International Series in Operations Research & Management Science

Volume 347

The book series **International Series in Operations Research and Management Science** encompasses the various areas of operations research and management science. Both theoretical and applied books are included. It describes current advances anywhere in the world that are at the cutting edge of the field. The series is aimed especially at researchers, advanced graduate students, and sophisticated practitioners.

The series features three types of books:

- Advanced expository books that extend and unify our understanding of particular areas.
- Research monographs that make substantial contributions to knowledge.
- Handbooks that define the new state of the art in particular areas. Each handbook will be edited by a leading authority in the area who will organize a team of experts on various aspects of the topic to write individual chapters. A handbook may emphasize expository surveys or completely new advances (either research or applications) or a combination of both.

The series emphasizes the following four areas:

Mathematical Programming: Including linear programming, integer programming, nonlinear programming, interior point methods, game theory, network optimization models, combinatorics, equilibrium programming, complementarity theory, multiobjective optimization, dynamic programming, stochastic programming, complexity theory, etc.

Applied Probability: Including queuing theory, simulation, renewal theory, Brownian motion and diffusion processes, decision analysis, Markov decision processes, reliability theory, forecasting, other stochastic processes motivated by applications, etc.

Production and Operations Management: Including inventory theory, production scheduling, capacity planning, facility location, supply chain management, distribution systems, materials requirements planning, just-in-time systems, flexible manufacturing systems, design of production lines, logistical planning, strategic issues, etc.

Applications of Operations Research and Management Science: Including telecommunications, health care, capital budgeting and finance, economics, marketing, public policy, military operations research, humanitarian relief and disaster mitigation, service operations, transportation systems, etc.

This book series is indexed in Scopus.

H. A. Eiselt • Vladimir Marianov
Editors

Uncertainty in Facility Location Problems

Springer

Editors
H. A. Eiselt
Faculty of Management
University of New Brunswick
Fredericton, NB, Canada

Vladimir Marianov
Department of Electrical Engineering
Pontificia Universidad Católica de Chile
and Instituto Sistemas Complejos de
Ingeniería (ISCI)
Santiago, Chile

ISSN 0884-8289 ISSN 2214-7934 (electronic)
International Series in Operations Research & Management Science
ISBN 978-3-031-32337-9 ISBN 978-3-031-32338-6 (eBook)
https://doi.org/10.1007/978-3-031-32338-6

This Springer imprint is published by the registered company Springer Nature Switzerland AG
The registered company address is: Gewerbestrasse 11, 6330 Cham, Switzerland

Paper in this product is recyclable.

"Where do I go from here?" asked the pilgrim. "This depends entirely on you," replied the peasant. "There is a trail that is pleasant except for the risk posed by wild animals, another on which you may encounter mudslides, another that is plagued by robbers, and yet another that has no water for a long stretch. Each trail carries its own risk. What are you best prepared for?" The pilgrim pondered the situation. "Whichever road I am going to take," he asked, "will it get me to the same destination?" "No," replied the peasant, "you will end up at the place that the trail leads you to. Just have faith and be prepared for everything."
Jesús Peregrino

Preface

When faced with the daunting task of writing a Preface to this volume, we were faced with a number of choices: write a piece that surveys the use and technique of stochastic location modeling, simply introduce the authors in this volume to our readers, or follow Mazo de la Roche in his introduction to the volume *Northern Lights*, in which he wrote

> After scanning the table of contents, I find that I have not read many of these stories. However, this does not much matter, as the names of the authors are assurance enough of the interest of what they write. They are Canadians and they carry the weight of their responsibility with assurance.

Having actively planned and solicited chapters from colleagues from all over the world and having them refereed by other colleagues does not permit us to take the last option.

The project, initiated by the ever-active and pleasant Dr. Camille Price, sounded immediately intriguing. All of us who teach the subject or are even remotely associated with it have learned long ago that most location problems are on the strategic level. This means they tend to be big-money, long-term decisions that need to be made. And, the longer the time frame of a decision, the higher the degree of uncertainty. This is where this volume comes in. We subsequently asked colleagues working in the field for their contribution. Contrary to our expectations, our suggestion was met with much enthusiasm. After the usual vetting and refereeing process, we ended up with 15 contributions, which fall into four categories.

Naturally, the first category includes those contributions that have taken a step back, and describe the sources of uncertainty, the risk, the imprecision, and similar features. The contribution by Murray addresses this very problem. His main thesis is to address uncertainty not by way of different modeling types, but as something that gets lost or introduces doubt in the different stages involved in finding good locations, and to convert the original problem with all of its subtleties and imponderabilities into a more formal, but inherently simpler, model. Bronfman, Paredes-Belmar, Marianov, and Eiselt utilize the framework of location and transport of hazardous materials to discuss the criteria that can be used to deal with their effects on population, environment, and property, as well as taking into account public, companies' and regulating agencies' concerns. These criteria include risk, hazard, and exposure time, combined in different ways. Marianov and Méndez-Vogel focus on customer-related uncertainties, what their origin is, how these uncertainties affect the choices the customers do, how to model these choices, and how to use these models in facility location. They describe the uncertainties coming from product heterogeneity, lack of or imperfect information, taste for variety, and compulsive behavior. Consumer behavior, in turn, has effects on the location chosen by firms.

The second category groups chapters that deal with protection against (and reaction to) acts of nature, and different adversaries: attackers or competitors. The first such chapter, by Bayram, Kara, Saldanha-da-Gama, and Yaman, focuses on what has been called humanitarian logistics, understood as emergency evacuation planning and management in the event of an act of nature, such as hurricanes, floods, earthquakes, and similar disasters. The authors discuss hedging against uncertainties in shelter location, and approaches to evacuation traffic management, integrating both in a stochastic model. Tammy Drezner addresses competitive facility location problems and pinpoints the uncertainties in the attraction function utilized in the well-known Huff user-choice model. She discusses different forms of the attraction function, the estimation of model parameters, uncertainty-based objectives, and some refinements of the probabilistic Huff model, such as leader-follower models, lost demand, and cannibalization. She wraps up the chapter by applying the Huff model to the p-median and its obnoxious version, the hub location problem, and a multiple-server location problem. The piece by Chicoisne, Ordóñez, and Castro addresses the location of defender's resources in Stackelberg security games with risk aversion. These games are known for their high complexity, as the leader or defender has to solve its location problem in such a way as to preempt the best attacker's strategy. They analyze different risk models (among others, maximum expected disutility, bounded distortion risk, plain risk minimization, value at risk, and conditional value at risk), and how these models are included in the leader's problem to obtain tractable models. In the chapter by Heckmann, Nickel, and Saldanha-da-Gama, location analysis is integrated into supply chain analysis, to obtain risk-aware decisions. Risk, in this case, refers to a disruption of the supply chain due to unexpected events, including strikes, floods, pandemics, and similar. The main idea is to maintain the supply chain efficiency (cost minimization) and effectiveness (service level). A stochastic facility location model is proposed that embeds uncertainty in a time horizon, as well as the possible attitudes of the decision-makers toward risk. To end this group, the chapter by Church addresses the advanced facility planning and modeling for resilience and protection against nature- or human-based disasters. Different approaches are described, including r-interdiction and fortification for resilience, to tri-level problems with a leader (defender) and followers (attackers) that act in two stages. He also describes, among others, approaches in which the loss of efficiency is minimized after disruptive events, or the weaknesses of a network are minimized and the network is made safer.

The third part of the book comprises contributions that deal with various aspects that involve the response time from facilities to customers or congested facilities. The chapter by Stratman, Boutilier, and Albert discusses problems of emergency medical services, which include the availability of ambulances (which may be busy responding to other calls), uncertain response time, and equally uncertain demand. The authors discuss systems in different countries as well as relocation models in countries with different income levels. The piece by Aboolian and Karimi deals with the location of public facilities under uncertainly, particularly congestion. It pinpoints the main difference between public and private facilities as the objective and identifies welfare and its proxies as an objective function. Formulations from the provider's and the consumers' perspective are provided. The work by Zvi Drezner starts with the concept of gradual coverage model that, in contrast to the original coverage models, allows for the degree of coverage being not only binary. Different versions of gradual cover models are discussed, before a stochastic version of the gradual covering model is presented. In his "directional covering models," each demand point is represented by a circle, and then the gradual coverage is defined and calculated. An extensive series of tests concludes the chapter. The piece by Shehadeh and Snyder deals with equity in the delivery of health care. In particular, the chapter discusses a number of possibilities to define and model "equity," followed by introducing stochasticity of demand, costs, and travel times. The chapter examines the proposed approaches in an example with real data in Pennsylvania.

The fourth and last part of this book comprises contributions that deal with methods and approaches for the solution of location models that involve uncertainty at some level. Taherkhani and Alumur examine and discuss hub location problems with stochastic demand and transportation costs and develop a number of different models along with their solution techniques. The authors describe stochastic and robust formulations of the problems. Escudero and Monge apply a multistage and a scenario approach for different uncertain parameters. They then use a decomposition matheuristic to solve the problem. Finally, Albareda-Sambola, Fernández, and Saldanha-da-Gama discuss the facility location problem with Bernoulli demand. Based on known heuristics, they then develop a tailor-made solution approach for the problem.

Finally, it is our pleasure to thank the many people who have made this volume a reality. First and foremost there are, of course, the authors, who have contributed their most recent work in the area. Then there is the aforementioned Dr. Price, who came up with the idea for this volume, and last, but certainly not least, there are the many helpers at Springer-Verlag, such as Mrs. Chockalingam, Ms. Yan, Ms. Su, and Ms. Prakash, whom we would like to thank for their active and moral support. Without them, this volume would never have seen the light of day. With great sadness did we hear a short while ago that our colleague, good friend, and contributing author to this volume, Tammy Drezner, has passed on. Our heartfelt condolences go out to her husband Zvi and daughter Taly Dawn. She will be missed.

Fredericton, NB, Canada H. A. Eiselt
Santiago, Chile Vladimir Marianov
January 2023

Contents

Part III Facility-Customer Response Time and Congested Facilities

Contributors

Robert Aboolian College of Business Administration, California State University San Marcos, San Marcos, CA, USA

Maria Albareda-Sambola Departamento d'Estadística i Investigació Operativa, Universitat Politècnica de Catalunya, Terrassa, Spain

Laura A. Albert Department of Industrial and Systems Engineering, University of Wisconsin-Madison, Madison, WI, USA

Sibel A. Alumur Department of Management Sciences, University of Waterloo, Waterloo, ON, Canada

Vedat Bayram Department of Industrial Engineering, TED University, Ankara, Turkey

Justin J. Boutilier Mechanical Engineering, College of Engineering, University of Wisconsin – Madison, Madison, WI, USA

Andrés Bronfman Engineering Sciences Department, Universidad Andres Bello, Santiago, Chile

Daniel Castro Industrial Engineering Department, Universidad de Chile, Santiago, Chile

Renaud Chicoisne Clermont Auvergne INP, LIMOS, Aubière, France

Richard L. Church University of California, Santa Barbara, CA, USA

Tammy Drezner College of Business and Economics, California State University-Fullerton, Fullerton, CA, USA

Zvi Drezner College of Business and Economics, California State University-Fullerton, Fullerton, CA, USA

H. A. Eiselt Faculty of Management, University of New Brunswick, Fredericton, NB, Canada

Laureano F. Escudero Area of Statistics and Operations Research, Universidad Rey Juan Carlos, URJC, Móstoles (Madrid), Spain

Elena Fernández Universidad de Cádiz, Departamento de Estadística e Investigación Operativa, Puerto Real, Spain

Iris Heckmann CamelotITLab, Innovative Technologies Lab, Cologne, Germany

Bahar Y. Kara Department of Industrial Engineering, Bilkent University, Ankara, Turkey

Majid Karimi College of Business Administration, California State University San Marcos, San Marcos, CA, USA

Vladimir Marianov Department of Electrical Engineering, Instituto Sistemas Complejos de Ingeniería (ISCI), Pontificia Universidad Católica de Chile, Santiago, Chile

Gonzalo Méndez-Vogel Pontificia Universidad Católica de Chile, Santiago, Chile

Juan F. Monge Center of Operations Research, Universidad Miguel Hernández, UMH, Elche (Alicante), Spain

Alan T. Murray Department of Geography, University of California at Santa Barbara, Santa Barbara, CA, USA

Stefan Nickel Karlsruhe Institute of Technology (KIT), Karlsruhe, Germany

Fernando Ordóñez Industrial Engineering Department, Universidad de Chile, Santiago, Chile

Germán Paredes-Belmar School of Industrial Engineering, Pontificia Universidad Católica de Valparaíso, Valparaíso, Chile

Francisco Saldanha-da-Gama Departamento de Estaística e Investigacão Operacional e Centro de Matemática, Aplicações Fundamentais e Investigação Operacional, Faculdade de Ciências, Universidade de Lisboa, Lisbon, Portugal

Karmel S. Shehadeh Department of Industrial and Systems Engineering, Lehigh University, Bethlehem, PA, USA

Lawrence V. Snyder Department of Industrial and Systems Engineering, Lehigh University, Bethlehem, PA, USA

Eric G. Stratman Industrial & Systems Engineering, University of Wisconsin-Madison, Madison, WI, USA

Gita Taherkhani Quinlan School of Business, Loyola University, Chicago, IL, USA

Hande Yaman Faculty of Economics and Business, ORSTAT, KU Leuven, Leuven, Belgium

Part I
Sources of Uncertainty, Risk, and Imprecision

Sources of Uncertainty in Location Analysis

Alan T. Murray

Abstract This chapter provides an overview of uncertainty in location analysis, highlighting different ways sources of error can be introduced. The chapter intentionally deviates from past efforts discussing uncertainty in location analysis that emphasize particular model types, such as risk, robust, and stochastic approaches. The rationale is that defining characteristics of uncertainty suggest that doubt is key. Accordingly, doubt can be found in a range of identified categories, including understanding of problem/issue, abstraction, model specification, attribute(s), location, spatial properties, solution, and implementation. The modeling implications for select categories are illustrated in various ways in order to highlight the spatial and aspatial implications. The intent is to make future avenues for investigation more comprehensive and ultimately ensure that uncertainty is addressed in a rigorous fashion.

Keywords Analytics · Abstraction · MAUP · Error

1 Introduction

There are many variants of the idea that uncertainty is ever present in our daily lives, with quotes and sayings like "... nothing can be said to be certain, except death and taxes ..." by Benjamin Franklin and "the only certainty is uncertainty" echoed by many in different ways. In facility location modeling uncertainty is pervasive due to problem ambiguity, measurement, data collection/process, computational challenges, etc. However, the significance and importance of modeling is to provide insight, facilitate decision making, inform policy, etc., putting associated uncertainty into perspective. Of course, the implication is that humans and/or decision-making processes will somehow reconcile associated uncertainties. As with all modeling

A. T. Murray (✉)
Department of Geography, University of California at Santa Barbara, Santa Barbara, CA, USA
e-mail: amurray@ucsb.edu

H. A. Eiselt, V. Marianov (eds.), *Uncertainty in Facility Location Problems*,
International Series in Operations Research & Management Science 347,
https://doi.org/10.1007/978-3-031-32338-6_1

efforts, the hope is that fundamentally important components of a problem are sufficiently reflected and relatively certain in associated model(s), making derived insights meaningful in various ways. This has arguably been the overarching perspective in facility location modeling, with a range of models developed, applied, studied, extended, enhanced, and rediscovered to address a wide variety of problem contexts and nuanced considerations. Uncertainty too has been acknowledged and addressed in different ways, but could be considered rather narrowly interpreted, with efforts instead largely focusing on model capabilities, solution potential, and planning support insights.

In this chapter, I offer my own view of uncertainty in facility location modeling, outlined in Fig. 1, suggesting understanding of problem/issue, abstraction, model specification, attribute(s), location, spatial properties, solution, and implementation. This view extends the general summary in Murray (2003) where existing location model research incorporating uncertainty is put into categories of objective function(s), solution approaches, distance measure, demand location error, attribute

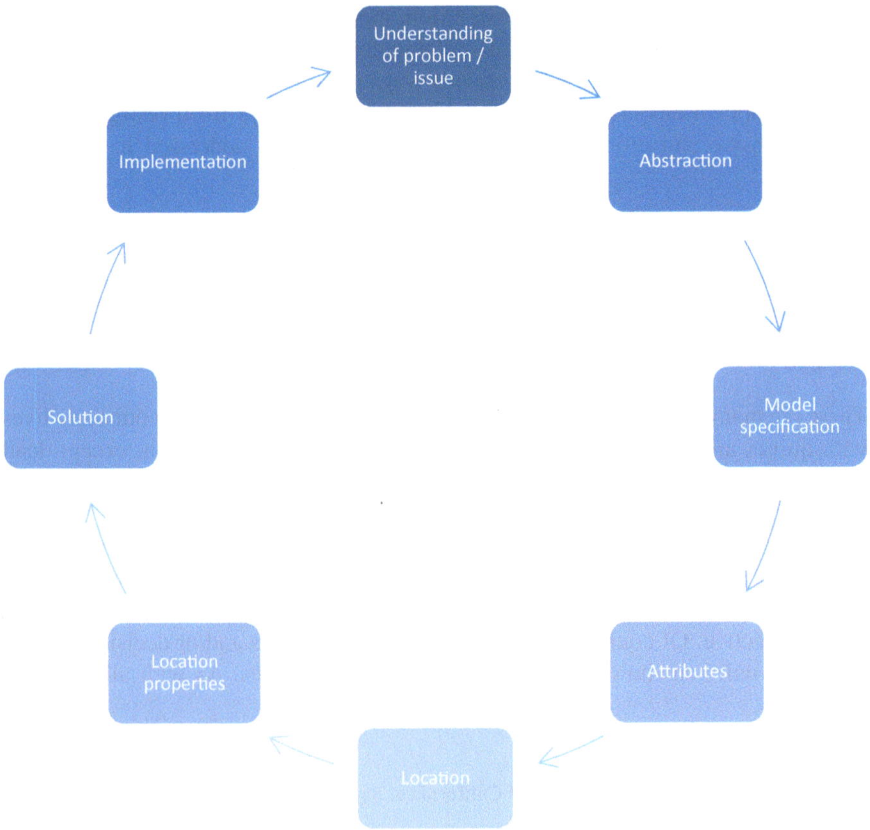

Fig. 1 Categories of uncertainty in location analysis

accuracy, representation, and site placement. The perspective reflected in Fig. 1 can be contrasted with the more common takes on uncertainty offered in reviews by Snyder (2006), Correia and Saldanha-da-Gama (2019), and other related topics found in Laporte et al. (2019). Snyder (2006) discusses uncertainty in facility location models by categorizing problems as risk, stochastic optimization, and robust optimization. Related themes can be found in Correia and Saldanha-da-Gama (2019), noting topics of congestion, uncertain parameters, robust, stochastic, and chance-constrained approaches. Further highlighted is future work involving multi-stage stochastic models, algorithm development, and scenario generation. While not considered as uncertainty topics per se, a number of chapters in Laporte et al. (2019) are focused on uncertainty related themes, including multi-criteria, multi-period, and competitive approaches.

This chapter is meant to be more practically oriented, emphasizing both spatial and aspatial elements of location analysis where uncertainty arises. Categories of note in the next section include understanding of problem/issue, abstraction, model specification, attribute(s), location, spatial properties, solution, and implementation. This is followed by specific examples highlighting modeling implications. The intent of this chapter is to make future avenues for investigation more comprehensive and ultimately ensure that uncertainty is addressed in a rigorous fashion. The chapter ends with discussion and concluding comments.

2 Uncertainty

An understanding of uncertainty is something that we all generally possess, at least to some degree. Merriam-Webster defines uncertain as "not known beyond doubt" or "not clearly identified or defined." Given this, anything under this broad umbrella of uncertainty is a potentially important and significant topic for any analysis approach. Since location science has emphasized models, it is not surprising then that reviews of uncertainty would focus on different modeling approaches, such as those by Snyder (2006) and Correia and Saldanha-da-Gama (2019). However, the emphasis on model categories (e.g., risk, stochastic, robust optimization) likely ignores important fundamental issues and challenges that must be fully considered and rationalized.

Understanding of the problem or issue being addressed is the first item noted in Fig. 1 associated with elements contributing to uncertainty in location analysis. This serves as a beginning because it is where we invariably start when undertaking location analysis. There would appear to be at least three ways in which uncertainty could be introduced as a result of understanding: limited grasp of the relevant issues and concerns; incorrect interpretation of the relevant issues and concerns; and the wrong method and/or model is proposed/applied. The distinction between these is subtle. A limited understanding is meant to denote the situation where all facets of the location analysis context may not be known or fully appreciated, often with respect to planning, management, policy, etc. A different nuance is an incorrect understanding, typically where certain aspects of the problem are interpreted as more significant or central, but this could go beyond this. Finally, the wrong approach being identified and/or applied for whatever reason is another possibility. Along these lines, one could imagine that a p-median problem is proposed, but the actual problem of interest is a min-max problem or a coverage problem. Similarly there are many cases where a center is sought, but a centroid is identified instead (Murray et al. 2020). Such a situation is not inconceivable given access to location analysis in commercial packages like ArcGIS and TransCAD (see Murray et al. 2019) as well as through open-source software (see Chen et al. 2021), where a user can easily select a location analysis approach irrespective of whether it is appropriate and valid for the given context of analysis.

The abstraction aspect of uncertainty indicated in Fig. 1 reflects that all models are an approximation to some observed system, where the most important elements of the system are incorporated into the model in some way. This is equivalent to what Marianov (2022) terms "imperfect modeling." Accordingly, all models are an abstraction of an actual process or situation. However, it goes further than this because models invariably rely on data inputs, and such data is also an abstraction based on observed attributes, objects, interactions, etc. Recognizing and accepting these two important realities, a number of abstraction distinctions are well recognized in location analysis. The most common is a continuous space model and a discrete space model. This has to do with the object or facility being located, and the idea that potential locations for siting are anywhere in a region (continuous) or limited to a finite and distinct number (discrete). Accordingly, there is a long tradition of developing, solving, and applying both continuous and discrete location models to address a range of planning and analysis situations. Potential facility sites are but one facet of the continuous-discrete abstraction distinction. Other aspects include temporal context (static/cross-sectional vs. dynamic/longitudinal) and whether demand is conceived to be continuously distributed or occurring at discrete (and finite) locations. There are more abstraction-oriented features that could be discussed, but are reviewed under other categories of uncertainty due to characteristic features.

The third source of uncertainty in Fig. 1 is model specification. A beginning point for uncertainty is recognizing that there are many commonly relied upon ways to communicate a problem. These include descriptions, flowcharts, pseudo-code, and mathematical equations. The significance is that the different forms may lend themselves to some degree of ambiguity. Descriptions, flowchart, and pseudo-code, as an example, may omit an important detail or characteristic in the process of summary and generalization. It would be fair to say that mathematical specification is the most clear and precise, assuming that it accurately reflects what is actually being done. Other aspects of uncertainty in model specification could include an incorrect objective(s), missing constraints, logical inconsistences, etc. A final aspect of model specification where uncertainty could arise is the modifiable areal unit problem (MAUP). This is a recognized issue in location and geographic application of quantitative methods, where the results of model application vary based on a change of spatial scale or unit definition. Such a topic could also be included and discussed under other uncertainty categories in Fig. 1, such as abstraction and location, but is included here because model specification that is frame independent (not impacted by MAUP) is an elusive goal with a growing number of attempts to address this (see Murray 2005, 2018).

Attributes as a source of uncertainty in Fig. 1 reflect the common treatment in operations research of model inputs, or attributes, as being known and absent of error. In general such inputs are observed or estimated. For example, an often-utilized source of expected demand is total population, taken from an official census. However, it is well known that such data have limitations, including that they are residential population counts, undercount the actual population, and are biased. Their usage, therefore, could be problematics in certain circumstances. Other attribute-oriented information may similarly be problematic due to human or sensor error/bias in their creation or processing. Issues of concern include accuracy, precision, validity, reliability, sampling process, intended usage, etc. (Murray & Grubesic, 2012). An additional potential source of uncertainty with respect to a given attribute is that it is often created through a combining process, such as addition, subtraction, etc. of two or more attributes. Uncertainty arises because such an approach may mask what is ultimately an underlying multi-objective problem, where the individual attributes should be independently optimized. Accordingly, solution of a single combined attribute would result in a single optimal solution, but individual attributes would reflect Pareto trade-off solutions. Thus, only one of potentially many solutions are identified, yet these unidentified trade-off solutions are equally valid. They are missed through an unintentional masking process that combines attributes in advance, rather than treating them independently. Another rather common source of uncertainty arises through the use of interpolated attributes. Interpolation (or extrapolation) is a process of estimating an attribute value based on a sample of observed values. Thus, estimates are derived at other (unobserved) locations based on some interpolation/extrapolation method, making use of observed sample attribute values. There are many examples of such data, including temperature, rainfall, and air quality.

Another potential source of uncertainty noted in Fig. 1 is location, recognizing that data creation, processing, and manipulation can contribute to whether or not geographic position is accurate. This is a topic that has received perhaps the most attention in geography and GIS areas, though perhaps not as much in the context of location analysis. The creation of location, or geographic, information is often accomplished through a process of digitizing existing basemaps, application of geocoding, or reliance on GPS (Church & Murray, 2009). Human or automated digitizing relies on data layers or physical maps from which location is extracted. The reality is that all data layers and maps have limited positional accuracy. This is due to how they were created, but also cartographic license and design. As a result, they may be very accurate in the location of reported information, within centimeters, or very inaccurate, off by hundreds of meters (or more). This ignores, of course, the added element of human involvement in the process that may also contribute to potential error. Another source of location information is through a process known as geocoding, where an address is used to produce an associated coordinate reference on the surface of the Earth. Such a process relies on a database of street segments, street names, address ranges, etc., a type of look-up process, except there is invariably interpolation and street offset involved as well as default location used when an address cannot be found in the database. While beyond the scope of this discussion, suffice to say that there are many ways in which error and inaccuracy could be resident in associated location data relying on a geocoding process. What about data generated through the use of GPS? While seemingly accurate, it relies on satellites (and base stations) to determine positional location. As a result, atmospheric conditions, weather, buildings, mountains, etc. can interfere with signal communication, thereby impacting the spatial precision of derived location. Often not widely acknowledged, location data may be 2-dimensional, sometimes the by-product of a projection of an original 3-dimensional data source. The process of projecting 3-dimensional location to 2 dimensions cannot be accomplished without introducing some source of error. A final observation regarding uncertainty in location data is that such data is often the by-product of formal manipulation, including aggregation and simplification. Aggregation involves combining neighboring points or polygons to create a fewer number of observations, generally intended to reduce the computational of spatial analytical methods, such as optimization. Worth noting is that location data could be disaggregated as well, the opposite of aggregation, with a range of associated uncertainty issues that accompany such manipulation, including interpolation. Simplification is manipulation that involves representing a spatial object in a less complicated manner. One example is taking a polygon and representing its location as a coordinate pair, such as a centroid. Another example is a street, highway, or freeway and representing it as a simple line, the centerline. This discussion is simply to highlight that data associated with spatial location is far from perfect due to various issues.

The spatial properties (spatial relationships) source of uncertainty noted in Fig. 1 indicates that geographic attributes play a significant role in location analysis. Often these are topographical and/or topological, having to do with proximity and arrangement. One possibility for error arises through adjacency, a property reflecting that two locations neighbor each other. Adjacency is commonly defined by two areas sharing a common boundary or point, but other extended interpretations are possible. Thus, there could be variability due to adjacency definition as well as object location and position. Distance too is a spatial property, and could introduce error in various forms. The most obvious is that there are different ways to measure distance, including network, Euclidean, rectilinear, and l_p, as well as sensitivities to underlying location that arise in assessing distance between two objects. Other spatial properties that can be defined in differing ways or impacted by location and topological variability include connectivity, contiguity, compactness, and shape, among others, and each have been central to location analysis efforts and optimization. A final category of spatial properties to point out are cover sets and buffers. Cover sets are generally associated with coverage location problems, reflecting those demand objects that can be suitably served by a facility sited at a particular location. Depending on the assessment criteria and method, such sets could be uncertain in various ways, not to mention partial service considerations.

Somewhat related is the notion of a buffer, reflecting a topographical transformation of a location or object using a specified distance or travel time threshold.

Model solution as a source of uncertainty in Fig. 1 may be a surprising category. However, there are actually a number of ways that uncertainty can be introduced through a solution process. An exact method is one that produces a guaranteed optimal solution, but only if optimality criteria and conditions are met. Mixed-integer programming (MIP) problems are common in location analysis, with the branch and bound being a widely relied upon solution technique. An appealing aspect of linear programming combined with branch and bound, as an example, is that an optimality gap can be established once a feasible solution has been identified. In fact, many difficult MIP problem instances may terminate with a remaining optimality gap, offering only a bound on solution quality with some potential for a better solution. Uncertainty arises when a solution is not understood to be optimal within the stipulated conditions, or perhaps sub-optimal but within a threshold of a theoretical bound. Another common form of problem solution involves the use of a heuristic process. By definition, a heuristic is an ad hoc method or technique that produces a solution to an optimization problem with unknown or unproven optimality bounds or quality characteristics. As a result, a heuristically identified solution may be of high quality, or low quality. However, if a user, analyst, decision maker, etc. is not aware that a heuristic is used, then this is problematic as there exists uncertainty about solution quality. There are, in fact, many software packages providing access to exact and heuristic methods with essentially no communication of the approach used to solve, nor conditions under which the results should be interpreted. One could add as well that parameterization too may be a source of uncertainty, where a change in method parameters may produce better (or worse) result. While perhaps not an issue from a strict theoretical bound perspective, the existence of alternative (or multiple) optima may well be a source of uncertainty in that there are other solutions that could be considered, perhaps differing in significant ways. Finally, problems that are inherently multi-objective may not be interpreted and communicated in this manner, as noted above, but also that some methods may not be able to identify all Pareto or non-dominated solutions, creating some uncertainty about the actual trade-offs that exist.

The last category to be discussed in Fig. 1 is uncertainty due to implementation. Of course, models are viewed as an aid to management, planning, policy, and decision making more generally. Given this, it is understandable that implementation may deviate from the model prescribed results. However, an important question is how does this impact the overall solution quality. Is the solution now sub-optimal, infeasible, or degraded in some manner? Technically speaking, an optimal solution must be implemented as prescribed in order to maintain its theoretical properties. Any deviation is potentially significant. Of course, implementation itself is full of challenges, involving reconciliation of model abstraction with the realities of on-the-ground interpretation. This is precisely why there is invariably deviation from optimization prescriptions, yet this does raise concerns and creates associated uncertainty.

The significance of identifying and discussing the categories in Fig. 1 is that there is much potential for uncertainty in many different ways, but also a complex interaction or co-mingling of these individual items is inevitable. Table 1 offers a more detailed summary of Fig. 1, highlighting the various issues raised that could impact uncertainty. Summarized in this manner, it is undeniable that co-mingling is taking place, but more important that the implications are not at all understood. Not included in Fig. 1 or Table 1, but certainly worth noting are issues of omission, including attributes and objects in data as well as objectives and

Table 1 Uncertainty in location analysis

Uncertainty	Sources
Understanding	→ Limited grasp of relevant issues and concerns → Incorrect interpretation of relevant issues and concerns → Wrong method and/or model proposed/applied
Abstraction	→ Data → Continuous space (potential facility sites) → Discrete space (potential facility sites) → Static/cross-sectional → Dynamic/longitudinal → Continuous space (demand) → Discrete space (demand)
Model specification	→ Communication (e.g., descriptions, flowcharts, pseudo-code, mathematical equations) → Incorrect objective(s), missing constraints, logical inconsistences → Modifiable areal unit problem (e.g., spatial scale, unit definition) → Frame independence
Attribute(s)	→ Measurement error → Bias → Sampling → Human error → Masking of multiple attributes/objectives → Interpolation/extrapolation
Location	→ Data creation (e.g., digitizing, GPS, geocoding, cartographic license, design) → Processing (e.g., map projection) → Manipulation (e.g., simplification, aggregation, disaggregation)
Spatial properties	→ Adjacency/neighbor → Distance → Contiguity/connectivity → Cover sets and buffers
Solution	→ Exact (e.g., optimality gap) → Heuristic → Parameterization → Multiple / alternative optima → Pareto optima
Implementation	→ Modification → Adjustment → Inability to implement precisely

constraints. Similarly, the suitability of proxy measures and metrics likely could be more explicitly considered as well.

3 Modeling Implications

The detailed categories of uncertainty (Fig. 1 and Table 1) reflect a range of potential issues that could impact interpretation and significance in any location analysis and modeling effort to support planning, management, and policy development. Space limitations make it difficult to delve into specific instances of each category, but this section offers representative examples of how uncertainty is introduced in analysis. In particular, I draw upon experience in my own research and application of location analytics to highlight aspects of associated uncertainty that arise in abstraction, model specification, location, spatial properties, and solution. While equally important understanding, attributes, and implementation are left to the reader to contemplate.

3.1 Abstraction Uncertainty Implications

As introduced above, abstraction is a necessary component of any modeling effort, and permeates many aspects of location analysis. Abstraction is the idea that both data and a model approximate an associated system, process, or decision-making context. I offer one example in location analysis associated with the intent to cover or serve a region with a minimal level of resources. Consider the following notation:

i= index of demand objects
λ_i= area of demand i
S= service standard (time or distance)
f_i= coverage function for demand i with respect to service standard S
Φ= region of analysis
j= index of facilities providing service
Δ= set of facilities

The decision variables in this case are associated with where to site associated facilities:

$$\left(\alpha_j, \beta_j\right) = \text{location of facility } j$$

This is a classic location analysis problem. If it is assumed that facilities are permissible to site anywhere in continuous space, then the following formulation represents this particular problem:

$$\text{Minimize } |\Delta| \tag{1}$$

$$\text{Subject to} \iint f_i(\Delta)\ d\Phi \geq \lambda_i \quad \forall i \tag{2}$$

$$(\alpha_j, \beta_j) \in \Delta \quad \forall j \tag{3}$$

The objective, (1), is to site the fewest number of facilities possible. Constraints (2) require that the area of each demand object is to be completely covered or served within the standard S. Constraints (3) signify that location sites be in the region of analysis.

Associated formulations and analysis of coverage along these lines using (1)–(3) can be found in Wei and Murray (2015) and Church and Murray (2009). The challenge is that an efficient configuration of facilities is sought in order to serve, or cover, each demand object i, with an area of λ_i (could also be a length if the object is a line). Thus, through the combination of individual or combined coverage provided by the set of sited facilities, each demand must be served.

Readers familiar with location analysis may recognize the continuous space siting coverage problem in (1)–(3) as equivalent in intent to the location set cover problem detailed in Toregas et al. (1971), as well as Berge (1957) and Edmonds (1962). A distinction is that potential facility locations are finite and discrete, identified a priori. A closer look and comparison is facilitated by the introduction of the following additional notation:

j = index of potential facility sites
Ψ_i = set of facilities that suitably cover demand i

Consider as well these additional decision variables:

$$X_j = \begin{cases} 1 \text{ if facility located at site } j \\ 0 \text{ otherwise} \end{cases}$$

Thus, in contrast to (1)–(3), specific potential locations for siting facilities are known in advance, and finite in number. Accordingly, a decision variable can be defined for each location to represent the siting decision, eliminating the need to track location in continuous space with the coordinate pair decision variables, (α_j, β_j). With this, a discrete space location problem can be structured.

$$\text{Minimize} \sum_j X_j \tag{4}$$

$$\text{Subject to} \sum_{j \in \Psi_i} X_j \geq 1 \quad \forall i \tag{5}$$

$$X_j = \{0, 1\} \quad \forall j \tag{6}$$

The objective, (4), is to minimize the number of facilities necessary. Constraints (5) stipulate that each demand must be suitably served by at least one sited facility. Constraints (6) impose binary restrictions on decision variables.

Both models seek the same outcome, to identify the number and location of necessary facilities to cover all demand. However, they are doing so in very different ways, as (1)–(3) focuses on continuously distributed demand whereas (4)–(6) assumes demand is discrete. This is not particularly surprising or novel as most problems can be described, structured, and/or formalized in alternative ways, but the abstraction process in this case results in a fundamentally different spatial optimization model. The implications are many, including potential methods to solve each problem but also the findings that can be expected from each model. Since they are different, it should not be a surprise that different outcomes may well result. It is therefore in this context that abstraction is noted as a source of uncertainty, and is at the essence of what Murray (2018) demonstrates in a more expanded manner.

3.2 Model Specification Uncertainty Implications

Model specification was characterized as varying in terms of approach taken, including description, flowchart, pseudo-code, and mathematical equations. Consider the example of the description found in ArcGIS for a location-allocation problem, one of seven basic alternatives:[1] "MAXIMIZE_COVERAGE—This option solves the fire station location problem. It chooses facilities such that all or the greatest amount is within a specified impedance cutoff." Adding to this in ArcGIS is the option of including capacity restrictions on facilities. The first sentence likely fails to communicate any meaningful problem or model characteristic as many different location analysis approaches have been utilized to address fire station siting issues. The second sentence is a little more insightful, at least to those working in the area of location analysis with familiarity of the maximal and set covering location problems, but the description is ambiguous at best raising issues of uncertainty. Xu et al. (2020) show that what is implemented in ArcGIS is the maximal covering location problem (or capacitated maximal covering location problem when the capacity option is elected), with an assumption that potential facility locations are discrete and identified in advance. While only the above description is provided in ArcGIS, a mathematical formulation is possible. Offered here, using the previous notation, is the capacitated maximal covering location problem, based on the following additional notation:

[1] ArcGIS is arguably the leading commercial GIS (geographic information system) software package, produced by Esri (https://www.esri.com/). The different problem types are minimize weighted impedance, maximize coverage, maximize coverage and minimize facilities, maximize attendance, maximize market share, target market share, and maximize capacitated coverage.

δ_i = expected service amount associated with demand i
θ_j = service capacity of facility j
p = number of facilities to locate

Additional decision variables are necessary for tracking demand allocation to facilities, as follows:

$$Y_{ij} = \begin{cases} 1 \text{ if demand } i \text{ is served by facility } j \\ 0 \text{ otherwise} \end{cases}$$

The above ArcGIS description can then be formally stated as:

$$\text{Maximize} \sum_i \sum_{j \in \Psi_i} \delta_i Y_{ij} \tag{7}$$

$$\text{Subject to} \sum_{j \in \Psi_i} Y_{ij} \leq 1 \quad \forall i \tag{8}$$

$$\sum_j X_j = p \tag{9}$$

$$\sum_i \delta_i Y_{ij} \leq \theta_j X_j \quad \forall j \tag{10}$$

$$\begin{aligned} X_j &= \{0, 1\} \quad \forall j \\ Y_{ij} &= \{0, 1\} \quad \forall i, j \in \Psi_i \end{aligned} \tag{11}$$

The objective, (7), indicates an intent to maximize total demand covered within the service standard. Constraints (8) limit allocation to at most one facility. Constraint (9) requires p facilities to be sited. Constraints (10) impose capacity restrictions on sited facilities, ensuring that no facility is allocated more than θ_j total demand. Finally, binary restrictions are noted in constraints (11). This is a well-known location model, the capacitated maximal covering location problem (see Church and Murray 2018).

It would be extremely difficult to know with certainty that the above description offered in ArcGIS actually corresponds to the formulation given in (7)–(11). Regardless, the approach has seen broad and growing use and application, as detailed in Xu et al. (2020). This, however, is the essence of uncertainty arising in model specification, where a complete understanding and appreciation may not be possible with certain approaches accessible through commercial or open-source software. Further, it opens the door to questionable or inappropriate use and application in practice.

3.3 Location Uncertainty Implications

Irrespective of the methods used to generate geographic information, there is invariably associated positional error in the location of objects. Most geographic information systems conceive of spatial objects as points, lines (polylines), or polygon objects. While considerable information and analysis makes use of raster data, it is generally only used to represent a continuously distributed attribute or phenomenon, not objects per se. Given the reliance on spatial objects, consider their representation in technical or formal terms. A point is a coordinate pair, $\{(x, y)\}$. A line (or polyline) is a collection of sequenced coordinate pairs, $\{(x_1, y_1), (x_2, y_2), (x_3, y_3), \ldots, (x_k, y_k)\}$, with consecutive coordinates connected by a line segment, often assumed to be a straight line. A polygon is a collection of sequenced coordinate pairs, $\{(x_1, y_1), (x_2, y_2), (x_3, y_3), \ldots, (x_k, y_k)\}$, with consecutive coordinates connected by a line segment and the condition that $(x_k, y_k) = (x_1, y_1)$, indicating a closure of the object to form a polygon. For completeness, each object is illustrated in Fig. 2. As shown, the assumption is that the coordinate locations are known with certainty, and accurately represent the associated object. In reality, however, positional location is uncertain and line segments can only approximate continuous variation.

One implication of location uncertainty due to data generation processes is that any coordinate reference or associated line segment may be off by ε units. One can visualize such error, or uncertainty, for each object type. Figure 3 depicts an ε band, or buffer, for the point, line, and polygon objects shown in Fig. 2. Accordingly, the actual location of the point, as an example, could be anywhere in the uncertainty region shown in Fig. 3a. Similarly, the line boundary could be anywhere in the uncertainty region given in Fig. 3b. Finally, the polygon boundary could deviate within the uncertainty area shown in Fig. 3c.

The manipulation of geographic data is also a source of potential locational uncertainty. Consider the 295 polygons shown in Fig. 4a. It is very common that polygons such as these are manipulated to make them easier to evaluate and/or use in a location model. One such approach is simplification, where objects are often represented as a center or centroid. An example is a polygon $\{(x_1, y_1), (x_2, y_2), (x_3, y_3), \ldots, (x_k, y_k)\}$ simplified as a point, e.g., $(x^*, y^*) = \left(\sum_l x_l \big/ k, \sum_l y_l \big/ k\right)$. The result of such a process is shown in Fig. 4b, indicting the 295 centers, or centroids in this case, that represent a simplification of the more spatially complex polygon objects given in Fig. 4a. Clearly such a process of simplification introduces some level of uncertainty in location since the center could be considered for essentially any location within a polygon. Another aspect of manipulation is aggregation. This is a process where two or more neighboring polygons, as an example, are combined to form a single new polygon, with interior boundaries removed. Consider again the 295 polygons in Fig. 4a that have undergone aggregation to form the five new polygons shown in Fig. 4c. Notice that only exterior boundaries remain for each of the new polygons in Fig. 4c. The by-product of such a process clearly obscures and removes underlying spatial detail and variability, creating locational uncertainty. Often the rationale for aggregation is to

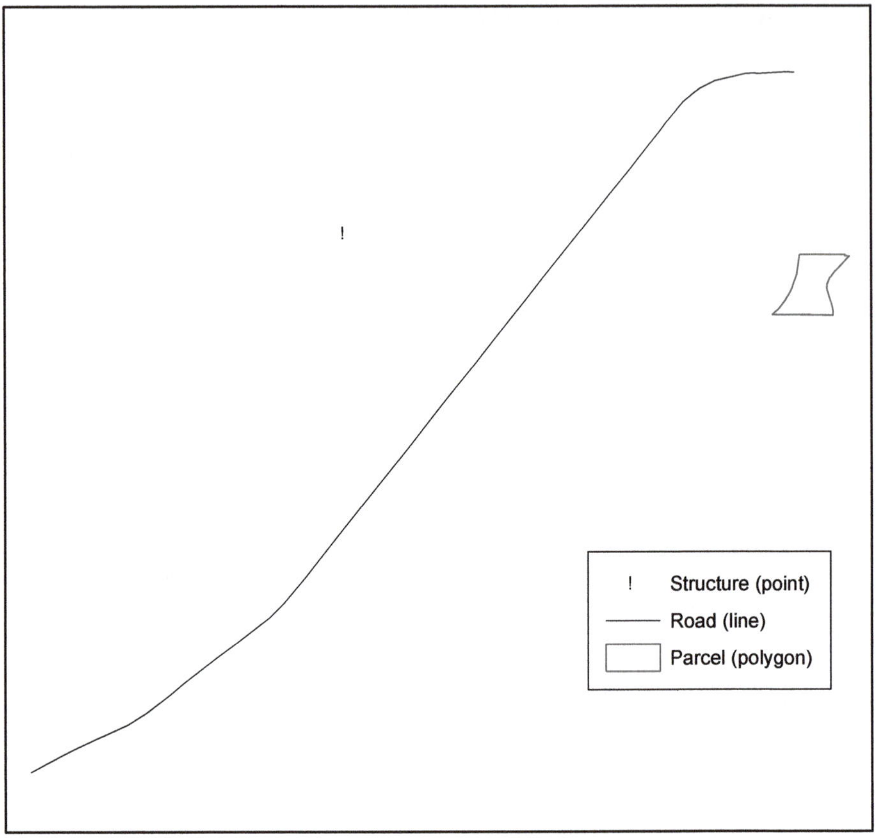

Fig. 2 Common location analysis objects (points, lines, polygons)

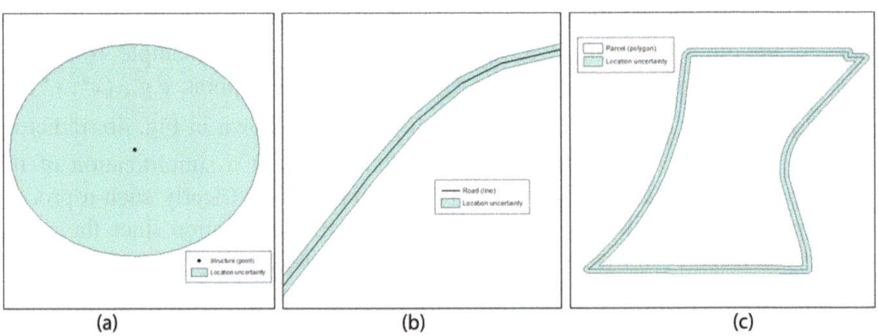

Fig. 3 Location uncertainty in common object boundaries. (**a**) Uncertainty in point location, (**b**) uncertainty in line location, and (**c**) uncertainty in polygon location

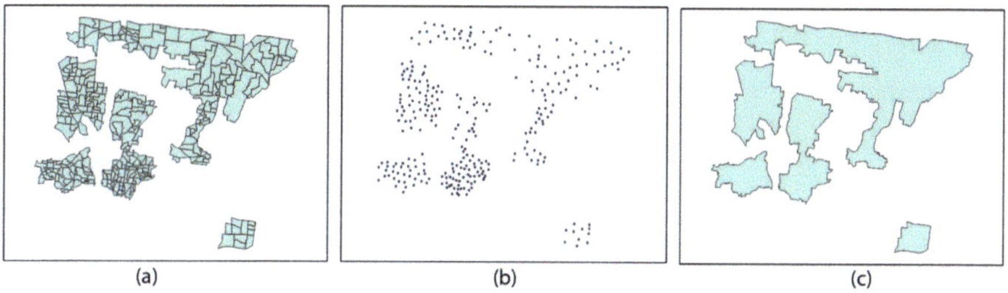

Fig. 4 Location manipulation. (**a**) Original polygons, (**b**) point simplification, and (**c**) aggregation of polygons

enhance computational capabilities in model solution, yet this is invariably being done at the expense of locational precision and certainty.

3.4 Spatial Properties Uncertainty Implications

The uncertainty in spatial properties (and spatial relationships) is not particularly surprising given the broader location uncertainty in geographic data illustrated previously. Nevertheless, location analysis generally focuses on spatial properties of various sorts in models. That is, the spatial properties are often central to the intent of optimization, or reflect primary restrictions and constraining conditions. This is especially true for the set covering models noted previously, continuous space siting in (1)–(3) and discrete space siting in (4)–(6). What are the important spatial properties in this case? Clearly the coverage functions f_i and the cover sets Ψ_i each rely on the service standard S in their respective models. Often coverage is based on distance or travel time, yet there are many different metrics that can be used. In the case of distance, there is Euclidean, rectilinear, l_p, and network travel. Not only are these different in strict mathematical terms, but the spatial footprint and length of travel can vary significantly. As an example, consider the shortest network travel path of 2.7 miles in Fig. 5 to that of Euclidean distance measuring 1.27 miles. This is significant, and for obvious reasons given the travel limitations evident in Fig. 5. A critical question, however, is how does this impact coverage assessment, particularly when uncertainty arises due to distance and proximity. Murray and Grubesic (2012) offer an expanded discussion of this topic, but clearly context and the nature of service are particularly important for both interpretation and utilization in location analysis.

Of course, there are many other types of spatial relationships that are critical in location analysis. Consider the relatively well-understood notion of adjacency. Figure 6 illustrates planning units from which adjacent units can be observed. In particular, look closely at unit 238. The set of adjacent units based on a shared point or boundary in this case would be {221, 225, 226, 234, 239, 240, 241, 242}. This assumes, of course, that the boundaries are error free, or precise.

Fig. 5 Contrasting Euclidean distance to network travel

Shown in Fig. 7, however, is the actual location uncertainty of the boundary (or specifically, imprecision), which appears to have significant implications for adjacency in this case, particularly with respect to unit 238. Inspection of Fig. 7 suggests that unit 238 may only be adjacent to {221, 234, 239, 240} with certainty, leaving other units uncertain depending on the actual boundary position. Boundary

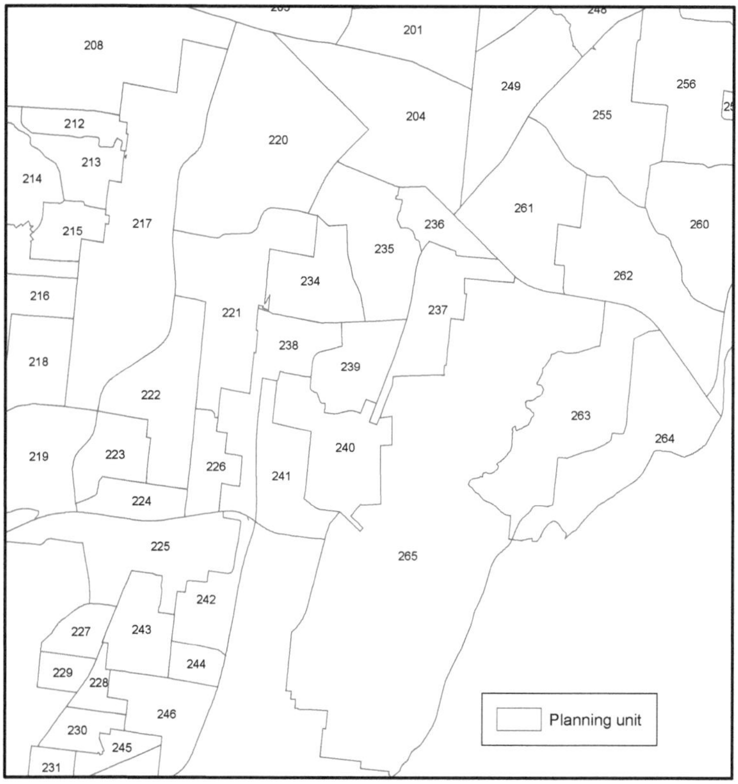

Fig. 6 Planning units with adjacent/neighboring units

imprecision along these lines can be attributed to many factors, including digitizing equipment, weather and atmospheric conditions, terrain, human error, etc. but also data processing, cleaning, management, and manipulation. Wei and Murray (2012, 2018) utilized information along these lines to derive probabilities of adjacency certainty for inclusion in a location model, offering one approach for addressing spatial uncertainty in a structured manner.

3.5 Solution Uncertainty Implications

As noted previously, model solution is likely an unexpected source of uncertainty, yet there are actually many opportunities for this to create uncertainty. The technical details associated with MIP approaches, such as an optimality gap, may well be beyond the expertise of many. Nevertheless, communicating these facts is important when an associated solution is not confirmed to be optimal. Perhaps the bigger challenge is communication encountered in location analysis, when map-based

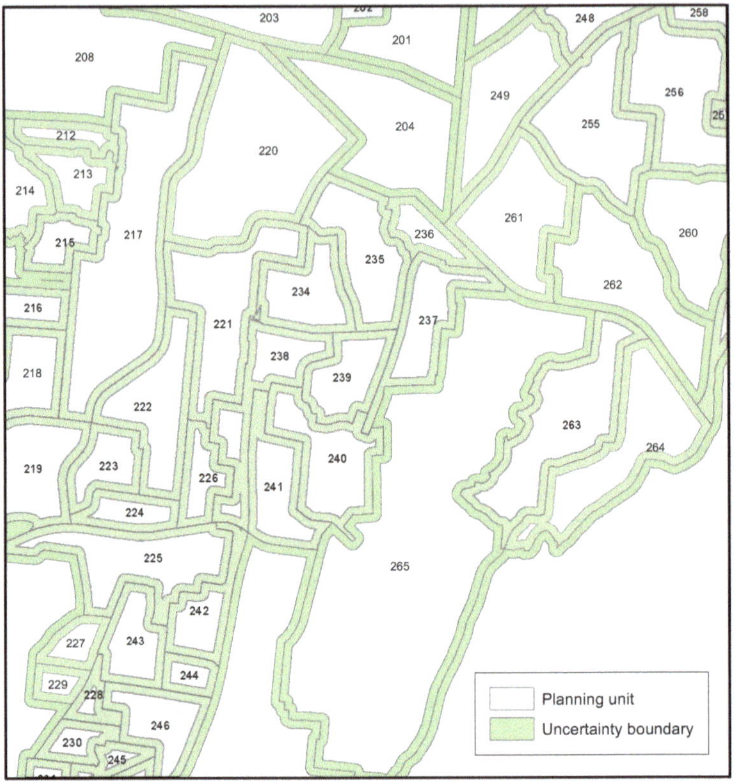

Fig. 7 Uncertainty associated with adjacency/neighboring spatial relationships

figures are relied upon to summarize recommendations and findings derived from optimization models. Consider the solution shown in Fig. 8 identified using the ArcGIS version 10.5 location-allocation function described previously, accessed through the Network Analysis toolbox. As noted above, the formal specification of the model is reflected in (7)–(11). Xu et al. (2020) indicate that a heuristic is used for solution. The depicted location analysis examines the Special Supplemental Nutrition Program for Women, Infants, and Children in the Santa Barbara area, seeking the best locations for this federally funded program to provide nutrition, healthy foods, breastfeeding education, and health care service to the region. Demand corresponds to 2070 census blocks, totaling 200,450 people (e.g., $\sum_i \delta_i =$ 200,450). Potential facility sites are identified in advance, totaling 82 potential sites. Travel and access are via the road network, with a service coverage standard assumed to be 5 miles. In this instance, three ($p = 3$) facilities are considered, with a capacity of $\theta_j =$ 64,135. The solution shown in Fig. 8 is capable of covering, or serving, 170,144 people within the 5-mile travel distance standard.

In contrast to the heuristic solution, Fig. 9 depicts the optimal solution for this problem instance. This solution was obtained using Xpress, and is proven to be

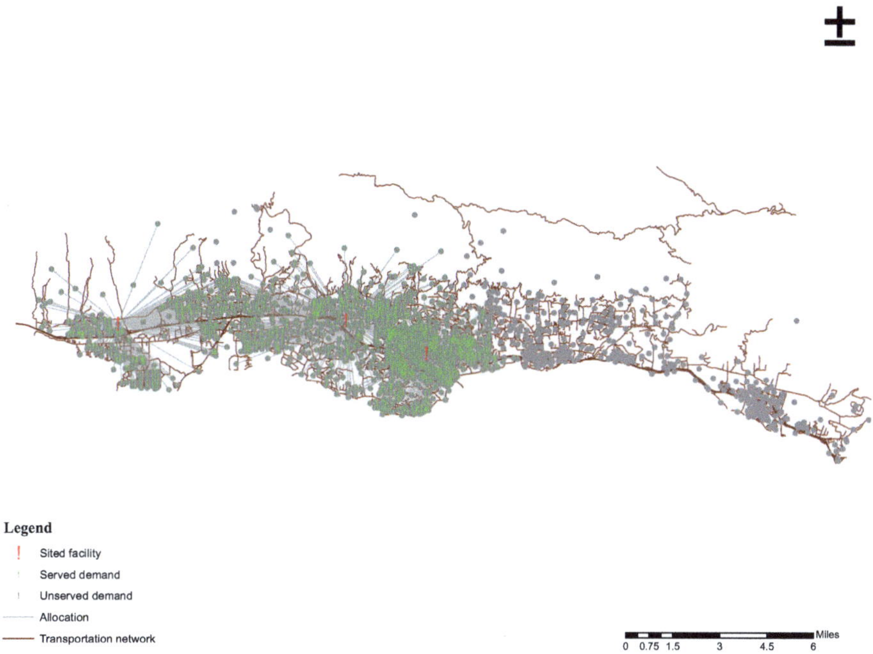

Fig. 8 ArcGIS capacitated "MAXIMIZE_COVERAGE" results

within 0.001% of optimal. Noteworthy is that 182,749 people can now be covered within 5 min, indicating that the ArcGIS heuristic solution is 6.90% less than the optimum. To achieve this, the facility siting and allocation differ in significant ways. Since access is incredibly important for social service like this, such an improvement is noteworthy. But the important point here is that challenges associated with communication, and in particular solution quality, clearly exist.

4 Discussion

There is much more that could be said and demonstrated regarding uncertainty in location analysis. This chapter offers one perspective, with supporting examples to illustrate particular instances that can be observed. At a minimum, there likely is a more general issue of effective communication, but perhaps ill-advised usage and application of location analysis as well. Returning to the contrast between the heuristic results produced by ArcGIS in Fig. 8 compared with exact results shown in Fig. 9, a few issues are worth highlighting. The output of analysis carried out in ArcGIS, as an example, offers no communication of potential suboptimality with the utilized heuristic (Fig. 8), nor is this true for the exact results shown in Fig. 9. It is likely critical that any ArcGIS location-allocation solution be considered

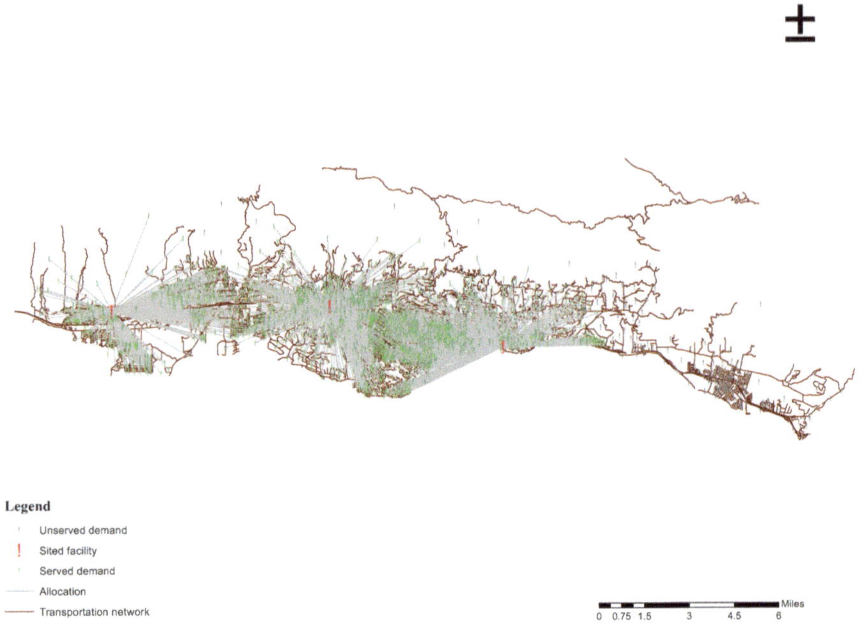

Fig. 9 Capacitated MCLP, (7)–(11), optimal results

uncertain with respect to optimality, as it may or may not be good given the very nature of a heuristic. Yet there is an abundant and growing literature detailed in Xu et al. (2020) applying this particular model that essentially fails to recognize any limitations with derived findings. Further, many often refer to ArcGIS heuristic results as "optimal," which is clearly incorrect. Murray et al. (2019) highlight this as well for a different model in ArcGIS, indicating that it is not an isolated incident but rather something more widespread and pervasive. Indeed, it is likely the by-product of easy-to-use software like ArcGIS, offering access to range of functions, methods, procedures, and models, making it possible to misuse them in various ways. And as noted previously, the offered documentation is generally lacking with respect to the precise model or solution approach, making it very much "black box" in nature. This is in contrast to the location modeling literature where problems are explicitly detailed and solution method limitations very well understood.

The included examples of uncertainty implications are admittedly limited, with additional nuances emphasized as well as examples associated with understanding, model specification, and attributes that could have been explored. While some overlap with themes of risk, stochastic optimization, and robust optimization appears in Fig. 1/Table 1, important nuances of uncertainty would appear to be omitted in previous reviews of uncertainty in location analysis.

5 Conclusions

This chapter provided an overview of uncertainty in location analysis. It was noted that the categories of uncertainty outlined in Fig. 1 and Table 1 intentionally deviate from past reviews of uncertainty in location analysis. Such past reviews have focused on particular types of models, such as risk, robust, and stochastic approaches. The reason and rationale for viewing this differently is to better account for the defining characteristics of uncertainty centered on doubt. In this way, understanding of problem/issue, abstraction, model specification, attribute(s), location, spatial properties, solution, and implementation all contribute to and have major implications for uncertainty. A number of examples illustrating modeling implications for select categories were detailed, making it evident how uncertainty arises. The intent is to make future avenues for investigation more comprehensive, and ultimately ensure that uncertainty is addressed in a rigorous fashion.

References

Berge, C. (1957). Two theorems in graph theory. *Proceedings of the National Academy of Sciences of the United States of America, 43*(9), 842–844.

Chen, H., Murray, A. T., & Jiang, R. (2021). Open-source approaches for location cover models: Capabilities and efficiency. *Journal of Geographical Systems*.

Church, R. L., & Murray, A. T. (2009). *Business site selection, location analysis and GIS*. Wiley.

Church, R. L., & Murray, A. (2018). *Location covering models*. Springer.

Correia, I., & Saldanha-da-Gama, F. (2019). Facility location under uncertainty. In G. Laporte, S. Nickel, & F. Saldanha da Gama (Eds.), *Location science* (pp. 185–213). Springer.

Edmonds, J. (1962). Covers and packings in a family of sets. *Bulletin of the American Mathematical Society, 68*(5), 494–499.

Laporte, G., Nickel, S., & Saldanha da Gama, F. (2019). *Location science*. Springer.

Marianov, V. (2022). User related uncertainties in facility location. In H.A. Eiselt, & V. Marianov (Eds.), *Uncertainty in facility location problems*.

Murray, A. T. (2003). Site placement uncertainty in location analysis. *Computers, Environment and Urban Systems, 27*(2), 205–221.

Murray, A. T. (2005). Geography in coverage modeling: Exploiting spatial structure to address complementary partial service of areas. *Annals of the Association of American Geographers, 95*(4), 761–772.

Murray, A. T. (2018). Evolving location analytics for service coverage modeling. *Geographical Analysis, 50*(3), 207–222.

Murray, A. T., & Grubesic, T. H. (2012). Spatial optimization and geographic uncertainty: Implications for sex offender management strategies. In M. Johnson (Ed.), *Community-based operations research* (pp. 121–142). Springer.

Murray, A. T., Xu, J., Wang, Z., & Church, R. L. (2019). Commercial GIS location analytics: Capabilities and performance. *International Journal of Geographical Information Science, 33*(5), 1106–1130.

Murray, A. T., Church, R. L., & Feng, X. (2020). Single facility siting involving allocation decisions. *European Journal of Operational Research, 284*(3), 834–846.

Snyder, L. V. (2006). Facility location under uncertainty: A review. *IIE Transactions, 38*(7), 547–564.

Toregas, C., Swain, R., ReVelle, C., & Bergman, L. (1971). The location of emergency service facilities. *Operations Research, 19*(6), 1363–1373.

Wei, R., & Murray, A. T. (2012). An integrated approach for addressing geographic uncertainty in spatial optimization. *International Journal of Geographical Information Science, 26*(7), 1231–1249.

Wei, R., & Murray, A. T. (2015). Continuous space maximal coverage: Insights, advances and challenges. *Computers & Operations Research, 62*, 325–336.

Wei, R., & Murray, A. T. (2018). Spatial uncertainty challenges in location modeling with dispersion requirements. In *Spatial analysis and location modeling in urban and regional systems* (pp. 283–300). Springer.

Xu, J., Murray, A., Wang, Z., & Church, R. (2020). Challenges in applying capacitated covering models. *Transactions in GIS, 24*(2), 268–290.

Risk, Hazard, and Exposure Time in Hazmat Location and Routing

Andrés Bronfman, Germán Paredes-Belmar, Vladimir Marianov, and H. A. Eiselt

Abstract Hazardous materials such as fuel, solvents, organic waste from hospitals, used batteries, explosives, and nuclear waste need to be transported to and from the facilities that use, produce, and dispose of them. Managing these transports requires a design that alleviates negative effects of these activities, such as the loss of lives, environmental damage, and the destruction of property. Despite the large body of literature addressing numerous aspects regarding hazardous materials, there is no clear consensus on how potential adverse effects should be measured when optimizing facility location and route design. Our analysis commences with a look at the primary stakeholders in these activities: the population that is potentially affected by transportation, the firms that pay for it, and the government regulator, whose task is to protect the population at large. This chapter proposes two new indicators related to these activities, which are easy to compute, avoid the use of unreliable very low probability estimations, take care of the regulatory agencies and public concern, and, in our view, are more understandable to the public. Mathematical programming problems that integrate criteria for all stakeholders are formulated and solved. The methodology is then applied to a real case in order to determine an optimal transport route for the transport of hazardous materials in and out of the city of Santiago, Chile.

A. Bronfman
Engineering Sciences Department, Universidad Andres Bello, Santiago, Chile
e-mail: abronfman@unab.cl

G. Paredes-Belmar
School of Industrial Engineering, Pontificia Universidad Católica de Valparaíso, Valparaíso, Chile
e-mail: german.paredes@pucv.cl

V. Marianov (✉)
Department of Electrical Engineering, Pontificia Universidad Católica de Chile, Instituto Sistemas Complejos de Ingeniería (ISCI), Santiago, Chile
e-mail: marianov@ing.puc.cl

H. A. Eiselt
Faculty of Management, University of New Brunswick, Fredericton, NB, Canada
e-mail: haeiselt@unb.ca

H. A. Eiselt, V. Marianov (eds.), *Uncertainty in Facility Location Problems*, International Series in Operations Research & Management Science 347,
https://doi.org/10.1007/978-3-031-32338-6_2

Keywords Hazardous materials transportation · Network design · Hazard · Risk minimization

1 Introduction

Worldwide, large quantities of hazardous materials are transported through urban areas between facilities that use it, produce it, or dispose of it as industrial waste, some of them located within these areas, e.g., gas stations. This activity entails considerable hazard due to accidental release (spills, fires, explosions, etc.). A classification of these materials can be found in EPA (2012).

Hazard is defined in the Britannica Dictionary as "a source of danger" (The Britannica Dictionary, 2022). It is the potential of an undesirable event *without regard of how likely it is*. Risk, on the other hand, as defined by Prince2 (2009), is "An uncertain event or set of events that, should it occur, will have an effect on the achievement of objectives. A risk is measured by a combination of the probability of a perceived threat or opportunity occurring, and the magnitude of its impact on objectives." In other words, risk includes a *probability* or *likelihood* weighted by a *consequence* or *impact* (i.e., an expected value). Hazard and risk are sometimes confused in common language, but they are different because hazard does not involve probabilities.

Dealing with situations involving hazardous materials is difficult, as there are several stakeholders involved: the main "players" in the game are regulatory agencies, the industry, and the public. Regulatory agencies intervene to reduce the *average risk* to the public and the environment, i.e., to minimize the *expected consequence* of incidents averaged over the whole region, and must do so without threatening the economic viability of the activity being regulated. With a few exceptions, risk is computed considering that all people, buildings, or vulnerable environment within a certain distance of a facility or a segment of the route are affected in some negative way.

Firms handling hazardous materials and operators involved in their transportation have a different point of view. Their concern is minimizing location and transport costs, within the frame established by the regulatory agencies.

The third point of view is that of the public. The general public typically opposes any activity involving hazardous materials in its neighborhood, seeking to minimize hazard. The main reason for the general public to choose hazard and not risk as a criterion is the public's inability to determine the probability, its inability to interpret it, and the main concern about the possibility of an accident and its impact on their own lives. In fact, the announcement that a chemical plant will be open or nearby routes will be used for transportation of corrosive liquids, for example, will generate strong public opposition, because both are perceived as very hazardous by the public, in spite of the fact that the risk to which this population will be exposed is usually extremely low, due to the very small probability of an accident. For instance, the US Department of Transportation (2022) states that in 2021, while there were 23,705 incidents in 2021 (highway, water, rail, and air, almost all of them highway), the total number of fatalities in that year was 2. In other words, when dealing with hazardous activities, people are not concerned with, or do not understand, probabilities and thus the resulting risk; they just consider hazard. In their minds, sooner or later an accident could happen, whose likelihood does not matter much. Moreover, people are not interested in averages, but in what can happen to them, which is assumed to be the worst case. Accidents can have very significant consequences on human lives or health, environmental damage, or costly damage, which is what the public sees. In the case of undesirable facilities, hazard is (correctly) perceived by the public as decreasing with their distance to the facility, so they prefer these facilities as well as the transport routes for the hazardous materials being as far as possible from them (Hung & Wang, 2011). In order to express the sentiment of the public, we use the public understanding of hazard: the potential of something bad happening, whose effects decrease with distance. The hazard is then a function of the distance between a populated point and the point that generates the hazard. To compute the individual hazard to which a route segment exposes a populated point, the hazard function is integrated over the length of the segment of a transportation route that lies within a safety distance of a population center. This individual hazard is then weighted by the population.

Cost and average risk have been dealt with profusely in the literature and practice: definitions of location and transportation cost are standard, and estimators of average risk have been calculated in various ways and used as an estimate of the disutility imposed on the population. Erkut et al. (2007) identify nine different models of risk for hazardous materials routing, each using its own method of combining the likelihood or probability of an incident on route segments with the associated potential consequences, or one of these factors alone. Mohri et al. (2022) offer an overview of hazardous materials transportation problems and present a complete table of what they call "risk measures," the great majority of them using some combination of consequences and probabilities.

Besides the fact that probabilities are not well understood by the public, a further drawback of using them is that unwanted events occur with very low probabilities and extremely severe consequences (not unlike airplane crashes, terrorist attacks, meltdowns of nuclear power plants, or similar catastrophic, but highly unlikely events). The estimates of the probabilities of unwanted events in specific points or route segments, computed using past history, are highly unreliable, as the rare occurrence of events means that there is not enough past data to achieve adequate precision. Furthermore, these rare events may have been caused by conditions that have no relation to the specific point or route and change in time. The product of an unreliable low probability times a large number indicating consequence is also unreliable.

Given the unreliability of probability estimations, as far as the regulator's objective of risk is concerned, we propose a measure that assesses the adverse effects in hazardous materials transportation. In particular, we consider the time length of exposure of the population to the hazardous material(s), which can also be seen as a proxy of probability. Clearly, the longer a part of the population is exposed to a dangerous activity, the higher the likelihood of something bad happening to them.

The firms' costs are straightforward and no proxy is needed. However, it is necessary to annualize the one-time location costs and the recurring transportation costs.

The remainder of the chapter is organized as follows: Sect. 2 reviews the literature related to the location and routing of hazardous materials. Section 3 introduces the estimators of hazard and period of exposure. The formulation of models using these new objectives is contained in Sect. 4, while Sect. 5 is devoted to a real case in Santiago. Section 6 summarizes the findings of this chapter and points at several future research directions.

2 Literature Review

The hazardous materials transportation problem has been widely studied, especially in the operations research field; see, e.g., Ditta et al. (2019), Holeczek (2019), Ma et al. (2020), and Mohri et al. (2022). One of the main concerns in hazardous materials transport research is the minimization of some estimator of the adverse effects resulting from a possible release of the material during its transportation. In terms of these estimators, ReVelle et al. (1991) minimize exposed population; Saccomanno and Chan (1985) and Abkowitz et al. (1992) minimize incident probability; Pijawka et al. (1985), Batta and Chiu (1988), Alp (1995), and Erkut and Verter (1995) minimize the product of incident probability and incident consequence; Sivakumar et al. (1993), Sivakumar et al. (1995), and Sherali et al. (1997) minimize the expected consequence given that an accident occurs on the route; Erkut and Ingolfsson (2000) propose diverse objectives: minimization of the maximum population exposure; simultaneous minimization of expected value and the variance of the number of people affected by an accident within a circle around the event, with both factors represented as attributes of each route link; and minimization of the expected disutility, defined as $u(X) = \exp.(\alpha X)$ where X is the population affected and $\alpha > 0$ a constant that measures catastrophe aversion. Abkowitz et al. (1992) minimize perceived risk imposed by a link, measured as pC^q where p is the probability of an incident on a link, C the incident consequence, and q a risk preference parameter; Erkut and Ingolfsson (2005) assume that the occurrence of an incident terminates a trip so that a new shipment must be sent to satisfy the original demand, and thus use the total expected consequence of all the necessary trips. Finally, Holeczek (2021) studies different risk models, presenting a detailed analysis of the impact of load-dependent or load-independent risk models.

The above objectives are used in various approaches for modeling hazardous materials transportation. For example, some works recognize the multiple actors involved in decision making and the multi-objective nature of the hazardous materials routing problem, such as Zografos and Davis (1989), Marianov and ReVelle (1998), Zero et al. (2019), Bula et al. (2019), and Li and Leung (2011).

Considering the relationship between the carrier and the regulatory agency is a concern considered by Kara and Verter (2004), Erkut and Gzara (2008), Verter and Kara (2008), Bianco et al. (2009), and Bruglieri et al. (2014). Another group of contributions also addresses the issue of population risk equity, among them Gopalan et al. (1990), Lindner-Dutton et al. (1991), Carotenuto et al. (2007), and Caramia et al. (2010), who develop models that consider equity in the spatial distribution of risk along the generated routes. Finally, Abkowitz et al. (1990), Lepofsky et al. (1993), Lovett et al. (1997), Chang et al. (1997), Brainard et al. (1996), Frank et al. (2000), Chen et al. (2008), and Kim et al. (2011) use GIS tools to support the calculation, comparison, and visualization of the attributes of alternative routes, as well as to compare different risk modeling techniques and serve as a decision support system for hazardous materials transport.

An extensive literature review has addressed the location of obnoxious and hazardous facilities; see, e.g., Erkut and Neuman (1989), Church and Drezner (2022), Cappanera et al. (2003), Melachrinoudis (2011), Daskin (2011), and Colebrook and Sicilia (2013). In what follows, we focus on current studies of integrated location and routing models for hazardous materials. Different problems have been addressed in relation to transportation of hazardous materials. Zografos and Samara (1989) presented a mixed programming model to minimize the risks of hazardous waste transportation, travel times, and disposal risks to determine the location of waste treatment facilities and establish the associated shipping routes. ReVelle et al. (1991) minimized transportation risks and the risks perceived by the population. Current and Ratick (1995) considered the transportation costs of a unit of hazardous materials and the variable costs at the facilities, minimizing the risks and incorporating equity in distributing the risks. Helander and Melachrinoudis (1997) presented an integrated model for the location of a facility, minimizing the expected number of accidents along multiple hazardous materials transportation routes. Giannikos (1998) considered the total operation cost, total perceived risk, equitable risk distribution, and equitable distribution of the disutility caused by hazardous facility operation as objectives. Samanlioglu (2013) developed a programming model with three criteria: minimizing the total transportation cost of hazardous materials and waste, as well as the fixed cost of treatment, disposal, and recycling centers; minimizing the total transportation risk, measured as the population exposed along those routes; and minimizing the total risk of the population located around treatment and disposal centers. Zhao and Zhao (2010) focused on the diversity of waste types and treatment technologies, the compatibility and capacity of treatment technologies, and disposal centers. Asgari et al. (2017) addressed the obnoxious waste location-routing problem by considering different types of waste and several treatment technologies. They developed an optimization model that minimizes the cost of undesirable treatment and disposal facilities, and the risk of transporting hazardous materials.

Rabbani et al. (2018) considered the incompatibility between hazardous waste in their multi-objective industrial hazardous waste location-routing problem. The authors simultaneously minimized the total cost, transportation risk, and site risk. They used the exposed population along the routes as a measure of the transportation risk, while the site risk was measured through the product of the amount of waste available at each facility and the number of people within a threshold distance of it. Ghaderi and Burdett (2019) presented a two-stage stochastic programming model for a bi-modal hazardous materials location-routing problem. Road and rail transportation are considered. The risk is on the arcs of transportation and in the transfer nodes of the bi-modal network. They implement three algorithms: sample average approximation, maximum likelihood sampling algorithm, and a combination of both. Ziaei and Jabbarzadeh (2021) solve a similar problem of Ghaderi and Burdett (2019) applied to gasoline transportation. They minimize risk (arcs and locations) and cost, addressing uncertainties in parameters of cost and risk and CO_2 emissions. Hassanpour et al. (2021) solve a hazardous materials location-routing problem considering edge unavailability (random edge disruptions), time-dependent parameters, and time windows. A robust optimization approach is used to solve the problem. They minimize cost and risk functions. The risk is considered in transportation and in locations.

In all the above contributions, both for the hazardous materials transport problem and for the location-routing problem, the consequence or the risk associated with the hazardous materials transportation is always expressed as the risk posed by each link of the route to its surroundings, as opposed to the risk to which are exposed the population centers, possibly because this approach decreases the number of variables. If two or more links or materials affect a single center, however, the magnitude of the effect over that particular center will not be captured when a route is designed following these approaches. This was recognized by List and Mirchandani (1991), Erkut and Verter (1995), and Bronfman et al. (2015) and recently in Fontaine et al. (2020). In List and Mirchandani (1991), the risk associated with each route and population point is defined as a function of an integral, although they do not propose any specific functional form. The total risk posed by a given route is the sum of the risks each point on it poses to the various population centers. In their case study, the authors do not use this estimator but the expected fatalities. Furthermore, their formulation requires that the candidate routes be explicitly enumerated, and the risk posed by using each one of them be calculated. As it stands, it can be used only for choosing routes, not designing them. In a very complete work by Erkut and Verter (1995), a first model assumes population distributed at points (populated points) in the plane, surrounded by a danger area. The risk to which a populated point is exposed is computed as the product of the length of the route segment that falls within its danger area, and the population of the center and the probability of an incident (release of hazardous materials). Their second model assumes population distributed continuously and uniformly over the plane. A route segment has a rectangular hazard area around it, with a width of twice the reach of an incident (which, in turn, depends on the material being transported). For this representation to be valid, the whole area is decomposed so that all route segments are straight (making it perfect for vector representations in geographical information systems), and the population density is uniform around each route segment. An individual within the rectangle is exposed to a risk that is computed as in the first model, and the risk is integrated over the rectangle and assigned to the segment as an attribute of it. Both models are applied to the selection of one of a set of existing routes.

A different approach was followed by Bronfman et al. (2015), who addressed the problem of hazardous materials routing in urban areas. They maximize the population-weighted distance from the route to its closest vulnerable point in the region, where vulnerable points are hospitals and clinics, schools, senior homes, and, in general, sites that are difficult or slow to evacuate. Each such point has a hazard circle around it and, ideally, the route must not cross any of these circles. They propose an exact model and a reduction technique for the number of variables and constraints, as well as an optimal polynomial time heuristic to reduce the total length of such crossings, which minimizes the likelihood of possible undesired effects on the vulnerable points. In some sense, they use crossing length as a proxy of probability.

Hazard has not been used as an objective for hazardous materials transportation, although it has been studied in relation to dangerous activities or natural events. In those cases, it has been recognized that the hazard a population is exposed to is a function of the distance, an observation that accords with the perception of the general public for hazardous facilities (Hung & Wang, 2011; Elliott et al., 1999; Brody et al., 2004; Lima, 2004), earthquakes (Lindell & Perry, 2000), hurricanes (Arlikatti et al., 2006), and flooding (Wachinger et al., 2013; Miceli et al., 2008; Heitz et al., 2009; Brilly et al., 2005). Saccomanno and Shortreed (1993), Jonkman et al. (2003), and Fernández et al. (2000) also point to this fact in their argument that the possible consequences for the population in the case of a hazardous materials spill incident vary as a function of the distance from the event.

The next section develops estimators to measure adverse effects on the population. Our approach in some sense resembles that of List and Mirchandani (1991), although we provide explicit functions of hazard, and we formulate a model that allows designing a route, as opposed to choosing one. It is also related to Erkut and Verter (1995) and Bronfman et al. (2015).

3 Proposed Estimators of Adverse Effects

Each population or vulnerable center is represented as a point k in a plane around which a circular hazard zone of radius λ_t is defined, as shown in Fig. 1, where t is the index of the material being dealt with. For the moment and for the sake of clarity, we do not use the subscript. The links of the route consist of straight-line sections of it. Segments (a, b) of link (i, j) and (d, e) of link (j, l) are the parts of a hypothetical route that expose the population to hazard and are therefore denoted the *exposure segments*. Different materials have different hazard zone radiuses and therefore different exposure segments.

Figure 1 depicts how a population center can be affected by more than one link, especially in urban areas. By expressing adverse effects as attributes of the affected center rather than of a link, we can account for the aggregate effect of all links on a

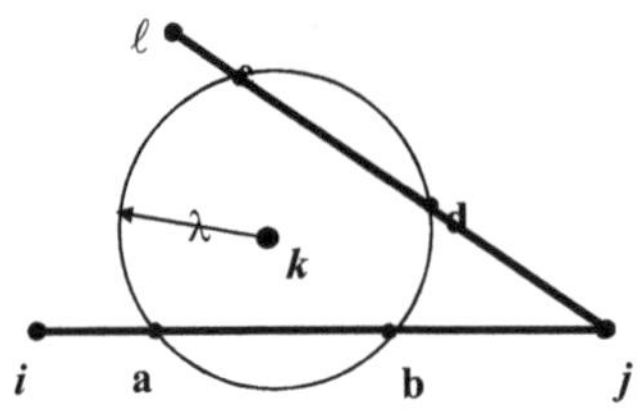

Fig. 1 Population at point k, with its circular hazard zone and exposure segments (a, b) and (d, e)

given center. The *hazard* imposed on k is a non-decreasing function of the individual hazard values imposed by each of the two link segments (a, b) and (e, d). In this chapter, we will use a simple sum of individual hazards to express the total hazard. (Depending on the case, that may underestimate the hazard due to an interaction effect: loads of household chemicals such as ammonia and bleach will, if mixed together, produce chloramines, which are much more of a health risk than ammonia and bleach by themselves). The *period of exposure* of the population of center k is the sum of the times during which it is exposed due to the use for hazardous materials transport of either link.

Expanding to the multi-product case, the hazard imposed on k by shipments of hazardous materials t is the sum of the hazard values imposed by each shipment t on its respective exposure segment. Also, the period of exposure of the point k is the sum of the times during which it is exposed to the use of the exposure segments of each shipment with hazardous materials t.

Note that, in most cases, a vulnerable center can be represented by a point on the plane. If the population to be protected is continuously distributed over the region, aggregation errors may be treated as in Sadigh and Fallah (2009) and Francis et al. (2004). Methods for reducing aggregation errors have been proposed by Current and Schilling (1990) and Emir-Farinas and Francis (2005).

3.1 *Hazard at a Population Center*

Let a transport network be represented by a directed graph $G(N, A)$, where N is the set of nodes and A the set of links. To derive a formal expression for the concept of hazard exposure, let $f^{\,k}(x)$ be the hazard to each individual in population center k emanating from a point x on link (i, j). The function $f^{\,k}(x)$ is assumed to be non-increasing in the distance $r^{\,k}(x)$ between x on the link (i, j) and k, and the form of the function depends on the type of material being transported. Then let f_{ij}^k be the hazard each individual in k is exposed to by the use of exposure segment (a, b) of link (i, j). To determine the value of f_{ij}^k, we divide the exposure segment (a, b) into a finite number $n = \frac{|b-a|}{\Delta x}$ of intervals of equal length Δx (see Fig. 2). Each interval represents a separate hazard to k that depends on the distance between them. Thus, the contribution to the hazard to k of a hazardous materials vehicle traveling each interval Δx in (a, b) is given by $f^{\,k}(x)\Delta x$. Summing the hazard represented by each

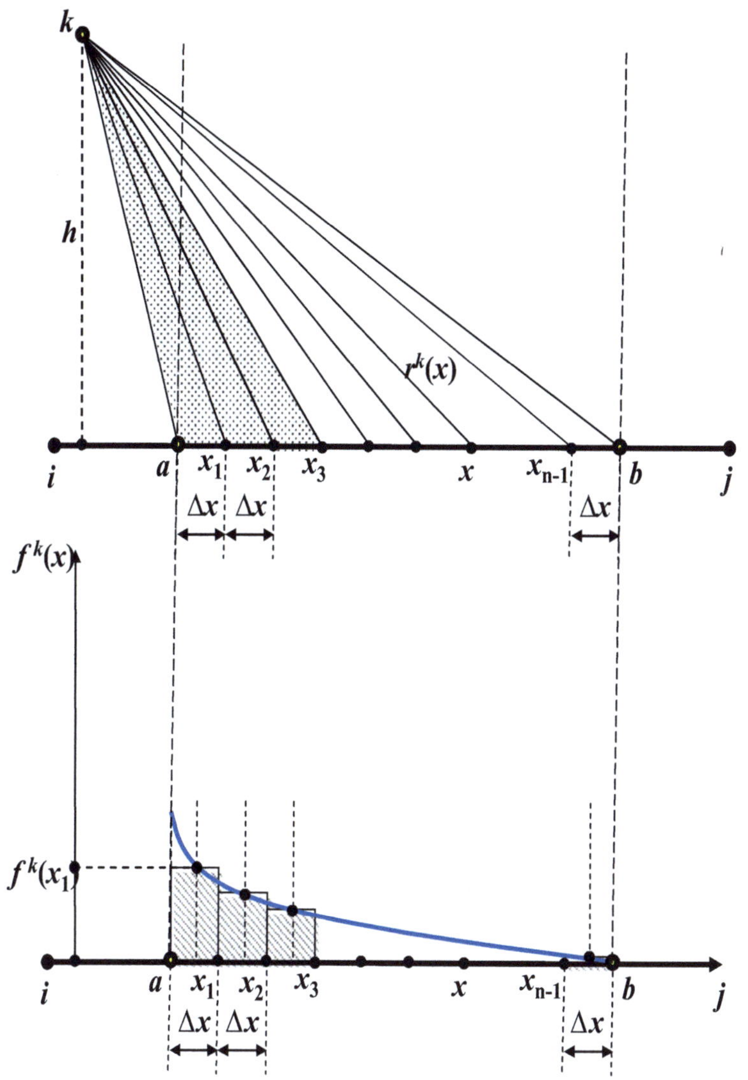

Fig. 2 Calculation of the hazard k is exposed to by use of segment (a, b) of link (i, j)

interval in segment (a, b), we obtain the following approximation:

$$f^k(x_1)\,\Delta x + f^k(x_2)\,\Delta x + \cdots + f^k(x_n)\,\Delta x = \sum_{q=1}^{n} f^k\left(x_q\right)\Delta x.$$

The value of f_{ij}^k is the limit of this sum as Δx tends to 0. Thus,

$$f_{ij}^k = \lim_{\Delta x \to 0} \sum_{q=1}^{n} f^k\left(x_q\right) \Delta x = \int_a^b f^k(x)dx. \tag{1}$$

The hazard function $f^k(x)$ can take various forms. Some of such forms used in modeling real situations of hazardous materials dispersion involve quadratic and exponential functions, as follows.

Example 1 Hazard is inversely proportional to the square of the Euclidean distance between the population unit and the location of the hazardous materials vehicle:

$$f^k(x) = \frac{1}{r^k(x)^2 + \varepsilon^2} \tag{2}$$

where $\varepsilon \geq 0$ is a constant that ensures $f^k(x)$ is not undefined when $r^k(x) = 0$.

Substituting (2) into (1) and solving the integral, we obtain

$$f_{ij}^k = \frac{1}{\sqrt{h^2+\varepsilon^2}}\left[\arctan\left(\frac{b}{\sqrt{h^2+\varepsilon^2}}\right) - \arctan\left(\frac{a}{\sqrt{h^2+\varepsilon^2}}\right)\right] \tag{3}$$

where h is the distance between population unit k and link (i, j), measured along a line that is perpendicular to the link.

Example 2 Hazard is an exponential function of the square of the Euclidean distance between the population unit and the location of the hazardous materials vehicle:

$$f^k(x) = e^{-\theta\left[r^k(x)\right]^2}$$

Substituting this expression into (1) and solving the integral, we obtain

$$f_{ij}^k = \frac{\sqrt{\pi}}{2\sqrt{\theta}}\operatorname{erf}\left[\sqrt{h^2+x^2}\sqrt{\theta}\right] \Rightarrow$$
$$f_{ij}^k = \frac{\sqrt{\pi}}{2\sqrt{\theta}}\left[\operatorname{erf}\left[\sqrt{h^2+b^2}\sqrt{\theta}\right] - \operatorname{erf}\left[\sqrt{h^2+a^2}\sqrt{\theta}\right]\right]$$

where *erf* (z) is the integral of the normal distribution:

$$\operatorname{erf}(z) = \frac{2}{\sqrt{\pi}}\int_0^z e^{-t^2}dt$$

Example 3 Hazard is a logistic function, that is, the exponential of the Euclidean distance between the population unit and the location of the hazardous materials vehicle:

$$f^k(x) = \frac{1}{1 + e^{(\alpha + \beta r^k(x))}},$$

where α and $\beta > 0$ are parameters to be estimated. Substituting this expression into (1) and solving the integral, we obtain

$$f_{ij}^k = \left(\sqrt{h^2 + b^2} - \frac{\log\left(e^{\left(\alpha + \sqrt{h^2+b^2}\right)} + 1 \right)}{\beta} \right) - \left(\sqrt{h^2 + a^2} - \frac{\log\left(e^{\left(\alpha + \sqrt{h^2+a^2}\right)} + 1 \right)}{\beta} \right).$$

This function can model different intensities of hazard to represent the transport of different types of hazardous materials. The smaller the value of α, the greater the hazard the population is exposed to, and the larger is the value of β, the greater is the decrease in the hazard as the distance to the link increases.

Our proxy for risk is the total population-weighted exposure time F_W^k facing a population center k due to the use of a route W for hazardous materials transport. It is given by the following formula, where G^k is the population of center k:

$$F_W^k = \sum_{(i,j) \in W} f_{ij}^k G^k \tag{4}$$

If the travel speed of hazardous materials transport over the link segment (a, b) in Fig. 2, assumed to be constant over the segment's entire length, is doubled, the vehicle's travel time on (a, b) will be reduced by half. Although the hazard to which k is exposed does not change, the period of exposure of the population does. Including the additional indicator, we propose next, can capture this effect.

Obviously, in urban areas, there are multiple factors that make the speed over a network arc not constant over time (vehicle congestion, traffic accidents, weather conditions, etc.), and there is extensive literature that addresses the time dependency in vehicle routing problems; see Malandraki and Daskin (1992) and Ichoua et al. (2003). However, to our knowledge, there are no proposed exposure time indicators for hazardous materials transport in the literature.

3.2 Period of Exposure of the Population

The period of exposure of the population (hereafter simply "period of exposure") depends on the length of the route segments that intercept the hazard zone of the population center k, and on the speed s_{ij} of the hazardous materials vehicles over each link (i, j). Thus, the period of exposure t_{ij}^k for k due to the use of link segment $(i, j) \in A$ is given by

$$t_{ij}^k = \ell_{ij}^k / s_{ij}, \tag{5}$$

where ℓ_{ij}^k is the length of the segment of link (i, j) which exposes population center k. Assuming s_{ij} is uniform over each link, the period of exposure T_W^k for k due to the use of route W to transport a load of hazardous materials is given by the following formula:

$$T_W^k = \sum_{(i,j)\in W} t_{ij}^k \tag{6}$$

4 Hazardous Materials Routing Models with Multiple OD Pairs/Multiple Materials

In what follows we formulate two models for using and comparing two different objectives consisting of the indicators proposed in the previous section. These objectives can be easily combined with cost objectives.

The first model, M_1, is a bi-objective model that minimizes a linear convex combination of a normalized expression of overall population-weighted hazard (an objective of interest to the public) and the population-weighted period of exposure time (an objective of interest to the regulating agency). We also investigate two special versions of model M1: First, there is the bi-objective model M1*, which minimizes risk (population-weighted exposure time) and transportation cost (the objectives of the regulator and the firm, respectively). The second version is the model M2*, which is another bi-objective model that minimizes population-weighted hazard and transportation costs (i.e., the public and the firms' objectives. We then formulate and solve model M2, which minimizes the total hazard, but puts upper bounds on individual hazard and individual periods of weighted exposure. As such, it addresses the concerns of the public at large while guaranteeing that no single individual is affected too much.

Let us now N^q denote the set of hazardous materials shipments between the origin-destination pair $q \in Q$. Note that different materials are treated as different origin-destination pairs. If the same physical origin-destination pair, say $q = (a, b)$, requires transportation of two different materials, in the model there will be two different "virtual origin-destination pairs" $q_1 = (a_1, b_1)$ and $q_2 = (a_2, b_2)$, both referring to physical origin-destination pair (a, b). The hazard radiuses will be possibly different (Beneventti et al., 2019). We define the following binary variables:

$$x_{ij}^{tq} = \begin{cases} 1 \text{ if arc } (i, j) \text{ is used for shipment } t \in N^q \\ \quad \text{between the origin-destination pair } q \in Q \\ 0 \text{ otherwise} \end{cases}$$

The first model is formulated as follows:

$$M_1 : \text{Min} \sum_{i=1}^{2} \delta_i \left[\frac{f_i - I_i}{AI_i - I_i} \right] \tag{7}$$

subject to

$$f_1 = \sum_{k \in K} \sum_{(i,j) \in A} \sum_{q \in Q} \left[\sum_{t \in N^q} \left(f_{ij}^k x_{ij}^{tq} \right) G^k \right] \tag{8}$$

$$f_2 = \sum_{k \in K} \sum_{(i,j) \in A} \sum_{q \in Q} \left[\sum_{t \in N^q} \left(t_{ij}^k x_{ij}^{tq} \right) G^k \right] \tag{9}$$

$$\sum_{\{j/(i,j) \in A\}} x_{ij}^{tq} - \sum_{\{j/(j,i) \in A\}} x_{ji}^{tq} = \begin{cases} 1 & \text{if} \quad i = O^q \\ -1 & \text{if} \quad i = D^q \ \forall i \in N, \forall q \in Q, \forall t \in N^q \\ 0 & \text{otherwise} \end{cases} \tag{10}$$

$$x_{ij}^{tq} \in \{0, 1\} \quad \forall (i, j) \in A, \forall q \in Q, \forall t \in N^q \tag{11}$$

Expression (7) corresponds to a normalized linear combination of expressions (8, 9), which are the population-weighted hazard and population-weighted period of exposure objectives, respectively. In (7), I_i is the best (lowest) possible value of objective f_i, and AI_i is its worst (highest) value. By normalizing the objectives, we avoid scaling problems. Each objective is multiplied by a weight factor $\delta_i \in [0, 1]$, with $\delta_1 + \delta_2 = 1$ which is changed in successive runs of the problem; to find an approximation of the efficient frontier, see Cohon (1978). Constraint set (10) represents flow conservation while (11) defines the nature of the variables.

Note that problem M_1 is separable by origin-destination pairs. As in this model, the adverse effects (or perceptions) are aggregated over the whole network; this model does not take into account the fact that for a particular center, the hazard or the period of exposure can be very high. Our second model M_2 addresses the issue. In M_2, one of the objectives is minimized. Alternatively, it is possible to minimize both objectives together or only one of them. Without loss of generality, we have chosen to minimize the total hazard. The problem M_2 can then be written as follows:

$$M_2 : \text{Min} \sum_{k \in K} \sum_{(i,j) \in A} \sum_{q \in Q} \left[\sum_{t \in N^q} \left(f_{ij}^k x_{ij}^{tq} \right) G^k \right]$$

subject to: (10)–(11)

$$\left[\sum_{(i,j) \in A} \sum_{q \in Q} \sum_{t \in N^q} \left(f_{ij}^k x_{ij}^{tq} \right) \right] G^k \leq \beta^k \quad \forall k \in K \tag{12}$$

$$\left[\sum_{(i,j) \in A} \sum_{q \in Q} \sum_{t \in N^q} \left(t_{ij}^k x_{ij}^{tq} \right) \right] G^k \leq \alpha^k \quad \forall k \in K, \tag{13}$$

where β^k and α^k can be set by the decision maker to represent different "protection levels," e.g., for centers k of different vulnerability.

5 Application

The models were applied to the real case of the transport of hazardous industrial solid waste (HW) between five origin-destination pairs in the city of Santiago, Chile (see Fig. 3 and Table 1).

For this case, we use a single hazardous material, but the extension to multiple materials is trivial, as each origin-destination pair can be either a different material or a different shipment, or both. The data regarding the road network and vulnerable centers are the same as those used in Bronfman et al. (2015). They consist of 6681 links, 2212 nodes, and 244 vulnerable centers (schools with over a thousand seventy students) populated by 386,254 people (students), distributed as shown in Fig. 3. The hazard zone radius of a hazardous material incident is taken to be $\lambda = 800$ m. This distance, about half a mile, is commonly chosen as the boundary line for hazardous materials. For each network link, the data include its length, travel speed for different times of day (morning peak, evening peak, and off-peak period), and

Table 1 Hazardous materials shipments by origin-destination pair, at morning peak period

Origin-destination pair	Shipments
O1-D1	2
O2-D2	1
O3-D3	1
O4-D4	3
O5-D5	1

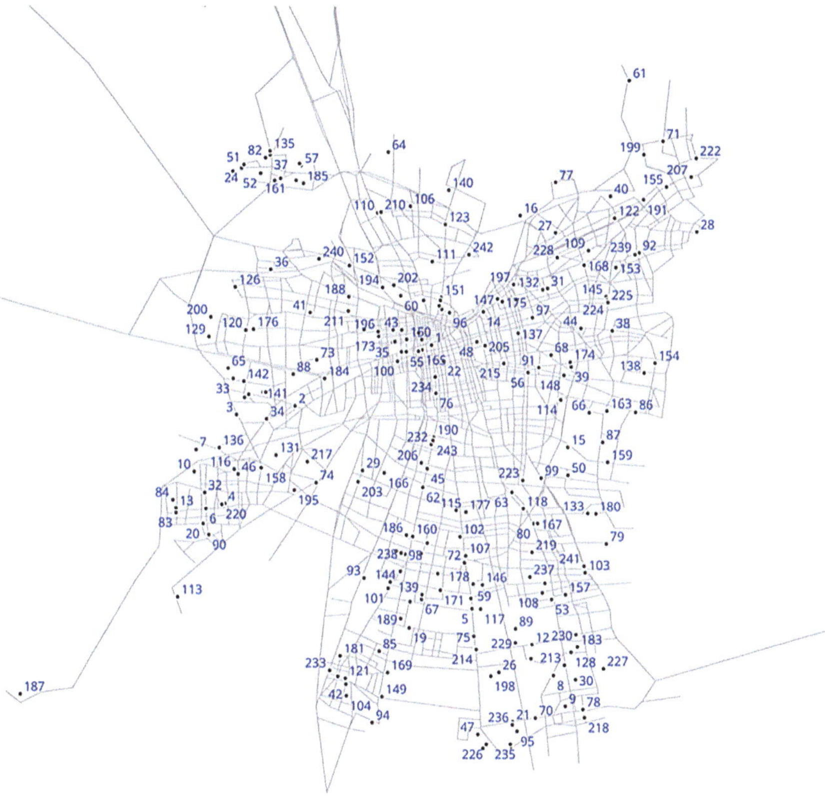

Fig. 3 Transport network and 244 schools with over a one thousand seventy students (vulnerable centers)

geographic coordinates. The transport of hazardous materials is evaluated during the morning peak period because students are at school at these times. This is done to avoid a worst-case scenario. The students in each school are assumed to be concentrated at its center. We identified the intersections of links with the hazard circles (exposure segments) of each school center k using simple geometry and the open-source geographical information system QGIS, version 3.12.3. We then applied Eqs. (3) and (5) to evaluate the hazard and period of exposure for each population center due to hazardous materials transport on each network link. The hazard function was assumed to be the inverse of the square of the distance as in Eq. (2) above, with $\varepsilon = 10^{-10}$. To calculate the hazard exposure for each k, f_{ij}^{k} as given by Eq. (3) was divided by $\theta = \text{Max}\ \{\theta^k | \, k \in K\}$, where $\theta^k = \max\limits_{(i,j)\in U^k} \left\{ f_{ij}^{k} G^k \right\}$ and U^k is the set of links $(i, j) \in A$ with segments within the hazard circle of k. The resulting hazard values are dimensionless. The instance was solved on a personal computer running Ubuntu 12.04 LTS with a 3.40 GHz Intel® Core™ i7-

2600 processor and 16 GB of RAM. The models were coded and solved using AMPL Cplex 12.5.

5.1 Results for M_1, M_{1*}, and M_{1**}

This subsection solves M_1 for different values of the weight δ_1 and approximates the efficient frontier. As this version of the problem considers only the public and regulating agency points of view, we then analyze the effect of considering each one of the new objectives on the transportation costs—the transportation company concern—represented by the total distance traveled $\sum_{(i,j)\in A} \sum_{q\in Q} \left[\sum_{t\in N^q} \ell_{ij} x_{ij}^{tq} \right]$, where l_{ij} is the length of arc (i, j). The bi-objective model M_{1*} uses as objectives the total weighted exposure time and the transportation cost, while the bi-objective model M_{1**} trades off the total hazard imposed on the population against the total transportation cost.

The values of I_i and AI_i shown in Tables 2–4 were obtained by solving each bi-objective model with extreme values of the weights δ_1. Tables 2–4 and Figs. 4, 5, and 6 show the efficient frontier approximations for the three versions of M_1. Also shown are the corresponding values of δ_1.

Table 2 shows how, going from $\delta_1 \approx 1$ to $\delta_1 \approx 0$ in M_1, the total hazard goes from 21.82 to 52.87, an increase of 2.4 times, while the population-weighted period of exposure decreases from 2874 to 1222 person-hours, a reduction of 57%. Good compromise solutions can be found in the efficient frontier; e.g., the hazard can be reduced from its maximum at 52.9 to only 32.4, in return for a small increase in time of exposure (from 1222 to 1286 person-hours).

Table 3 shows that a reduction of a 38% in the total transportation cost corresponds to an increase in the population-weighted period of exposure from 1221.5 a 12,340.3 person-hours, more than 10 times. Again, if transportation cost is

Table 2 Approximation of the efficient frontier for M_1

δ_1	Hazard	Period of exposure (person-hours)
≈ 0.0	$52.9 = AI_1$	$1221.5 = I_2$
0.1	36.7	1231.7
0.2	34.4	1248.0
0.3	32.4	1286.3
0.4	32.4	1286.3
0.5	32.4	1286.3
0.6	23.2	1819.2
0.7	22.5	1900.9
0.8	22.5	1900.9
0.9	22.3	1943.3
≈ 1.0	$21.8 = I_1$	$2784.4 = AI_2$

Table 3 Approximation of the efficient frontier for M_{1*}

δ_1	Period of Exposure (person-hours)	Length (km)
≈ 0.0	$12{,}340.3 = AI_2$	$201.94 = I_3$
0.1	8690.7	203.1
0.2	6575.1	205.6
0.3	4577.3	211.0
0.4	4113.8	213.1
0.5	2043.2	224.6
0.6	2008.3	224.9
0.7	1833.7	227.1
0.8	1479.9	233.8
0.9	1444.7	234.9
≈ 1.0	$1221.5 = I_2$	$278.60 = AI_3$

Table 4 Approximation of the efficient frontier for M_{1**}

δ_1	Hazard	Length (km)
≈ 0.0	$1502.6 = AI_2$	$201.94 = I_3$
0.1	270.0	202.5
0.2	256.3	202.7
0.3	223.2	203.4
0.4	167.5	205.6
0.5	143.4	207.2
0.6	143.4	207.2
0.7	78.2	217.4
0.8	65.1	220.3
0.9	43.2	228.6
≈ 1.0	$21.8 = I_2$	$311.0 = AI_3$

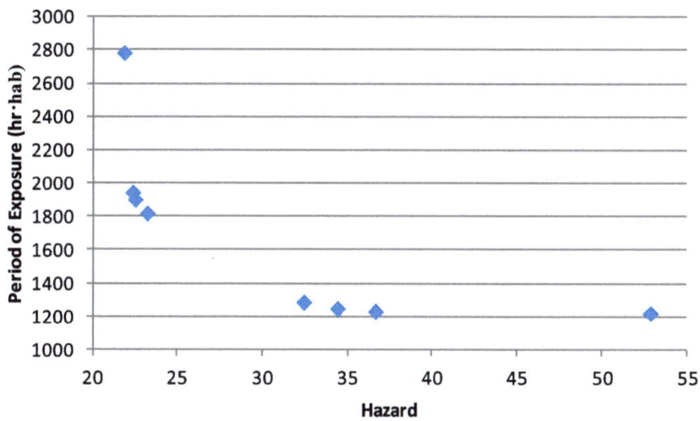

Fig. 4 Approximation of the efficient frontier for M_1

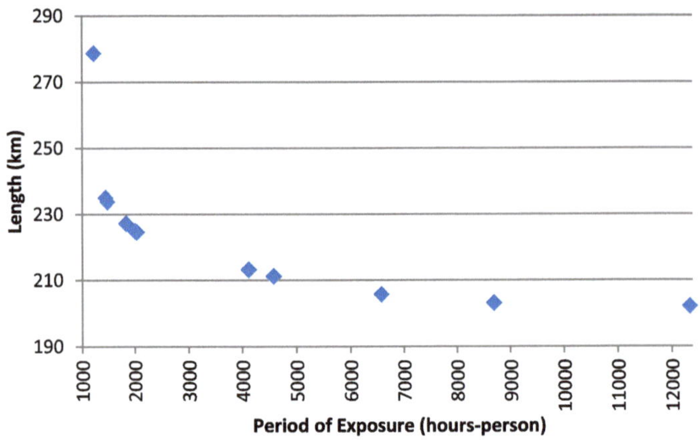

Fig. 5 Approximation of the efficient frontier for M_{1*}

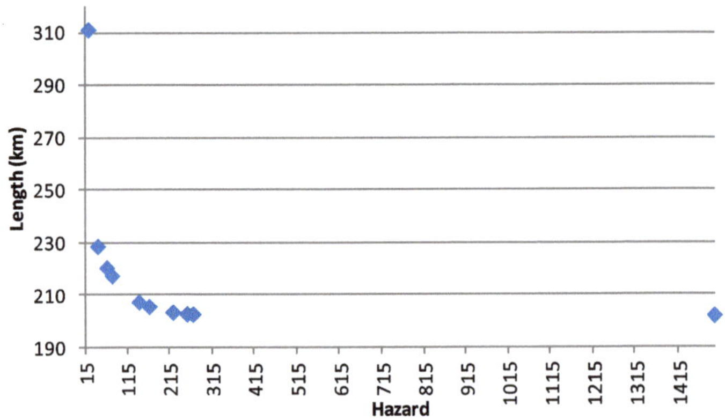

Fig. 6 Approximation of the efficient frontier for M_{1**}

increased from its minimum value by only a 4.5%, the period of exposure decreases to a 37% of its initial value ($\delta_1 = 0.3$).

Finally, Table 4 shows how an increase of a 54% of the transportation cost corresponds to an increase in hazard from 21.8 to 1502.6, equivalent to 68.9 times. A small increase of the transportation cost of a 7.1% reduces hazard in 19 times.

Figures 7a, b, and c show the transportation paths for the extreme values of δ_1 for each bi-objective model. Origins and destinations are marked in the figures, except for O3, which is out of the limits of the drawings. The hazard areas marked in gray are those intersected by the route, for one or more hazardous materials shipments.

Finally, these three tables allow choosing a strategy of a good compromise between hazard, risk, and transportation cost.

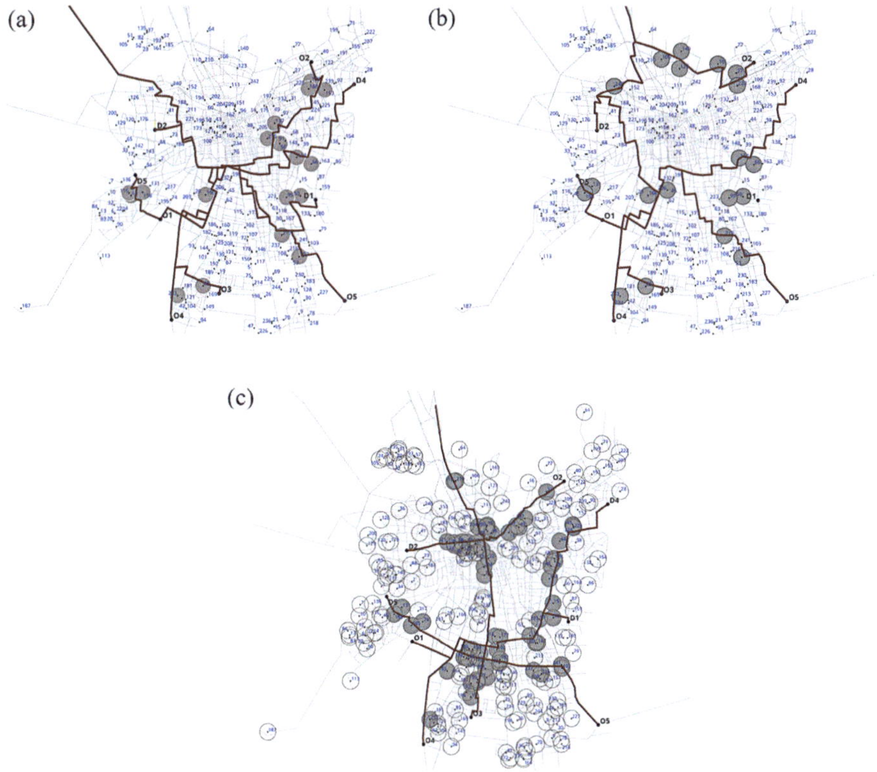

Fig. 7 Hazardous materials flows. (**a**) Model M_1 with $\delta_1 \approx 1$, (**b**) model M_1 with $\delta_1 \approx 0$, (**c**) model M_{1*} and M_{1**} with $\delta_1 \approx 1$ (shortest paths)

5.2 *Model M_2*

5.2.1 Analysis of M_2 for Different Values of α^k and β^k

We solved M_2 for different values of α^k and β^k, setting the value of β^k at a value that was sufficiently high ($\beta^k = 30 \ \forall\ k \in K$), so as to leave constraint (12) inactive. The parameter α^k was assigned values in the range (132.23; 414.99), in steps representing successive increments of 10% over the previous value. There is no feasible solution for $\alpha^k < 132.23$ person-hours, $\forall\ k \in K$, and for $\alpha^k > 414.99$ person-hours, $\forall\ k \in K$, we obtain the unconstrained solution for minimum hazard. Table 5a and Fig. 8a show the results. In this instance, an increase of 33% in the maximum period of exposure of each populated or vulnerable center (132.23 to 176.00 person-hours) reduces the total hazard by 25.45%. Table 5b and Fig. 8b show the results of a similar exercise when leaving α^k fixed at 460 person-hours and changing now the value of β^k by steps of 10% starting from its minimum feasible

Table 5 Hazard and total risk for different values of α^k and β^k

(a) M_2 with $\beta^k = 30\ \forall\ k \in K$ and different values of α^k			
α^k	Hazard	Risk (person-hours)	CPU time
132.23	33.41	2621.61	441.52
145.45	26.85	2091.55	13.62
160.00	26.21	2250.79	44.38
176.00	24.91	2191.14	9.59
193.60	23.77	2366.81	5.09
212.96	23.00	2215.55	4.60
234.26	23.00	2258.52	7.24
257.68	22.39	2297.96	2.83
283.45	22.18	2602.25	6.09
311.80	22.18	2602.25	3.38
342.97	21.87	2691.68	1.58
377.27	21.87	2691.68	1.77
414.99	21.82	2784.42	1.26
(b) M_2 with $\alpha^k = 460$ person-hours $\forall\ k \in K$ and different values of β^k			
β^k	Hazard	Risk (person-hours)	CPU time
2.20	24.83	3807.32	8.25
2.42	23.65	3382.72	2.58
2.66	23.60	3475.46	2.20
2.93	23.60	3475.46	2.47
3.22	23.60	3475.46	3.12
3.54	23.60	3475.46	2.80
3.90	23.18	3096.51	2.66
4.29	22.71	3129.94	2.11
4.72	22.71	3129.94	2.17
5.19	22.71	3129.94	2.80
5.71	22.52	2734.28	2.49
6.28	21.82	2784.42	1.08
6.90	21.82	2784.42	1.06

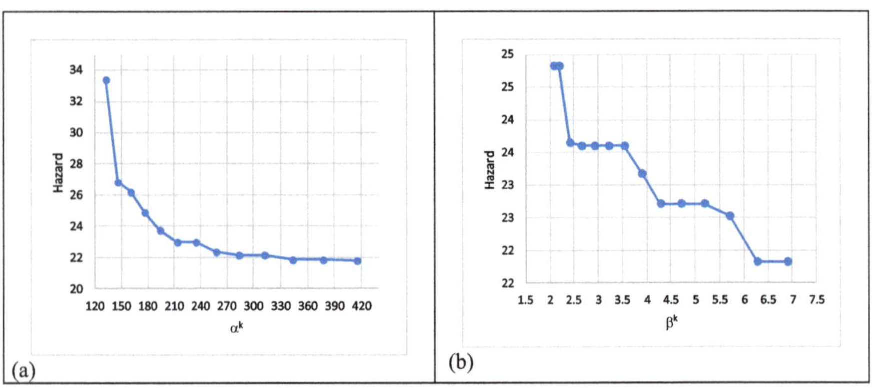

Fig. 8 Hazard and total risk for different values of α^k and β^k. (**a**) Varying α^k and $\beta^k = 30\ \forall\ k \in K$; (**b**) varying β^k and $\alpha^k = 460$ person-hours $\forall\ k \in K$

value. In this case, a tighter constraint on the individual hazard does not increase total hazard by a significant amount (just 12.12%).

5.2.2 Effects of Constraining Hazard and Period of Exposure at Individual Points

The effects of incorporating the constraints on individual hazard and period of exposure are shown in Tables 6 and 7 and Fig. 9.

While Fig. 9 displays solutions of M_2, Tables 6 and 7 show the results of M_2, compared with the results of M_1 for $\delta_1 \approx 1$ and $\delta_1 \approx 0$. The first column of Table 6 displays the identification of each vulnerable center exposed to one or more arc segments of the hazardous materials routes. The second column shows the number of students in each school. The third, fourth, and fifth columns display the hazard and, in parentheses, the population-weighted period of exposure of each school for model M_1 with $\delta_1 \approx 1$, M_1 with $\delta_1 \approx 0$, M_2 without constraints (13), and M_2, respectively. The chosen values of β^k and α^k are indicated in the top of each column.

Table 7 shows an apparent dominance of model M_1 over M_2 in terms of exposed students and affected schools. However, when the results are analyzed by vulnerable center, as in Table 6, hazard and population-weighted period of exposure are concentrated in a few vulnerable centers. For example, when $\delta_1 \approx 1$, schools 114 (1478 students) and 219 (1132 students) concentrate the 40.5% of the total hazard and are exposed during long periods (318.1 and 411.2 person-hours, respectively). In this case, the average period of exposure per person is 2.57 min.

When M_1 is solved with $\delta_1 \approx 0$, schools 16, 50, and 114 (representing a 9.1% of the students) concentrate the 43.3% of the total period of exposure. However, the average period of exposure per person drops to 1.13 min, a decrease of 56.1%. At the same time, school 158 is exposed to the 12.7% of the total hazard during a 1.9% of the total weighted period of exposure.

When considering model M_2 without constraints (13), the hazard is shared among more schools, and none of them is overexposed. However, the period of exposure can increase significantly for some centers, e.g., 56, 137, and 168, which together increase from an 8.4% to a 23.5% of the total period of exposure, when compared with M_1 with $\delta_1 \approx 1$. The average period of exposure per student also increases to 3.52 min. When constraints (13) are incorporated, this increase in period of exposure is controlled.

Naturally, there is no free improvement of the individual indicators: the imposed limit on the individual hazard and period of exposure, in this case, is obtained at the expense of an increase of 52.4% in the aggregated hazard and an increase of 169.4% in the total period of exposure, as well as an increase of 62.8% in the average period of exposure per student. Also, the number of exposed schools and total number of exposed students increase.

These results are mainly due to the chosen tight values for $\alpha^k = 160$ person-hours and $\beta^k = 2.2$. Recall that the smallest value that α^k and β^k can take are 130 person-

Table 6 Values obtained for M_1 with $\delta_1 \approx 1$, M_1 with $\delta_1 \approx 0$, and M_2, broken down by vulnerable exposed center

Vulnerable center	No. of students	M_1 with $\delta_1 \approx 1$ (Minimum hazard)	M_1 with $\delta_1 \approx 0$ (Min period of exposure)	M_2 without (13) and $\beta^k = 2.2$	M_2 with $\beta^k = 2.2$ and $\alpha^k = 160$ (person-hours)
16	2612	0 (0)	4.14 (136.5)	0 (0)	0 (0)
27	2094	0 (0)	0.73 (18.8)	0 (0)	0 (0)
39	1940	0 (0)	0 (0)	0 (0)	2.02 (153.6)
44	1922	0 (0)	0 (0)	0 (0)	0.57 (94.0)
46	1900	1.01 (274.7)	0 (0)	1.01 (274.7)	0.22 (120.8)
48	1831	0 (0)	0 (0)	1.58 (264.6)	1.66 (101.0)
50	1820	0.88 (210.5)	5.49 (128.2)	0.88 (210.5)	1.61 (128.8)
56	1750	0.69 (71.5)	0 (0)	1.38 (259.2)	0.80 (67.1)
63	1683	0 (0)	0 (0)	0 (0)	0.44 (47.4)
66	1668	0.39 (31.0)	0.39 (31.0)	0.13 (10.3)	0.13 (154.7)
68	1664	0 (0)	0 (0)	1.30 (113.9)	1.30 (113.9)
85	1602	2.01 (94.0)	2.01 (94.0)	2.01 (94.0)	2.01 (94.0)
91	1580	0 (0)	0 (0)	0 (0)	0.03 (4.6)
99	1547	0.32 (150.0)	0.87 (88.0)	1.26 (197.4)	0.32 (150.0)
106	1495	0 (0)	0.38 (16.4)	0 (0)	0 (0)
109	1487	0.92 (70.2)	0 (0)	0.92 (70.2)	0.92 (70.2)
114	1478	6.17 (318.1)	17.1 (263.9)	2.06 (106.0)	2.06 (87.6)
116	1454	0.72 (153.2)	0 (0)	0.72 (153.2)	0.19 (100.0)
118	1440	0 (0)	0 (0)	0 (0)	0.55 (65.7)
122	1414	0 (0)	0 (0)	0 (0)	0.41 (34.4)
123	1412	0 (0)	0.45 (13.3)	0 (0)	0 (0)
131	1379	0 (0)	0.51 (13.0)	0 (0)	0.51 (13.0)
140	1359	0 (0)	0.15 (7.1)	0 (0)	0 (0)
137	1365	0.25 (129.6)	0 (0)	0.64 (300.2)	1.82 (138.1)
145	1326	0 (0)	0 (0)	0.86 (71.1)	0.43 (35.5)
148	1320	0 (0)	0 (0)	0 (0)	2.00 (144.9)
153	1310	0.58 (80.0)	0 (0)	1.37 (146.1)	0.39 (33.0)
157	1295	1.36 (77.5)	1.36 (77.5)	1.36 (77.5)	1.36 (77.5)
158	1290	0.66 (79.2)	6.71 (22.7)	0.66 (79.2)	1.63 (61.7)
166	1264	0.07 (329.7)	0.27 (22.1)	0.07 (329.7)	0.31 (111.5)
168	1255	0.02 (31.9)	0 (0)	0.34 (335.7)	0.16 (151.9)
174	1237	0 (0)	0 (0)	0 (0)	0.83 (74.9)
179	1221	0.13 (45.2)	0 (0)	0.13 (45.2)	0.13 (45.2)
205	1173	0 (0)	0 (0)	0.73 (95.7)	0.45 (37.1)
206	1172	0 (0)	2.02 (51.1)	0 (0)	0 (0)
215	1137	0.89 (180.2)	0 (0)	1.55 (255.1)	2.06 (145.9)
216	1137	0 (0)	0 (0)	0 (0)	0.35 (92.0)
219	1132	2.66 (411.2)	4.25 (86.3)	1.77 (271.1)	0.89 (131.0)
223	1128	0 (0)	0 (0)	0 (0)	0.69 (78.6)

(continued)

Table 6 (continued)

Vulnerable center	No. of students	M_1 with $\delta_1 \approx 1$ (Minimum hazard)	M_1 with $\delta_1 \approx 0$ (Min period of exposure)	M_2 without (13) and $\beta^k = 2.2$	M_2 with $\beta^k = 2.2$ and $\alpha^k = 160$ (person-hours)
224	1128	0 (0)	0 (0)	0 (0)	0.30 (71.0)
225	1123	0 (0)	0 (0)	0 (0)	0.63 (66.3)
228	1110	0 (0)	0.93 (18.3)	0 (0)	0 (0)
233	1099	2.09 (46.7)	2.09 (46.7)	2.09 (46.7)	2.09 (46.7)
237	1089	0 (0)	0 (0)	0 (0)	0.98 (147.4)
240	1085	0 (0)	3.06 (86.6)	0 (0)	0 (0)
Total	64,927	21.,82 (2784.4)	52.87 (1221.5)	24.82 (3807.3)	33.26 (3291.1)

Table 7 Values obtained for M_1 with $\delta_1 \approx 1$, M_1 with $\delta_1 \approx 0$, and M_2

Attribute	M_1 with $\delta_1 \approx 1$ (Minimum hazard)	M_1 with $\delta_1 \approx 0$ (Min period of exposure)	M_2 without (13) ($\beta^k = 2{,}2$)	M_2 ($\beta^k = 2{,}2$ and $\alpha^k = 160$ person-hours)
Exposed students	27,074	27,913	33,068	52,588
% Exposed students	7.0%	7.2%	8.6%	13.6%
Total affected vulnerable centers	19	19	23	37
CPU Time (seconds)	1.13	1.08	9.89	121.30

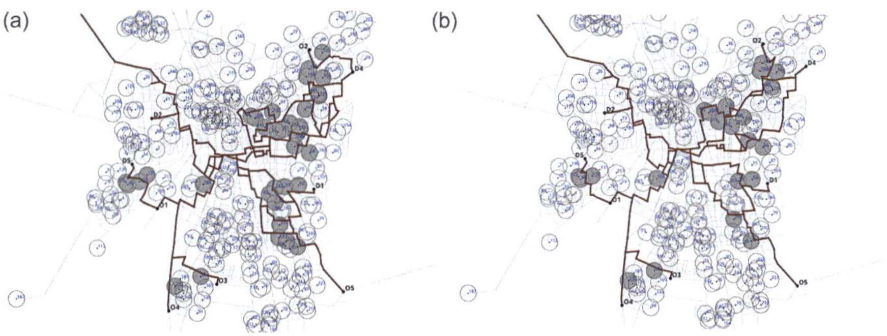

Fig. 9 Hazardous materials flows (**a**) Model M_2, $\beta^k = 2.2$ and $\alpha^k = 160$ (person·hr) $\forall\, k \in K$; (**b**) model M_2 without (13), $\beta^k = 2.2\ \forall\, k \in K$

hours and 2.2, respectively. For higher values of these parameters, the observed increases in hazard and total weighted exposure period will naturally be lower.

The point here is, however, that the decision maker can find an adequate compromise between total hazard and weighted period of exposure and individual values of both, i.e., equity of exposure, while keeping transportation costs within reasonable values. The models we propose are a useful tool for evaluating each strategy.

6 Conclusions and Future Research

The chapter presents an approach to the hazardous materials transport route design problem that can be applied to real-world situations. In our approach, the population is distributed at discrete points or centers in a plane, with a circle of radius λ around each one determining zone, in which the population is negatively affected. The major players—the regulating agency, the firms, and the general population—are identified and their objectives are delineated.

We then define a general hazard function, and add a weighted period of exposure function, both independent of incident probabilities. We then formulate models, in which the objectives of the individual players/stakeholders are traded off against each other.

We then apply the proposed methodology to a real instance of hazardous materials transport in Santiago, Chile. The results demonstrate that hazard exposure is an objective when it is minimized together with the weighted period of exposure. Both objectives can be traded off against transportation costs.

Trading off total hazard and total weighted period of exposure in a two-objective model exposes some of the vulnerable centers to high levels of both hazard and period of exposure. Consequently, we propose a second model that minimizes the total hazard subject to limits on the hazard and exposure period on each individual population center. We conclude that the incorporation of such thresholds can control the maximum hazard and population-weighted period of exposure for each population center, naturally at the expense of increased total hazard and total exposure period of the population. An adequate compromise can be easily explored by the decision maker.

The proposed objectives can be combined with other objectives. Probabilities of events can be included in the models if desired and reliable estimates are available. If speed statistics are known over the network, they can be used to improve the estimations of periods of exposure. The models can be solved for different times of the day, to consider the different distributions of the population throughout the day. Notice the trade-off: higher speed means shorter exposure time, but, at the same time, a higher risk of an incident per time unit.

Yet other possibilities opened up by the proposed approach of representing the undesirable effects of hazardous materials transport as attributes of population centers rather than network links would be to include emergency response center locations and hazardous materials routing as a combined factor in hazardous materials transport network design.

It is interesting to note that in the context of the transportation of hazardous materials, an alternative to reducing the possible consequences in some cases is to divide a shipment in smaller amounts and route each part through a different route. It remains of interest to analyze the trade-off between exposing a few points to a higher consequence against exposing many points to a lower consequence.

Acknowledgments This work was in part supported by FONDECYT grant 1220047; grants ANID PIA/PUENTE AFB220003; and Research Center for Integrated Disaster Risk Management (CIGIDEN) ANID/FONDAP/15110017.

References

Abkowitz, M., Cheng, P., & Lepofsky, M. (1990). Use of geographic information systems in managing hazardous materials shipments. *Transportation Research Record, 1261*, 35–43.

Abkowitz, M., Lepofsky, M., & Cheng, P. (1992). Selecting criteria for designating hazardous materials highway routes. *Transportation Research Record, 1333*, 30–35.

Alp, E. (1995). Risk-based transportation planning practice: Overall methodology and a case example. *Infor, 33*(1), 4–19.

Arlikatti, S., Lindell, M. K., Prater, C. S., & Zhang, Y. (2006). Risk area accuracy and hurricane evacuation expectations of coastal residents. *Environment and Behavior, 38*(2), 226–247.

Asgari, N., Rajabi, M., Jamshidi, M., Khatami, M., & Farahani, R. Z. (2017). A memetic algorithm for a multi-objective obnoxious waste location-routing problem: A case study. *Annals of Operations Research, 250*(2), 279–308.

Batta, R., & Chiu, S. (1988). Optimal obnoxious paths on network: Transportation of hazardous materials. *Operations Research, 36*, 84–92.

Beneventti, D., Bronfman, A., Paredes-Belmar, G., & Marianov, V. (2019). A multi-product maximin hazmat routing-location problem with multiple origin-destination pairs. *Journal of Cleaner Production, 240*, 118193.

Bianco, L., Caramia, M., & Giordani, S. (2009). A bilevel flow model for hazmat transportation network design. *Transportation Research Part C, 17*, 175–196.

Brainard, J. S., Lovett, A. A., & Parfitt, P. (1996). Assessing hazardous waste transport risks using a GIS. *Geographical Information Systems, 10*(7), 831–849.

Brilly, M., Polic, M., Castelli, F., & Caragliano, S. (2005). Public perception of flood risks, flood forecasting and mitigation. *Natural Hazards and Earth System Sciences, 5*(3), 345–355.

Brody, S. D., Peck, B. M., & Highfield, W. E. (2004). Examining localized patterns of air quality perception in Texas: A spatial and statistical analysis. *Risk Analysis, 24*(6), 1561–1574.

Bronfman, A., Marianov, V., Paredes-Belmar, G., & Lüer-Villagra, A. (2015). The maximin HAZMAT routing problem. *European Journal of Operational Research, 241*(1), 15–27.

Bruglieri, M., Cappanera, P., & Nonato, M. (2014). The gateway location problem: Assessing the impact of candidate site selection policies. *Discrete Applied Mathematics, 165*, 96–111.

Bula, G. A., Afsar, H. M., González, F. A., Prodhon, C., & Velasco, N. (2019). Bi-objective vehicle routing problem for hazardous materials transportation. *Journal of Cleaner Production, 206*, 976–986.

Cappanera, P., Gallo, G., & Maffioli, F. (2003). Discrete facility location and routing of obnoxious activities. *Discrete Applied Mathematics, 133*(1–3), 3–28.

Caramia, M., Giordani, S., & Iovanella, A. (2010). On the selection of k routes in multiobjective hazmat route planning. *Journal of Management Mathematics, 21*, 239–251.

Carotenuto, P., Giordani, S., & Ricciardelli, S. (2007). Finding minimum and equitable risk routes for hazmat shipments. *Computers and Operations Research, 34*(5), 1304–1327.

Chang, N.-B., Lu, H. Y., & Wei, Y. L. (1997). GIS technology for vehicle routing and scheduling in solid waste collection systems. *Journal of Environmental Engineering, 123*(9), 901–910.

Chen, Y.-W., Wang, C.-H., & Lin, S.-J. (2008). A multi-objective geographic information system for route selection of nuclear waste transport. *Omega, 36*(3), 363–372.

Church, R. L., & Drezner, Z. (2022). Review of obnoxious facilities location problems. *Computers and Operations Research, 138*, 105468.

Cohon, J. L. (1978). *Chapter 6: Techniques for generating noninferior solutions. Multiobjective Programming and Planning*. Academic Press.

Colebrook, M., & Sicilia, J. (2013). *Hazardous facility location models on networks. Handbook of OR/MS Models in Hazardous Materials Transportation* (pp. 155–186). Springer.
Current, J., & Ratick, S. (1995). A model to assess risk, equity and efficiency in facility location and transportation of hazardous materials. *Location Science, 3*(3), 187–201.
Current, J. R., & Schilling, D. A. (1990). Analysis of errors due to demand data aggregation in the set covering and maximal covering location problems. *Geographical Analysis, 22*(2), 116–126.
Daskin, M. S. (2011). *Network and discrete location: Models, algorithms, and applications*. Wiley.
Ditta, A., Figueroa, O., Galindo, G., & Yie-Pinedo, R. (2019). A review on research in transportation of hazardous materials. *Socio-Economic Planning Sciences, 68*, 100665.
Elliott, S. J., Cole, D. C., Krueger, P., Voorberg, N., & Wakefield, S. (1999). The power of perception: Health risk attributed to air pollution in an urban industrial neighbourhood. *Risk Analysis, 19*(4), 621–634.
Emir-Farinas, H., & Francis, R. L. (2005). Demand point aggregation for planar covering location models. *Annals of Operations Research, 136*(1), 175–192.
EPA. (2012). *Hazardous waste listings: A user-friendly reference document*. September 2012. Available online at https://www.epa.gov/sites/production/files/2016-01/documents/hw_listref_sep2012.pdf, last accessed on 11/30/2022.
Erkut, E., & Gzara, F. (2008). Solving the hazmat transport network design problem. *Computers and Operations Research, 35*, 2234–2247.
Erkut, E., & Ingolfsson, A. (2000). Catastrophe avoidance models for hazardous materials route planning. *Transportation Science, 34*(2), 165–179.
Erkut, E., & Ingolfsson, A. (2005). Transport risk models for hazardous materials: Revisited. *Operations Research Letters, 33*, 81–89.
Erkut, E., & Neuman, S. (1989). Analytical models for locating undersireble facilities. *European Journal of Operational Research, 40*, 275–291.
Erkut, E., & Verter, V. (1995). A framework for hazardous materials transport risk assessment. *Risk Analysis, 15*(5), 589–601.
Erkut, E., Tjandra, S. A., & Verter, V. (2007). Hazardous materials transportation. Chapter 9. In C. Barnhart & G. Laporte (Eds.), *Handbooks in operations research and management science* (pp. 539–621). Elsevier.
Fernández, J., Fernández, P., & Pelegrín, B. (2000). A continuous location model for siting a non-noxious undesirable facility within a geographical region. *European Journal of Operational Research, 121*(2), 259–274.
Fontaine, P., Crainic, T. G., Gendreau, M., & Minner, S. (2020). Population-based risk equilibration for the multimode hazmat transport network design problem. *European Journal of Operational Research, 284*(1), 188–200.
Francis, R. L., Lowe, T. J., & Tamir, A. (2004). Demand point aggregation for location models. In Z. Drezner & H. W. Hamacher (Eds.), *Facility location: Application and theory* (pp. 207–230). Springer.
Frank, W. C., Thill, J.-C., & Batta, R. (2000). Spatial decision support system for hazardous material truck routing. *Transportation Research Part C, 8*, 337–359.
Ghaderi, A., & Burdett, R. L. (2019). An integrated location and routing approach for transporting hazardous materials in a bi-modal transportation network. *Transportation Research Part E: Logistics and Transportation Review, 127*, 49–65.
Giannikos, I. (1998). A multiobjective programming model for locating treatment sites and routing hazardous wastes. *European Journal of Operational Research, 104*(2), 333–342.
Gopalan, R., Kolluri, K. S., Batta, R., & Karwan, M. H. (1990). Modeling equity of risk in the transportation of hazardous materials. *Operations Research, 38*(6), 961–975.
Hassanpour, S. T., Ke, G. Y., & Tulett, D. M. (2021). A time-dependent location-routing problem of hazardous material transportation with edge unavailability and time window. *Journal of Cleaner Production, 322*, 128951.
Heitz, C., Spaeter, S., Auzet, A.-V., & Glatron, S. (2009). Local stakeholders' perception of muddy flood risk and implications for management approaches: A case study in Alsace (France). *Land Use Policy, 26*(2), 443–451.

Helander, M. E., & Melachrinoudis, E. (1997). Facility location and reliable route planning in hazardous material transportation. *Transportation Science, 31*(3), 216–226.
Holeczek, N. (2019). Hazardous materials truck transportation problems: A classification and state of the art literature review. *Transportation Research Part D: Transport and Environment, 69*, 305–328.
Holeczek, N. (2021). Analysis of different risk models for the hazardous materials vehicle routing problem in urban areas. *Cleaner Environmental Systems, 2*, 100022.
Hung, H. C., & Wang, T. W. (2011). Determinants and mapping of collective perceptions of technological risk: The case of the second nuclear power plant in Taiwan. *Risk Analysis, 31*(4), 668–683.
Ichoua, S., Gendreau, M., & Potvin, J.-Y. (2003). Vehicle dispatching with time-dependent travel times. *European Journal of Operational Research, 144*(2), 379–396.
Jonkman, S. N., van Gelder, P. H. A. J. M., & Vrijling, J. K. (2003). An overview of quantitative risk measures for loss of life and economic damage. *Journal of Hazardous Materials, 99*, 1–30.
Kara, B., & Verter, V. (2004). Designing a road network for hazardous materials transportation. *Transportation Science, 38*(2), 188–196.
Kim, M., Miller-Hooks, E., & Nair, R. (2011). A geographic information system-based real-time decision support framework for routing vehicles carrying hazardous materials. *Journal of Intelligent Transportation Systems, 15*(1), 28–41.
Lepofsky, M., Abkowitz, M., & Cheng, P. (1993). Transportation hazard analysis in an integrated GIS environment. *Journal of Transportation Engineering, 119*(2), 239–254.
Li, R., & Leung, Y. (2011). Multi-objective route planning for dangerous goods using compromise programming. *Journal of Geographical Systems, 13*(3), 249–271.
Lima, M. L. (2004). On the influence of risk perception on mental health: Living near an incinerator. *Journal of Environmental Psychology, 24*(1), 71–84.
Lindell, M. K., & Perry, R. W. (2000). Household adjustment to earthquake hazard a review of research. *Environment and Behavior, 32*(4), 461–501.
Lindner-Dutton, L., Batta, R., & Karwan, M. H. (1991). Equitable sequencing of a given set of hazardous materials shipments. *Transportation Science, 25*(2), 124–137.
List, G. F., & Mirchandani, P. B. (1991). An integrated network/planar multiobjective model for routing and siting for hazardous materials and wastes. *Transportation Science, 25*, 146–156.
Lovett, A. A., Parfitt, J. P., & Brainard, J. S. (1997). Using GIS in risk analysis: A case study of hazardous waste transport. *Risk Analysis, 17*(5), 625–633.
Ma, C., Zhou, J., Xu, X. D., Pan, F., & Xu, J. (2020). Fleet scheduling optimization of hazardous materials transportation: A literature review. *Journal of Advanced Transportation*, Article #4079617. Available online at doi:https://doi.org/10.1155/2020/4079617, last accessed on 11/30/2022.
Malandraki, C., & Daskin, M. S. (1992). Time dependent vehicle routing problems: Formulations, properties and heuristic algorithms. *Transportation Science, 26*(3), 185–200.
Marianov, V., & ReVelle, C. (1998). Linear, non-approximated models for optimal routing in hazardous environment. *Journal of the Operational Research Society, 49*, 157–164.
Melachrinoudis, E. (2011). The location of undesirable facilities. In H. A. Eiselt & V. Marianov (Eds.), *Foundations of location analysis* (pp. 207–239). Springer.
Miceli, R., Sotgiu, I., & Settanni, M. (2008). Disaster preparedness and perception of flood risk: A study in an alpine valley in Italy. *Journal of Environmental Psychology, 28*(2), 164–173.
Mohri, S. S., Mohammadi, M., Gendreau, M., Pirayesh, A., Ghasemaghaei, A., & Salehi, V. (2022). Hazardous material transportation problems: A comprehensive overview of models and solution approaches. *European Journal of Operational Research, 302*(1), 1–38.
Pijawka, K. D., Foote, S., & Soesilo, A. (1985). Risk assessment of transporting hazardous material: Route analysis and hazard management. *Transportations Research Record, 1020*, 1–6.
Prince2. (2009) *Prince2 glossary of terms*. Available online at https://www.stakeholdermap.com/prince2/prince2-glossary-R-records.html, last accessed on 11/30/2022.

Rabbani, M., Heidari, R., Farrokhi-Asl, H., & Rahimi, N. (2018). Using metaheuristic algorithms to solve a multi-objective industrial hazardous waste location-routing problem considering incompatible waste types. *Journal of Cleaner Production, 170*, 227–241.

ReVelle, C., Cohon, J., & Shobrys, D. (1991). Simultaneous sitting and routing in the disposal of hazardous wastes. *Transportation Science, 25*, 138–145.

Saccomanno, F. F., & Chan, A. (1985). Economic evaluation of routing strategies for hazardous road shipments. *Transportation Research Record, 1020*, 12–18.

Saccomanno, F. F., & Shortreed, J. H. (1993). Hazardous material transport risk: Societal and individual perspectives. *Journal of Transportation Engineering, 119*, 177–188.

Sadigh, A. N., & Fallah, H. (2009). Demand point aggregation analaysis for location models. Chapter 22. In R. Zanjirani Farahani & M. Hekmatfar (Eds.), *Facility location* (pp. 523–534). Physica.

Samanlioglu, F. (2013). A multi-objective mathematical model for the industrial hazardous waste location-routing problem. *European Journal of Operational Research, 226*(2), 332–340.

Sherali, H. D., Brizendine, L. D., Glickman, T. S., & Subramanian, S. (1997). Low probality - high consecuence considerations in routing hazardous material shipments. *Transportation Science, 31*(3), 237–251.

Sivakumar, R. A., Batta, R., & Karwan, M. H. (1993). A network-based model for transporting extremely hazardous materials. *Operation Research Letters, 13*(2), 85–93.

Sivakumar, R. A., Batta, R., & Karwan, M. H. (1995). A multiple route conditional risk model for transporting hazardous materials. *Information Systems and Operational Research, 33*(1), 20–33.

The Britannica Dictionary. (2022). Available online at https://www.britannica.com/dictionary/hazard, last accessed on 12/1/2022.

U.S. Department of Transportation. (2022) *Pipeline and Hazardous Materials Safety Administration*. Office of Hazardous Material Safety. Available online at https://portal.phmsa.dot.gov/analytics/saw.dll?Portalpages&PortalPath=%2Fshared%2FPublic%20Website%20Pages%2F_portal%2F10%20Year%20Incident%20Summary%20Reports, last accessed on 11/30/2022.

Verter, V., & Kara, B. Y. (2008). A path-based approach for hazmat transport network design. *Management Science, 54*(1), 29–40.

Wachinger, G., Renn, O., Begg, C., & Kuhlicke, C. (2013). The risk perception paradox—Implications for governance and communication of natural hazards. *Risk Analysis, 33*(6), 1049–1065.

Zero, L., Bersani, C., Paolucci, M., & Sacile, R. (2019). Two new approaches for the bi-objective shortest path with a fuzzy objective applied to HAZMAT transportation. *Journal of Hazardous Materials, 375*, 96–106.

Zhao, J., & Zhao, J. (2010). Model and algorithm for hazardous waste location-routing problem. In J. Zhang, L. Xu, X. Zhang, & M. Jian (Eds.), *ICLEM 2010: Logistics for sustained economic development: Infrastructure, information, integration* (pp. 2843–2849). American Society of Civil Engineers.

Ziaei, Z., & Jabbarzadeh, A. (2021). A multi-objective robust optimization approach for green location-routing planning of multi-modal transportation systems under uncertainty. *Journal of Cleaner Production, 291*, 125293.

Zografos, K. G., & Davis, C. F. (1989). Multi-objective programming approach for routing hazardous materials. *Journal of Transportation Engineering, 115*(6), 661–673.

Zografos, K. G., & Samara, S. (1989). A combined location-routing model for hazardous waste transportation and disposal. *Transportation Research Record, 1245*, 52–59.

Customer-Related Uncertainties in Facility Location Problems

Vladimir Marianov and Gonzalo Méndez-Vogel

Abstract In many situations, customers choose the facilities they want to interact with. One possible objective of the facility managers is to maximize the number of customers who use their facilities. In order to achieve this objective, they will need to make decisions regarding the features of their facilities, such as product variety, parking space, ambiance, prices, and, not least, the location of these facilities, particularly relative to the location of customers and possible competitors. To make their facilities attract as many customers as possible, the firms need to know what makes customers behave the way they do. Unfortunately for the firms, customer behavior is uncertain. This chapter examines the sources of customer-related uncertainty. These include the occurrence of unplanned purchases, the taste for variety—given product heterogeneity, imperfect information available to the customers about product and store features, and imperfect information on customers available to decision-makers. The effects of these uncertainties on customers' behavior are also described: purchases distributed among all competitors, comparison shopping, multipurpose trips, and price and feature search. This behavior results in facility locations different from those obtained using models that do not consider uncertainty. In particular, we do see more agglomeration. The chapter then describes some models that include customer probabilistic choice rules and demonstrates how these rules can be integrated into facility location models.

Keywords Uncertain customer behavior · Imperfect information · Customer choice rules · Competitive facility location

V. Marianov (✉)
Department of Electrical Engineering, Instituto Sistemas Complejos de Ingeniería (ISCI), Pontificia Universidad Católica de Chile, Santiago, Chile
e-mail: marianov@ing.puc.cl

G. Méndez-Vogel
Pontificia Universidad Católica de Chile, Santiago, Chile
e-mail: ggmendez@uc.cl

H. A. Eiselt, V. Marianov (eds.), *Uncertainty in Facility Location Problems*, International Series in Operations Research & Management Science 347, https://doi.org/10.1007/978-3-031-32338-6_3

1 Introduction

There are facilities that operate without the need of interacting with humans except for maintenance, e.g., meteorological radars or remote weather stations. Their location is decided considering mainly factors that are related to the efficient measurement of the meteorological phenomena. Most of the facilities, however, are expected to have physical interaction with their users. When facilities and users interact, one of the most important factors—if not the most, in deciding their location—is the provision of the best possible service to those who use these facilities. This includes locational factors, such as closeness to demand, good access, and a good general environment, and non-locational features such as offering good products or fast service. Overall, an understanding is required of what the users need or desire, which drives users' or customers' behavior.

The importance of customers' behavior in facility location depends on who decides the assignment of users to facilities. It can be a decision-maker associated with the facility (we call it "allocation") or the users themselves ("user choice").

Examples of allocation happen in the provision of the Internet through fiber, in an ambulance service, or in deliveries from a fulfillment center or warehouse. In these cases, it is the provider who decides the locations of the facilities and which users will be served by which facility. Users do not need to choose or even know what facility serves them as long as the service is adequately provided. The location is determined based mostly on cost or timely service. In the allocation case, the user-related uncertainties are generally limited to the demand: when and how much will each user require. In general, having a good statistical representation of the demand is enough to design allocation systems with an adequate level of service.

On the contrary, in a user choice system, customers are free to choose what, where, and when to obtain a service or good. They are the ones who choose which facilities to use, when to do so, and what and how much of it to obtain. This applies to non-competitive environments, such as a network of voluntary vaccination centers and libraries, but, far more importantly, to competitive settings. One of the best examples of competitive environments is retail, as customers can usually choose to purchase a product among different mutual substitute alternatives, and in any of the competitors' facilities. When there is competition, it is vital for competitors to know what the users want, when, where, and how much of it they want, and what attributes of products and facilities (including location) will make the customers prefer one competitor over the others, i.e., the customers' decision-making processes (Radu, 2022). This knowledge has to be put into *rules* of how customers choose, so these rules can be used in location decisions. Generally, these decisions will be oriented to attract as many customers as possible or maximize the profit obtained from their purchases. In this chapter, we will use retail as our context.

The expected utility model (Mongin, 1998; Coto-Millán, 2003) in its current interpretation considers that customers are rational, they are utility maximizers, and they have full and correct information. The utility is a quantitative holistic indicator that includes a number of dimensions and can have different functional forms. When choosing between alternatives, the utility of each one of them is combined with a function of the probability of that alternative being chosen, and the expectation is computed. Schoemaker (1980) provides an extended treatment of the subject, including attitudes toward risk. Several model variants are presented, including the prospect theory model by Kahneman and Tversky (1979). Customers possess a certain amount of a dummy good (money), which they will use to acquire different services and goods, assigning preference to those whose utility is higher until there are no more products with a utility that is higher than that of the dummy good, or the dummy good is all spent. Note that as more units of an item are acquired, the successive marginal utility can decrease.

What is called "goods" in expected utility theory could go from a piano to lettuce, a haircut, or a plane ticket. Unless explicitly stated otherwise, we will refer in this chapter mostly to goods that are within a specific category: they are substitutes of each other, i.e., goods that serve the same basic purpose but have differences in their secondary features. Facility location has dealt with products that are either *homogeneous* (identical substitutes of each other, except for price) or *heterogeneous*—imperfect mutual substitutes (differ in secondary characteristics). When products are substitutes for each other, even imperfect ones, choosing one of them on a shopping trip excludes purchasing any of the remaining ones: one either chooses to travel by car or public transport and purchases only one brand of dog food at a time. As customers can pick only a single option, choices are discrete.

The expected utility theory assumes that customers are, in general, rational. However, this is not necessarily the case. In fact, from the point of view of an observer, customers behave in a seemingly random way. The single most important fact that makes them do so is that the available products are differentiated or heterogeneous. This difference is what drives the customers to make choices, as we will see in the next section. If the products, e.g., pairs of shoes, were all identical, the choice would boil down to deciding where to get the shoes at the least full price, that is, the price of the product plus the travel (or delivery) cost. Since the products in practice are heterogeneous, their differences make customers choose one pair of shoes over another one in the same store, and, when the attributes of the shoes are sufficiently different between two competing stores, choose the pair of shoes offered by one store over that in the competing store.

Secondly, customers' behavior changes over time. It is, to a point, unpredictable, although rarely *completely* so, as we humans act within physical limitations and, in most cases, for an individual, similar stimuli result in similar reactions, belonging to the same limited set of possibilities. Customers do not use all the possible product differentiation factors in their decisions, and the set of factors, and their relative importance, change over time. Different customers use different information sets, and they are influenced by different factors (Radu, 2022): demographics (age, gender, culture), psychological factors (perceptions, attitudes toward a marketing message, being tired of the same product), and social factors (income, education level, family and friends, social media). In addition, the effect of some of the factors is observable, but not of all of them. In synthesis, there is uncertainty about what users will do, and what will be their choice when faced with different alternatives. And this is exactly what firms need to know to best locate their facilities.

In this chapter, we address the influence of customer (or user)-related uncertainties on the optimal facility locations. We describe these uncertainties, analyze how to model their effects on customer choice, and use these models (customer choice rules) when optimizing facility locations. Our analysis is oriented toward this single end, as opposed to addressing customer behavior in depth, as psychologists or behavioral economists would do, e.g., Solomon (2017), or Foxall (2005). Neither do we discuss other kinds of uncertainties involved in facility location (see Murray, this volume) nor how to address general uncertainty in optimization (Snyder, 2006; Correia & Saldanha-da-Gama, 2019).

We refer in this chapter to purchases made in physical stores, although most of the uncertainties are also present in online shopping. We chose to focus on physical stores because even though a big share of retail purchases is made online, this has not made and will not make brick-and-mortar stores disappear in the foreseeable future. Even more, physical sales are on the rise (Business Insider, 2021; McCall, 2021; Schnure, 2021; Sheth, 2021; Hübner et al., 2022), and it is more important than ever to locate physical stores and choose the products to offer so to attract customers (Gauri et al., 2021).

The remainder of the chapter will deal with sources of uncertainty, customer behavior driven by uncertainties, user choice rules with uncertainty, facility location models considering customer-related uncertainties, and conclusions.

2 Sources of Uncertainty

2.1 *Planned and Unplanned Purchases*

There are purchases that are planned, which respond to needs that are previous to starting any shopping trip, and are the result of reasoning and a positive decision based on the (limited) information the customer has about the required product or products. On the other hand, there are unplanned purchases, usually in response to some state of mind, time pressure, occasional discounts, in-shop marketing, or other stimuli that a shopper receives during a shopping or any other type of trip. This is the dominating type of purchase during leisure or entertainment shopping. A significant percentage of purchases are driven by impulse. Impulse purchases are triggered by a desire to buy a product that is attractively displayed, conveniently placed on a shelf, and accessible, or while the customer is window shopping, or because of promotions in the store during the customer's stay in it, or driven by the general environment in a store or mall inviting to purchase, and some other factors related to the customer's emotions and, hence, unconscious (Jamal & Lodhi, 2015). Figures for the percentage of purchases made impulsively range from 20% of all purchases (Bell et al., 2013) to some percentage between 40% and 80% depending on the category of product (Amos et al., 2014), who also report that more than 87% of American adults admit to engaging in impulse purchases, and impulse purchases account for more than 50% of grocery purchases. A very recent survey shows that in the USA, 73% of adults said most of their purchases tend to be spontaneous (Tronier, 2022). Unfortunately, very few authors (Massara et al., 2014) distinguish between impulse versus opportunistic purchases. Opportunistic purchases are unplanned, but they have a rational trigger: they are driven by real needs that become conscious when the product is in view. We will later refer to one of the behaviors that are partly a result of opportunistic purchases: multipurpose shopping.

Most of the remaining causes of uncertainty in this section are applicable to both planned and unplanned shopping, unless stated the contrary.

2.2 *Product and Facility Heterogeneity or Differentiation*

For many years since the seminal work of Hotelling (1929), it was common in the facility location literature to consider that stores and products were completely homogeneous.[1] However, all products are heterogeneous. They can be differentiated from each other. Chamberlin (1933) states that a product is differentiated if there is any significant basis for distinguishing what different stores offer, which suggests that the customer "purchases an experience" involving both the product and the store. Moreover, this difference can be real or can be in the mind of the customer, and includes the secondary features of the product, such as product name, package, color, quality, pattern, size, fabric, price, style, and so on, and also the context (store) in which it is being sold, as location, the assortment of products, ambiance, courtesy and appearance of the personnel, availability of parking space, the neighborhood and the existence of other stores in the vicinity, and even attachment to someone in the store. What this means is that virtually all tuples (product, store) have differences between them, even if belonging to the same firm. If two sellers offer products that are sufficiently differentiated, there will also be a reason for customers to prefer one seller over the other.

Product differentiation *is the main underlying reason for uncertainties in customers' behavior*. If stores were identical, and products were undifferentiated, customers would not have choices other than picking the closest place at which to obtain whatever is what they need—which is a choice rule that many location models actually assume. It may also be a trigger of unplanned purchases when a customer discovers a product type that is new and he particularly likes.

For a long time, market researchers, economists, psychologists, and anthropologists (Li, 2015) have tried to understand what attributes determine how brick-and-mortar purchasers choose where and what to buy, in the case of planned

[1] Hotelling explicitly refers to product heterogeneity when firms "locate" in a space of customer preferences. However, when addressing geographical location, he uses homogeneous products.

or unplanned purchases. As mentioned, the secondary features of the heterogeneous products have a strong influence. From the point of view of the facilities, Attri and Jain (2018), in a field study, determined that the most important factors influencing the choice of a facility are store atmospherics, customer demographics, social, and psychological factors, marketing communications, service, retail outlet perceptions, availability of time, and display at the store. They cite research on the subject going back to the 1960s. Other studies include retailers' assortment, pricing, and promotional policies (Fox et al., 2004).

2.3 *Taste for Variety*

In planned purchases, as stores and products are heterogeneous, customers can choose among all the stores and products the one they like best, and this depends significantly on the features of the store and the products being offered. Many of the popular models of customer behavior consider that customers possess all the required information and take into account what they consider to be the relevant attributes of the product and facility, and rationally weigh each attribute to reach a decision on what alternative to choose.

Although the classical expected utility theory considers so, the human brain is not capable of taking into account (and processing) too many factors simultaneously (Kahneman & Tversky, 1979; Thaler, 2015), and customers must choose what manageable subset of relevant attributes they will consider in the decision making. The attributes perceived as the most relevant by the customer, and their weights, change over time. As an extreme example, in purchasing a medicine, a customer may consider its price in different stores, the travel cost, maybe the brand of medicine she prefers, and her time availability to go to a store, among other factors. However, the same customer, in urgent need of medicine, will choose the fastest way to get it, no matter what the price or time availability is. Customers can also forget to take some attributes into account or make an error in estimating their importance (Anderson et al., 1992).

Finally, there are factors that are out of the awareness of the customer, unconscious factors, that have been found to play an important role (Fitzsimons et al., 2002). In this category may fall personality traits, compulsive behavior, some marketing actions, and even social trends. As mentioned, a significant part of all purchases is driven by impulse or are opportunistic.

As a result, the same customer can show a preference for different models of shoes on shopping trips made at different times, even a short time apart, under apparently identical conditions (Loomes et al., 2009). Or choose not to dine at the same restaurant on two occasions. This apparent inconsistency has been called *taste for variety*. Note, though, that there are cases in which customers prefer to stick to a store or a brand, as happens with beer or soda drinkers, or (at least some) pub or fast-food-goers. This is *loyalty*. Furthermore, customers have some memory: they can purchase because they tried a certain product in the past and it did (or did not) well.

2.4 Imperfect Information Available to the Customer

Note that in describing taste for variety or diversity, the implicit assumption is that the customers have all the necessary information on stores and products. This is not so. Customers have *imperfect or incomplete information*. Due to imperfect information, a customer may be undecided about his willingness to pay for a non-essential product until he visits a store in which the product is displayed, and its features can be evaluated. Furthermore, because of product heterogeneity, taste for variety, and incomplete information on the attributes or the products in different stores, which he cannot acquire by web search, the customer cannot be sure which product, among all imperfect mutual substitutes, he will prefer at that particular time.

Some authors consider that taste for variety and imperfect information are the two factors of uncertainty that are intrinsic to the customers. Urbany et al. (1989), through a factor analysis of empirical data, found these two factors being significantly predominant. However, other researchers add *evaluation uncertainty* (Shiu et al., 2011), which is the lack of confidence the customer has in his own ability to correctly evaluate choices, even having full information. Evaluation uncertainty influences the willingness to search for missing information: the higher the evaluation uncertainty, the more likely a customer is to keep searching for his ideal product.

Underlying these sources of uncertainty is also *credibility* (of the information) which negatively impacts the available information, part of it becoming useless for the customer. An example of non-credible information is online reviews of a product, written by company associates. See Shiu et al. (2011) for a more detailed treatment of these factors and an empirical study of their importance.

In addition to the previous uncertainties, the product portfolio available to the customers at stores changes over time, and there may be temporary unavailability of certain products at a store, which may be unknown to the customer. Neither can the customer predict the conditions in which she will make use of the good, e.g., when planning a vacation trip for some time in the future (Loomes et al., 2009). There also could be situations in which customers may be uncertain about the future availability of a product, or their available budget at future times, which could make them overstock, changing a regular pattern of purchases.

Konishi (2005) calls this seemingly random behavior of the customers, due to their apparent uncertainty as to what alternative to choose, *taste uncertainty*, which includes all the factors that make the customer behave in a seemingly semi-random way.

Later in this chapter, we will describe the consequences of all these uncertainties.

2.5 Imperfect Information of the Decision-Maker on Customers

A different source of uncertainty is the lack of knowledge of the decision-makers on customers' actions. This imperfect and incomplete information is related to both the rules the customers use to choose and the value they assign to different factors. This uncertainty comes from the imperfect *observation and measurement* or quantification of customer choice behavior. Decision-makers need to know how customers choose the best locations and attributes of their stores. This requires observing and measuring customers' actions and inferring from these actions what attributes are relevant to them, as well as the weights that customers assign to each one of the attributes. Observations must be made of large groups of customers (as there is *taste dispersion* among customers), over periods of time, as different individuals behave differently, each one having her own assessment of the important attributes and their weights, and each of them having possibly different information about the market (Hausman & Newey, 2016) and different tastes.

Unfortunately, observations are *incomplete,* and measurements are *imprecise*. In fact, only some of the attributes' effects are observable. This happens, e.g., with price and travel cost. But there are a number of store and product attributes whose effect is not observable, e.g., the attraction a customer feels for someone that works at a particular store or his preference for a scenic route to go to the store, as somewhat extreme examples.

Neither is observable the sequence in which customers make some choices of store and product. Choices can be made simultaneously or sequentially. Suppose a customer makes a trip to purchase a pair of shoes. He may change his mind in the middle of a trip, or he can remember that he also needs a jacket and engage in a multipurpose trip (O'Kelly, 1981; Eaton & Lipsey, 1982).

Furthermore, not all relevant actions of a customer can be observed. A firm can keep track of the purchases of a particular individual in its stores, but it cannot know when and how many times the same individual chooses a different chain to purchase substitute products, as it is extremely unlikely to have stores sharing this information. Moreover, it cannot know what kept that individual from purchasing on some occasions. This makes the measurements imprecise, in the sense that the parameters have necessarily large confidence intervals.

The incompleteness and impreciseness of observations added to the *intrinsic* uncertainties in customers' choices, i.e., their own uncertainty as to what alternative to choose, make the customer behavior seem random. Randomness implies unpredictable behavior. Most choice models assume, implicitly or explicitly, that there are deterministic and random components in customers' choices.

Note, finally, that when customers choose an alternative, they compute what in their brains is a deterministic value for the utility of the different alternatives. They assign some values to all the relevant parameters and, based on their evaluation, their choice is the best, from a deterministic point of view. However, as the factors that are considered and their weights are not observable, the utility for an external observer can be described as having a deterministic component (the observable factors) and a random component (the unobservable factors).

In general, customer uncertainties and observation uncertainties are treated as one phenomenon ("uncertainty"), as most of the time, it is not possible to separate them.

3 Effects of Uncertainties on Customer and Firm Behavior

3.1 Purchases Are Distributed Among All Available Stores

The Hotelling (1929) setting for geographical competition, involving product homogeneity, was profusely dominant until the 1970s. It is still in use for analyses, as it allows us to obtain insights into a variety of problems. However, there are goods that must be purchased periodically, and other goods that wear out, and must be replaced and, in practice, customers make many trips to purchase from different facilities. Another assumption in Hotelling's analysis is that there is only one homogeneous product. This assumption leads to considering only shopping trips that have one stop at the chosen store, which is always the same.

Heterogeneity was introduced in the interaction of firms and customers by Papageorgiou and Thisse (1985), to explain facility agglomeration in a linear and in a circular market. Their starting point is the assumption that, as products are heterogeneous and there is customers' taste dispersion and taste for variety, customers will not always choose the store that offers the least full price, but the purchases will be distributed among all the available facilities. In their view, the previous results of agglomeration were due to the bounded market, which makes facilities agglomerate at the center as that is the position in which they can attract more customers in equilibrium. This argument of Hotelling is destroyed when the market becomes a circle, as in this case there is no agglomeration. Marianov and Eiselt (2016) provide a similar argument: in Hotelling's setting, two competing facilities locate at the center of a line at the equilibrium because that is the location at which each one of them maximizes the demand for which the facility of the firm is the closest, and that is what they call a "weak force of agglomeration," but there are other forces that are more significant, related to heterogeneity as multipurpose and comparison shopping. What Papageorgiou and Thisse (1985) do is dispute the well-established notion that full price is the only drive for purchases and replace it with the notion that customers prefer diversity, although the full price is indeed one of the important factors. Their market is divided into small areas, and each area can house some stores. They assume that during a standard period, the customers gather information on stores until they have full information.[2] Once they have full information, the relative frequency of their visits to a particular area, say *j*, is given

[2] Actually, the gathering of information can be more permanent, which result in search behavior.

by the ratio between the number of visits to area j and the number of visits to all areas. Furthermore, the number of visits to any area is strictly decreasing and strictly concave in distance. Also, and here is the importance of heterogeneity in customers and products, they argue that a larger number of stores in an area j attract more total visits because the area offers a larger variety. Their conclusion is that agglomeration of firms and households is because it decreases the total transportation costs, increases the frequency of visits, and hence, the total market. Firm agglomeration, common in practice, is explained by a higher volume of sales due to the taste for variety, rather than due to Hotelling's explanation.

De Palma et al. (1985) use a linear market to show that the agglomeration principle holds when products and customers are sufficiently heterogeneous (again, taste for variety and differences between customers), expressing utility as a sum of a deterministic component and a random component:

They clarify that the random component accounts for the unobservability of the taste and valuations of the customers for the different attributes of the heterogeneous products. This does not mean the behavior of the customers is irrational but merely non-observable. They assume that the customers are distributed along a line, and stores can locate anywhere on that line. By allowing each customer to change his taste randomly, they represent the choice behavior of the customers by a logit choice rule and obtain two important conclusions. First, uncertainty, represented as a random component of the utility, makes disappear the formerly abrupt discontinuities in Hotelling's model, replaced by customers' smooth changes of facility preference when any of the parameters (price, location) changes. Secondly, agglomeration follows, for two or more facilities, and competition stabilizes, although a larger heterogeneity is required to maintain agglomeration when the number of firms increases.

3.2 Search Behavior, Comparison Shopping, Multipurpose-Shopping Trips. Firms Agglomerate

The uncertainties described above make customers unsure about which of the available mutual substitute products, let us say shirts, is closer to their ideal at the time of their shopping trip. Taste uncertainty, lack of information, evaluation uncertainty, and credibility of the available information, added to selectivity (the fact that not all the products in the set are acceptable to the customer), result in the need of visiting two or more stores, to try the products on, and to check on the features and availability, until they find a product that is acceptable for their taste or decide that the cost of further search exceeds the benefits of increased information (if they believe that the newly acquired information will be reliable and useful for choice purposes) and the expected marginal increase in the utility of a more suitable product. Search costs and patterns are well described by Anderson and Renault (1999). Because of economies of scale in transportation, all the stores are visited on the same trip, and the shopping trips become multiple-stop. This behavior is called *comparison shopping*. Note that if products are only slightly different, or commodities, there is no strong need for comparison, unless the customers are extremely selective. However, a higher differentiation between products or a very high range of available options (e.g., shoes) increases the drive for comparison (Krider & Putler, 2013; Fischer & Harrington Jr, 1996).

Comparison shopping has a significant impact on the location of retail stores. Eaton and Lipsey (1979) analyze the effect of comparison shopping on the location of several competitors located in a linear, bounded market. Customers are distributed uniformly, and they must visit exactly two stores, to compare their products each time they engage in a shopping trip. As in Hotelling's setting, the stores cannot co-locate but they can locate at an arbitrarily short distance from each other. They show that stores locate next to each other in triplets, even though this increases their distance from some of their customers, because the triplet attracts customers from farther away locations, due to the need to compare. The model was extended by Stahl (1982) to the case in which customers search for their preferred product from several mutually imperfect substitutes. The search criterion is optimizing the product characteristics, and the customers can visit more than two stores. Stahl assumes n possible products and n types of customers, uniformly distributed over a linear market, each type i having $m_i \leq n$ products that are acceptable, i.e., customers are selective. There are several marketplaces at which the stores agglomerate in different numbers. Customers visit only one marketplace and can search in all its stores at no cost. Note that this implies a two-step choice procedure: choose first the marketplace and then the store. The utility of a trip to a marketplace increases with the number of stores in it, and therefore, the market radius of a marketplace increases with the number of stores. A higher selectivity makes customers avoid trips to stand-alone stores and prefer larger marketplaces even if they are farther away. In conclusion, comparison shopping draws stores to agglomerate, e.g., in food courts, or the wedding dress shopping hub in Brooklyn (Hoo, 2018). Wolinsky (1983) finds conditions for a marketplace to attract more customers than a close-by stand-alone store. In his setting, a stand-alone store attracts customers from its neighborhood and becomes a monopoly, while a marketplace with several stores attracts customers from longer distances and the stores share the purchases. The conditions are established for a customer to prefer a comparison-shopping trip over a trip to a single store. In addition to taste uncertainty, Konishi (2005) justifies customers' attraction to marketplaces by a lower price expectation of customers, who believe that agglomeration increases competition and, hence, lowers prices. However, as Stahl shows, prices are not always lower in this setting. For more details, see Marianov and Eiselt (2016).

Taste uncertainty, impulse shopping, and lack of information also intervene in the case of bundles or multipurpose shopping. This activity consists in making a trip during which a customer purchases more than one product. The strongest driving force for this type of trip is economies of scale in transportation. Customers make up a shopping list and decide to take a trip to a marketplace that hopefully offers all the products in the list, or at least, is located in such a way that all the products can be purchased from nearby stores. However, uncertainties and lack of information frequently make a customer change his or her mind during the trip and either acquire less or more products than those on the previously made list. The shopping list is not always the same (it is uncertain), because the stock of different products that the customer has at home changes every time. Furthermore, the presence of other services in the shopping places, e.g., coffee shops, fast-food restaurants (conveniently located for the customers to do comparison shopping), car washing facilities, or other stores, can make customers use these services even if it was not planned. A single-stop trip can become a multi-stop trip. Again, this multipurpose shopping behavior implies a search for the products in the list, which makes it attractive for the customer to patronize marketplaces with more stores offering different products and even services, increasing the utility of such places and pushing firms to co-locate their stores. Lancaster (1966) proposed what he calls a "new approach to customer theory," in which instead of assuming that the products are what generates utility to the purchasers, it is the characteristics of the products that gives rise to utilities, which immediately suggests that customers not necessarily are looking for a single product, but possibly for a set of products that possess a set of characteristics. He analyzes groups of products and their relationship as complements or substitutes, which, in turn, leads to multipurpose and comparison shopping.

Regarding comparison shopping, Marianov et al. (2020) proposed and solved the follower location problem in a duopoly with comparison shopping. They do not force customers to necessarily visit both competitors' stores but assume that the probability of finding a suitable product increases when customers visit one store of each competitor. This increases the utility obtained by such action, which results in the co-location of the stores of the competitors and larger marked radii of the clusters.

Multipurpose shopping has also been addressed in the literature (Eaton & Lipsey, 1982; O'Kelly, 1981; Marianov et al., 2018; Khapugin & Melnikov, 2019; Méndez-Vogel et al., 2022).

4 Most Representative User Choice Rules Addressing Uncertainty

4.1 Brief Overview of the Fundamental Deterministic Rules

Hotelling (1929) recognizes that in practice, there is product/store heterogeneity, which makes customers take into consideration attributes like "his model of doing business is more to their liking, or because he sells other articles which they desire, or because he is a relative or a fellow Elk or Baptist, or on account of some difference in service or quality, or for a combination of reasons." However, for finding the location equilibrium of two competing stores, he assumes that the only attribute that customers in the linear market consider is the full price of a product. Customers and products are assumed to be homogeneous. Each customer simply chooses the store that minimizes the full price he observes. This choice rule is fully deterministic, with binary decisions: choose one or the other. Infinitesimal changes in price or location can change the choice, which is one of the handicaps of this rule. Hotelling rule was profusely used by researchers in marketing, economics, transportation, and retail until the 1970s, and continues to be used nowadays because of its simplicity. Hotelling's rule has been called "binary" rule or "winner-take-all" rule, as an infinitesimal difference between the utilities of two alternative stores makes a customer to prefer only, and always, the highest utility one.

In parallel, Reilly (1929), a marketing specialist, made an early attempt of finding the "relative" trade areas of marketplaces, using a rule based on the formula of gravity. Hence, the name "gravity rule." In his setting, customers in a small city located between two large cities a and b distribute their purchases among them, and the businesses attracted by each large city from the small one are in the proportion

$$\frac{B_a}{B_b} = \left(\frac{P_a}{P_b}\right)^N \left(\frac{D_b}{D_a}\right)^n$$

where

B_a = the sales volume that city a attracts from the small city
B_b = the sales volume that city b attracts from the small city
P_a = Population of city a
P_b = Population of city b
D_a = Distance from city a to the small city
D_b = Distance from city b to the small city

The parameters n and N need to be estimated using actual data. Reilly argues in favor of setting $N = 1$, and in his experiments, he finds the mode for n in the range 1.51–2.5. Most of researchers using this rule assign a value of 2, which is the value in Newton's gravity formula.

He recognizes that customers tend to open charge accounts in several stores rather than in only one, which would indicate either that there is taste dispersion or that the customers acquire different products in different stores. Note, though, that the formulation does not deal with utilities or individual customers, but only with trade areas, sales volume and trade area proportions, and the factors affecting these two are populations and distances. The model does not include any random factors.

Still in the deterministic playground, Hakimi (1990) argued that, in practice, the purchasing power of a customer is not captured completely by the least full price or the closest facility, as in the binary case, but it is divided among facilities proportionally to their relative distance to the customer ("proportional" rule)

$$w_{ij} = \frac{w/f\left(d_{ij}\right)}{\sum_{k \in R} 1/f\left(d_{ik}\right)}$$

where

w_{ij} = portion of customers at i that use the facility at j
$f(d_{ij})$ = nondecreasing, non-negative function of the distance
R = set of all facilities available to the customers at i.

Furthermore, he analyzes the "partially binary" rule, which assumes that faced with two or more competitors, each with several stores, a customer will use the proportional rule, but consider only the closest facility of each competitor. This rule makes sense especially when all the facilities of a competitor offer the same product and have homogeneous features, as in the case of franchises. This rule has been extended by Fernández et al. (2017) by using utilities that are mainly positive, instead of disutility (distance), and by Lin and Tian (2021) to the limited choice set rule, which considers not only the closest facility of each competitor but a possibly larger set, not containing all the facilities.

Note that, although the proportional rule does not explicitly address uncertainty, it gives the idea that there are factors other than the deterministic (distance, in this case), that have an influence on the behavior of the customers, and that make them distribute their purchases among several stores. Furthermore, it gives a usable expression for Luce's (1957) choice axiom, described in the next subsection.

4.2 Choice Rules Explicitly Assuming Uncertainty: Proportional and Gravity

Thirty years after Hotelling and Reilly, Luce (1957) stated the choice axiom in the field of psychology, including a formula that he recognizes as closely related to the logit analysis that started in the 1830s (see Cramer, 2002). The logit or logistic curve was first used to model human growth and some chemical reactions and to represent variability in human responses. This axiom indicates implicitly that the choices of people are not (at least completely) deterministic, and he proposes a formula for the probability of an agent choosing, say a, over all other possible alternatives. One of the parts of the axiom proposes a proportional choice rule that measures the probability of making a choice or choosing an option a as the ratio

$$P_{Ra} = v_a / \sum_{b \in R} v_b$$

where P_{Ra} = probability of choosing a among all alternatives in set R and v_a is "the response strength associated with response a"; see Luce (1977). This response strength is naturally dependent on the context, so this rule is not applicable to product or store choice as it is. In our context, the response strength must be determined, and it could be a utility, an attraction level, or any other pertinent indicator.

Huff (1963) formulated what appears to be the first probabilistic choice rule in the field of retail analysis. In his setting, customers choose between shopping centers, which are "complementary and competing agglomerations of retail firms which are geographically contained." Huff's rule looks very similar to that of Luce (1977), but he proposes an explicit expression for what Luce calls the "response strength." This expression comes from the gravity rule of Reilly (1929), with two important differences: Huff's analysis recognizes the probabilistic nature of choice, and it focuses on the utility of an action for segments of customers ("statistical units"), rather than sales volume. Referring to the use of Reilly's formula to determine the breaking point between two cities (the point at which the volumes of the trade from the smaller, intermediate city are equal), he states that the breaking point can be interpreted as the "0.5 *probability* position between two cities." Using the probabilistic concept, the sales volume proportion that is captured by the city a can be written using Reilly's formula as

$$\frac{B_a}{B_a + B_b} = \left(\frac{P_a}{D_a^n}\right) / \left[\frac{P_a}{D_a^n} + \frac{P_b}{D_b^n}\right]$$

which resembles the Luce model, in which the left-hand side represents the probability of the demand in the small town purchasing at a, and the population divided by a power of the distance is the "strength." Instead, Huff computes what would be the strength as "the utility of a shopping center to a customer," and estimates its numerator using the number of items of the type that the customer needs that are available at the shopping center, an intuition for multipurpose shopping (for which, in a further step, he uses the square footage of the center, as a proxy). Huff uses the travel time raised to a power in the denominator, as a longer travel time decreases the utility for the customers. Furthermore, he states that customers do not know a priori whether their needs will be satisfied at a particular shopping center as they make their decisions under uncertainty (lack of information, taste uncertainty), but they do know that "the greater the number of items carried by such (shopping) centers, the greater is the customer's expectation that his shopping trip will be successful."

Finally, Huff's formula for the utility u_{ij} of a shopping center at j to all customers located at i is:

$$u_{ij} = S_j / T_{ij}^n$$

where

S_j = square footage of the shopping center,
T_{ij} = travel time from i to j, and
n = parameter to be estimated.

And the probability P_{ij} of all customers located at a point i of patronizing a shopping center located at j is:

$$P_{ij} = \frac{S_j / T_{ij}^n}{\sum_k S_k / T_{ik}^n}$$

as before, n is a parameter to be estimated according to the class of products.

Huff's rule has been profusely used and continues in use in the field of facility location (Drezner, 2019), using different indicators for the attraction of a shopping center, an area, or a store. Among these, are the origin and destination of trips and, in the marketing field, a product of attributes of a brand raised to a power (Nakanishi & Cooper, 1974). Other distance (or travel time) functions have also been used, as a negative exponential (Hodgson, 1981).

Parallel to Huff, a different, deterministic gravity rule was being used since the 1940s that estimated the number of trips T_{ij} from zone i to zone j. Its expression is

$$T_{ij} = k \frac{O_i D_j}{d_{ij}^2}$$

where

O_i denotes the total number of trips originating at i,
D_i counts the total number of trips with j as the destination,
d_{ij} measures the distance between i and j, and
k is a constant.

This rule had no other rationality than its resemblance with the gravity formula. Wilson (1967) provided a theoretical justification for this gravity formula using an analogy to statistical mechanics, obtaining the following form:

$$T_{ij} = \frac{A_i B_j O_i D_j e^{-\beta c_{ij}}}{\left[\sum_j B_j D_j e^{-\beta c_{ij}}\right]\left[\sum_i A_i O_i e^{-\beta c_{ij}}\right]},$$

where the parameters A_i and B_j are balancing factors that enforce $\sum_j T_{ij} = O_i$ and $\sum_i T_{ij} = D_j$. Instead of using the inverse of the distance to a power as in Huff's rule, a new (exponential) function is introduced of c_{ij}, the "impedance" or generalized cost of traveling from i to j. In this version of Wilson's formula, $\{T_{ij}\}$ is interpreted as a distribution of the total number of trips between all origins and destinations. Note that the parameters of this rule are estimated using aggregated data on origins and destinations. Variations of this model have been used for spatial interaction in general (Fotheringham & O'Kelly, 1989).

4.3 *Choice Rules Explicitly Assuming Uncertainty: Random Utility Models (RUM)*

McFadden (1974) proposed the first rule that explicitly considers that the utility of a customer has two components: a deterministic or observable component and a random component. The deterministic component includes all these features of the store and product that have an observable effect on the behavior of a customer, while the random component takes care of all the uncertainties and factors that the customer considers but are not known to an observer.

In its simplest form, the rule is called the multinomial logit (MNL) rule and it assumes the following expression for the utility of a customer i of choosing option j:

$$u_{ij} = \sum_k \beta_{jk} X_{jk} + \varepsilon_{ij} = v_{ij} + \varepsilon_{ij}$$

where v_{ij} and ε_{ij} are the deterministic (observable) and random (non-observable) parts of the utility, respectively. In the observable part of the utility, several attributes of the product and the store can be included. X_{jk} is the value of attribute k at store j and β_{jk} is the weight of the attribute on the decision of the customer, and it is the same for all customers, although it can be specialized for customer segments.

Assuming that ε_{ij} are IID, Gumbel distributed, McFadden finds the following expression for the probability p_{ij} of a customer i using the store at j

$$p_{ij} = \frac{e^{v_{ij}}}{\sum_{k \in R} e^{v_{ik}}}$$

De Palma et al. (1985) use the Hotelling settings on a line to significantly extend Hotelling's result. They use the same expression for the utility as the MNL, in which the only observable attributes are the valuation of the product by the customer (or reservation price), the price of the product, and the distance between the point at which the customer is sited and the store at which the product is offered. They also assume that the choice rule is MNL. They prove that, when products and customers are sufficiently heterogeneous (product heterogeneity, taste dispersion, and uncertainty), the principle of minimum differentiation of Hotelling holds and does not have the abrupt changes generated by infinitesimal changes in price or distance. Furthermore, they prove that beyond a threshold of heterogeneity (a large enough standard deviation of the random component), $n \geq 2$ firms agglomerate at the center of the line. For details, see Marianov and Eiselt (2016).

Although the rule was developed for an individual customer, it can be used for a set of customers and its parameters estimated using data from all customers in a set (Anas, 1983). It is interesting to note that Anas (1983) finds that, when predicting origin-destination trips, the multinomial Logit model, derived by McFadden (1974) from utility maximization and likelihood maximization for estimating its parameters, using small-sample disaggregated data, is equivalent to the gravity model derived by Wilson (1967) from statistical entropy minimization (information maximization), and estimated using aggregated data. Anas first estimates the parameters of an MNL model using both likelihood maximization and entropy minimization, aggregated over a small sample of customers, and proves that both methods result in the same parameters of the MNL model. Next, he minimizes entropy to derive an aggregated version of MNL, i.e., a model for a set of customers. By adequately representing the aggregated parameters in the minimization, he arrives at the Wilson (1967) formula.

An aspect that has been criticized of the MNL rule is that it does not consider the correlation between alternatives. In fact, take the case of two competing franchises, each with stores that are identical to each other except for their location. The MNL rule is not capable of including this fact and it will consider all stores as distinct options, when they are not. This phenomenon has been represented by the red bus-blue bus example (Ortúzar & Willumsen, 2022), in which commuters have two available alternatives for travel: their car or a red bus. They will distribute in a certain proportion among the two alternatives, say one-half each. If a new bus line is added, with the exact same bus type, route, and frequencies as the red bus, but the buses are blue, the MNL will assume that the new line is a new, distinct alternative and the proportions of commuters using a red bus, a blue bus, and car will be one-third each. It is similar to assuming that by painting half of the buses of a different color will change the preference for that mode of transportation from 50% to 66%.

There are several extensions of the MNL that take care of the correlation, being the nested logit a representative one (Williams, 1977; Daly & Zachary, 1978). In the nested logit, customers choose first the "nest," e.g., the bus over the car in a certain proportion, and later, they will choose the options within the nest, e.g., the red bus or the blue bus. As an example, assume that there are two franchises as described above, and each franchise has several stores. The expression of the utility is now

$$u_{ij} = v_{ij} + \varepsilon_{if} + \varepsilon_{ij|f}$$

in which there are two random components: ε_{if} is related to the uncertainties in the choice of the franchise f, and $\varepsilon_{ij\,|\,f}$ is the random component related to the uncertainty in the choice of a store j given that franchise f has been chosen. Both choices are assumed to follow the MNL rule, $\varepsilon_{ij\,|\,f} \sim \text{Gumbel}(0, \lambda)$ and $\varepsilon_{if} + \varepsilon_{ij\,|\,f} \sim \text{Gumbel}(0, \mu)$. Then, the expression of the probability of choosing store j according to the Nested Logit rule is:

$$p_{ij} = p_{if}\, p_{ij|f} \frac{\left(\sum\limits_{j \in F} e^{\lambda v_{ij}}\right)^{\frac{\mu}{\lambda}}}{\left(\sum\limits_{j \in F} e^{\lambda v_{ij}}\right)^{\frac{\mu}{\lambda}} + \left(\sum\limits_{k \in R} e^{\lambda v_{ik}}\right)^{\frac{\mu}{\lambda}}} \frac{e^{\lambda v_{ij}}}{\sum\limits_{j \in F} e^{\lambda v_{ij}}}$$

in which F is the set of stores of franchise f, the practice is to set $\mu = 1$, and λ is known as the similarity factor. From a more general point of view, nests represent options that are correlated, and in our example, a franchise could have more than one style of store, i.e., more than one nest.

Other extensions of the MNL are the mixed logit (McFadden & Train, 2000; see also Ortúzar & Willumsen, 2022) which considers that the weights on the attributes are random, the constrained MNL (Martínez et al., 2009), and the partially binary logit rule (Méndez-Vogel et al., 2022).

All the rules revised in this section assume single-stop shopping, but they can be modified to include comparison shopping and multipurpose shopping. Further details on choice rules can be found in Eiselt et al. (2019).

5 Integrating User Choice Rules with Uncertainty in Facility Location Models

We present a general competitive facility location model in which any probabilistic choice rule can be incorporated. We restrict ourselves to the case in which there are two competitors, offering mutual substitute products. One of them is already located (the leader) and the second one looking for the best locations for its n^F stores (the follower). Let I, J, and K be the set of customers, the set of candidate locations, and the stores located by the leader, respectively. Each customer has a buying power of w_i. Let us define the variables x_j and p_{ij} (x). x_j takes the value of 1 if a store is located in j, and 0 otherwise. $p_{ij}(x)$ is the probability that customer i purchases the product from the follower's store located at j given the location vector x. The following general model maximizes customers' capture by the follower:

$$\text{Max} \sum_{i \in I} w_i \sum_{j \in J} p_{ij}(x) \tag{1}$$

$$\text{s.t.} \quad \sum_{j \in J} x_j = n^F \tag{2}$$

$$x_j \in \{0, 1\} \quad \forall j \in J \tag{3}$$

$$p_{ij}(x) \in [0, 1] \quad \forall i \in I, j \in J \tag{4}$$

The probability p_{ij} (x) is replaced according to the customers' choice rule. In general, a nonlinear objective follows. Using the Huff rule, the model becomes

$$\sum_{i \in I} w_i \sum_{j \in J_L} \frac{S_j x_j / T_{ij}^n}{\sum_{k \in J_L^*} S_k / T_{ik}^n + \sum_{k \in J_L} S_k x_k / T_{ik}^n} \tag{5}$$

s.t. (2) and (3)where J_L^*, J_F are the set of located leader's stores and the set of follower's candidate locations.

If the rule is the multinomial logit, the model becomes

$$\text{Max} \sum_{i \in I} w_i \sum_{j \in J} \frac{e^{v_{ij}} x_j}{\sum_{k \in J_L^*} e^{v_{ik}} + \sum_{k \in J_F} e^{v_{ik}}} \tag{6}$$

s.t. (2) and (3)

For non-essential products, a non-purchase option can be included, whose utility must be determined exogenously and added as an additional option for the customers in the denominator of the expressions.

The models with objectives (5) and (6) can be linearized as in Aros-Vera et al. (2013). Tighter linearizations as well as reformulations of the multinomial logit model that make the problem suitable for branch and cut or cutting-plane methods have been presented in Benati and Hansen (2002), Haase and Müller (2014), Kress and Pesch (2016), Freire et al. (2016), Ljubić and Moreno (2018), Mai and Lodi (2020), Altekin et al. (2021), and references therein. Models using gravity rules have been presented by multiple authors; see Eiselt et al. (2019) and Drezner (2019).

In terms of performance, the best-known solution approaches for the multinomial logit rule are in Ljubić and Moreno (2018) and Mai and Lodi (2020). Ljubić and Moreno proposed a branch and cut approach based on submodular cuts and outer-approximation cuts. Mai and Lodi (2020) proposed a multi-cut outer-approximation approach in a cutting-plane fashion. Both methods require a separation algorithm to dynamically generate the cuts since these linear formulations use a big number of constraints (one for each possible solution). An easier implementation is possible using the new conic reformulation of Altekin et al. (2021).

It is worth mentioning that in practice, the customer does not always consider all available alternatives, as there may be some that are not sufficiently attractive to be considered. There are techniques for eliminating these alternatives in the multinomial logit case (see Bierlaire et al., 2009; Ortúzar & Willumsen, 2022), and in a general proportional rule, as in Lin and Tian (2021), who solve this problem using a branch and cut approach based on generalized Benders decomposition. A limiting case is when the choice is restricted to only one store of each competitor, the one that provides the highest utility. This is the case of competing franchises in which all stores have the same features and products. This is the partially binary rule, and it has been successfully used by, e.g., Fernández et al. (2017). A partially binary logit rule has been recently proposed by Méndez-Vogel et al. (2022).

One of the important remaining challenges is the implementation of models that consider the correlation between alternatives, especially in the case of franchises. The action of purchasing at one of the stores of a firm is correlated with the action of purchasing at another store of the same firm, as in the example of the red bus/blue bus problem: a significant number of features of the combination product-store that are relevant for a customer are the same, irrespective of the store of the firm chosen by this customer. In fact, the product offered is the same, or the service offered is very similar. There might even be a correlation between alternatives of different franchises if there are similarities between them. The nested logit rule aims at including these types of correlation. However, it has never been possible to propose exact methods in the location field because the concavity of the model has not been assured (Dam et al., 2022).

Finally, in relation to other behaviors triggered by uncertainty, a model that considers comparison shopping in a deterministic setting has been proposed by Marianov et al. (2020), and models taking into account multipurpose shopping by Marianov et al. (2018) and Lüer-Villagra et al. (2022), while a probabilistic version of the follower problem can be found in Méndez-Vogel et al. (2022).

6 Conclusions

In this chapter, we discuss the uncertainty in customers' behavior and its effect on how customers choose a product or a store. We describe the sources of uncertainty. We then describe the most relevant families of probabilistic customer choice models and end by demonstrating how these choice models can be included in facility location models.

The most important factor driving uncertainty is the heterogeneity of products and stores. Due to the heterogeneity, customers need to decide between different products that serve the same purpose but have different secondary features. In addition to heterogeneity, most of the time there is some lack of information on the products, their price, and availability, which complicates and adds randomness to the choice process. These two factors, as well as the emotional or psychological state of the customers, make customers change their decisions in time, not be trustful about the information they receive, and in general, behave in a seemingly random way. From the point of view of decision-makers, an added layer of uncertainty comes from the fact that they cannot observe each decision by the customers, which makes perfect modeling impossible. All these uncertainties are incorporated into choice models either indirectly, by assuming a probability of a customer patronizing a store and product that depends on their characteristics, or explicitly assuming that the customers' utilities for their different actions have random components.

Finally, we use these customer choice rules when optimizing facility locations, by including them in facility location models.

Important challenges remain for the facility location community in the application of increasingly complex customer choice rules in facility location models. Modeling sequential choices and unplanned purchases and finding adequate solution methodology for the resulting location models are among these challenges.

Acknowledgments We gratefully acknowledge the support by grants FONDECYT 1220047, ANID PIA/PUENTE AFB220003, and Ph.D. scholarship from CONICYT-PFCHA/Doctorado Nacional/2019-21190765.

References

Altekin, F. T., Dasci, A., & Karatas, M. (2021). Linear and conic reformulations for the maximum capture location problem under multinomial logit choice. *Optimization Letters, 15*(8), 2611–2637.

Amos, C., Holmes, G. R., & Keneson, W. C. (2014). A meta-analysis of customer impulse buying. *Journal of Retailing and Customer Services, 21*(2), 86–97.

Anas, A. (1983). Discrete choice theory, information theory and the multinomial logit and gravity models. *Transportation Research Part B, 17*(1), 13–23. https://doi.org/10.1016/0191-2615(83)90023-1

Anderson, S. P., & Renault, R. (1999). Pricing, product diversity, and search costs: A Bertrand–Chamberlin–Diamond model. *RAND Journal of Economics, 30*, 719–735.

Anderson, S. P., De Palma, A., & Thisse, J.-F. (1992). *Discrete choice theory of product differentiation*. MIT Press.

Aros-Vera, F., Marianov, V., & Mitchell, J. E. (2013). P-hub approach for the optimal park-and-ride facility location problem. *European Journal of Operational Research, 226*(2), 277–285.

Attri, R., & Jain, V. (2018). A study of factors affecting customer shopping behavior. *IUP Journal of Marketing Management, 17*(1), 38–52.

Bell, D. R., Corsten, D., & Knox, G. (2013). *Unplanned category purchase incidence: Who does it, how often, and why*. https://repository.upenn.edu/marketing_papers/301. Last accessed on 01/30/2023.

Benati, S., & Hansen, P. (2002). The maximum capture problem with random utilities: Problem formulation and algorithms. *European Journal of Operational Research, 143*(3), 518–530.

Bierlaire, M., Hurtubia, R., & Flötteröd, G. (2009). A comparative analysis of implicit and explicit methods to model choice set generation. *9th Swiss Transport Research Conference*, Switzerland. http://www.strc.ch/2009/Bierlaire.pdf. Last accessed on 01/30/2023.

Business Insider. (2021). *The pandemic might have saved brick-and-mortar shopping as restrictions lift and customers head back to stores*. https://www.businessinsider.com/pandemic-saved-brick-and-mortar-stores-walmart-and-target-2021-5. Last accessed on 01/30/2023.

Chamberlin, E. H. (1933). *Theory of monopolistic competition*. Harvard University Press.

Correia, I., & Saldanha-da-Gama, F. (2019). Facility location under uncertainty. In G. Laporte, S. Nickel, & F. Saldanha da Gama (Eds.), *Location science* (pp. 185–213). Springer.

Coto-Millán, P. (2003). Theory of utility and customer behaviour: A comprehensive review of concepts, properties and the most significant theorems. In *Utility and production. Contributions to economics*. Physica.

Cramer, J. S. (2002). *The origins of logistic regression*. Tinbergen Institute Working Paper No. 2002-119/4, Available at SSRN: https://ssrn.com/abstract=360300 or https://doi.org/10.2139/ssrn.360300. Both last accessed on 01/30/2023.

Daly, A., & Zachary, S. (1978). Improved multiple choice models. In D. Hensher & Q. Dalvi (Eds.), *Identifying and measuring the determinants of mode choice*. Teakfields.

Dam, T. T., Ta, T. A., & Mai, T. (2022). Submodularity and local search approaches for maximum capture problems under generalized extreme value models. *European Journal of Operational Research, 300*(3), 953–965.

De Palma, A., Ginsburgh, V., Papageorgiu, Y., & Thisse, J.-F. (1985). Heterogeneity and taste dispersion: The principle of minimum differentiation holds under sufficient heterogeneity. *Econometrica, 53*, 767–782.

Drezner, T. (2019). Gravity models in competitive facility location. In H. A. Eiselt & V. Marianov (Eds.), *Contributions to location analysis. International series in operations research and management science* (pp. 253–275). Springer.

Eaton, B. C., & Lipsey, R. G. (1979). Comparison shopping and the clustering of homogeneous firms. *Journal of Regional Science, 19*, 421–435.

Eaton, B. C., & Lipsey, R. G. (1982). An economic theory of central places. *The Economic Journal, 92*, 56–72.

Eiselt, H. A., Marianov, V., & Drezner, T. (2019). Competitive location models, Chapter 14. In G. Laporte, S. Nickel, & F. Saldanha da Gama (Eds.), *Location science* (2nd ed., p. 644p). Springer.

Fernández, P., Pelegrín, B., Lančinskas, A., & Žilinskas, J. (2017). New heuristic algorithms for discrete competitive location problems with binary and partially binary customer behavior. *Computers and Operations Research, 79*, 12–18.

Fischer, J. H., & Harrington, J. E., Jr. (1996). Product variety and firm agglomeration. *The Rand Journal of Economics*, 281–309.

Fitzsimons, G. J., Hutchinson, J. W., Williams, P., Alba, J. W., Chartrand, T. L., Huber, J., Kardes, F. R., Menon, G., Raghubir, P., Russo, J., Shiv, B., & Tavassoli, N. T. (2002). Non-conscious influences on customer choice. *Marketing Letters, 13*(3), 269–279. https://doi.org/10.1023/A:1020313710388

Fotheringham, A. S., & O'Kelly, M. E. (1989). *Spatial interactions: Formulations and applications*. Kluwer Academic Publishers.

Fox, E. J., Montgomery, A. L., & Lodish, L. M. (2004). Customer shopping and spending across retail formats. *The Journal of Business, 77*(S2), S25–S60. https://www.jstor.org/stable/10.1086/381518

Foxall, G. (2005). *Understanding customer choice*. Palgrave Macmillan.

Freire, A. S., Moreno, E., & Yushimito, W. F. (2016). A branch-and-bound algorithm for the maximum capture problem with random utilities. *European Journal of Operational Research, 252*(1), 204–212.

Gauri, D. K., Jindal, R. P., Ratchford, B., Fox, E., Bhatnagar, A., Pandey, A., Navallo, J. R., Fogarty, J., Carr, S., & Howerton, E. (2021). Evolution of retail formats: Past, present, and future. *Journal of Retailing, 97*(1), 42–61. https://doi.org/10.1016/j.jretai.2020.11.002

Haase, K., & Müller, S. (2014). A comparison of linear reformulations for multinomial logit choice probabilities in facility location models. *European Journal of Operational Research, 232*(3), 689–691.

Hakimi, S. (1990). Locations with spatial interactions: Competitive locations and games. In P. B. Mirchandani & R. L. Francis (Eds.), *Discrete location theory* (pp. 439–478). Wiley.

Hausman, J. A., & Newey, W. K. (2016). Individual heterogeneity and average welfare. *Econometrica, 84*(3), 1225–1248.

Hodgson, M. J. (1981). A location–allocation model maximizing customers' welfare. *Regional Studies, 15*(6), 493–506.

Hoo, F. S. (2018). *The one-stop shopping area for all your 'Brooklyn bride' needs*. Available online at https://fashionista.com/2018/08/brooklyn-brides-bridal-shops-third-avenue. Last accessed on 01/30/2023.

Hotelling, H. (1929). Stability in competition. *The Economic Journal, 39*(153), 41–57.

Hübner, A., Hense, J., & Dethlefs, C. (2022). The revival of retail stores via omnichannel operations: A literature review and research framework. *European Journal of Operational Research, 302*, 799–818. https://doi.org/10.1016/j.ejor.2021.12.021

Huff, D. L. (1963). A probabilistic analysis of shopping center trade areas. *Land Economics, 39*(1), 81–89.

McCall, B. (2021). Bricks and mortar retailing far from dead – but now it's just part of the mix. *The Irish Times*https://www.irishtimes.com/special-reports/future-of-retail/bricks-and-mortar-retailing-far-from-dead-but-now-it-s-just-part-of-the-mix-1.4645578. Last accessed on 01/30/2023.

Jamal, M., & Lodhi, S. (2015). Customer shopping behavior in relation to factors influencing impulse buying: A case of superstores in Karachi, Pakistan. *International Journal of Scientific and Research Publications, 5*(12), 41–59.

Kahneman, D., & Tversky, A. (1979). Prospect theory: An analysis of decisions under risk. *Econometrica, 47*, 263–262.

Khapugin, S., & Melnikov, A. (2019). Local search approach for the medianoid problem with multi-purpose shopping trips. In M. Khachay, Y. Kochetov, & P. Pardalos (Eds.), *Mathematical optimization theory and operations research* (MOTOR 2019. Lecture Notes in Computer Science) (Vol. 11548). Springer. https://doi.org/10.1007/978-3-030-22629-9_23

Konishi, H. (2005). Concentration of competing retail stores. *Journal of Urban Economics, 58*(3), 488–512. https://doi.org/10.1016/j.jue.2005.08.005

Kress, D., & Pesch, E. (2016). Competitive location and pricing on networks with random utilities. *Networks and Spatial Economics, 16*(3), 837–863.

Krider, R. E., & Putler, D. S. (2013). Which birds of a feather flock together? Clustering and avoidance patterns, of similar retail outlets. *Geographical Analysis, 45*, 123–149.

Lancaster, K. J. (1966). A new approach to customer theory. *Journal of Political Economy, 74*(2), 132–157.

Li, M. (2015). Convenience and online consumer shopping behavior: A business anthropological case study based on the contingent valuation method. *Anthropologist, 21*(1,2), 8–17.

Lin, Y. H., & Tian, Q. (2021). Branch-and-cut approach based on generalized benders decomposition for facility location with limited choice rule. *European Journal of Operational Research, 293*(1), 109–119.

Ljubić, I., & Moreno, E. (2018). Outer approximation and submodular cuts for maximum capture facility location problems with random utilities. *European Journal of Operational Research, 266*(1), 46–56.

Loomes, G., Orr, S., & Sugden, R. (2009). Taste uncertainty and status quo effects in customer choice. *Journal of Risk and Uncertainty, 39*, 113–135. https://doi.org/10.1007/s11166-009-9076-y

Luce, R. D. (1957). *A theory of individual choice behavior*. Defense Technical Information Center Report. May 1957. https://apps.dtic.mil/sti/pdfs/AD0130718.pdf. Last accessed on 01/30/2023.

Luce, R. D. (1977). The choice axiom after twenty years. *Journal of Mathematical Psychology, 15*, 215–233.

Lüer-Villagra, A., Marianov, V., Eiselt, H. A., & Méndez-Vogel, G. (2022). The leader multipurpose shopping location problem. *European Journal of Operational Research, 302*(2), 470–481.

Mai, T., & Lodi, A. (2020). A multicut outer-approximation approach for competitive facility location under random utilities. *European Journal of Operational Research, 284*(3), 874–881.

Marianov, V., & Eiselt, H. A. (2016). On agglomeration in competitive location models. *Annals of Operations Research, 246*, 31–55. https://doi.org/10.1007/s10479-014-1704-5

Marianov, V., Eiselt, H. A., & Lüer-Villagra, A. (2018). Effects of multipurpose shopping trips on retail store location in a duopoly. *European Journal of Operational Research, 269*, 782–792. https://doi.org/10.1016/j.ejor.2018.02.024

Marianov, V., Eiselt, H. A., & Lüer-Villagra, A. (2020). The follower competitive location problem with comparison-shopping. *Networks and Spatial Economics, 20*(2), 367–393. https://doi.org/10.1007/s11067-019-09481-6

Martínez, F., Aguila, F., & Hurtubia, R. (2009). The constrained multinomial logit: Asemi-compensatory choice model. *Transportation Research Part B: Methodological, 43*(3), 365–377. https://doi.org/10.1016/j.trb.2008.06.006

Massara, F., Melara, R. D., & Liu, S. S. (2014). Impulse versus opportunistic purchasing during a grocery shopping experience. *Marketing Letters, 25*, 361–372.

McFadden, D. (1974). Conditional logit analysis of qualitative choice behaviour. In P. Zarembka (Ed.), *Frontiers in econometrics* (pp. 105–142). Academic Press.

McFadden, D., & Train, K. (2000). Mixed MNL models for discrete response. *Journal of Applied Econometrics, 15*(5), 447–470.

Méndez-Vogel, G., Marianov, V., Lüer-Villagra, A., & Eiselt, H. A. (2022). Store location with multipurpose shopping trips and a new random utility customers' choice model. *European Journal of Operational Research*. https://doi.org/10.1016/j.ejor.2022.06.008
Mongin, P. (1998). Expected utility theory. In J. B. Davis, D. W. Hands, & M. Uskali (Eds.), *The handbook of economic methodology*. Edward Elgar.
Nakanishi, M., & Cooper, L. G. (1974). Parameter estimation for a multiplicative competitive interaction model: Least squares approach. *Journal of Marketing Research, 11*(3), 303–311.
O'Kelly, M. E. (1981). A model of the demand of retail facilities, incorporating multistop, multipurpose trips. *Geographical Analysis, 13*, 134–148.
Ortúzar, J. D., & Willumsen, L. G. (2022). *Modelling transport* (5th ed.). Wiley.
Papageorgiou, Y. Y., & Thisse, J.-F. (1985). Agglomeration as spatial interdependence between firms and households. *Journal of Economic Theory, 37*, 19–31.
Radu, V. (2022). *Customer behavior in marketing – patterns, types, segmentation*. Available online at https://www.omniconvert.com/blog/customer-behavior-in-marketing-patterns-types-segmentation/. Last accessed on 01/30/2023.
Reilly, W. J. (1929). *Methods for the Study of Retail Relationships* (Vol. 4). Bureau of Business Research Studies in Marketing.
Schnure, C. (2021). Brick-and-mortar retail is bouncing back. *Forbes*. Available online at https://www.forbes.com/sites/calvinschnure/2021/03/18/brick-and-mortar-retail-is-bouncing-back/?sh=30a76841d0f4. Last accessed on 01/30/2023.
Sheth, J. N. (2021). Future of brick-and-mortar retailing: how will it survive and thrive? *Journal of Strategic Marketing, 29*(7), 598–607. https://doi.org/10.1080/0965254X.2021.1891128
Shiu, E., Walsh, G., Hassan, L., & Shaw, D. (2011). Customer uncertainty revisited. *Psychology and Marketing, 28*(6), 584–607.
Schoemaker, P. J. H. (1980). *Experiments on decisions under risk: The expected utility hypothesis*. Martinus Nijho.
Snyder, L. V. (2006). Facility location under uncertainty: A review. *IIE Transactions, 38*(7), 547–564.
Solomon, M. R. (2017). *Customer behavior: Buying, having and being* (12th ed.). Pearson.
Stahl, K. (1982). Differentiate products, customer search, and locational oligopoly. *Journal of Industrial Economics, 31*, 97–113.
Thaler, R. H. (2015). *Misbehaving: The making of behavioral economics*. W W Norton & Co..
Tronier, R. M. (2022). *America's love for impulse spending is going strong in 2022*. https://money.slickdeals.net/surveys/slickdeals-impulse-spending-survey-2022/. Last accessed on 01/30/2023.
Urbany, J. E., Dickson, P. R., & Wilkie, W. L. (1989). Buyer uncertainty and information search. *Journal of Customer Research, 16*, 208–215.
Wilson, A. G. (1967). A statistical theory of spatial distribution models. *Transportation Research, 1*, 253–269.
Williams, H. C. (1977). On the formation of travel demand models and economic evaluation measures of user benefit. *Environment and Planning A, 9*, 285–344.
Wolinsky, A. (1983). Retail trade concentration due to customers' imperfect information. *Bell Journal of Economics, 14*, 275–282.

Part II
Models that Protect Against Acts of Nature, Attackers, and Competitors

Humanitarian Logistics Under Uncertainty: Planning for Sheltering and Evacuation

Vedat Bayram, Bahar Y. Kara, Francisco Saldanha-da-Gama, and Hande Yaman

Abstract This chapter focuses on a major area emerging in the context of humanitarian logistics: emergency evacuation planning and management. Two major aspects are covered: shelter site location and evacuation traffic assignment. Both are discussed separately before an integrated problem is considered. Throughout the chapter, uncertainty in the underlying parameters is assumed. The major sources of uncertainty analyzed are the demand for sheltering and capacity of the edges in the underlying network. Congestion issues emerge in this context that are also considered. Different paradigms for capturing uncertainty are considered for illustrative purposes, namely, robust optimization, chance-constrained programming, and stochastic programming.

Keywords Humanitarian logistics · Uncertainty · Shelter site location · Evacuation traffic assignment · Mathematical models

---

V. Bayram
TED University, Department of Industrial Engineering, Ankara, Turkey

University of Kent, Kent Business School, Department of Analytics, Operations and Systems, Canterbury, Kent, UK
e-mail: vedat.bayram@tedu.edu.tr; v.bayram@kent.ac.uk

B. Y. Kara (✉)
Department of Industrial Engineering, Bilkent University, Ankara, Turkey
e-mail: bkara@bilkent.edu.tr

F. Saldanha-da-Gama
Departamento de Estatística e Investigação Operacional e Centro de Matemática, Aplicações Fundamentais e Investigação Operacional, Faculdade de Ciências, Universidade de Lisboa, Lisbon, Portugal
e-mail: fsgama@ciencias.ulisboa.pt; faconceicao@fc.ul.pt

H. Yaman
ORSTAT, Faculty of Economics and Business, KU Leuven, Leuven, Belgium
e-mail: hande.yaman@kuleuven.be

H. A. Eiselt, V. Marianov (eds.), *Uncertainty in Facility Location Problems*, International Series in Operations Research & Management Science 347,
https://doi.org/10.1007/978-3-031-32338-6_4

1 Introduction

Many regions around the globe are vulnerable to disastrous events that can endanger human lives and property. Among such events, we find floods, earthquakes, fires, landslides, volcanic eruptions, etc.

Disaster operations management is the term used for the activities concerned with making decisions and planning for operations that can anticipate or react to a disaster. In this context, one can distinguish among different types of operations: pre-disaster and post-disaster operations. The former include (i) mitigation, i.e., actions taken to prevent and mitigate the consequences of a disaster, and (ii) preparedness, which seeks the elaboration of plans to provide a more efficient response when a disaster occurs. The latter are divided into (i) response operations –those starting immediately after the event to quickly provide the affected people with relief goods (water, food, medical care, shelter, etc.)– and (ii) recovery, which gathers the operations aiming at recovering all the damaged (infrastructures) to resume the normal functioning of the affected area.

The Operations Research and Management Science have a prominent role in the development of useful (and often decisive) decision-making tools to better plan for the above types of operations (alone or combined). Humanitarian logistics is the term used to designate the operations focusing on preparedness and response.

Upon the occurrence of a disaster or when a serious threat is foreseen (e.g., an approaching hurricane), the affected or threatened region or part of it may require evacuation. A specialized area of humanitarian logistics dealing with planning problems within this context is evacuation planning and management. It includes decisions related to traffic assignment along with routing, location of support facilities, and allocation decisions. Some relevant problems in this area are discussed in this chapter.

The goal of an emergency evacuation operation is to save lives by moving people out of the area affected by a disaster or under threat (Lindell et al., 2018). In accordance with the US Department of Homeland Security's (DHS) Federal Emergency Management Agency (FEMA) and US Department of Transportation's (DOT) Federal Highway Administration (FHWA) reports, annually 45–75 actual or prospective disasters require an evacuation and every 2–3 weeks an evacuation order is directed to 1,000 or more people (FHWA, 2007). Only in 2017, over eight million people were evacuated in the USA due to several types of disasters such as wildfires, hurricanes, and floods (DHS, 2019). Efficiently managing (assigning) evacuation traffic is regarded as a critical capability (USDHS, 2013) that must be attained as it is challenging to manage an unusual, sudden, and widespread surge in traffic demand far beyond the capacity of the existing road network over a large geographical area for an extended time that could span days. Failure to do so may lead to further losses (Thompson et al., 2017). Evacuation planning and management deals with this type of emergency. Two intertwined decisions often need to be made in this case: traffic assignment and facility location. The former ensures that if necessary the populations can successfully get to safe locations where aid/support can be provided and their normal lives can be resumed to the larger possible extent. The latter regards the support facilities whose location and establishment should be planned in advance to maximize the chances of providing an adequate help to the affected populations.

When it comes to locating facilities in the context of humanitarian logistics, we find different possibilities depending on their function. Dönmez et al. (2021) distinguish among six categories, namely, suppliers, distribution centers, points of distribution, shelters, field hospitals, and blood centers. When thinking of combining evacuation planning with facility location, we conclude that the facilities of interest are shelter areas since these are the facilities whose function is exactly to support the populations moved away from the disaster region.

Different from a humanitarian supply chain perspective, in an evacuation management setting, demand is related to the number of people or vehicles that will be evacuated from the area under possible disaster threat, and supply corresponds to the capacity of the road network to serve the evacuation demand. Among the supply and demand management strategies that are key to a successful evacuation operation are location-related decisions such as shelter location, contraflow, zone-based evacuate or shelter-in-place, and dynamic resource allocation decisions. The evacuation studies that account for uncertainty and also consider location decisions related to contraflow, zone-based evacuate or shelter-in-place, or dynamic resource allocation are rare (Bayram, 2016).

An important aspect of relevance when planning for pre- or after-disaster operations regards uncertainty. In the context of humanitarian logistics, different sources of uncertainty emerge, which again can be grouped into a few categories (see Dönmez et al. 2021 for further details): demand, supply, and network connectivity.

The above aspects combined set the motivation for the current chapter: to discuss how to capture uncertainty in optimization models aiming at support decision-making in sheltering location and evacuation planning and management.

The remainder of this chapter is organized as follows. In Sect. 2, we revisit the shelter site location problem with emphasis to the underlying decisions, assumptions, and objectives; in Sect. 3, we discuss different possibilities for embedding uncertainty when planning for the location of shelters. Section 4 focuses on traffic assignment decisions and the related efficiency measures assuming that a decision has already been made on the shelters to open. In Sect. 5, an integrated approach is presented and its relevance discussed. The final section presents an overview of the contents presented in the chapter.

2 The Shelter Site Location Problem

Depending on the type of the disaster, shelters are selected among existing facilities or constructed from scratch to protect a population from the impact of the disaster (ARC, 2002; FEMA, 2006, 2008). They can be high grounds and vertical structures in a flood or tsunami or safe places out of the reach of or strong facilities that can withstand the impact of hazards from tornadoes, hurricanes, or wildfires. Although protection of the evacuees is a priority, they may also provide the evacuees with food, water, medical care, and accommodation. The facilities to be fortified or locations where new shelters will be constructed are chosen before a disaster hits during the preparedness phase.

The shelter site location problem has been studied alone or combined with other decisions such as those related with evacuation planning. This is the case in Bayram et al. (2015), Kılcı et al. (2015), Kulshrestha et al. (2011), Li and Jin (2010), Li et al. (2011), and Li et al. (2012), to mention a few early works on the topic. Candidate locations for sheltering typically include parks, yards, schools, parking lots, etc. Such locations must be identified beforehand (see Kılcı et al. 2015 for many practical details).

Next we present a base modeling framework for the selection of shelter sites. Let I denote the set of potentially affected populations (or populations that need to be protected in advance to an approaching threat) and J the potential set of locations for the shelter areas. The demand of population $i \in I$ is denoted by d_i, and the capacity of shelter area $j \in J$ is denoted by q_j. To derive a mathematical model for the problem, we consider two sets of decision variables. For $j \in J$, y_j is a binary variable equal to one iff shelter site j is selected; for $i \in I$ and $j \in J$, x_{ij} is equal to one if the affected population i is accommodated in site j and zero otherwise. Note that the number of shelters to locate is not known beforehand. In fact, it is endogenous and resulting from different aspects still to be discussed. Let us denote $(\mathbf{x}, \mathbf{y}) = \big((x_{ij})_{i\in I,\, j\in J}, (y_j)_{j\in J}\big)$ and by $f(\mathbf{x}, \mathbf{y})$ the measure(s) of interest to optimize. An integer optimization model for the problem can be conceptually stated as follows:

$$\text{opt} \quad f(\mathbf{x}, \mathbf{y}), \tag{1}$$

$$\text{s. t.} \quad \sum_{j\in J} x_{ij} = 1, \qquad i \in I, \tag{2}$$

$$\sum_{i\in I} d_i x_{ij} \le q_j y_j, \qquad j \in J, \tag{3}$$

$$y_j \in \{0, 1\}, \qquad j \in J, \tag{4}$$

$$x_{ij} \in \{0, 1\}, \qquad i \in I,\ j \in J. \tag{5}$$

In the above model, Constraints (2) ensure that all potential populations requiring sheltering are assigned to a shelter site, whereas (3) guarantee that the capacity of the shelter sites is respected. Looking into the constraints of the above model, we observe those typically adopted in a capacitated facility location problem with single assignment. In fact, in the shelter site location problem, each potentially affected population is fully assigned to one and only one location.

The above model can be enriched by ensuring that each population is allocated to the closest open shelter. If we denote by ℓ_{ij} the traveling time or distance between the site of population i and the shelter site j, then the closest assignment can be ensured by using the following constraints (see Espejo et al. 2012 and Wagner & Falkson 1975):

$$\sum_{s\in J:\, \ell_{is}>\ell_{ij}} x_{is} + y_j \leq 1, \quad i \in I,\ j \in J. \tag{6}$$

Imposing closest assignment may turn out to be too strict in the presence of capacity constraints and may result in a deterioration in the system efficiency. To maintain a certain level of efficiency without sacrificing people's willingness to comply, we can define a tolerance, say λ, and accept that a population can be allocated to shelters whose distances are not larger than $(1+\lambda)$ times the distance to the closest open shelter:

$$\sum_{s\in J:\, \ell_{is}>(1+\lambda)\ell_{ij}} x_{is} + y_j \leq 1, \quad i \in I,\ j \in J. \tag{7}$$

Note that the above constraints are valid under the assumption that the decision-maker is responsible for the evacuation plan or, at least, can decide the shelter sites each population is allocated to. Nevertheless, by ensuring such constraints, the final decision will certainly be close to what the populations would do by themselves—to patronize an open shelter site close to their homes.

What remains to be discussed is the objective function. In the shelter site location problem, the objectives of interest are different from what we observe in facility location models emerging in an economic context or even in the context of public facilities location. In the problem we are investigating, the potential locations are assessed beforehand in terms of their aptitude for the function to accomplish—sheltering. The result depends on features such as distance to a hospital, electrical infrastructure, sanitary system, etc. Again, the interested reader can refer to Kılcı et al. (2015) for all details. Nevertheless, such an aptitude is measured by means of a weight in the interval (0, 1]. In particular, for every $j \in J$, a value w_j is found. The closer to 1, the better suited the site is for locating shelters.

A primary objective to consider in the shelter site location problem is fairness, i.e., a so-called Rawlsian approach is sought. By fairness in this context, we mean to put the focus on the least advantaged victims upon the occurrence of a disaster. A surrogate for such objective is a function (to be maximized) accounting for the minimum value of the weights across the open shelters:

$$f(\mathbf{x}, \mathbf{y}) = W_{\min} = \min_{j\in J: y_j=1} w_j. \tag{8}$$

Therefore, maximizing $f(\mathbf{x}, \mathbf{y})$ is a natural goal in our problem. Such objective can be straightforwardly linearized. Nonetheless, as pointed out by Kınay et al. (2019), the above objective does not ensure that the best locations are selected (i.e., locations with the highest weights), and thus the authors concluded that a "pure" Rawlsian objective may not guarantee the best use of the available resources. This motivates another objective—the average weight of the selected shelters:

$$f(\mathbf{x}, \mathbf{y}) = W_{\text{AVG}} = \frac{\sum_{j \in J} w_j y_j}{\sum_{j \in J} y_j}. \tag{9}$$

to maximize. Again, this objective can be linearized in a standard way (Williams 2013). It must be pointed out that the previous objectives are a consequence of the fairness concept that we are adopting. Selecting shelters according to the previous criteria does not guarantee that the total distance traveled is minimized, which is an aspect of relevance in the context of a disaster. Then, a third objective function can be of interest:

$$f(\mathbf{x}, \mathbf{y}) = \text{ADT} = \frac{\sum_{i \in I, j \in J} \ell_{ij} d_i x_{ij}}{\sum_{i \in I} d_i}. \tag{10}$$

to minimize. As above, ℓ_{ij} is denoting the distance or travel time between population $i \in I$ and the shelter site location $j \in J$. When the three above objective functions are of interest, then the problem should be cast in the context of goal programming (if a hierarchy exists between the objectives) or in the context of vectorial optimization (if no hierarchy exists)—see Kınay et al. (2019) for all details.

Once evacuation management authorities complete their plans as to which shelter sites to open, which evacuation zone to assign to which shelter, and how to route each zone to their assigned shelters, public education campaigns through various means should be started to inform and educate the public and to increase compliance rates of vulnerable populations to evacuation orders (DHS, 2019; FEMA, 2021). The education campaigns should make sure people know their evacuation zones, whether they will evacuate or shelter-in-place, which shelter to evacuate to, which route(s) to use, and other critical information needed. And several warning messages from multiple channels such as mobile phones, television, radio, and social media should be disseminated to each evacuation zone with more detailed information throughout the evacuation management process.

3 Hedging Against Uncertainty in the Shelter Site Location Problem

A major issue of concern when planning in advance for humanitarian logistics operations regards the inherent uncertainty underlying the problems. In fact, it is the magnitude of an event (its absolute strength) and the intensity (how seriously it affects each population) that determine the demand for sheltering. Note that this is the case no matter we are protecting populations against a possible threat or we are rescuing people after a catastrophe. Several researchers looked into hedging against uncertainty in the shelter site location problem: Kınay et al. (2018), Kınay et al.

(2019), Mostajabdaveh et al. (2019), and Ozbay et al. (2019). In this section, we discuss how uncertainty can be captured in the problem.

Several paradigms have been proposed for capturing uncertainty in an optimization problem. These depend on the specific problem being considered as well as on the exact knowledge we have about the uncertainty. When no probabilistic quantification exists (either because it could not be determined or because it is irrelevant due to the decision-maker's goal), one typically resorts to robust optimization. Different robustness measures have been proposed such as the maximum regret (to be minimized). Additionally, when a given cumulative distribution function can be associated with the uncertain vector of parameters, one can take advantage from stochastic programming models and techniques. Chance-constrained programming is a sub-topic within stochastic programming and emerges when we face a setting in which the decision-maker is satisfied with a solution that satisfies some constraints with some given probability. Next we discuss some of the above paradigms in the context of the shelter site location problem.

When demand for sheltering is now known in advance, the vector $\xi = (d_1, \ldots, d_{|I|})$ is uncertain. Let ω designate one possible scenario for the demand. In this case, we consider demands given by $\xi_\omega = (d_{1\omega}, \ldots, d_{|I|\omega})$. If we know beforehand that this scenario will occur, then the problem we should solve is the following one that we call P_ω:

$$\text{opt} \quad f(\mathbf{x}_\omega, \mathbf{y}_\omega), \tag{11}$$

$$\text{s. t.} \quad \sum_{j \in J} x_{ij\omega} = 1, \qquad i \in I, \tag{12}$$

$$\sum_{i \in I} d_{i\omega} x_{ij\omega} \le q_j y_{j\omega}, \qquad j \in J, \tag{13}$$

$$\sum_{s \in J : \ell_{is} > (1+\lambda)\ell_{ij}} x_{is\omega} + y_{j\omega} \le 1, \qquad i \in I,\ j \in J, \tag{14}$$

$$y_{j\omega} \in \{0, 1\}, \qquad j \in J, \tag{15}$$

$$x_{ij\omega} \in \{0, 1\}, \qquad i \in I,\ j \in J. \tag{16}$$

In the above problem, variables $x_{ij\omega}$ $(i \in I,\ j \in J)$ and $y_{j\omega}$ $(j \in J)$ emphasize that both in terms of the shelter sites to open and in terms of the allocation of the potentially affected populations to those sites, we seek a solution for a specific demand scenario. Note that closest assignment can be imposed by setting $\lambda = 0$ in Constraints (14).

In what follows, we denote by $V(\mathrm{P}_\omega)$ the optimal value of the above problem.

Under uncertainty, we do not know the exact scenario that will be observed. Thus, a decision must be made here and now about the shelter sites to select without complete information about the future. Suppose that we can foresee a finite set of scenarios; denote by Ω the corresponding index set. Note that we can keep assuming that the allocation of the populations to the open shelters depends on the scenarios. It is the sheltering location solution that should be decided upon beforehand and should be the same no matter the occurring scenario (non-anticipativity decision).

One possible way to tackle the problem under the above conditions consists of planning for a complete risk-averse decision-maker: find the best solution for the worst-case scenario. Unfortunately, in the context of the objective functions presented above, it is not clear what exactly the worst-case scenario means. Note also that the worst-case scenario is dependent on the exact objective adopted. For instance, the scenario corresponding to the largest total demand (which sounds a really bad one) is not necessarily the worst when it comes to measuring the performance of the system using $W_{\min}$. Therefore, an adequate model for the problem would require including the scenario sorting in the model (so that the worst can be identified). This easily leads to a cumbersome mathematical structure. A more interesting alternative is to minimize the maximum regret across all scenarios.

For a set of selected shelter sites, the regret in a scenario is the difference between the value of the solution in case that scenario occurs and the best possible value under that scenario (which corresponds to implementing the optimal shelter site locations for that scenario). Formally, the regret of a solution, say $\hat{\mathbf{y}}$, in scenario $\omega \in \Omega$ is given by $R(\hat{\mathbf{y}}, \omega) = |V(\mathrm{P}_\omega \mid \hat{\mathbf{y}}) - V(\mathrm{P}_\omega)|$, with $V(\mathrm{P}_\omega \mid \hat{\mathbf{y}})$ denoting the optimal value of problem (11)–(16) fixing $\mathbf{y}$ equal to $\hat{\mathbf{y}}$. The use of $|.|$ in the above expression has to do with the fact that we are considering a general objective function (to minimize or maximize) and also with the fact that the regret of a solution is commonly accepted as being a non-negative quantity with the value zero indicating that the solution is optimal for the scenario.

We can now look for the solution that minimizes the maximum regret, which is a solution to the following problem:

$$\min \quad \nu, \tag{17}$$

$$\text{s. t.} \quad \nu \geq |V(\mathrm{P}_\omega \mid \mathbf{y}) - V(\mathrm{P}_\omega)|, \qquad \omega \in \Omega, \tag{18}$$

$$\sum_{j \in J} x_{ij\omega} = 1, \qquad i \in I,\ \omega \in \Omega, \tag{19}$$

$$\sum_{i \in I} d_{i\omega} x_{ij\omega} \leq q_j y_j, \qquad j \in J, \omega \in \Omega, \tag{20}$$

$$\sum_{s \in J: \ell_{is} > (1+\lambda)\ell_{ij}} x_{is\omega} + y_j \leq 1, \qquad i \in I,\ j \in J,\ \omega \in \Omega, \tag{21}$$

$$y_j \in \{0, 1\}, \qquad j \in J, \tag{22}$$

$$x_{ij\omega} \in \{0, 1\}, \qquad i \in I,\ j \in J,\ \omega \in \Omega. \tag{23}$$

In the above model, the objective function (17) together with constraints (18) ensures the adequate computation of the minmax regret. The other constraints were already explained before. In particular, closed assignment constraints are a particular case of (21) by setting $\lambda = 0$.

The discussion presented in this section so far relies on the assumption that no probabilistic quantification is considered for the uncertainty. Robust optimization models and techniques can then come into play for supporting decision-making. This research direction has been pursued in the literature as done by Sun et al. (2021) who proposed a bi-objective robust optimization model considering two injury levels for the people affected by a disaster. In that work, temporary facilities (e.g., shelter sites) are to be located as part of the decisions to make. Yahyaei and Bozorgi-Amiri (2019) make use of robust optimization to design a relief network that includes locating shelter sites for temporary accommodation and selecting supportive distribution centers for supplying the shelter sites. The authors consider interval uncertainty for the amount of affected people in each population center and also assume that some supportive distribution centers may be disrupted by the same event affecting the populations to rescue.

In addition to the lack of probabilistic quantification of the demand, in the above contents, we are assuming that the capacity constraints are hard constraints—no solution violating them is acceptable. Next, we complement the above contents by addressing these issues.

Let us assume now that the demand vector $\boldsymbol{\xi} = \left(d_1, \ldots, d_{|I|}\right)$ is actually a random vector and that a cumulative distribution function (CDF) has been estimated for it (e.g., using historical data). This allows casting the problem as a stochastic programming problem. Nevertheless, the exact paradigm of interest depends on the specific problem being considered. First, we note that the randomness of $\boldsymbol{\xi}$ makes constraints (3) in models (1)–(6) no longer well-defined. Again we can resort to a "fat solution" by devising a plan that works no matter the occurring scenario. Nevertheless, next we consider an alternative.

Suppose that we have soft capacity constraints—if the capacity constraints are "slightly" violated, the solution is still acceptable. We could consider a penalty associated with surplus demand in the shelters and seek to minimize its expected value. This would define another objective in the problem. However, in that case, we would be assuming that it is undesirable to have surplus, which is not what we want. In fact, we are assuming the setting in which, up to a certain extent, having surplus is not an issue. In this case, we can resort to chance-constrained programming. We can define thresholds γ_j $(j \in J)$ corresponding to an acceptable probability of exceeding the capacity of shelter site j and replace the capacity constraints (3) by

$$\mathbb{P}_{\boldsymbol{\xi}}\left[\sum_{i \in I} d_i x_{ij} \leq q_j y_j\right] \geq 1 - \gamma_j, \quad j \in J. \tag{24}$$

The use of chance constraints allows us to further enrich the model by also imposing a minimum throughput that justifies opening a shelter site. In practice, it may not be acceptable that a shelter site is opened no matter the amount of demand it will possibly accommodate. Let β be the minimum threshold for the utilization rate of a shelter site. Since demand is stochastic, again, a minimum utilization rate for the shelter sites is not meaningful but a probabilistic constraint can be considered:

$$\mathbb{P}_{\xi}\left[\sum_{i\in I} d_i x_{ij} \geq \beta q_j y_j\right] \geq 1-\zeta_j, \quad j \in J. \tag{25}$$

In the above inequality, ζ_j is the user-defined probability that the minimum threshold is not satisfied in shelter site $j \in J$. A new model emerges:

$$\begin{aligned} \text{opt} \quad & (1) \\ \text{s. t.} \quad & (2),\ (6),\ (4),\ (5), \\ & (24),\ (25), \\ & x_{ij} \leq y_j, \qquad\qquad i \in I,\ j \in J. \end{aligned} \tag{26}$$

Note that constraints (26) are redundant when the capacity constraints are not probabilistic. However, now we must impose them to make sure that allocations are only made to open shelter sites.

The probabilistic constraints raise some mathematical challenges. In particular, we must find a tractable deterministic counterpart (or at least a good approximation for it). In practice, there are a few aspects that help us in finding such counterpart: (i) the number of demand points is large compared to the number of shelter sites opened. This means that a solution will typically consist of assigning many demand points to each shelter site. (ii) Often, the demand points result themselves from an aggregation—sum of many small demands. (iii) Demands are independent. In fact there are many examples of disastrous events that fully affect one population but not a nearby one (e.g., with floods or a volcanic eruption). Unfortunately, in other situations, there may exist some degree of correlation. Next, we proceed assuming independence since the existence of correlation renders our models intractable.

The above arguments justify invoking the Central Limit Theorem for approximating the probability distribution of the demand allocated to a shelter site. Kınay et al. (2018) take advantage from these facts to tackle the problem considering $f(\mathbf{x}, \mathbf{y}) = W_{\min}$. Kınay et al. (2019) follow the same reasoning to cast the problem in a multi-objective setting—the three objective functions revisited in the previous section were used.

To make this chapter self-contained, we briefly review the elements underlying the approximate linear counterpart model. Let us define $\mu_i = \mathbb{E}[d_i]$ and $\sigma_i^2 = \text{Var}[d_i]$, $i \in I$. Additionally, consider $\Gamma = \sqrt{\sum_{i\in I}\sigma_i^2}$ and denote by z_α the α-quantile of the standard normal distribution. Under the above assumptions, the chance constraints can be approximated by

$$\sum_{i\in I}\frac{\mu_i}{\Gamma}x_{ij} + z_{1-\gamma_j}\sum_{m=1}^{n}\lambda_{jm}b_m \leq \frac{q_j}{\Gamma}y_j, \qquad j \in J, \tag{27}$$

$$\sum_{i\in I}\frac{\mu_i}{\Gamma}x_{ij} + z_{\zeta_j}\sum_{m=1}^{n}\lambda_{jm}b_m \geq \frac{\beta q_j}{\Gamma}y_j, \qquad j \in J, \tag{28}$$

$$\sum_{i\in I}\frac{\sigma_i^2}{\Gamma^2}x_{ij} = \sum_{m=1}^{n}\lambda_{jm}b_m^2, \qquad j \in J, \tag{29}$$

$$\sum_{m=1}^{n}\lambda_{jm} = y_j, \qquad j \in J, \tag{30}$$

$$\lambda_{jm} \geq 0, \qquad j \in J,\ m = 1,\ldots,n, \tag{31}$$

$$(\lambda_{j1},\ldots,\lambda_{jn}) \text{ SOS2}, \qquad j \in J. \tag{32}$$

In this model, n refers to the division of the [0, 1] interval considering n break points. This number is user-defined. Variables λ are actually auxiliary variables that help in identifying the exact segment of the piecewise linear function that is used. This also explains why they define a special ordered set of type 2 (SOS2): at most, two λ variables can be positive, and if exactly two are positive, then they must be consecutive. The larger the number of breakpoints considered, the better the resulting model is as an approximate counterpart. However, it also becomes larger and thus potentially more difficult to tackle. The interested reader can refer to Kınay et al. (2018) for all details.

An alternative to using the above probabilistic constraints consists of casting the problem as a two-stage stochastic programming problem: the shelter sites are decisions to implement here and now, whereas the allocation of the affected populations to the open shelters is implemented after demand is revealed (i.e., when an actual threat appears or a catastrophic event occurs, depending on the specific setting). A consequence of this two-stage decision process is that the total capacity associated with the open shelters may turn out to be not enough for the observed demand. In this case, we must consider surplus demand at one or several shelters or, similarly, we consider capacity shortage at the shelters. Let us focus on the objective function $f(\mathbf{x}, \mathbf{y}) = W_{\min}$ to illustrate this approach. Assuming a risk-neutral decision-maker, the problem can be formulated as follows:

$$\min \quad -W_{\min} + \mathcal{Q}(\mathbf{y}), \tag{33}$$

$$\text{s. t.} \quad W_{\min} \leq w_j y_j + M(1 - y_j), \quad j \in J, \tag{34}$$

$$y_j \in \{0, 1\}, \quad j \in J, \tag{4}$$

with $\mathcal{Q}(\mathbf{y}) = \mathbb{E}_{\boldsymbol{\xi}}[Q(\mathbf{y}, \xi)]$ and

$$Q(\mathbf{y}, \xi) = \min \quad h(\mathbf{x}(\xi)), \tag{35}$$

$$\text{s. t.} \quad \sum_{j \in J} x_{ij}(\xi) = 1, \qquad i \in I, \tag{36}$$

$$\sum_{i \in I} d_i(\xi) x_{ij}(\xi) \leq q_j y_j + \psi_j(\xi), \qquad j \in J, \tag{37}$$

$$\sum_{s \in J: \ell_{is} > (1+\lambda)\ell_{ij}} x_{is}(\xi) + y_j \leq 1, \qquad i \in I,\ j \in J, \tag{38}$$

$$x_{ij}(\xi) \in \{0, 1\}, \qquad i \in I,\ j \in J, \tag{39}$$

$$\psi_j(\xi) \geq 0, \qquad j \in J. \tag{40}$$

The first-stage problem seeks a shelter site selection maximizing the minimum weight across the selected shelters plus a future "consequence" from that decision. The latter is represented by the recourse function $\mathcal{Q}(\mathbf{y})$. In the second-stage problem, the new variables $\psi_j(\xi)$ represent a minimum value for the capacity shortage at shelter $j \in J$. Constraints (37) help in finding the shortages. The other second-stage constraints have a straightforward meaning given the previous contents.

Note that we have one model of type (35)–(40) for every possible realization, ξ, of the random vector $\boldsymbol{\xi}$. The notation used for the second-stage parameters highlights this.

A natural second-stage objective is the average capacity shortage at the shelter sites, i.e., considering

$$h(\mathbf{x}(\xi)) = \frac{\sum_{j \in J} \psi_j(\xi)}{\sum_{j \in J} y_j}. \tag{41}$$

This is a nonlinear objective function that, nonetheless, can be easily linearized. To do so, we can replace the fraction with a new variable, say $\psi_{\text{AVG}}(\xi)$, and then define variables $\tau_j(\xi) = \psi_{\text{AVG}}(\xi) \times y_j$, $j \in J$. Now we set

$$h(\mathbf{x}(\xi)) = \psi_{\text{AVG}}(\xi), \tag{42}$$

and add the following constraints to the second-stage problem:

$$\tau_j(\xi) \leq \overline{\psi_{\text{AVG}}}(\xi) \times y_j, \qquad j \in J,$$

$$\tau_j(\xi) \leq \psi_{\text{AVG}}(\xi), \qquad j \in J,$$

$$\tau_j(\xi) \geq \psi_{\text{AVG}}(\xi) - \left[(1 - y_j) \times \overline{\psi_{\text{AVG}}}(\xi)\right], \qquad j \in J,$$

$$\sum_{j\in J} \tau_j(\xi) = \sum_{j\in J} (\psi_j(\xi) \times y_j),$$

$$\tau_j(\xi) \geq 0, \qquad j \in J,$$

$$\psi_{\text{AVG}}(\xi) \geq 0.$$

$\overline{\psi_{\text{AVG}}}(\xi)$ denotes an upper bound on the average capacity shortage.

The major drawback of the above objective function $h(\mathbf{x}(\xi))$ is that the magnitude of the involved values easily becomes much different from that of $W_{\min}$, which is in (0, 1). This means that a decision will likely be too much influenced by the second-stage decision, which may lead to misleading results. This drawback can be overcome by considering the relative capacity shortage at the open shelters and by defining the second-stage objective function as the maximum relative capacity shortage. The second-stage problem can be written as follows:

$$Q(\mathbf{y}, \xi) = \min \quad \varphi(\xi), \tag{43}$$

$$\text{s. t.} \quad \varphi(\xi) \geq \left[\frac{\sum_{i\in I} d_i(\xi) x_{ij}(\xi)}{q_j} - 1\right] - M(1 - y_j), \quad j \in J, \tag{44}$$

$$\sum_{j\in J} x_{ij}(\xi) = 1, \qquad i \in I, \tag{36}$$

$$\sum_{s\in J:\ell_{is}>(1+\lambda)\ell_{ij}} x_{is}(\xi) + y_j \leq 1, \qquad i \in I,\ j \in J, \tag{38}$$

$$x_{ij}(\xi) \in \{0, 1\}, \qquad i \in I,\ j \in J, \tag{39}$$

$$0 \leq \varphi(\xi) \leq 1. \tag{45}$$

In the above model, constraints (44) are those calling for a detailed explanation. Due to the big M, these constraints are activated if $y_j = 1$; otherwise, they can be discarded. In case shelter site j is set open (i.e., $y_j = 1$), then

$$\frac{\sum_{i\in I} d_i(\xi) x_{ij}(\xi)}{q_j} - 1$$

represents the relative capacity shortage at the shelter site with respect to the total capacity. A negative value indicates no shortage (the capacity is above the demand allocated to the shelter). A value greater than one indicates that the demand assigned to the shelter is more than the double of its capacity—situation that in practice we certainly wish to avoid. Thus, we impose $\varphi(\xi)$ to be smaller than one (constraints (45)), which means that we do not accept a capacity shortage at a shelter greater than or equal to the capacity itself. Of course this is always feasible only if the potential overall capacity at the shelter sites is at least equal to half of the demand in all scenarios. In the end, constraints (44) ensure that the objective function represents the maximum relative capacity shortage.

If we have a finite number of scenarios as before, indexed in a set Ω, then we can write the extensive form of the deterministic equivalent. Denoting by π_ω the probability that scenario $\omega \in \Omega$ occurs with $\pi_\omega \geq 0$ and $\sum_{\omega \in \Omega} \pi_\omega = 1$, the full problem becomes

$$\min \quad -W_{\min} + \sum_{\omega \in \Omega} \pi_\omega \varphi_\omega, \tag{46}$$

$$\text{s. t.} \quad W_{\min} \leq w_j y_j + (1 - y_j), \qquad j \in J, \tag{34}$$

$$\varphi_\omega \geq \left[\frac{\sum_{i \in I} d_{i\omega} x_{ij\omega}}{q_j} - 1 \right] - M(1 - y_j), \qquad j \in J,\ \omega \in \Omega, \tag{47}$$

$$\sum_{j \in J} x_{ij\omega} = 1, \qquad i \in I,\ \omega \in \Omega, \tag{48}$$

$$\sum_{s \in J : \ell_{is} > (1+\lambda)\ell_{ij}} x_{is\omega} + y_j \leq 1, \qquad i \in I,\ j \in J,\ \omega \in \Omega, \tag{49}$$

$$y_j \in \{0, 1\}, \qquad j \in J, \tag{4}$$

$$x_{ij\omega} \in \{0, 1\}, \qquad i \in I,\ j \in J,\ \omega \in \Omega, \tag{50}$$

$$0 \leq \varphi_\omega \leq 1, \qquad \omega \in \Omega. \tag{51}$$

The previous discussion is suitable for a risk-neutral decision-maker. If this is not the case, then capturing the future outcome by an expected value is not adequate. Due to its mathematical properties, the α-conditional value-at-risk (α-CVaR) emerges as a possibility. This is a popular way to account for risk aversion (see, e.g., Shapiro 2021). To make this chapter self-contained, we provide the essential details.

Given a shelter site solution $\mathbf{y}$, the α-CVaR that we denote by $\Psi_\alpha(\mathbf{y})$ is given by the expected value of the objective function for the $(1-\alpha) \times 100\%$ worst scenarios (for that shelter site solution), i.e., it is the expected cost conditional to the scenarios whose value exceeds a certain threshold, say $\eta_\alpha(\mathbf{y})$. The latter is in fact the value-at-risk (associated with shelter site solution $\mathbf{y}$), that is, the α-quantile of the (random) objective function.

It is worth noticing that α-CVaR contains the expected cost as a particular case. In fact, when $\alpha = 0$, all scenarios become involved in the evaluation of $\Psi_\alpha(\mathbf{y})$.

In general, unless an analytical representation for $\eta_\alpha(\mathbf{y})$ can be derived, it is very difficult to find a solution of minimum α-CVaR. For a given solution $\mathbf{y}$, it is already difficult to express $\Psi_\alpha(\mathbf{y})$ since $\eta_\alpha(\mathbf{y})$ is involved in its definition. One possibility (see, e.g., Rockafellar & Uryasev 2000, 2002) is to consider the function

$$\Phi_\alpha(\mathbf{y}, \eta) = \eta + \frac{1}{1-\alpha}\mathbb{E}\left[(\mathcal{R}(\mathbf{y}; \boldsymbol{\xi}) - \eta)\,|\mathcal{R}(\mathbf{y}; \boldsymbol{\xi}) > \eta\right],$$

where $\mathcal{R}(\mathbf{y}; \boldsymbol{\xi})$ is the random variable representing the optimal value of the problem when we fix the shelter site solution, $\mathbf{y}$.

If uncertainty can be captured by finite set of scenarios indexed in a set, say Ω, then the previous function reduces to

$$\Phi_\alpha(\mathbf{y}, \eta) = \eta + \frac{1}{1-\alpha}\sum_{\omega\in\Omega}\left(\mathcal{R}(\mathbf{y}; \boldsymbol{\xi}_\omega) - \eta\right)^+ \pi_\omega.$$

In this case, the α-CVaR of $\mathbf{y}$ can be computed as

$$\Psi_\alpha(\mathbf{y}) = \Phi_\alpha(\mathbf{y}, \eta(\mathbf{y})) = \min_\eta \Phi_\alpha(\mathbf{y}, \eta).$$

The problem of finding a feasible vector $\mathbf{y} \in \{0, 1\}^{|J|}$ with the smallest α-CVaR value can formally stated as

$$\min_{\mathbf{y}\in\{0,1\}^{|J|}} \Psi_\alpha(\mathbf{y}) = \min_{\mathbf{y}\in\{0,1\}^{|J|}} \Phi_\alpha(\mathbf{y}, \eta(\mathbf{y})) = \min_{\mathbf{y}\in\{0,1\}^{|J|},\, \eta} \Phi_\alpha(\mathbf{y}, \eta)$$

Finally, we can formulate the problem as follows:

$$\min \quad \eta + \frac{1}{1-\alpha}\sum_{\omega\in\Omega}\pi_\omega \rho_\omega, \tag{52}$$

$$\text{s. t.} \quad \rho_\omega \geq \left(-W_{\min} + \sum_{\omega\in\Omega}\pi_\omega\varphi_\omega\right) - \eta, \quad \omega \in \Omega, \tag{53}$$

$$\rho_\omega \geq 0, \qquad \omega \in \Omega, \tag{54}$$

$$W_{\min} \leq w_j y_j + (1 - y_j), \qquad j \in J, \tag{34}$$

$$\varphi_\omega \geq \left[\frac{\sum_{i\in I} d_{i\omega}x_{ij\omega}}{q_j} - 1\right] - M(1 - y_j), \quad j \in J,\ \omega \in \Omega, \tag{47}$$

$$\sum_{j\in J} x_{ij\omega} = 1, \qquad i \in I,\ \omega \in \Omega, \tag{48}$$

$$\sum_{s\in J:\ell_{is}>(1+\lambda)\ell_{ij}} x_{is\omega} + y_j \leq 1, \qquad i \in I,\ j \in J,\ \omega \in \Omega, \tag{49}$$

$$y_j \in \{0, 1\}, \qquad j \in J, \tag{4}$$

$$x_{ij\omega} \in \{0, 1\}, \qquad i \in I,\ j \in J,\ \omega \in \Omega, \tag{50}$$

$$0 \leq \varphi_{\omega} \leq 1, \qquad \omega \in \Omega. \tag{51}$$

In line with the above approach, Ozbay et al. (2019) propose a three-stage stochastic programming model for a shelter site location problem aiming at hedging against the consequences of an earthquake. The authors consider the real circumstance in which an aftershock occurs which, sometimes, has consequences not smaller than the main event. For this case, the authors consider shelter site selection made in two stages with the allocation of affected populations to the shelter sites adapted to the aftershock (if advantageous).

We end this section by emphasizing again the contents presented in terms of shelter site location are valid both when we are planning for sheltering populations threatened by an upcoming disaster (e.g., a hurricane) and populations that have suffered from a catastrophic event (e.g., a landslide or a flood).

4 Evacuation Traffic Assignment Approaches

In this section, we assume that shelter location and allocation decisions are already given, and we deal with assigning evacuation traffic to routes in the network to optimize a system or a user (evacuee) objective or a combination of both.

The models proposed in the evacuation planning and management literature are mostly extensions of existing traffic assignment models (Bayram, 2016). These models can mainly be categorized with respect to the traffic assignment approach adopted, i.e., system optimal (SO), user equilibrium (UE), nearest allocation (NA), and constrained system optimal (CSO). In SO approach, evacuation traffic is distributed evenly throughout the road network to reduce congestion effects and to minimize total evacuation time. In UE, however, evacuees act selfishly and minimize their individual evacuation times under the assumption that they have perfect information regarding the road network and the traffic conditions. An equilibrium condition is reached when no traveler can improve his/her travel time by changing routes (Sheffi, 1985). NA approach assigns evacuees to nearest safe destinations. And finally, CSO relaxes the requirement of nearest assignment but ensures that evacuees are assigned to acceptable fair paths only (Jahn et al., 2005; Bayram et al., 2015). For basic theories and models in traffic management problems, the reader is referred to Sheffi (1985).

Consider a directed network $G = (N, A)$, where N is the set of nodes and A is the set of arcs (road segments) in the network. Each arc a is associated with a convex travel time function $t_a(f_a)$, when the traffic flow on that arc is f_a. Let $I \subset N$ be the set of origin (demand) nodes from where the population at risk is to be evacuated, and $J \subset N$ is the set of destination nodes (safe shelter sites) evacuees are assigned. For $i \in I$ and $j \in J$, let d_{ij} be the evacuation traffic demand (trip rate), v_{ij}^k be the flow assigned to route k, and P_{ij} be the set of all routes from origin i to destination

j. Below is the UE formulation (Sheffi, 1985):

$$\min \quad \sum_{a \in A} \int_0^{f_a} t_a(w)dw, \tag{55}$$

$$\text{s. t.} \quad \sum_{k \in P_{ij}} v_{ij}^k = d_{ij}, \qquad i \in I,\ j \in J, \tag{56}$$

$$f_a = \sum_{i \in I,\ j \in J} \sum_{k \in P_{ij}: a \in k} v_{ij}^k, \qquad a \in A, \tag{57}$$

$$v_{ij}^k \geq 0 \qquad k \in P_{ij}, i \in I,\ j \in J. \tag{58}$$

The objective function is the sum of the integrals of the travel time functions over arcs. Constraints (56) ensure that the flow on all paths connecting each origin-destination pair i-j is equal to the evacuation demand for that origin-destination. Constraints (57) compute traffic flow on every arc, and constraints (58) define the variable domains.

For the SO formulation, the objective function is replaced with

$$\sum_{a \in A} t_a(f_a) f_a. \tag{59}$$

Given a path k, let $l_k = \sum_{a \in k} l_a$ be its normal length, where l_a is the normal length (geographical distance, free-flow travel time, travel time in UE) of arc a. For the CSO model, objective function is the same as that of SO, and the set of acceptable fair paths from origin i to destination j is defined as $P_{ij}^{\lambda} = \{k \in P_{ij} : l^k \leq (1+\lambda)l_{ij}^*\}$, where $l_{ij}^* = \min_{k \in P_{ij}} l^k$ is the normal length of the shortest path between $i \in I$ and $j \in J$ and $\lambda \geq 0$ is a tolerance/fairness factor. In other words, the CSO model can be obtained by setting $v_{ij}^k = 0$ for all $i \in I$, $j \in J$, and $k \in P_{ij} \setminus P_{ij}^{\lambda}$ in the SO model. When $\lambda = 0$, the CSO model is the same as the UE/NA traffic assignment model, where only UE travel times/shortest paths can be used. When $\lambda = \infty$, a model for the SO traffic assignment is obtained.

The majority of the evacuation models try to minimize total evacuation time, network clearance time, maximum latency, and total risk or maximize the number/percentage of people evacuated up to a specific time. Whichever goal is pursued, it is important to consider the congestion effect in the model. This is achieved through travel time functions, which are also referred to as link performance functions, link capacity functions, volume-delay curves, link impedance functions, and latency functions. These functions represent travel time on a road segment as convex, positive, and monotonically increasing functions of traffic flow, i.e., as traffic flow (congestion) on a road segment increases, travel speed decreases and hence travel time increases. Among these functions, Greenshields' function (Greenshields et al., 1935), Bureau of Public Roads (BPR) function (TAM, 1964), and Davidson's function (Davidson, 1966) are the most commonly used ones.

Greenshields' function defines the relationship between speed and density and is given below:

$$\vartheta(k) = \vartheta_0 \left(1 - \frac{k}{k_j}\right), 0 \le k \le k_j,$$

where $\vartheta(k)$ is the speed, ϑ_0 is the free-flow speed, and k and k_j are the density and jam density of the road segment, respectively.

BPR function is represented with the following mathematical expression:

$$t(f) = t^0 \left(1 + \alpha \left(\frac{f}{c}\right)^{\beta}\right),$$

where $t(f)$ is the travel time on the road segment; f and c are the amount of traffic assigned to the road segment and the capacity of road segment, respectively; and t^0 is the theoretical free-flow travel time when there exist no vehicles on the road segment. The parameters $\alpha \ge 0$ and $\beta \ge 0$ are the tuning parameters defined in accordance with the road characteristics, which generally are taken as 0.15 and 4, respectively (TAM, 1964).

Davidson's function (Davidson, 1966) is formulated as

$$t(\rho) = t^0 \left(1 + \delta \frac{\rho}{1 - \rho}\right),$$

where $t(\rho)$ is the travel time on the road segment, δ is a delay parameter, and $\rho = f/c$ is the degree of saturation.

5 Planning for Shelter Locations for an Effective Evacuation Management: An Integrated View

Bayram (2016) points out that not considering evacuation traffic assignment and shelter location decisions simultaneously would lead to suboptimal solutions as the selection of safe shelter locations has a direct impact on traffic assignment. Most of the models found in the evacuation literature that consider shelter location decisions are bi-level models, i.e., they decide on the locations and number of shelters in an SO manner at the upper level while they assign evacuees to routes and to shelters in a UE manner at the lower level. There are also single-level models that optimize shelter location and traffic assignment decisions in an SO or CSO manner (see, e.g., Abdelgawad & Abdulhai 2009; Xie & Turnquist 2009, as well as Bayram 2016 and the references therein).

Like in any other area of humanitarian logistics and disaster management, evacuation operations are planned under uncertain information about the future.

The sources of uncertainty include evacuation demand due to unpredictability of time, exact location, and the impact of the disaster (which may also impact the existing shelters) along with the human behavior. The road network may also lose some capacity due to debris, flooding, landslides, and damages. Not accounting for uncertainty may result in further losses (O'Driscoll et al., 2005; Lindell and Prater, 2007). In the literature focusing on evacuation planning and management under uncertainty, the problems that also include shelter location decisions are cast either as bi-level models (Shen et al., 2008; Kulshrestha et al., 2011; Li et al., 2012) or as single-level models (Bayram & Yaman, 2018b,a) as described earlier. While Shen et al. (2008) consider the uncertainty in demand and disruption in shelters, Kulshrestha et al. (2011) and Li et al. (2012) consider only demand uncertainty and disruption in shelters, respectively. Bayram and Yaman (2018a,b) are the only studies that consider demand and capacity uncertainty and disruption in shelters simultaneously.

In this section, we extend the CSO model defined in Sect. 4 with the purpose of ensuring a fair assignment of evacuees to shelters and to routes. Since the CSO model is a generalization of UE/NA models, UE and NA model versions can easily be obtained from it as described. The problem consists of deciding where to locate p shelters and how to assign evacuees to safe shelters and to routes leading to the shelters so as to minimize the total evacuation time. There may be reasons for limiting the number of shelters that can be open such as an endogenous budget and/or personnel constraints.

The problem can be cast as a two-stage stochastic programming problem: in the first stage, a decision is made about the shelter site locations to open; the recourse actions comprise the shelter and traffic assignment decisions.

We assume that uncertainty can be captured by a finite set of scenarios, indexed in set Ω with each scenario specifying the values of all uncertain parameters. As before, π_ω denotes the probability associated with scenario $\omega \in \Omega$. In particular, we assume that the potential shelter site locations may be disrupted upon the occurrence of a disaster. Hereafter, we consider the following notation:

$d_{i\omega}$,	demand at origin $i \in I$ under scenario $\omega \in \Omega$, i.e., number of passenger vehicles that will be evacuated under the scenario.
$J_\omega \subseteq J$,	set of potential shelters that are not disrupted in scenario $\omega \in \Omega$.
$c_{a\omega}$,	(possibly degraded) capacity of road segment a under scenario $\omega \in \Omega$; $0 \leq c_{a\omega} \leq c_a$, for all $a \in A$.
$A_\omega \in A$,	set of usable road segments, i.e., the set of segments such that $c_{a\omega} > 0$, $a \in A_\omega$.
$P_{ij\omega}$,	set of alternative paths from demand point $i \in I$ to shelter $j \in J$ under scenario $\omega \in \Omega$.

A shelter site $j \in J$ is reachable from demand point $i \in I$ under scenario $\omega \in \Omega$ if this shelter is not disrupted and if there exists a route with usable arcs from i to j under this scenario (i.e., $P_{ij\omega} \neq \emptyset$). Accordingly, we can define $\overline{J}_{i\omega}$ as the set of reachable shelters for demand point $i \in I$ under scenario $\omega \in \Omega$. For feasibility purposes, we assume that there exists at least one reachable shelter for every origin $i \in I$ under every scenario $\omega \in \Omega$.

To ensure fairness, the evacuation planning authority may not be willing to assign an evacuee to a path whose normal length is more than $(1 + \lambda)$ times the normal length of a shortest path to the closest open and usable shelter under a given scenario. This implies that some evacuees may be assigned to an open shelter within this fairness level although it might not be the nearest one.

We define $P^{\lambda}_{ij\omega} = \{k \in P_{ij\omega} : l^k \leq (1 + \lambda)l^*_{ij\omega}\}$. This is the set of acceptable and usable paths from origin i to destination j under scenario $\omega \in \Omega$ considering a fairness level λ. $l^*_{ij\omega}$ denotes the length of a shortest path from i to j under scenario $\omega \in \Omega$. To compute the above sets, geographical distances, free-flow travel times $t^0_{a\omega}$, or travel times in UE solution can be used. For a given origin $i \in I$ and a scenario $\omega \in \Omega$, the set of acceptable paths is defined based on the length of a shortest path from node i to the closest open and usable shelter. Since the shelters that will be open is not known a priori, the actual set of acceptable paths is also not known. Nevertheless, we know that this set is a subset of the union of $P^{\lambda}_{ij\omega}$ over all potential shelters $j \in J$.

To model mathematically the integrated problem, we keep using a binary variable y_j equal to 1 iff a shelter is located/opened at node $j \in J$. These are the first-stage decision variables. As for the second-stage variables, we consider $v^k_{ij\omega}$ representing the fraction of the demand of origin $i \in I$ that uses path $k \in P^{\lambda}_{ij\omega}$ to shelter $j \in J_{\omega}$ under scenario $\omega \in \Omega$. Finally, $f_{a\omega}$ is the amount of traffic on arc $a \in A_{\omega}$ under scenario $\omega \in \Omega$.

The stochastic constrained system optimal (SCSO) evacuation model (Bayram & Yaman, 2018b,a) is the following:

$$\min \sum_{\omega \in \Omega} \pi_{\omega} \sum_{a \in A_{\omega}} t_a(f_{a\omega}) f_{a\omega}, \tag{60}$$

$$\text{s.t.} \sum_{j \in J} y_j = p, \tag{61}$$

$$\sum_{j \in J_{\omega}} \sum_{k \in P^{\lambda}_{ij\omega}} v^k_{ij\omega} = 1, \qquad \forall i \in I,\ \omega \in \Omega, \tag{62}$$

$$\sum_{j \in \overline{J}_{i\omega}} y_j \geq 1, \qquad \forall i \in I,\ \omega \in \Omega, \tag{63}$$

$$\sum_{k \in P^{\lambda}_{ij\omega}} v^k_{ij\omega} \leq y_j, \qquad \forall i \in I,\ \omega \in \Omega,\ j \in J_{\omega}, \tag{64}$$

$$\sum_{s \in J_{\omega}} \sum_{k \in P^{\lambda}_{is\omega} : l^k > (1+\lambda) l^*_{ij\omega}} v^k_{is\omega} + y_j \leq 1, \quad \forall i \in I,\ \omega \in \Omega,\ j \in J_{\omega}, \tag{65}$$

$$f_{a\omega} = \sum_{i \in I} \sum_{j \in J_\omega} \sum_{k \in P^\lambda_{ij\omega}: a \in k} d_{i\omega} v^k_{ij\omega} \qquad \forall \omega \in \Omega,\ a \in A_\omega, \tag{66}$$

$$v^k_{ij\omega} \geq 0, \qquad \forall \omega \in \Omega,\ k \in \bigcup_{i \in I,\ j \in J_\omega} P^\lambda_{ij\omega}, \tag{67}$$

$$y_j \in \{0, 1\}, \qquad \forall j \in J. \tag{68}$$

The objective function (60) accounts for the expected total evacuation time spent by the evacuees in the network. Constraint (61) limits the number of shelters open to this pre-specified number p. Constraints (62) ensure that for every scenario, the demand of every origin is assigned to a shelter as well as to a route leading to that shelter. Constraints (63) guarantee at least one open and reachable shelter for each demand point under each scenario. Constraints (64) prevent assigning demand to a non-open shelter. Constraints (65) ensure that if the shelter is open and usable under some scenario, then the demand routed to that shelter should use a path with a length that respects the imposed fairness level. The equalities (66) account for the traffic on every arc under each scenario. Finally, Constraints (67) and (68) define the domain of the decision variables.

Note that in the above model, Constraint (61) can be replaced with a budget constraint if the data on the associated costs are available.

In some applications, the evacuation management authority may require all the evacuees from the same origin to be allocated to the same shelter although allowing the traffic to be distributed between alternative routes connecting the origin to the shelter. To enable having separate control levels over the assignment of demand to shelters and to alternative paths, we define an additional set of decision variables: for $i \in I$, $j \in J$, and $\omega \in \omega$, $x_{ij\omega}$ is equal to one if origin i is assigned to shelter j under scenario $\omega \in \Omega$ and zero otherwise. Using these variables, the new condition can be embedded in the model by adding

$$\sum_{k \in P^\lambda_{ij\omega}} v^k_{ij\omega} = x_{ij\omega}, \qquad i \in I,\ j \in J,\ \omega \in \Omega. \tag{69}$$

These constraints impose that the demand originated at i can only be routed using paths connecting i and the destination shelter.

Another variant of the base model emerges when shelters are capacitated as we considered in Sects. 2 and 3. In this case, we need to add

$$\sum_{i \in I} \sum_{k \in P^\lambda_{ij\omega}} d_{i\omega} v^k_{ij\omega} \leq q_j y_j, \qquad \omega \in \Omega,\ j \in J_\omega, \tag{70}$$

to the model. As before, q_j stands for the capacity of shelter $j \in J$. Unfortunately, due to the inclusion of (70), the stochastic problem no longer has (relatively) complete recourse, i.e., there may be some first-stage feasible solution for which no second-stage feasible completion exists for some scenarios. Still, we can try to reduce second-stage infeasibility by replacing (70) with

$$\sum_{j \in \overline{J}_{i\omega}} q_j y_j \geq d_{i\omega}, \qquad \forall i \in I,\ \omega \in \Omega, \tag{71}$$

and imposing

$$\sum_{j \in \overline{J}_{\omega}} q_j y_j \geq \sum_{i \in I} d_{i\omega}, \qquad \omega \in \Omega, \tag{72}$$

with $\overline{J}_{\omega} = \cup_{i \in I} \overline{J}_{i\omega}$.

The SCSO evacuation problem is NP-hard even when there is a single scenario, $\alpha = 0$ (congestion parameter used in expression (60)), and G is bipartite. In fact, this specification reduces the problem to the classical p-median facility location problem.

The SCSO evacuation problem also generalizes the SO and UE/NA traffic assignment approaches when geographical distances and UE travel times are used as the length of a path. When $\lambda = 0$, we have the UE/NA model, and when $\lambda = \infty$, we obtain a model for the SO traffic assignment.

Finally, the SCSO evacuation problem generalizes the congested facility location problem (Desrochers et al., 1995; Fischetti et al., 2016) where the congestion costs at facilities can be modeled by splitting facility nodes into arcs with convex congestion costs.

6 Conclusions

This chapter discussed two major classes of problems stemming from the preparedness phase in humanitarian logistics: shelter site location and evacuation traffic assignment. Different models were discussed throughout the chapter to highlight different assumptions, underlying conditions, decisions to make, and sources of uncertainty. Above all, we conclude that the existing knowledge gathered allows building progressively more comprehensive models that hopefully can better support authorities when it comes to planning in advance for evacuation of populations either due to a catastrophic event or due to a serious threat foreseen.

It is important to note that the problems discussed in this chapter involve people supposedly affected by a disaster. This is a setting in which human behavior becomes of major relevance. In many situations, one cannot expect the affected populations to act in a purely rational way. For instance, part of the affected people may simply not wait for being rescued or may not look for sheltering in the best possible available location. The models discussed neglect this unpredictable behavior. The relevance of capturing such source of uncertainty in humanitarian logistics is an interesting topic for debate. We refer the reader to Bayram (2016) for more details on evacuee behavior analyses and how they can be incorporated into evacuation modeling approaches.

Another aspect of relevance regards the given distribution function assumed in the context of stochastic programming and chance-constrained programming. Often, there is some "ambiguity" in such distribution. Therefore, the stability of the solutions obtained using a specific distribution is certainly an interesting research direction. The models presented do not reflect any possible deviations between the probability distribution adopted and the real one. Again, this is an interesting research line to pursue.

In any case, the problems and models discussed in this chapter are certainly of help when it comes to finding solutions that hedge against uncertainty in humanitarian logistics, namely, when planning for sheltering and evacuation.

References

Abdelgawad, H., & Abdulhai, B. (2009). Emergency evacuation planning as a network design problem: a critical review. *Transportation Letters, 1*, 41–58.

ARC. (2002). Standards for hurricane evacuation shelter selection, ARC 4496. Technical report. American Red Cross.

Bayram, V. (2016). Optimization models for large scale network evacuation planning and management: a literature review. *Surveys in Operations Research and Management Science, 21*(2), 63–84.

Bayram, V., Tansel, B.Ç., & Yaman, H. (2015). Compromising system and user interests in shelter location and evacuation planning. *Transportation Research Part B: Methodological, 72*, 146–163.

Bayram, V., & Yaman, H. (2018a). Shelter location and evacuation route assignment under uncertainty: A benders decomposition approach. *Transportation Science, 52*(2), 416–436.

Bayram, V., & Yaman, H. (2018b). A stochastic programming approach for shelter location and evacuation planning. *RAIRO-Operations Research, 52*(3):779–805.

Davidson, K. (1966). A flow travel time relationship for use in transportation planning. In *Australian Road Research Board (ARRB) Conference*, 3rd, 1966, Sydney (Vol. 3)

Desrochers, M., Marcotte, P., & Stan, M. (1995). The congested facility location problem. *Location Science, 3*(1), 9–23.

DHS. (2019). Planning considerations: Evacuation and shelter-in-place guidance for state, local, tribal, and territorial partners. US Department of Homeland Security. https://www.fema.gov/sites/default/files/2020-07/planning-considerations-evacuation-and-shelter-in-place.pdf.

Dönmez, Z., Kara, B. Y., Karsu, Ö., & Saldanha-da-Gama, F. (2021). Humanitarian facility location under uncertainty: Critical review and future prospects. *Omega, 102*, 102,393.

Espejo, I., Marín, A., & Rodríguez-Chía, A. M. (2012). Closest assignment constraints in discrete location problems. *European Journal of Operational Research, 219*, 49–58.

FEMA. (2006). Risk management series, safe rooms and shelters: Protecting people against terrorist attacks, FEMA 453. Technical report. Federal Emergency Management Agency.

FEMA. (2008). Design and construction guidance for community safe rooms, FEMA (p. 361, 2nd edn.) Technical report. Federal Emergency Management Agency.

FEMA. (2021). Improving public messaging for evacuation and shelter-in-place, findings and recommendations for emergency managers from peer-reviewed research. US Department of Homeland Security Federal Emergency Management Agency. https://www.fema.gov/sites/

default/files/documents/fema_improving-public-messaging-for-evacuation-and-shelter-in-place_literature-review-report.pdf.

FHWA. (2007). Using highways for no-notice evacuations. Federal Highway Administration, Routes to Effective Evacuation Planning Primer Series FHWA-HOP-08-003. https://www.fema.gov/media-library-data/1564165488078-09ab4aac641f77fe7b7dd30bad21526b/Planning_Considerations_Evacuation_and_Shelter-in-Place.pdf.

Fischetti, M., Ljubić, I., & Sinnl, M. (2016). Benders decomposition without separability: A computational study for capacitated facility location problems. *European Journal of Operational Research, 253*(3), 557–569.

Greenshields, B., Channing, W., Miller, H., et al. (1935). A study of traffic capacity. In: *Highway Research Board Proceedings, National Research Council (USA), Highway Research Board* (vol. 1935).

Jahn, O., Möhring, R. H., Schulz, A. S., & Stier-Moses, N. E. (2005). System-optimal routing of traffic flows with user constraints in networks with congestion. *Operations Research, 53*(4), 600–616.

Kınay, O., Kara, B., Saldanha-da-Gama, F., & Correia, I. (2018). Modeling the shelter site location problem using chance constraints: A case study for Istanbul. *European Journal of Operational Research, 270*(1), 132–145.

Kılcı, F., Kara, B. Y., & Bozkaya, B. (2015). Locating temporary shelter areas after an earthquake: A case for Turkey. *European Journal of Operational Research, 243*, 323–332.

Kınay, O., Saldanha-da-Gama, F., & Kara, B. (2019). On multi-criteria chance-constrained capacitated single-source discrete facility location problems. *Omega, 83*, 107–122.

Kulshrestha, A., Wu, D., Lou, Y., & Yin, Y. (2011). Robust shelter locations for evacuation planning with demand uncertainty. *Journal of Transportation Safety & Security, 3*(4), 272–288.

Li, A. C., Xu, N., Nozick, L., & Davidson, R. (2011). Bilevel optimization for integrated shelter location analysis and transportation planning for hurricane events. *Journal of Infrastructure Systems, 17*(4), 184–192.

Li, A. C. Y., Nozick, L., Xu, N., & Davidson, R. (2012). Shelter location and transportation planning under hurricane conditions. *Transportation Research Part E: Logistics and Transportation Review, 48*(4), 715–729.

Li, L., & Jin, M. (2010) Sheltering planning and management for natural disasters. In: *Proceedings of THC-IT 2010—Texas Hurricane Center for Innovative Technology*. Texas, USA: University of Houston.

Lindell, M. K., Murray-Tuite, P., Wolshon, B., & Baker, E. J. (2018). *Large-scale evacuation: The analysis, modeling, and management of emergency relocation from hazardous areas*. CRC Press.

Lindell, M. K., & Prater, C. S. (2007). Critical behavioral assumptions in evacuation time estimate analysis for private vehicles: Examples from hurricane research and planning. *Journal of Urban Planning and Development, 133*(1), 18–29.

Mostajabdaveh, M., Gutjahr, W. J., & Salman, F. S. (2019). Inequity-averse shelter location for disaster preparedness. *IISE Transactions, 51*, 809–829.

O'Driscoll, P., Wolf, R., & Hampson, R. (2005). The evacuation worked, but created a highway horror. USA Today (pp. 712–716). http://usatoday30.usatoday.com/news/nation/2005-09-25-evacuation-cover_x.htm

Ozbay, E., Özlem, Ç., & Kara, B. Y. (2019). Shelter site location under multi-hazard scenarios. *Computers & Operations Research, 106*, 102–118.

Rockafellar, R., & Uryasev, S. (2000). Optimization of conditional value-at-risk. *Journal of Risk, 2*, 21–41.

Rockafellar, R., & Uryasev, S. (2002). Conditional value-at-risk for general loss functions. *Journal of Banking & Finance, 26*, 1443–1471.

Shapiro, A. (2021). Tutorial on risk neutral, distributionally robust and risk averse multistage stochastic programming. *European Journal of Operational Research, 288*, 1–13.

Sheffi, Y. (1985). *Urban transportation networks: Equilibrium analysis with mathematical programming methods*. Englewood Cliffs, NJ: Prentice-Hall, Inc.

Shen, Z. J. M., Pannala, J., Rai, R., & Tsoi, T. S. (2008). Modeling transportation networks during disruptions and emergency evacuations. Technical report. Modeling Transportation Networks During Disruptions and Emergency Evacuations.
Sun, H., Wang, Y., & Xue, Y. (2021). A bi-objective robust optimization model for disaster response planning under uncertainties. *Computers & Industrial Engineering, 155*, 107,213.
TAM. (1964). Traffic Assignment Manual. Bureau of Public Roads, US Department of Commerce.
Thompson, R. R., Garfin, D. R., & Silver, R. C. (2017). Evacuation from natural disasters: A systematic review of the literature. *Risk Analysis, 37*(4), 812–839.
USDHS. (2013). National response framework. Department of Homeland Security, WD.
Wagner, J. L., & Falkson, L. M. (1975). The optimal nodal location of public facilities with price-sensitive demand. *Geographical Analysis, 7*, 69–83.
Williams, H. (2013). *Model building in mathematical programming* (5th edn.). Chichester: Wiley.
Xie, C., & Turnquist, M. A. (2009). Integrated evacuation network optimization and emergency vehicle assignment. *Transportation Research Record, 2091*, 79–90.
Yahyaei, M., & Bozorgi-Amiri, A. (2019). Robust reliable humanitarian relief network design: an integration of shelter and supply facility location. *Annals of Operations Research, 283*, 897–916.

Stochastic Components of the Attraction Function in Competitive Facilities Location

Tammy Drezner

Abstract In this chapter, we briefly review basic competitive facilities location models and discuss many extensions that have a stochastic component. Examples include the minimax regret objective, the probability of not meeting a threshold, and the leader-follower game. The most widely used user choice rule is the one applied in the probabilistic gravity model also referred to as the Huff model. There are p facilities located in the area. The probability that a customer selects a particular facility to patronize is a function of all facilities' attractiveness levels and travel distances. We discuss the assessment of the attractiveness level of competing facilities which is based on stochastic analysis. We also present other non-competitive location models that apply the gravity rule in the formulation of their model.

Keywords Competitive facilities location · Stochastic models · Minimax regret · Leader-follower · Gravity model

1 Introduction

The competing facilities location problem is the location of one or more facilities among existing competing facilities. The facilities attract demand generated by customers in the area. The most common objective is to maximize the market share captured by the new facilities. Over the years, many ways of estimating the market share captured by each facility were proposed. It is assumed that customers divide their buying power among facilities according to the facilities' attractiveness and their distance relative to other facilities. Once the market share attracted by one or more facilities can be estimated, a procedure for finding the best locations for the new facilities can be constructed.

T. Drezner (✉)
College of Business and Economics, California State University-Fullerton, Fullerton, CA, USA
e-mail: tdrezner@fullerton.edu; zdrezner@fullerton.edu

H. A. Eiselt, V. Marianov (eds.), *Uncertainty in Facility Location Problems*,
International Series in Operations Research & Management Science 347,
https://doi.org/10.1007/978-3-031-32338-6_5

Recent reviews of competitive facilities location models are Berman et al. (2009), Eiselt (2011), Drezner (2019), Kress and Pesch (2012), Pelegrín et al. (2018), Marianov et al. (2020), Lederer (2020), and Marianov and Eiselt (2016).

2 Probabilistic Models

There are several models that assume that the market share captured by a facility is determined by the probability that customers are attracted to that facility.

2.1 *The Probabilistic Gravity Model*

The most widely used competitive model is the probabilistic gravity model, which is termed in many papers as the gravity model or the Huff model. Reilly (1931) proposed the gravity model where the area between facilities is partitioned according to the physical law of gravity and all customers in such an area patronize the facility determined by the gravity rule. The probabilistic gravity model, sometimes referred to as the "Huff" model, was proposed by Huff (1964, 1966). According to the probabilistic gravity model, the probability that a customer patronizes a facility is proportional to its attractiveness and declines according to a distance decay function. The basic probabilistic gravity model is based on p competing facilities and n demand points that exist in an area (Drezner, 1994b). A distance decay function $f(d, \lambda)$ with a parameter λ, depending on the retail category, is defined. For example, the distance decay function for grocery stores is different from the one for shopping malls.

In the original gravity model (Reilly, 1931), it is assumed that the distance decay parallels gravity decay and thus $f(d) = \frac{1}{d^2}$. Huff (1964, 1966) suggested a decay function $f(d, \lambda) = \frac{1}{d^\lambda}$. Other distance decay functions include: exponential decay $f(d, \lambda) = e^{-\lambda d}$ (Wilson, 1976; Hodgson, 1981), $f(d) = e^{-1.705d^{0.409}}$ (Bell et al., 1998), and a logit function (Drezner et al., 1998b). Based on a real dataset, Drezner (2006) showed that exponential decay $f(d, \lambda) = e^{-\lambda d}$ fits the data better than a power decay $f(d, \lambda) = \frac{1}{d^\lambda}$. It is well recognized that the decay function varies across retail categories. For example, for the decay function $f(d, \lambda) = \frac{1}{d^\lambda}$, it was found that $\lambda = 3$ for grocery stores (Huff, 1966), $\lambda = 3.191$ for clothing stores (Huff, 1964), $\lambda = 2.723$ for furniture stores (Huff, 1964), and $\lambda = 1.27$ for shopping malls (Drezner, 2006).

Let:

B_i	be the buying power at demand point i for $i = 1, \ldots, n$
d_{ij}	be the distance between demand point i and facility j
A_j	be the attractiveness of facility j based on its features, without considering distances d_{ij}, for $j = 1, \ldots, p$
$f(d, \lambda)$	be the distance decay function
λ	be the parameter of the distance decay function
M_j	be the expected market share captured by facility j

The estimated market share captured by facility j according to the gravity model is:

$$M_j = \sum_{i=1}^{n} B_i \frac{A_j f(d_{ij}, \lambda)}{\sum\limits_{k=1}^{p} A_k f(d_{ik}, \lambda)} . \quad (1)$$

where the distance decay function $f(d, \lambda)$ is the same for all competing facilities in the same retail category. Note that some models assume a decay function $f(d)$ without a parameter λ.

2.2 *Random Utility*

The random utility rule (Leonardi & Tadei, 1984; Drezner & Drezner, 1996) is an extension of the utility rule (Drezner, 1994a). In the utility rule, the utility function is defined as a weighted sum of attributes minus the distance. The weight of each attribute is determined by a customer's opinion survey. Hotelling (1929) proposed that competitors compete by charging different mill prices and customers select the facility that provides the lowest total price of mill price plus the cost of travel. Many early competitive location models assumed that each customer patronizes the closest facility which is the case that all the weights are zero (Hakimi, 1981; Drezner, 1982; Hakimi, 1983, 1986, 1990; ReVelle, 1986; Ghosh & Rushton, 1987; Serra & ReVelle, 1995).

In the random utility model, the utility function weights, except for the distance, are assumed to be randomly distributed. Each customer patronizes the facility with the largest utility according to his assessment of the parameters. Therefore, not all customers residing at the same demand point patronize the same facility.

2.3 *Cover-Based Model*

Launhardt (1885) and Fetter (1924) coined the term "economic law of market areas." This concept was formalized by defining a "radius of influence," which is at the core of central place theory (Lösch, 1954; Christaller, 1966). According to central place theory, there is a maximum "range of the good," depending on retail category, that customers are willing to travel to obtain the good. ReVelle (1986) coined the term "sphere of influence." Drezner et al. (2011, 2012) proposed that each competing facility has a sphere of influence determined by its attractiveness level. More attractive facilities have a larger sphere of influence. The buying power spent by a customer in the sphere of influence of several facilities is divided among the competing facilities. The buying power of a customer located outside all the spheres of influence is lost. Drezner et al. (2020a) refined the model by assuming that patronage does not drop abruptly at the radius of influence, but declines gradually near that radius.

3 Estimating Model Parameters

3.1 Distance Correction

In most location models, it is assumed that demand is generated at demand "points." In reality, demand is generated in neighborhoods, and not all residents in a neighborhood reside at the same distance from a facility. The distances between demand points and the facility follow some probability distribution such as a uniform distribution. Demand generated in an area (for non-competitive location models) is investigated in, for example, Wesolowsky and Love (1971). Listing all individual customers is impractical. One way, termed the aggregation problem, is to partition the set of demand points to subsets and replace each subset by a single point (Hodgson & Neuman, 1993; Plastria & Vanhaverbeke, 2007; Francis et al., 2009).

Drezner and Drezner (1997) proposed a distance correction to the gravity model. They found that if the area of a demand "point" is A and the distance to a facility from the center of the area (the demand point) is d, then the corrected distance to be used in the gravity model is about $\sqrt{d^2 + 0.24A}$.

Drezner and Drezner (1997) used an example problem of 100 demand points in a square of size 10 by 10 with 7 existing facilities. Each demand point has an area of 1. The market share captured by the new facility is plotted in Drezner and Drezner (1997) and depicted in Fig. 1. On the left, the surface plot of the "standard" gravity model using $f(d) = \frac{1}{d^2}$ as the decay function is depicted. In the middle, the market share captured when demand is continuous in the 10 by 10 square is shown. On the right, the market share surface using a decay function of $f(d) = \frac{1}{d^2+0.24}$ (distance correction) is depicted. When demand is generated at demand "points," there are many local maxima at various locations. The surface on the right is "smooth" and very close to the continuous surface with two local maxima.

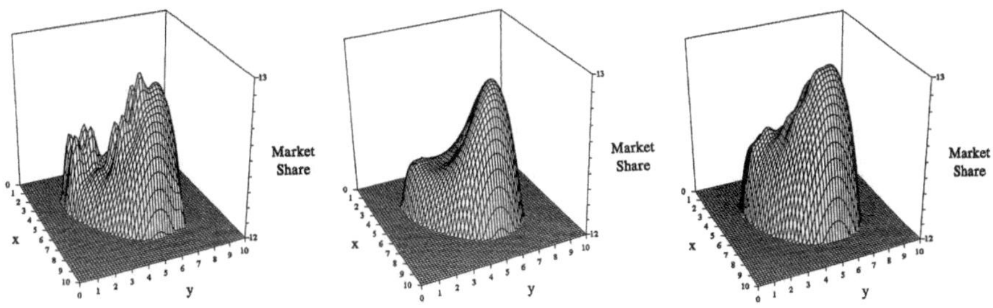

Demand generated at points. Continuous demand in the square. Using distance $\sqrt{d^2 + 0.24}$.

Fig. 1 Discrete and continuous market share surfaces

3.2 On the Attractiveness Level of Competing Facilities

The models for estimating the captured market share (except for the proximity rule) rely on a good estimate of the facilities' attractiveness levels. Therefore, estimating the attractiveness of a facility is an important component required for a successful implementation of the models.

Nakanishi and Cooper (1974) suggested to determine a list of properties and calculated the attractiveness of a facility as a product of these properties' values, each raised by a power. Many researchers (e.g., Bell et al., 1998; Jain & Mahajan, 1979; Schuler, 1981; Timmermans, 1988; Drezner et al., 1998a) conducted public opinion surveys to determine the attributes affecting the attractiveness of the competing facilities and then establish their attractiveness.

Properties that were found by opinion surveys to affect attractiveness include:

Supermarkets: price (Prosperi & Schuler, 1976); price, freshness, availability, convenience, quality service, parking (Schuler, 1981); choice range for daily/non-daily goods, price for daily/non-daily goods, parking (Timmermans, 1988); store image, layout, appearance, accessibility, service, employee composition (Jain & Mahajan, 1979); cleanliness, brands I like, better produce, low prices (Drezner, 1994a); cost of products (Bell et al., 1998).

Clothing: parking availability, choice range (Timmermans, 1982).

Central Business District: price, visual appearance, reputation, range of goods, shopping hours, atmosphere, design, service (Downs, 1970).

Shopping Malls: variety of stores, mall appearance, favorite brand names (Drezner et al., 1998a). They tested six more attributes that were found non-significant: mall prices, distance to mall, adequate parking, mall safety, food court/restaurants, and movies/entertainment.

Drezner and Drezner (2002) suggested to apply the available data of buying power at communities and the reported market share captured by facilities to estimate the attractiveness levels of the competing facilities by a least square model. The best attractiveness levels that yield market shares as close as possible to the reported market shares are found. This method does not require any public opinion surveys.

Drezner (2006) estimated the attractiveness levels of 10 shopping malls in Orange County, California, by analyzing data obtained from a survey of 3,112 intercepted customers. Customers were not asked about their "opinion" on attributes of the shopping mall. They were asked only about their residence zip code and whether they came from home. 1,660 intercepted customers came from home, and their information was used in the analysis. Two distance decay functions were tested: power decay $\frac{1}{d^{\lambda}}$ and exponential decay $e^{-\lambda d}$. By defining the attractiveness levels of the malls and λ as variables, she compared the best fit between the expected number of customers from each zip code to the actual number. In conclusion, exponential decay provided better fit to the data and is recommended as the preferred decay function.

Drezner et al. (2020b) proposed that each facility has a different distance decay function rather than a multiplicative attractiveness level. As the distance increases, the decay in patronage by more attractive facilities is slower than the decay by less attractive facilities. The distance decay parameter can be estimated by a simple survey of intercepted customers inquiring only about the origin of their trip, and no opinion survey, as was suggested in Drezner (2006). No modifications are required in order to apply existing solution algorithms to the new model. The effectiveness and accuracy of the new approach is demonstrated using the Drezner (2006) dataset.

Drezner et al. (2022) further refined the gravity model by extending the modification proposed in Drezner et al. (2020b). When specifying the distance decay function, the basic gravity model and its variants use actual distance. But in reality, travel time to a retail outlet is only a fraction of the time spent on shopping trips. They propose the introduction of an extra distance parameter α. The resulting market share formulation is:

$$M_j = \sum_{i=1}^{n} B_i \frac{f(d_{ij} + \alpha, \lambda_j)}{\sum\limits_{k=1}^{p} f(d_{ik} + \alpha, \lambda_k)} \,. \tag{2}$$

For $f(d, \lambda) = e^{-\lambda d}$, which is the recommended decay function (Drezner, 2006), the formula (2) for the market share is:

$$M_j = \sum_{i=1}^{n} B_i \frac{e^{-\lambda_j (d_{ij}+\alpha)}}{\sum\limits_{k=1}^{p} e^{-\lambda_k (d_{ik}+\alpha)}} \,. \tag{3}$$

The difference between (3) and (2) is the additional extra distance α. If $\alpha = 0$, formulation (3) is equivalent to (2). Formulation (3) extends the original gravity model (1) in two ways. First, every facility may have a different parameter λ_j, and second, A_j is replaced by $e^{\lambda_j \alpha}$. Drezner et al. (2022) found empirically by linear regression on the dataset of Drezner (2006) that indeed $A_j \approx e^{-\alpha\lambda_j}$ for $\alpha = 6.71$ with p-value of 0.0025.

Drezner et al. (2018) suggested a model where attractiveness levels are not constants but follow some probability distribution with a mean and variance. All models assume a given attractiveness level obtained by surveys or other approaches and apply the gravity model based on these values. These attractiveness levels are actually the means of the distribution. Drezner et al. (2018) showed that the "effective" attractiveness level is lower than the mean and the decrease is proportional to the variance divided by the mean. The increase in market share by increasing the attractiveness by Δ is lower than the loss in the market share by decreasing the attractiveness by Δ. Therefore, for a better estimate of the captured market share, the attractiveness level should be replaced by the effective one.

4 Uncertainty-Based Objectives

4.1 Minimax Regret Criterion

Drezner (2009) incorporated future market conditions into the gravity model for the retail facility location. Future market conditions were analyzed as a set of possible scenarios. The best location for a new retail facility is at a location where the market share captured at that location is as close to the maximum as possible regardless of which future scenario takes place. Each scenario may also span different time horizons. The objective is the minimax regret which is used in other models of location analysis, for example, Daskin et al. (1997), Puerto et al. (2009), and Averbakh and Berman (2000).

Suppose that there are K possible scenarios, $k = 1, \ldots, K$. For each scenario, we can calculate the market share $M_k(X)$ at location X. The maximum buying power that can be captured according to each scenario, $M_k^* = \max_X \{M_k(X)\}$, is calculated. The minimax regret objective $R(X)$, to be minimized by selecting the best location X, is then:

$$R(X) = \max_{k=1,\ldots,K} \left\{M_k^* - M_k(X)\right\}$$

Drezner (2009) applied a multi-start heuristic approach to find M_k^* and minimize $R(X)$ for the location of one facility in the plane. Exact algorithms that can find the optimal solution (Drezner & Suzuki, 2004; Hansen et al., 1981) can be implemented.

4.2 The Threshold Objective

Drezner et al. (2002) suggested a different objective for competitive location models. Rather than the objective of maximizing the total buying power attracted by a chain, there is a minimum buying power threshold T to be met. If the chain fails to attract the buying power T, the company fails. The proposed objective is minimizing the probability that the company fails to meet the threshold. The threshold concept has been employed in financial circles as a form of insurance on a portfolio, either to protect the portfolio or to protect a firm's minimum profit, for example, Jacobs and Levy (1996); Olsen (1997); Johansson et al. (1999).

In competitive facility location, let the buying power at demand point $1 \leq i \leq n$ be distributed according to some distribution with a mean of μ_i and a standard deviation σ_i. The buying powers of two demand points i and j are correlated with a correlation coefficient r_{ij}. The total buying power attracted by the chain has a mean of μ and a standard deviation σ (see Drezner et al. (2002) for detailed calculations). The objective function is to minimize $p(X) = P\left(Z \leq \frac{T-\mu}{\sigma}\right)$. Any cumulative distribution is monotonically increasing; thus, minimizing $p(X)$ is equivalent to minimizing $\frac{T-\mu}{\sigma}$.

This problem was solved heuristically in Drezner et al. (2002). It is possible to solve it optimally using BTST (Drezner & Suzuki, 2004) or BSSS (Hansen et al., 1981). Drezner and Drezner (2011b) replaced the weighted sum objective of the p-median problem by the objective of minimizing the probability of exceeding a threshold of the weighted sum of distances.

5 Refinements of the Probabilistic Gravity Model

5.1 *Leader-Follower Models*

Drezner and Drezner (2017) provide a review of the leader-follower model. Other papers on the topic are Plastria and Vanhaverbeke (2008); Küçükaydın et al. (2012).

There are two well-researched two players' simultaneous and sequential games, Nash equilibrium (Nash, 1951) and the leader-follower game, which is also termed the von Stackelberg equilibrium (Stackelberg, 1934) and in voting theory is known as Simpson's problem (Simpson, 1969). In the Nash equilibrium game, no player can improve his objective when the other player does not change his strategy. In many cases, no equilibrium exists (Pelegrín et al., 2018; Bhadury & Eiselt, 1995; Eiselt & Bhadury, 1998). In the leader-follower game, the leader adopts a strategy, and the follower adopts his best strategy knowing the leader's strategy. The follower's goal is to maximize his objective function, while the leader's goal is to maximize his objective function *following* the follower's action.

Early contributions to the Nash equilibrium location problems include Hotelling (1929), Lerner and Singer (1937), Eaton and Lipsey (1975), and Wendell and McKelvey (1981). The leader-follower location problem was introduced to competitive location models by Hakimi (1981), and published in Hakimi (1983, 1986, 1990), for location on network nodes using the premise that each customer patronizes the closest facility; see also Hansen and Labbè (1988).

Drezner (1982) analyzed two competitive location models in the plane. One is the location of a new facility that will attract the most buying power from an existing facility (the follower's problem). The other is the location of a facility that will secure the most buying power against the best location of a competing facility to be set up in the future (the leader's problem). The proximity rule using Euclidean distances is assumed.

Let n demand points be located in the plane. A buying power $b_i > 0$ is associated with demand point i for $i = 1, \ldots, n$. The leader locates his facility at X, and the follower locates his facility at Y. Customers will patronize the follower's facility Y if the Euclidean distance between the customer and Y is less than the distance between the customer and X. Two problems are considered:

The follower's problem: Given the location of an existing facility X serving the demand points, find a location for a new facility Y that will attract the most buying power from demand points.

The leader's problem: Find a location for X such that it will retain the most buying power against the best possible location for the follower's facility Y.

Drezner and Zemel (1992) considered the following problem: a large number of customers are spread uniformly over a given region $A \subseteq \mathbb{R}^2$. What configuration of facilities that cover the area will best protect against a future competing facility? The proximity rule is assumed, i.e., each customer patronizes the closest facility. There are three evenly spread configurations that cover the whole $\mathbb{R}^2$ plane with equilateral polygons depicted in Fig. 2: a triangular grid, a square grid, and an hexagonal grid (beehive). No other cover of the plane by identical equilateral polygons exists. Drezner and Zemel (1992) found that the solution to the problem of covering the whole $\mathbb{R}^2$ plane is the hexagonal pattern. Then they analyzed the finite area problem and found bounds on the difference between the configurations as the number of facilities increases.

Let A be the area attracted by each facility (the area of the triangle, square, or hexagon). It is shown in Drezner and Zemel (1992) that:

- For the triangular grid, the competitor's facility can attract a maximum of $\frac{2}{3}A$.

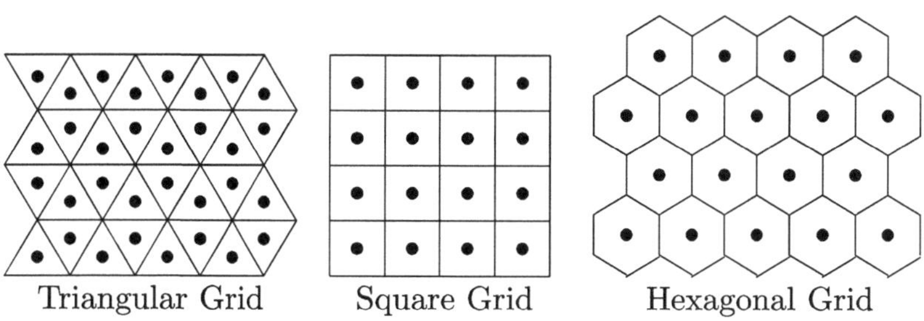

Fig. 2 Various configurations

- For a square grid, the competitor's facility can attract a maximum of $\frac{9}{16}A = 0.5625A$.
- For an hexagonal grid, the competitor's facility can attract a maximum of $0.5127A$.

The hexagonal pattern provides the best protection from a future competitor. It is interesting that for hexagonal and square grids, the competitor captures at least half of A at any point in the plane.

Hexagonal pattern is optimal for many location problems with numerous facilities covering a large area, for example:

- Packing the largest number of circles in an area (Coxeter, 1973; Hilbert & Cohn-Vossen, 1932; Szabo et al., 2007)
- p-median (Okabe & Suzuki, 1997), p-center (Suzuki & Drezner, 1996), and p-cover (Drezner & Suzuki, 2010)
- p-dispersion (Locatelli & Raber, 2002; Maranas et al., 1995; Nurmela & Oestergard, 1999)
- Equalizing the load covered by facilities (Suzuki & Drezner, 2009)

It is also the preferred arrangement for a beehive in nature which has developed over the years in the evolutionary process.

Drezner and Drezner (1998) proposed three heuristic approaches for finding a good solution to the leader-follower model where market share is estimated by the gravity model: brute force, pseudo-mathematical programming, and gradient search. For complete details, see Drezner and Drezner (1998).

Drezner et al. (2015) investigated a leader-follower competitive location model incorporating facilities' attractiveness (design) subject to limited budgets for both the leader and follower. The competitive model is based on the concept of cover (Drezner et al., 2011). The leader and the follower each has a budget to be spent on the expansion of their chains either by improving their existing facilities or by constructing new ones. The objective of the leader is to maximize his market share following the follower's reaction. The follower's problem is identical to the three problems analyzed in Drezner et al. (2012) because market conditions are fully known to the follower. A branch and bound algorithm and a tabu search (Glover, 1977, 1986; Glover & Laguna, 1997) were proposed in Drezner et al. (2012) for the solution of each of these three strategies. For complete details, the reader is referred to Kalczynski (2019).

5.2 *Lost Demand*

Customers may choose a substitute product if the product they are looking for is located too far. For example, if potential customers are interested in a Chinese restaurant but the closest one is too far, they may choose a non-Chinese restaurant which is close by or eat at home. This issue was observed by Lösch (1954) and

Christaller (1966) who developed central place theory and defined the radius of influence in deterministic models.

In the cover models (discussed in Sect. 2.3), the lost demand is automatically addressed. If a demand point is outside the sphere of influence of all facilities, its demand is lost.

In the gravity model, the total demand is assigned to the facilities and no demand is lost. By Eq. (1), $\sum_{j=1}^{p} M_j = \sum_{i=1}^{n} B_i$. This is also true for Eqs. (2) and (3). Drezner and Drezner (2008) proposed to adjust the buying power B_i by the distances from demand point i to the p facilities and then apply the gravity model. They defined a decline of $e^{-\lambda_j d_{ij}}$ for a given set of λ_j for $j = 1, \ldots, p$ in the buying power spent at facility j. The total buying power at demand point i spent at all facilities is

$$B_i \left\{ 1 - \prod_{j=1}^{p} \left[1 - e^{-\lambda_j d_{ij}} \right] \right\}$$

Therefore, the total lost demand by all demand points is:

$$\sum_{i=1}^{n} B_i \prod_{j=1}^{p} \left[1 - e^{-\lambda_j d_{ij}} \right]$$

Drezner and Drezner (2012) added a "dummy" facility to the list of competitors. The dummy facility has no actual location, but $d_{ij} = D$ for some distance D. The distance D represents a reasonable distance customers are willing to travel to patronize a facility. The total buying power attracted by the dummy facility is the lost demand. The standard gravity model or any extension of it can be applied. There is no need to develop specific solution methods for solving the gravity model and variations of it.

5.3 *Cannibalization*

Marketers commonly use a definition of cannibalization that focuses on a company eating into its own market by introducing a new product to an existing product line or an established brand (product line extension and multi-brand strategies) at the expense of the old brand. In such cases, overall company sales may not increase. This form of cannibalization is well recognized and well researched in the marketing literature. See, for example, Mazumdar et al. (1996), Chandy and Tellis (1998), Moorthy and Png (1992), Mason and Milne (1994), and Drezner (2011).

Another form of cannibalization occurs at the retail level of chain facilities, especially in the case of franchises. In this form of cannibalization, opening a new retail outlet in close proximity to an existing outlet, the new outlet cannibalizes the sales of the existing one. Schneider et al. (1998) report cases of lawsuits regarding cannibalization in fast-food franchise systems. This phenomenon is referred to as encroachment. A similar problem is observed and documented in the hospitality/lodging industry.

When a retail chain plans an expansion by building additional outlets, two not necessarily compatible objectives should be considered: (1) maximize the market share captured by the expanding chain and (2) minimize cannibalization of existing chain outlets so as not to gain too much market share at the expense of member outlets. This consideration is especially critical when the outlets are franchised and gain in market share at the expense of member franchisees may be damaging to the profitability of the whole chain.

Drezner (2011) formulated and solved the problem of maximizing the market share captured by the chain facilities subject to a given limit of cannibalization. The market share captured, and consequently the cannibalized portion of it, was calculated using the gravity model discussed in Sect. 2.1.

Plastria (2005) applied the utility function model (Drezner, 1994b) in which the optimal solution to maximizing market share is usually not unique, but there is an area in the plane such that a facility located at any point in that area attracts the same (maximal) market share. Plastria (2005) suggested to locate the facility at the point in that region that minimizes cannibalization, thus maintaining the maximum market share. When the gravity model is used, there is only one optimal solution point that maximizes chain's market share, and the planner must accept the cannibalization at that point if he or she does not wish to consider suboptimal location solutions regarding the market share captured.

Zeller et al. (1980) considered the market share captured by an expanding chain. The franchisor attempts to maximize the total market share of the chain (thus implicitly considers the cannibalization of existing outlets), while the new franchisee considers the market share captured by his new outlet. They conclude that the franchisee of a new store may choose a different location for his store than the franchisor. In reality, the franchisee has no input into the location decision, and thus his objective is ignored.

Ghosh and Craig (1991) developed the FRANSYS model for franchise system growth. Firms seeking to expand franchise distribution systems have to balance two incompatible goals, maximizing system revenues and minimizing the cannibalization of sales of existing outlets. The model uses two constraint types: (1) constraints that disallow new unit locations that do not provide a minimum revenue threshold for the new unit and (2) disallow new units that fail to either protect current revenue for existing units as a group or protect current revenue for each existing unit.

Fernández et al. (2007a) proposed a related model. Their model is a bi-objective model of maximizing profit while minimizing cannibalization. They consider the location of the new facility along with its attractiveness as a decision variable. The construction cost of the new facility is included in the profit function. In addition, they added constraints forbidding the location of a facility in the vicinity of demand points. All of these components lead to a complicated model that requires extensive data collection and relies on many modeling assumptions.

5.4 Location and Design

Combining the location decision with the facility design (treating the attractiveness level of the facility as a variable) was investigated, for example, in Aboolian et al. (2007), Drezner (1998), Fernández et al. (2007b), Plastria and Carrizosa (2004), and Toth et al. (2009).

Drezner (1998) also assumed that the facilities' attractiveness are variables. A budget is available for locating new facilities and for establishing the new facilities' attractiveness levels. The problem is determining the facilities' attractiveness levels within the available budget. It is solved by a gradient search when the budget constraint is kept as equality. Plastria and Vanhaverbeke (2008) combined the limited budget model with the leader-follower model.

The analysis in Drezner (1998) for various assumptions about the attractiveness as a function of the investment in the facility leads to some interesting insights:

1. For firms with a decreasing marginal return on investment curve, the fixed budget allocation solution with equally divided budget among the new facilities is very close to the optimal investment strategy.
2. For firms with a fixed (constant) marginal return on investment, the fixed budget allocation solution with equally divided budget works well and can be used if the computational effort required to obtain the flexible budget allocation solution is prohibitive.
3. For a rapidly increasing marginal return, one should consider opening only one new facility investing all the budget in it.
4. Mildly increasing marginal return leads to a middle-ground solution, and none of the extreme budget allocation strategies is appropriate. In this case, it is recommended to find the best budget allocation by applying the algorithm in Drezner (1998).

Aboolian et al. (2007) studied the problem of simultaneously finding the number of facilities, their location, and their design. For the problem with discrete design scenarios, the TLA (tangent line approximation) procedure (Aboolian et al., 2007) is applied.

Drezner et al. (2016) suggested a model assuming that the market can be partitioned into mutually exclusive sub-markets, for example, expanding a franchise around the world in New York, Paris, Tokyo, Beijing, etc. that customers residing in one sub-market patronize facilities only in that sub-market. Suppose that a procedure for finding the market share at each sub-market for a given budget allocated to the sub-market is available. The problem is then to determine the allocation of the budget among the sub-markets. A constraint that the sum of these individual budgets is equal to the available budget is added.

Three objectives were investigated: (i) maximizing the firm's profit, (ii) maximizing the firm's return on investment, and (iii) maximizing profit subject to a minimum threshold return on investment. Once the market share for a given budget at each individual market can be determined, the allocation of the budget among the markets is found by dynamic programming. For complete details, see Drezner et al. (2016).

6 Applying the Probabilistic Gravity Rule to Other Location Models

The probabilistic gravity rule can be applied to other commonly used non-competitive location objectives. Rather than assuming that a user gets services from the closest facility, he chooses a facility according to the gravity rule. The probability of patronizing a facility is proportional to the facility's attractiveness and to some decay function of the distance.

6.1 Gravity p-Median

In the standard p-median model (Daskin, 1995), it is assumed that each user travels to the closest facility. This implicitly implies that facility choice is centrally controlled or that all facilities charge the same price for the service. Drezner and Drezner (2007) proposed the gravity p-median model. It is assumed that users choose from among the facilities providing services according to the gravity rule rather than from the closest facility. Users consider facilities' attractiveness in their choice. Similar to the standard p-median problem, the objective is to minimize the sum of the expected weighted distances.

Brimberg et al. (2021) suggested a similar p-median model based on the idea that customers do not necessarily patronize the closest facility. A list of probabilities $P_1 \geq P_2, \ldots, \geq P_p$ that add up to 1 is constructed. The probability that a customer patronizes the closest facility is P_1. The probability he patronizes the second closest facility is P_2 and so on.

6.2 The Gravity Obnoxious p-Median Problem

Kalczynski and Drezner (2021) proposed and solved the obnoxious p-median problem. Each facility must be at least at a given distance D from all demand points, and the objective is the minimization of the sum of weighted distances of demand points to their closest facility.

Kalczynski and Drezner (2022) extended the Kalczynski and Drezner (2021) model and proposed three obnoxious p-median models where the facilities may have different sizes which are proportional to the number of customers patronizing the facility. One of the three models is based on the gravity rule. The probability that a customer patronizes a facility is proportional to a distance decay function to that facility. Facility j must be located at least a distance D_j from all demand points where D_j is proportional to the facility size (the volume of services provided by the facility) and the average of the D_j distances is a given distance D.

6.3 Gravity Hub Location

Drezner and Drezner (2001) applied the gravity rule to the hub location problem. A traveler needs to fly from one airport to another. Several potential hubs are available. If the origin or the destination is a hub airport, the traveler chooses a non-stop flight. Otherwise, the probability that a certain hub is selected is proportional to the hub's attractiveness (price, walking distance from the arrival gate to the connecting one, chance of inclement weather, etc.) and to a distance decay function such as the total travel distance (or time) raised to a given inverse power. Such a model can be generalized to selecting a sequence of two or more hubs.

6.4 Gravity Multiple Server

Drezner and Drezner (2011a) considered the gravity rule version of the multiple server location problem (Berman & Drezner, 2007). Total service time consists of travel time to the facility, waiting time in line, and service time. There is a given number of servers to be distributed among the facilities. Each facility acts as an M/M/k queuing system. In Drezner and Drezner (2011a), customers select a server with a probability proportional to its attractiveness and to a decay function of the distance, not necessarily the closest one. Two models are proposed: a stationary one and an interactive one. In the stationary model, it is assumed that customers do not consider the expected waiting time in line and service time at the facility in their facility selection decision simply because they do not know these values. In the interactive model, it is assumed that customers know the expected waiting time in line and service time at the facility and do consider them in their facility selection decision.

7 Summary and Suggestions for Future Research

In this chapter, we reviewed competitive location models which are part of the field of facility location. Facility location models investigate the location of one or more facilities to achieve a certain objective. In competitive location models, the objective is to attract as much buying power as possible from competitors' facilities by constructing new facilities and/or improving existing ones. A main component of such models is the estimation of how customers select the facility to patronize. Demand attracted by a facility depends on its attractiveness, on the buying power customers are planning to spend, and on the distance customers need to travel to get to the facility. What distinguishes different models is the assessment of the relationship between these factors and the market share captured. It is clear that higher attractiveness and buying power lead to higher market share, and a greater distance lowers the expected market share captured.

The gravity model (Reilly, 1931; Huff, 1964, 1966) estimates the probability of patronizing a facility by these three components. Other approaches include the proximity rule (customers patronize the closest facility), utility and random utility models, and cover-based models. One important implementation issue is the assessment of these components, especially the attractiveness level of a facility.

Many extensions to the basic models were investigated, for example, anticipating future changes in the market, considering lost demand due to long distances, and cannibalization of one's chain facilities. Optimal location of one facility can be found by branch and bound algorithms such as big square small square (Hansen et al., 1981) or big triangle small triangle (Drezner & Suzuki, 2004). Location of multiple facilities is usually solved heuristically by various approaches tailored to the specific model or meta-heuristic methods such as tabu search (Glover, 1977, 1986; Glover & Laguna, 1997), simulated annealing (Kirkpatrick et al., 1983), genetic algorithms (Holland, 1975; Goldberg, 2006), variable neighborhood search (Mladenović & Hansen, 1997; Hansen & Mladenović, 1997), and others.

There are many opportunities for future research: improving and fitting the models better to real circumstances and obtaining better estimates for attractiveness of facilities. There are many solution methods for multiple facilities location models and constrained models that can be improved by designing more efficient heuristic algorithms that will enable practitioners to solve larger problems.

References

Aboolian, R., Berman, O., & Krass, D. (2007). Competitive facility location and design problem. *European Journal of Operations Research, 182*, 40–62.

Averbakh, I., & Berman, O. (2000). Minmax regret median location on a network under uncertainty. *INFORMS Journal on Computing, 12*, 104–110.

Bell, D., Ho, T., & Tang, C. (1998). Determining where to shop: Fixed and variable costs of shopping. *Journal of Marketing Research, 35*(3), 352–369.

Berman, O., & Drezner, Z. (2007). The multiple server location problem. *Journal of the Operational Research Society, 58*, 91–99.

Berman, O., Drezner, T., Drezner, Z., & Krass, D. (2009). Modeling competitive facility location problems: New approaches and results. In M. Oskoorouchi (Ed.), *TutORials in Operations Research* (pp. 156–181). San Diego: INFORMS.

Bhadury, J., & Eiselt, H. (1995). Stability of Nash equilibria in locational games. *RAIRO-Operations Research, 29*, 19–33.

Brimberg, J., Maier, A., & Schöbel, A. (2021). When closest is not always the best: The distributed p-median problem. *Journal of the Operational Research Society, 72*, 200–216.

Chandy, R. K., & Tellis, G. J. (1998). Organizing for radical product innovation: The overlooked role of willingness to cannibalize. *Journal of Marketing Research, 35*, 474–487.
Christaller, W. (1966). *Central places in Southern Germany*. Englewood Cliffs, NJ: Prentice-Hall.
Coxeter, H. S. M. (1973). *Regular polytopes*. DoverPublications.
Daskin, M., Hesse, S., & Revelle, C. (1997). α-reliable p-minimax regret: A new model for strategic facility location modeling. *Location Science, 5*, 227–246.
Daskin, M. S. (1995). *Network and discrete location: Models, algorithms, and applications*. New York: John Wiley & Sons.
Downs, R. M. (1970). The cognitive structure of an urban shopping center. *Environment and Behavior, 2*, 13–39.
Drezner, T. (1994a). Locating a single new facility among existing unequally attractive facilities. *Journal of Regional Science, 34*, 237–252.
Drezner, T. (1994b). Optimal continuous location of a retail facility, facility attractiveness, and market share: An interactive model. *Journal of Retailing, 70*, 49–64.
Drezner, T. (1998). Location of multiple retail facilities with limited budget constraints – in continuous space. *Journal of Retailing and Consumer Services, 5*, 173–184.
Drezner, T. (2006). Derived attractiveness of shopping malls. *IMA Journal of Management Mathematics, 17*, 349–358.
Drezner, T. (2009). Location of retail facilities under conditions of uncertainty. *Annals of Operations Research, 167*, 107–120.
Drezner, T. (2011). Cannibalization in a competitive environment. *International Regional Science Review, 34*, 306–322.
Drezner, T. (2019). Gravity models in competitive facility location. In H. A. Eiselt & V. Marianov (Eds.), *Contributions to Location Analysis – In Honor of Zvi Drezner's 75th Birthday* (pp. 253–275). Springer.
Drezner, T., & Drezner, Z. (1996). Competitive facilities: Market share and location with random utility. *Journal of Regional Science, 36*, 1–15.
Drezner, T., & Drezner, Z. (1997). Replacing discrete demand with continuous demand in a competitive facility location problem. *Naval Research Logistics, 44*, 81–95.
Drezner, T., & Drezner, Z. (1998). Facility location in anticipation of future competition. *Location Science, 6*, 155–173.
Drezner, T., & Drezner, Z. (2001). A note on applying the gravity rule to the airline hub problem. *Journal of Regional Science, 41*, 67–73.
Drezner, T., & Drezner, Z. (2002). Validating the gravity-based competitive location model using inferred attractiveness. *Annals of Operations Research, 111*, 227–237.
Drezner, T., & Drezner, Z. (2007). The gravity p-median model. *European Journal of Operational Research, 179*, 1239–1251.
Drezner, T., & Drezner, Z. (2008). Lost demand in a competitive environment. *Journal of the Operational Research Society, 59*, 362–371.
Drezner, T., & Drezner, Z. (2011a). The gravity multiple server location problem. *Computers & Operations Research, 38*, 694–701.
Drezner, T., & Drezner, Z. (2011b). The Weber location problem: The threshold objective. *INFOR: Information Systems and Operational Research, 49*, 212–220.
Drezner, T., & Drezner, Z. (2012). Modelling lost demand in competitive facility location. *Journal of the Operational Research Society, 63*, 201–206.
Drezner, T., & Drezner, Z. (2017). Leader-follower models in facility location. In *Spatial interaction models* (pp. 73–104). Springer.
Drezner, T., Drezner, Z., & Kalczynski, P. (2011). A cover-based competitive location model. *Journal of the Operational Research Society, 62*, 100–113.
Drezner, T., Drezner, Z., & Kalczynski, P. (2012). Strategic competitive location: Improving existing and establishing new facilities. *Journal of the Operational Research Society, 63*, 1720–1730.
Drezner, T., Drezner, Z., & Kalczynski, P. (2015). A leader-follower model for discrete competitive facility location. *Computers & Operations Research, 64*, 51–59.

Drezner, T., Drezner, Z., & Kalczynski, P. (2016). The multiple markets competitive location problem. *Kybernetes, 45*, 854–865.
Drezner, T., Drezner, Z., & Kalczynski, P. (2020a). A gradual cover competitive facility location model. *OR Spectrum, 42*, 333–354.
Drezner, T., Drezner, Z., & Shiode, S. (2002). A threshold satisfying competitive location model. *Journal of Regional Science, 42*, 287–299.
Drezner, T., Drezner, Z., & Zerom, D. (2018). Competitive facility location with random attractiveness. *Operations Research Letters, 46*, 312–317.
Drezner, T., Drezner, Z., & Zerom, D. (2020b). Facility dependent distance decay in competitive location. *Networks and Spatial Economics, 20*, 915–934.
Drezner, T., Drezner, Z., & Zerom, D. (2022). An extension of the gravity model. *Journal of the Operational Research Society*. 73, 2732–2740.
Drezner, T., Marcouldies, G., & Drezner, Z. (1998a). Methods for comparing the attractiveness of shopping centers. In *Proceedings of the DSI Meeting, Las Vegas* (Vol. 2, pp. 1090–1092).
Drezner, Z. (1982). Competitive location strategies for two facilities. *Regional Science and Urban Economics, 12*, 485–493.
Drezner, Z., & Suzuki, A. (2004). The big triangle small triangle method for the solution of non-convex facility location problems. *Operations Research, 52*, 128–135.
Drezner, Z., & Suzuki, A. (2010). Covering continuous demand in the plane. *Journal of the Operational Research Society, 61*, 878–881.
Drezner, Z., Wesolowsky, G. O., & Drezner, T. (1998b). On the logit approach to competitive facility location. *Journal of Regional Science, 38*, 313–327.
Drezner, Z., & Zemel, E. (1992). Competitive location in the plane. *Annals of Operations Research, 40*, 173–193.
Eaton, B. C., & Lipsey, R. G. (1975). The principle of minimum differentiation reconsidered: Some new developments in the theory of spatial competition. *The Review of Economic Studies, 42*, 27–49.
Eiselt, H. A. (2011). Equilibria in competitive location models. In H. A. Eiselt & V. Marianov (Eds.), *Foundations of location analysis* (pp. 139–162). New York: Springer.
Eiselt, H. A., & Bhadury, J. (1998). Reachability of locational Nash equilibria. *Operations-Research-Spektrum, 20*, 101–107.
Fernández, J., Pelegrín, B., Plastria, F., & Tóth, B. (2007a). Planar location and design of a new facility with inner and outer competition: an interval lexicographical-like solution procedure. *Networks and Spatial Economics, 7*, 19–44.
Fernández, J., Pelegrin, B., Plastria, F., & Toth, B. (2007b). Solving a Huff-like competitive location and design model for profit maximization in the plane. *European Journal of Operational Research, 179*, 1274–1287.
Fetter, F. A. (1924). The economic law of market areas. *The Quarterly Journal of Economics, 38*, 520–529.
Francis, R. L., Lowe, T. J., Rayco, M. B., & Tamir, A. (2009). Aggregation error for location models: survey and analysis. *Annals of Operations Research, 167*, 171–208.
Ghosh, A., & Craig, C. S. (1991). FRANSYS: A franchise location model. *Journal of Retailing, 67*, 212–234.
Ghosh, A., & Rushton, G. (1987). *Spatial analysis and location-allocation models*. New York, NY: Van Nostrand Reinhold Company.
Glover, F. (1977). Heuristics for integer programming using surrogate constraints. *Decision Sciences, 8*, 156–166.
Glover, F. (1986). Future paths for integer programming and links to artificial intelligence. *Computers & Operations Research, 13*, 533–549.
Glover, F., & Laguna, M. (1997). *Tabu search*. Boston: Kluwer Academic Publishers.
Goldberg, D. E. (2006). *Genetic algorithms*. Delhi, India: Pearson Education.
Hakimi, S. L. (1981). On locating new facilities in a competitive environment. In *Presented at the ISOLDE II Conference*, Skodsborg, Denmark.

Hakimi, S. L. (1983). On locating new facilities in a competitive environment. *European Journal of Operational Research, 12*, 29–35.
Hakimi, S. L. (1986). p-Median theorems for competitive location. *Annals of Operations Research, 6*, 77–98.
Hakimi, S. L. (1990). Locations with spatial interactions: Competitive locations and games. In P. B. Mirchandani & R. L. Francis (Eds.), *Discrete location theory* (pp. 439–478). New York, NY: Wiley-Interscience.
Hansen, P., & Labbè, M. (1988). Algorithms for voting and competitive location on a network. *Transportation Science, 22*, 278–288.
Hansen, P., & Mladenović, N. (1997). Variable neighborhood search for the p-median. *Location Science, 5*, 207–226.
Hansen, P., Peeters, D., & Thisse, J.-F. (1981). On the location of an obnoxious facility. *Sistemi Urbani, 3*, 299–317.
Hilbert, D., & Cohn-Vossen, S. (1932). *Anschauliche geometrie*. Berlin: Springer. English translation published by Chelsea Publishing Company, New York (1956): Geometry and the Imagination.
Hodgson, M. J. (1981). The location of public facilities intermediate to the journey to work. *European Journal of Operational Research, 6*, 199–204.
Hodgson, M. J., & Neuman, S. (1993). A GIS approach to eliminating source C aggregation error in p-median models. *Location Science, 1*, 155–170.
Holland, J. H. (1975). *Adaptation in natural and artificial systems*. Ann Arbor, MI: University of Michigan Press.
Hotelling, H. (1929). Stability in competition. *Economic Journal, 39*, 41–57.
Huff, D. L. (1964). Defining and estimating a trade area. *Journal of Marketing, 28*, 34–38.
Huff, D. L. (1966). A programmed solution for approximating an optimum retail location. *Land Economics, 42*, 293–303.
Jacobs, B. I., & Levy, K. N. (1996). Residual risk: How much is too much? *Journal of Portfolio Management, 22*, 10–16.
Jain, A. K., & Mahajan, V. (1979). Evaluating the competitive environment in retailing using multiplicative competitive interactive models. In J. N. Sheth (Ed.), *Research in marketing* (Vol. 2, pp. 217–235). Greenwich, CT: JAI Press.
Johansson, F., Seiler, M. J., & Tjarnberg, M. (1999). Measuring downside portfolio risk. *The Journal of Portfolio Management, 26*, 96–107.
Kalczynski, P. (2019). Cover-based competitive location models. In H. A. Eiselt & V. Marianov (Eds.), *Contributions to location analysis – In Honor of Zvi Drezner's 75th birthday* (pp. 277–320). Springer.
Kalczynski, P., & Drezner, Z. (2021). The obnoxious facilities planar p-median problem. *OR Spectrum, 43*, 577–593.
Kalczynski, P., & Drezner, Z. (2022). The obnoxious facilities planar p-median problem with variable capacities. *OMEGA, 111*, 102639.
Kirkpatrick, S., Gelat, C. D., & Vecchi, M. P. (1983). Optimization by simulated annealing. *Science, 220*, 671–680.
Kress, D., & Pesch, E. (2012). Sequential competitive location on networks. *European Journal of Operational Research, 217*, 483–499.
Küçükaydın, H., Aras, N., & Kuban Altınel, İ. (2012). A leader–follower game in competitive facility location. *Computers & Operations Research, 39*, 437–448.
Launhardt, W. (1885). *Mathematische Begründung der Volkswirthschaftslehre*. W. Engelmann.
Lederer, P. J. (2020). Location-price competition with delivered pricing and elastic demand. *Networks and Spatial Economics, 20*, 449–477.
Leonardi, G., & Tadei, R. (1984). Random utility demand models and service location. *Regional Science and Urban Economics, 14*, 399–431.
Lerner, A. P., & Singer, H. W. (1937). Some notes on duopoly and spatial competition. *The Journal of Political Economy, 45*, 145–186.

Locatelli, M., & Raber, U. (2002). Packing equal circles in a square: a deterministic global optimization approach. *Discrete Applied Mathematics, 122*, 139–166.
Lösch, A. (1954). *The economics of location*. New Haven, CT: Yale University Press.
Maranas, C. D., Floudas, C. A., & Pardalos, P. M. (1995). New results in the packing of equal circles in a square. *Discrete Mathematics, 142*, 287–293.
Marianov, V., & Eiselt, H. A. (2016). On agglomeration in competitive location models. *Annals of Operations Research, 246*, 31–55.
Marianov, V., Eiselt, H. A., & Lüer-Villagra, A. (2020). The follower competitive location problem with comparison-shopping. *Networks and Spatial Economics, 20*, 367–393.
Mason, C. H., & Milne, G. R. (1994). An approach for identifying cannibalization within product line extensions and multi-brand strategies. *Journal of Business Research, 31*, 163–170.
Mazumdar, T., Sivakumar, K., & Wilemon, D. (1996). Launching new products with cannibalization potential: an optimal timing framework. *Journal of Marketing Theory and Practice, 4*, 83–93.
Mladenović, N., & Hansen, P. (1997). Variable neighborhood search. *Computers & Operations Research, 24*, 1097–1100.
Moorthy, K. S., & Png, I. P. (1992). Market segmentation, cannibalization, and the timing of product introductions. *Management Science, 38*, 345–359.
Nakanishi, M., & Cooper, L. G. (1974). Parameter estimate for multiplicative interactive choice model: Least squares approach. *Journal of Marketing Research, 11*, 303–311.
Nash, J. (1951). Non-cooperative games. *Annals of Mathematics, 54*, 286–295.
Nurmela, K. J., & Oestergard, P. (1999). More optimal packings of equal circles in a square. *Discrete & Computational Geometry, 22*, 439–457.
Okabe, A., & Suzuki, A. (1997). Locational optimization problems solved through Voronoi diagrams. *European Journal of Operational Research, 98*, 445–456.
Olsen, R. A. (1997). Investment risk: The experts' perspective. *Financial Analysts Journal, 53*, 62–66.
Pelegrín, B., Fernández, P., & García, M. D. (2018). Computation of multi-facility location Nash equilibria on a network under quantity competition. *Networks and Spatial Economics, 18*, 999–1017.
Plastria, F. (2005). Avoiding cannibalisation and/or competitor reaction in planar single facility location. *Journal of the Operational Research Society of Japan, 48*, 148–157.
Plastria, F., & Carrizosa, E. (2004). Optimal location and design of a competitive facility. *Mathematical Programming, 100*, 247–265.
Plastria, F., & Vanhaverbeke, L. (2007). Aggregation without loss of optimality in competitive location models. *Networks and Spatial Economics, 7*, 3–18.
Plastria, F., & Vanhaverbeke, L. (2008). Discrete models for competitive location with foresight. *Computers & Operations Research, 35*, 683–700.
Prosperi, D. C., & Schuler, H. J. (1976). An alternate method to identify rules of spatial choice. *Geographical Perspectives, 38*, 33–38.
Puerto, J., Rodríguez-Chía, A. M., & Tamir, A. (2009). Minimax regret single-facility ordered median location problems on networks. *INFORMS Journal on Computing, 21*, 77–87.
Reilly, W. J. (1931). *The law of retail gravitation*. New York, NY: Knickerbocker Press.
ReVelle, C. (1986). The maximum capture or sphere of influence problem: Hotelling revisited on a network. *Journal of Regional Science, 26*, 343–357.
Schneider, K. C., Johnson, J. C., Sleeper, B. J., & Rodgers, W. C. (1998). A note on applying retail location models in franchise systems: A view from the trenches. *Journal of Consumer Marketing, 15*, 290–296.
Schuler, H. J. (1981). Grocery shopping choices: Individual preferences based on store attractiveness and distance. *Environment and Behavior, 13*, 331–347.
Serra, D., & ReVelle, C. (1995). Competitive location in discrete space. In Z. Drezner (Ed.), *Facility location: A survey of applications and methods* (pp. 367–386). New York, NY: Springer.

Simpson, P. B. (1969). On defining areas of voter choice: Professor Tullock on stable voting. *The Quarterly Journal of Economics, 83*, 478–490.

Stackelberg, H. V. (1934). *Marktform und Gleichgewicht*. Vienne: Julius Springer.

Suzuki, A., & Drezner, Z. (1996). The *p*-center location problem in an area. *Location Science, 4*, 69–82.

Suzuki, A., & Drezner, Z. (2009). The minimum equitable radius location problem with continuous demand. *European Journal of Operational Research, 195*, 17–30.

Szabo, P. G., Markot, M., Csendes, T., & Specht, E. (2007). *New approaches to circle packing in a square: With program codes*. New York: Springer.

Timmermans, H. (1982). Consumer choice of shopping centre: an information integration approach. *Regional Studies, 16*, 171–182.

Timmermans, H. (1988). Multipurpose trips and individual choice behaviour: an analysis using experimental design data. In *Behavioural modelling in geography and planning* (pp. 356–367). Croom Helm.

Toth, B., Fernandez, J., Pelegrin, B., & Plastria, F. (2009). Sequential versus simultaneous approach in the location and design of two new facilities using planar Huff-like models. *Computers & Operations Research, 36*, 1393–1405.

Wendell, R., & McKelvey, R. (1981). New perspectives in competitive location theory. *European Journal of Operational Research, 6*, 174–182.

Wesolowsky, G. O., & Love, R. F. (1971). Location of facilities with rectangular distances among point and area destinations. *Naval Research Logistics Quarterly, 18*, 83–90.

Wilson, A. G. (1976). Retailers' profits and consumers' welfare in a spatial interaction shopping mode. In I. Masser (Ed.), *Theory and practice in regional science* (pp. 42–59). London: Pion.

Zeller, R. E., Achabal, D. D., & Brown, L. A. (1980). Market penetration and locational conflict in franchise systems. *Decision Sciences, 11*, 58–80.

Location and Strategies in Stackelberg Security Games with Risk Aversion

Renaud Chicoisne, Fernando Ordóñez, and Daniel Castro

Abstract In Stackelberg security games, a leader locates security resources to protect a set of targets from strategic adversaries that aim to attack these targets after observing the leader's strategy. In this setting, the leader decision problem is to optimize an uncertain reward that can take a discrete set of values with a probability distribution that depends on the decision variable.

We show how diverse risk aversion models of the leader decision problem can be formulated as tractable optimization problems, such as imposing a bound on the expected disutility, chance constraints, bounded distortion risk, and first- and second-order stochastic dominance constraints or optimizing a value at risk and conditional value at risk. We detail the resulting optimization problems and present computational results that show how the solution changes in two specific settings: (1) an entropic risk measure or value-at-risk minimization with a quantal response follower and (2) a prospect theory model with optimal follower response.

Keywords Stackelberg security games · Risk aversion · Quantal response · Convex optimization · Mixed-integer programming

1 Introduction

A Stackelberg game models the strategic interaction between a leader and one or more followers, where the leader decides on a strategy to maximize its utility knowing that followers will observe this strategy when deciding their own utility maximizing action (Von Stackelberg, 1952). In particular, Stackelberg game models have been used in security applications to represent the interaction between

R. Chicoisne
Clermont Auvergne INP, LIMOS, Clermont-Ferrand, France

F. Ordóñez (✉) · D. Castro
Industrial Engineering Department, Universidad de Chile, Santiago, Chile
e-mail: ordon@dii.uchile.cl

H. A. Eiselt, V. Marianov (eds.), *Uncertainty in Facility Location Problems*, International Series in Operations Research & Management Science 347,
https://doi.org/10.1007/978-3-031-32338-6_6

defenders (that act as the leader) and attackers (corresponding to followers) (Bier, 2007; Brown et al., 2006; Kar et al., 2017). We denote by Stackelberg security games (SSGs) a Stackelberg game where the leader is the defender that locates security resources to protect a subset of targets that can be attacked by one or more adversaries (followers) (Paruchuri et al., 2008; Jain et al., 2010). Such SSGs have been successfully deployed in real-world security applications to help locate the patrols conducted by the Los Angeles International Airport Police on the LAX airport and the US Federal Air Marshal Service on transatlantic flights (Jain et al., 2010), the LA Sheriff department on Los Angeles' subway system (Delle Fave et al., 2014), and the US Coast Guard on the ports and waterways in Boston and New York City (An et al., 2013).

In an SSG, both the defender and attacker receive a penalty or a reward depending on whether the defender strategy locates security resources on the target attacked by the follower strategies. Therefore, the players' utility functions depend on the strategies selected by the adversaries. Assuming that players use mixed strategies, i.e., a probability distribution over possible actions, the utility of a player for a given strategy is uncertain, depending on the outcome of the combined mixed strategies. Note that this means that the uncertainty of the utility functions depends on the decision variables.

Different expressions of the uncertain utility can be considered to solve these SSGs with decision variables that modify the probability distribution of the utility function. It is natural to consider that players, individually, optimize the expected value of these uncertain utility functions (Myerson, 2013; Paruchuri et al., 2008; Jain et al., 2010). In other words, players optimize the expectation of a reward that is stochastic due to the uncertainty of the adversary's strategy. In a security setting, however, the expected reward utility function does not always provide an accurate model of player interaction; see Camerer (1999). If an expected utility model is used, the adversary response can be misrepresented which can lead to less than optimal strategies. Also, by optimizing the expected utility, the outcome of catastrophic unlikely events is not explicitly considered. Doing so can provide mixed strategy solutions that are fragile or that have high likelihood of very bad outcomes. Both effects can be modeled with nonlinear distortion functions that transform the uncertain reward objective, such as prospect theory (Kahneman & Tversky, 1979), and risk measures (Artzner et al., 1999; Markowitz, 1952).

In this work, we investigate how to efficiently formulate and solve an SSG with decision variables that influence the uncertainty distribution of the utility function. We consider a single follower and a finite set of actions for each player. In particular, we focus on modeling risk-averse behavior with respect to the uncertainty due to the adversary's probability distribution over actions (i.e., its mixed strategy). We present different mathematical optimization formulations to represent chance constraints, perturbed utility functions, stochastic dominance, value at risk (VaR), and conditional value at risk (CVaR). Here we explore how to efficiently express these formulations and do not make a critical assessment on which model is preferable, since that depends on the application, the meaning of the utility function, and the decision-maker's risk attitude. We also present computational results for important examples that do not consider the expected reward utility function. In particular, we consider Stackelberg security game models where the leader either uses an *entropic risk measure* (Pratt, 1964) and a quantal response model (McKelvey & Palfrey, 1995) or a model that uses prospect theory (Kahneman & Tversky, 1979). We briefly describe these three concepts below.

An entropic risk measure amplifies the importance of outcomes that exceed a given threshold to model risk-adverse behavior against the attacker's probability over actions. The entropic risk measure of parameter $\alpha \geq 0$ of a random variable Y is defined by $\alpha \ln \mathbb{E}[e^{Y/\alpha}]$. While all outcomes are weighted, scenarios with a payoff larger than α contribute more to this measure. Therefore, the parameter α corresponds to a payoff value of risky outcomes and must be chosen carefully to tune the risk aversion level of the decision-maker.

The quantal response (QR) equilibrium model presented in McKelvey and Palfrey (1995) assumes that human adversaries do not behave rationally, sometimes selecting actions that do not maximize their utility. In this model, followers use a logit discrete choice model to decide between n possible actions, where action i (that gives a payoff U_i) is selected with probability

$$\mathbb{P}\left[\text{selecting action } i\right] = \frac{1}{\sum_{j=1}^{n} e^{\lambda U_j}} e^{\lambda U_i} ,$$

where the parameter λ represents a *degree of rationality*, with perfect rationality ($\lambda \to \infty$) or indifference ($\lambda = 0$) as special cases. The QR model has been used to model human behavior in various settings, including economics (Haile et al., 2008; Stahl & Wilson, 1994), game theory (Wright & Leyton-Brown, 2010), transportation engineering (Ben-Akiva & Lerman, 2018), marketing (Gensch & Recker, 1979), and security applications (Yang et al., 2011).

Prospect theory (Kahneman & Tversky, 1979) explicitly represents player biases, modeling risk-averse and risk-seeking behavior. It does so by considering perturbation functions on both the reward values and the probability distribution of possible outcomes. That is, if outcome i has a probability of occurrence p_i and payoff U_i, prospect theory proposes players perceive the following expected utility:

$$V(p, U) = \sum_{i=1}^{n} \pi(p_i) V(U_i) .$$

where $\pi(\cdot)$ and $V(\cdot)$ are perturbation functions with specific properties that model how players perceive both payoffs and the likelihood of occurrence. Prospect theory has contributed in economics (Tversky & Kahneman, 1986), politics (McDermott, 2004), online auctions (Brünner et al., 2019), and security (Yang et al., 2011) applications.

We note that some facility location models with random assignment of clients to facilities lead to optimization problems with uncertainty that also have probability distributions that depend on decision variables and are thus similar to the SSG presented. In particular, the SSG with the QR model has a similar structure to the maximum capture facility location problem with random utilities (Freire et al., 2016; Ljubić & Moreno, 2018). Both these problems can be seen as examples of a facility location model with multinomial logit choice probabilities (Haase & Müller, 2014).

In the next section, we present the SSG problem and fix the notation. Section 3 formulates an SSG problem for different risk aversion models. Section 4 presents the algorithms for computing VaR and CVaR with an uncertainty that depends on decision variables. We present some preliminary computational results in Sect. 7 and conclude the paper in Sect. 8.

2 Notation and Basic Assumptions

We begin introducing the Stackelberg security game considered, which is similar to the problem in Kiekintveld et al. (2009). The SSG assumes there is a finite set of targets denoted by $I = \{1, \dots n\}$. The attacker decides between n actions that indicate which target to attack. One of the targets can represent the decision not to attack. The defender actions determine where to locate security resources to protect or cover a subset of targets. A defender action, or pure strategy, $z \subset I$ indicates which targets are covered simultaneously and depends on physical constraints, such as the number of defender resources, capacity of defender resources, or target compatibility. Let Z denote the set of feasible defender actions. The payoff of each player depends only on whether the attacked target $i \in I$ is protected by the defender action $z \in Z$, denoted by $i \in z$, or not. Given actions $i \in I$ and $z \in Z$, the reward received by the defender (by the attacker) is either a reward $\bar{R}_i$ (a penalty P_i) if $i \in z$ or a penalty $\bar{P}_i$ (a reward R_i) if $i \notin z$. Here, $\bar{R}_i, R_i > 0$ and $\bar{P}_i, P_i < 0$. Therefore, under actions $i \in I$ and $z \in Z$, the utilities of the defender and attacker, respectively, are

$$u_D(i,z) = \begin{cases} \bar{R}_i & i \in z \\ \bar{P}_i & i \notin z \end{cases} \qquad u_A(i,z) = \begin{cases} P_i & i \in z \\ R_i & i \notin z \, . \end{cases}$$

We assume players decide on mixed strategies, or probability distributions over their set of actions, denoted by $y \in \mathcal{I} = \{y \in [0,1]^n : \sum_{i=1}^n y_i = 1\}$ and $q \in \mathcal{Z} = \{q \in [0,1]^{|Z|} : \sum_{z \in Z} q_z = 1\}$. Since player payoff only depends on whether the attacked target is protected or not, we consider the more succinct $x \in \mathcal{X} = \{x \in [0,1]^n : x_i = \sum_{z \in Z : i \in z} q_z,\ q \in \mathcal{Z}\}$. The set $\mathcal{X}$ is the projection on $[0,1]^n$ of the feasible mixed strategies of the defender, and, for $x \in \mathcal{X}$, the value x_i is the frequency with which target i is protected by a mixed strategy in $\mathcal{Z}$. The players' rewards as a function of the mixed strategies, denoted by $U_D(y,x)$ and $U_A(y,x)$ for the defender and attacker, respectively, are discrete random variables. For example, the defender utility equals $\bar{P}_i$ with probability $y_i(1-x_i)$ and equals $\bar{R}_i$ with probability $y_i x_i$. If Ψ and Ψ' denote statistics for the leader and follower utilities, we can write the problem that optimizes the leader utility as the following bilevel problem:

$$\begin{aligned} \max\ & \Psi\left(U_D(y,x)\right) \\ \text{s.t.}\ & x \in \mathcal{X} \\ & y = \operatorname{argmax}\ \Psi'(U_A(y,x)) \\ & \qquad \text{s.a.}\ \ y \in \mathcal{I}\,. \end{aligned} \tag{1}$$

The solution to this problem determines the strong Stackelberg equilibrium of the Stackelberg game, where the follower breaks ties in favor of the leader (Kiekintveld et al., 2009).

For any mixed strategy $x \in \mathcal{X}$, we let $y(x)$ denote the follower's best response, given by the solution to the subproblem in (1). Then the leader's disutility $D(x) = -U_D(y(x), x)$ is a discrete random variable that takes the value $-\bar{R}_i$ with probability $x_i y_i(x)$ and $-\bar{P}_i$ with probability $(1 - x_i)y_i(x)$. All the possible disutilities $\{-\bar{P}_i, -\bar{R}_i\}_{i\in\{1,\ldots,n\}}$ can be referred to as $\{V_v\}_{v\in\mathcal{V}}$, with $|\mathcal{V}| = 2n$ outcomes that do not depend on the decision variables x. Without loss of generality, we assume these values are sorted in increasing order: $V_1 \le V_2 \le \ldots \le V_{2n}$. However, the probabilities of these discrete outcomes $p_v(x) := \mathbb{P}[D(x) = V_v]$ depend on x.

Different forms of the best response $y(x)$ are due to the specifics of the subproblem being solved. In the classic Stackelberg setting, the statistic for the subproblem Ψ' is the expectation, making the subproblem a linear optimization problem, which has optimal pure strategies. Nonlinear statistics, such as variance or distortion functions—as in prospect theory—can generate a mixed strategy best response. A quantal response (QR) model of the follower replaces the second level problem with the assumption that a follower selects an alternative following the probability distribution

$$y_i(x) = \frac{e^{\lambda U_A(i,x)}}{\sum_{j=1}^{n} e^{\lambda U_A(j,x)}}.$$

If we assume that the utility statistic of the leader is the expected value, then $\Psi[-U_D(y(x), x)] = \mathbb{E}[D(x)] = \sum_{v\in\mathcal{V}} V_v p_v(x)$. We can then express the leader's optimization problem as

$$\min_{x\in\mathcal{X}} \sum_{v\in\mathcal{V}} V_v p_v(x)\,.$$

We show in the next section that, under reasonable conditions, this kind of problem and generalizations of the form

$$\min_{x\in\mathcal{X}} \{f_0(x) : f(x) \le 0\} \tag{2}$$

Table 1 Payoff matrix for the two targets, one defender example: each cell contains the utilities for [defender, attacker]

	Attack 1	Attack 2
Patrol 1	3, −1	−3, 1
Patrol 2	−1, 3	1, −3

can be tackled efficiently. The generalization considered is able to represent different methods to handle and model the uncertainty present in the leader's utility including chance constraints, risk distortion functions, and stochastic dominance constraints.

Consider now an example where the leader and follower play mixed strategies and, therefore, they induce a probability distribution on the outcomes. The example has two targets, a single patrol and a single attacker that, observing the mixed strategy of the leader, selects its target with a QR model with rationality factor $\lambda = 0.25$. The payoffs of this game (where the defender is the row player and the attacker is the column player) are given in Table 1.

The problem of maximizing the expected payoff in this simple setting can be written as follows:

$$\max_{x_1+x_2=1,\ x_1,x_2\geq 0} \frac{e^{x_1-0.75}(4x_1-1)+e^{x_2-0.25}(4x_2-3)}{e^{x_1-0.75}+e^{x_2-0.25}},$$

where x_i represents the frequency at which target $i = 1, 2$ is patrolled. The objective function attains its maximum value at $x^* = (0.505, 0.495)$ which in turn induces the adversary's quantal response into $y(x^*) = (0.622, 0.378)$. The mixed strategies determine the probabilities on the (discrete) set of outcomes: for example, the probability that the leader gets attacked on the non-patrolled target 2 will be

$$\begin{aligned}\mathbb{P}(2 \text{ not defended}) \cdot \mathbb{P}(2 \text{ attacked}) &:= (1-x_2^*) \cdot y_2(x^*) \\ &= (1-0.495) \cdot 0.378 = 0.19\,.\end{aligned}$$

3 Efficient Leader Problem Formulations

Here we present reformulations of (1) in the situation where there is a known follower best response $y(x)$ and the disutility function $D(x)$ takes $2n$ values that do not depend on x with probabilities that depend on x.

The formulations considered will aim to either maintain some risk measure of the disutility $D(x)$ under a given threshold—translated by some constraints $f(x) \leq 0$—or minimize a risk measure of $D(x)$, which translates into minimizing some function

$f_0(x)$. We will transform these different problem formulations to constraints over the set of decision variables $x \in \mathcal{X}$ of the form

$$\sum_{v \geq \bar{v}} p_v(x)\xi_v \leq \Xi\,, \tag{3}$$

for a real-valued vector $(\xi_v)_{v \in \mathcal{V}}$ such that $\xi_1 \leq \xi_2 \leq \ldots \leq \xi_{|\mathcal{V}|}$, some index $\bar{v} \in \mathcal{V}$, and a right-hand side $\Xi \in \mathbb{R}$.

Notice that we can assume that $\xi_v \geq 0$ for $v \geq \bar{v}$. If this is not the case, simply define $\zeta := \max_{v \geq \bar{v}}(-\xi_v)_+$ and construct the following non-negative vector $\xi'_v = \xi_v + \zeta$ if $v \geq \bar{v}$ and $\xi'_v = \zeta$ for $v \leq \bar{v} - 1$. Then, constraint (3) is equivalent to

$$\Xi + \zeta \geq \sum_{v \geq 1} p_v(x)\xi'_v\,.$$

Constraints of the form (3) are easy to solve if the dependency of x through the probability functions $p_v(x)$ forms convex constraints on $\mathcal{X}$. We now show situations where enforcing bounded risk of the leader can be modeled with type (3) constraints, for different choices of $\bar{v}$, ξ, and Ξ.

3.1 Maximum Expected Disutility

Given a reference disutility $\mathbb{E}[D(\widetilde{x})]$ coming from some known solution $\widetilde{x} \in \mathcal{X}$, we want to find some $x \in \mathcal{X}$ having an expected disutility that is no worse than the reference disutility from $\widetilde{x}$. In other words, x must satisfy the following constraint, $\mathbb{E}[D(x)] \leq \mathbb{E}[D(\widetilde{x})]$, which is by definition equivalent to the generic constraint (3) with $\Xi := \mathbb{E}[D(\widetilde{x})]$, $\xi_v := V_v$ for every $v \in \mathcal{V}$ and $\bar{v} := 1$, i.e.,

$$\sum_{v \in \mathcal{V}} p_v(x) V_v \leq \mathbb{E}[D(\widetilde{x})]\,.$$

3.2 Chance Constraints

Given a threshold value $\widetilde{V} \in \mathbb{R}$ and a tolerance $\epsilon \in [0, 1]$, a chance constraint (Mayer, 1992; Charnes & Cooper, 1959) on the disutility $D(x)$ bounds the likelihood that $D(x) \geq \widetilde{V}$ by ϵ, which means

$$\mathbb{P}\left[D(x) \geq \widetilde{V}\right] \leq \epsilon\,. \tag{4}$$

This constraint over $x \in \mathcal{X}$ is equivalent to the generic constraint (3) taking $\Xi = \epsilon$, $\xi_v = 1$ for every $v \in \mathcal{V}$ and $\bar{v} := \arg\min_{v \in \mathcal{V}} \{V_v : V_v \geq \widetilde{V}\}$, i.e.,

$$\sum_{v \geq \widetilde{v}} p_v(x) \leq \epsilon \,. \tag{5}$$

3.3 Bounded Distortion Risk

A distortion risk measure (Balbás et al., 2009) is a real-valued function ρ taking as argument a random variable Z that can be described as

$$\rho : Z \to d^{-1} \left(\mathbb{E}\left[d(Z) \right] \right),$$

where $d : \mathbb{R} \to \mathbb{R}$ is an increasing bijective disutility function. The entropic risk measure $Z \to \alpha \ln \mathbb{E}[e^{Z/\alpha}]$ of parameter $\alpha > 0$ is a particular distortion risk measure. A constraint that bounds a distortion risk is a constraint over $x \in \mathcal{X}$ so that the distortion risk is less than a given threshold $\widetilde{\rho}$, i.e.,

$$\rho(D(x)) \leq \widetilde{\rho} \,. \tag{6}$$

Constraint (6) is equivalent to $\mathbb{E}[d(D(x))] \leq d(\widetilde{\rho})$, i.e., $\sum_{v \in \mathcal{V}} p_v(x) d(V_v) \leq d(\widetilde{\rho})$, which is exactly the generic constraint (3) with $\Xi = d(\widetilde{\rho})$, $\xi_v = d(V_v)$ for every $v \in \mathcal{V}$ and $\bar{v} = 1$. Because d is increasing, we indeed have $\xi_1 \leq \xi_2 \leq \ldots \leq \xi_{|\mathcal{V}|}$.

3.4 First-Order Stochastic Dominance Constraints

Let $F_Z : t \to \mathbb{P}[Z \leq t]$ denote the cumulative distribution of a random variable Z. Given two random variables Z and T, Z is said to stochastically dominate T in the first order, $Z \succeq_{(1)} T$, if $F_Z(t) \geq F_T(t)$ for all $t \in \mathbb{R}$ (Dentcheva & Ruszczyński, 2004).

Given a reference random variable $D(\widetilde{x})$, we can write a constraint over $x \in \mathcal{X}$ such that $D(x)$ stochastically dominates $D(\widetilde{x})$ in the first order, i.e., $D(x) \succeq_{(1)} D(\widetilde{x})$. In our context where both random variables $D(x)$ and $D(\widetilde{x})$ have the same discrete support, this can be rewritten as follows: for every $\widetilde{v} \in \mathcal{V}$, we must have $F_{D(x)}(V_{\widetilde{v}}) \geq F_{D(\widetilde{x})}(V_{\widetilde{v}})$, i.e., $\sum_{v \leq \widetilde{v}} p_v(x) \geq \sum_{v \leq \widetilde{v}} p_v(\widetilde{x})$. In other words,

$$\sum_{v \geq \widetilde{v}+1} p_v(x) \leq 1 - \sum_{v \leq \widetilde{v}} p_v(\widetilde{x}) \qquad \forall \widetilde{v} \in \mathcal{V} \,. \tag{7}$$

The first-order stochastic dominance constraint $D(x) \succeq_{(1)} D(\widetilde{x})$ can thus be represented by the $|\mathcal{V}|$ constraints in (7) which are of type (3) with $\Xi = 1 - \sum_{v \leq \widetilde{v}} p_v(\widetilde{x})$, $\xi_v = 1$ for every $v \in \mathcal{V}$, and $\bar{v} = \widetilde{v} + 1$.

3.5 Second-Order Stochastic Dominance Constraints

The second-order cumulative distribution function of a random variable Z is given by

$$F_Z^{(2)}(\eta) := \int_{-\infty}^{\eta} F_Z(t)\mathrm{d}t\,.$$

Given two random variables Z and T, Z is said to stochastically dominate T in the second order, $Z \succeq_{(2)} T$, if $F_Z^{(2)}(\eta) \geq F_T^{(2)}(\eta)$ for all $\eta \in \mathbb{R}$, (Dentcheva & Ruszczyński, 2004).

Given a reference random variable $D(\widetilde{x})$, we want to enforce the fact that $D(x)$ stochastically dominates $D(\widetilde{x})$ in the second order, i.e., $D(x) \succeq_{(2)} D(\widetilde{x})$. A result from Dentcheva and Ruszczynski (2003) states that $D(x) \succeq_{(2)} D(\widetilde{x})$ is equivalent to

$$\mathbb{E}\left[(V_{\widetilde{v}} - D(x))_+\right] \geq \mathbb{E}\left[(V_{\widetilde{v}} - D(\widetilde{x}))_+\right] \qquad \forall \widetilde{v} \in \mathcal{V}\,.$$

We can rewrite this equivalently as

$$-\sum_{v \in \mathcal{V}} p_v(x)(V_{\widetilde{v}} - V_v)_+ \leq -\sum_{v \in \mathcal{V}} p_v(\widetilde{x})(V_{\widetilde{v}} - V_v)_+ \qquad \forall \widetilde{v} \in \mathcal{V}\,. \tag{8}$$

In consequence, the second-order stochastic dominance constraint $D(x) \succeq_{(2)} D(\widetilde{x})$ can be represented by the $|\mathcal{V}|$ constraints in (8) which are of type (3) with $\Xi = -\sum_{v \in \mathcal{V}} p_v(\widetilde{x})(V_{\widetilde{v}} - V_v)_+$, $\bar{v} = 1$, and $\xi_v = -(V_{\widetilde{v}} - V_v)_+$ for every $v \in \mathcal{V}$. Note that ξ_v are also in increasing order.

3.6 Some Difficult Risk Models

We say that constraint (3) is tractable if it describes a convex set on the decision variables or can be reasonably approximated with a handful of binary variables. While the previous examples show that the risk aversion constraints can be expressed in a tractable form, there are some examples for which it is not clear whether there is a tractable transformation or not.

For example, constraining the variance of $D(x)$ to be under a given threshold σ^2, i.e., $\mathbb{V}[D(x)] \leq \sigma^2$, boils down to

$$\sum_{v\in\mathcal{V}} p_v(x)V_v^2 - \left(\sum_{v\in\mathcal{V}} p_v(x)V_v\right)^2 \leq \sigma^2,$$

which is a complicated constraint for general probability functions $p_v(x)$. The same can be said about the upper semideviation $USD : Z \to \mathbb{E}[(Z - \mathbb{E}[Z])_+]$ where enforcing $USD(D(x)) \leq \widetilde{U}$ is equivalent to

$$\sum_{v\in\mathcal{V}} p_v(x)\left(V_v - \sum_{v'\in\mathcal{V}} p_{v'}(x)V_{v'}\right)_+ \leq \widetilde{U}$$

Both these constraints suggest non-convex constraints on the decision variables.

Interestingly, in a classic stochastic optimization setting where the probabilities are fixed and the uncertainty is affecting the payoffs alone, modeling chance constraints or using the value at risk turns the resulting problem NP-hard in general, whereas in our context, chance constraints are perfectly tractable computationally.

3.7 Risk Minimization

We now make use of all the machinery available for risk-inducing constraints in the context of minimizing risk measures. First, the generic problem (2) can be equivalently recast as

$$\begin{aligned} \min_{x\in\mathcal{X},\eta} \quad & \eta \\ \text{s.t.} \quad & \eta \geq f_0(x) \\ & f(x) \leq 0 \end{aligned}$$

For several of the aforementioned risk measures such as distortions in Sect. 3.2 or the probability of having a poor outcome in Sect. 3.3, f_i for any i can be of the form

$$\begin{aligned} f_0 : \quad & x \to \sum_{v\geqslant\bar{v}_0} \xi_v^0 p_v(x) \\ f_i : \quad & x \to \sum_{v\geqslant\bar{v}_i} \xi_v^i p_v(x) - \Xi_i \\ \text{with} \quad & 0 \leq \xi_1 \leq \xi_2 \leq \ldots \leq \xi_{|\mathcal{V}|} \end{aligned}$$

Following the ideas in Shieh et al. (2012), Chicoisne and Ordóñez (2016), we can iteratively make guesses about the optimal value η^* with a binary search: when fixing $\eta = \widetilde{\eta}$, the latter problem reduces to investigate whether there exists $x \in \mathcal{X}$ satisfying

$$\sum_{v \geqslant \bar{v}_i} \xi_v^i p_v(x) \leq \Xi_i, \qquad \forall i \geqslant 1$$

$$\sum_{v \geqslant \bar{v}_0} \xi_v^0 p_v(x) \leq \widetilde{\eta}.$$

In the end, considering a risk measure in the objective is not harder—modulo the binary search—than considering a constraint equivalent.

4 VaR and CVaR Minimization

In this section, we consider two classic risk measures: the value at risk and the conditional value at risk. It remains open if it is possible to express these risk models in a tractable form. However, we see below that it is possible to minimize them in our context.

4.1 *Value at Risk*

The objective in this subsection is to minimize the value at risk of parameter $\epsilon \in]0, 1[$ (VaR_ϵ) of the *disutility* of the defender. The value-at-risk-ϵ of a disutility random variable $D(x)$ is defined as $\text{VaR}_\epsilon(Z) := \inf_{t \in \mathbb{R}} \{t : F_Z(t) \geq 1 - \epsilon\}$. Because $D(x)$ has a discrete and finite probability distribution, the only values $\text{VaR}_\epsilon\,(D(x))$ can possibly take are the payoffs $(V_v)_{v \in \mathcal{V}}$. In consequence, we have that

$$\text{VaR}_\epsilon\,(D(x)) = \min_{\widetilde{v} \in \mathcal{V}} \left\{ V_{\widetilde{v}} : \sum_{v \leq \widetilde{v}} p_v(x) \geq 1 - \epsilon \right\}.$$

The problem of finding a defense strategy $x \in \mathcal{X}$ that minimizes $\text{VaR}_\epsilon\,(D(x))$ can then be cast as follows:

$$\min_{x \in \mathcal{X}, \widetilde{v} \in \mathcal{V}} \left\{ V_{\widetilde{v}} : \sum_{v \leq \widetilde{v}} p_v(x) \geq 1 - \epsilon \right\}. \tag{9}$$

After rearranging the minimizations in the latter problem (9), we obtain

$$\min_{\widetilde{v}\in\mathcal{V}}\left\{V_{\widetilde{v}}+\min_{x\in\mathcal{X}}\left\{0:\sum_{v\leq\widetilde{v}}p_v(x)\geq 1-\epsilon\right\}\right\}. \tag{10}$$

Notice that the inner problem (10) in x given $\widetilde{v}\in\mathcal{V}$ is a feasibility problem that only requires to check if there exists some $x\in\mathcal{X}$ such that $\sum_{v\leq\widetilde{v}}p_v(x)\geq 1-\epsilon$, which is equivalent to

$$\sum_{v\geq\widetilde{v}+1}p_v(x)\leq\epsilon.$$

Proposition 1 *The feasibility of the inner problem in x from problem* (10) *can be checked as follows: the inner problem in x is feasible iff the optimal objective value* $u_{\widetilde{v}}$ *of the following problem is lesser than or equal to* ϵ*:*

$$u_{\widetilde{v}}:=\min_{x\in\mathcal{X}}\sum_{v\geq\widetilde{v}+1}p_v(x). \tag{11}$$

The last problem simulates the fact that if the chosen $\widetilde{v}\in\mathcal{V}$ is associated with a value $V_{\widetilde{v}}$ that is too low to guarantee that $F_{D(x)}(V_{\widetilde{v}})\geq 1-\epsilon$ for at least one $x\in\mathcal{X}$, then it is an underestimator of the optimal objective value of the original problem (9). In consequence, the optimal objective value must lie strictly above $V_{\widetilde{v}}$, which allows us to eliminate from the candidates for the optimal objective value all the outcomes V_v such that $v\leq\widetilde{v}$.

On the other hand, if the chosen $\widetilde{v}\in\mathcal{V}$ is associated with a value $V_{\widetilde{v}}$ that guarantees that $F_{D(x)}(V_{\widetilde{v}})\geq 1-\epsilon$ for some $x\in\mathcal{X}$, then it is either an overestimator of the optimal objective value of the original problem (9) or the optimal objective value itself. In consequence, the optimal objective value must lie at $V_{\widetilde{v}}$ or under, which allows us to eliminate from the candidates for the optimal objective value all the outcomes V_v such that $v>\widetilde{v}$.

These observations suggest a binary search scheme iteratively looking for the index $v^*\in\mathcal{V}$ that corresponds to the true optimal value V_{v^*} of problem (9). We summarize the procedure in Algorithm 1 where the routine `solve`$(\widetilde{v})$ takes as argument an index $\widetilde{v}\in\mathcal{V}$ and returns a tuple $\left(x^{\widetilde{v}},u_{\widetilde{v}}\right)$ corresponding, respectively, to an optimal solution and the optimal value of problem (11).

Proposition 2 *Algorithm 1 returns an optimal solution for problem* (9) *by solving* $O\left(\log_2|\mathcal{V}|\right)$ *times a minimization problem* (11) *with different values of* $\bar{v}$.

Algorithm 1

```
Data: An instance of problem (9)
Result: An optimal solution x* for (9)
(x*, u):=solve(1);
if u ≥ 1 − ϵ then
    return x*;
(x*, u):=solve(|V|);
U := |V|; L := 1;
while U > L + 1 do
    v := ⌈(L + U)/2⌉;
    (x, u):=solve(v);
    if u ≥ 1 − ϵ then
        U := v; x* := x;
    else
        L := v;
return x*;
```

4.2 *Conditional Value at Risk*

The objective in this subsection is to minimize the conditional value at risk of parameter $\epsilon \in]0, 1[$ (CVaR_ϵ) of the *disutility* of the defender, defined as

$$\text{CVaR}_\epsilon(D(x)) := \inf_{t\in\mathbb{R}} \left\{ t + \epsilon^{-1}\mathbb{E}\left[(D(x) - t)_+\right] \right\}.$$

Furthermore, as shown in Rockafellar and Uryasev (2000), the minimum in ta is attained at $t^* = \text{VaR}_\epsilon\ (D(x))$ so that we also have the following alternative identity:

$$\text{CVaR}_\epsilon(D(x)) := \text{VaR}_\epsilon\ (D(x)) + \epsilon^{-1}\mathbb{E}\left[(D(x) - \text{VaR}_\epsilon\ (D(x)))_+\right].$$

We now want to determine an $x \in \mathcal{X}$ that minimizes the conditional value at risk of $D(x)$, which is modeled by the following optimization problem:

$$\omega^* := \min_{x\in\mathcal{X}, t\in\mathbb{R}} \left\{ t + \epsilon^{-1}\mathbb{E}\left[(D(x) - t)_+\right] \right\}. \tag{12}$$

Recalling that $D(x)$ follows a discrete probability distribution, we also have

$$\omega^* := \min_{x\in\mathcal{X}, t\in\mathbb{R}} \left\{ t + \epsilon^{-1} \sum_{v\in\mathcal{V}} p_v(x)\,(V_v - t)_+ \right\}. \tag{13}$$

In the previous section, we saw that $\mathrm{VaR}_\epsilon(D(x)) \in \mathrm{supp}(D(x)) = \{(V_v)_{v\in\mathcal{V}}\}$, meaning that (13) is equivalent to

$$\omega^* := \min_{x\in\mathcal{X},\widetilde{v}\in\mathcal{V}} \left\{ V_{\widetilde{v}} + \epsilon^{-1} \sum_{v\in\mathcal{V}} p_v(x)\,(V_v - V_{\widetilde{v}})_+ \right\}.$$

4.2.1 A Basic Algorithm

First, notice that for any optimal solution (x^*, t^*) of (13), we have

$$\omega^* := \min_{x\in\mathcal{X}} \left\{ t^* + \epsilon^{-1} \sum_{v\in\mathcal{V}} p_v(x)\left(V_v - t^*\right)_+ \right\}.$$

In consequence, we can "guess" the optimal value of t by fixing it to $V_{\widetilde{v}}$ for every $\widetilde{v} \in \mathcal{V}$ and then solve the corresponding problem in $x \in \mathcal{X}$:

$$\omega_{\widetilde{v}} := V_{\widetilde{v}} + \epsilon^{-1} \min_{x\in\mathcal{X}} \sum_{v\in\mathcal{V}} p_v(x)\,(V_v - V_{\widetilde{v}})_+ .$$

Because the outcomes are sorted in increasing order, the latter can be rewritten as

$$\omega_{\widetilde{v}} = V_{\widetilde{v}} + \epsilon^{-1} \underbrace{\min_{x\in\mathcal{X}} \sum_{v\geq\widetilde{v}+1} p_v(x)\,(V_v - V_{\widetilde{v}})}_{u_{\widetilde{v}}}, \tag{14}$$

whose optimal solution is denoted $x^{\widetilde{v}}$. Keeping track of the values $w_{\widetilde{v}}$, we find $\omega^* := \arg\min_{v\in\mathcal{V}} w_v$ and return x^{v^*} as an optimal solution of the original problem (13). The procedure has to solve $|\mathcal{V}| = 2n$ times problem (14). We summarize the procedure in Algorithm 2 where the routine $\texttt{solve}(\widetilde{v})$ takes as argument an index $\widetilde{v} \in \mathcal{V}$ and returns a tuple $\left(x^{\widetilde{v}}, u_{\widetilde{v}}\right)$ corresponding, respectively, to an optimal solution and the optimal value of (14).

4.2.2 An Improved Algorithm

Notice that Algorithm 2 requires to solve $2n$ optimization problems (14) with different values of $\widetilde{v}$, whereas minimizing VaR, only $O\left(\log_2 n\right)$ problems must be solved, as opposed to the classical optimization setting (where the uncertainty affects only the outcomes and the probabilities are constant) where minimizing VaR is NP-hard, whereas minimizing CVaR can be modeled via additional linear constraints and continuous variables. We now present a way to decrease the number of problems we need to solve.

Algorithm 2: Minimize CVaR

Data: An instance of problem (13)
Result: An optimal solution x^* for (13)

$v = 1$;
$w^* = +\infty$;
while $v \neq |\mathcal{V}|$ **do**
  (x, u):=`solve`(v);
  if $w^* > V_v + \epsilon^{-1}u$ **then**
    $x^* := x$;
    $w^* := V_v + \epsilon^{-1}u$;
  $v++$;
return x^*;

Proposition 3 *The function* $t \rightarrow t + \epsilon^{-1} \min_{x \in \mathcal{X}} \sum_{v \in \mathcal{V}} p_v(x)[V_v - t]_+$ *is continuous and piecewise concave with breakpoints* $(V_v)_{v \in \mathcal{V}}$. *Unfortunately, there is no guarantee that even the same function in its discrete form—i.e., restricting its domain to* $(V_v)_{v \in \mathcal{V}}$*—is convex. However, we can find a locally optimal solution for the original problem solving* $O\left(\log_2 n\right)$ *problems in* $x \in \mathcal{X}$ *with t fixed to some* V_v.

The last proposition allows us to return an upper bound that is hopefully better than just solving the problems in a sequential order. Together with the next proposition, we show how to prune values V_v without solving the problem they are associated with.

Proposition 4 *Recalling that* $VaR_\epsilon\,(D(x)) \leq CVaR_\epsilon\,(D(x))$, *each time we solve a problem with fixed* $t = V_{\tilde{v}}$, *we can eliminate from the list of candidates all the* V_v*s lying over* $w_{\tilde{v}}$ *as they cannot possibly produce a solution improving the current best objective value. Marking each* $v \in \mathcal{V}$ *when we solve its corresponding problem in x during the local minimization via binary search in v or when we eliminate it by bounds, we can accelerate the practical convergence of the first algorithm.*

Notice that in the worst case, we will solve at most $|\mathcal{V}| = 2n$ problems, which is no worse than using Algorithm 2. We summarize the procedure in Algorithm 3 where the routine `binary_search`$(\mathcal{V}^+)$ takes as argument a subset $\mathcal{V}^+ \subseteq \mathcal{V}$ of marked outcomes and returns a locally optimal solution x found by binary search with its objective value u and the outcome number v it is associated with. The routine also updates the set $\mathcal{V}^+$ with the previously nonmarked outcomes it visited during the binary search.

Algorithm 3: Improved CVaR Algorithm

Data: An instance of problem (13)
Result: An optimal solution x^* for (13)
$\mathcal{V}^+ := \emptyset$;
$w^* = +\infty$;
while $\mathcal{V}^+ \neq \mathcal{V}$ **do**
    (x, u, v):=`binary_search`$(\mathcal{V}^+)$;
    if $w^* > V_v + \epsilon^{-1}u$ **then**
        $x^* := x$;
        $w^* := V_v + \epsilon^{-1}u$;
        $\mathcal{V}^+ := \mathcal{V}^+ \cup \{v' \in \mathcal{V} : w^* \leq V_{v'}\}$
return x^*;

5 Quantal Response (QR)

5.1 Defining the Response Probabilities $p_v(x)$

Recalling that the expected utility of the attacker when the target i is attacked is $U_i(x_i) = x_i P_i + (1 - x_i)R_i$, if the attacker is not perfectly rational and follows a QR of rationality factor $\lambda > 0$ (McKelvey & Palfrey, 1995), the probability that target i is attacked is given by

$$y_i(x) = \frac{e^{\lambda U_i(x_i)}}{\sum_{j=1}^{n} e^{\lambda U_j(x_j)}}. \tag{15}$$

Defining $R := \max_{i \in \{1,\dots,n\}} R_i$, for theoretical complexity and computational tractability purposes, it is better (Chicoisne & Ordóñez, 2016) to divide by $e^{\lambda R}$ both the numerator and denominator in (15): $y_i(x) = e^{\lambda(U_i(x_i)-R)} / \sum_{j=1}^{n} e^{\lambda(U_j(x_j)-R)}$. Defining $\beta_i := e^{\lambda(R_i - R)} \geqslant 0$, $\gamma_i := \lambda(R_i - P_i) \geqslant 0$, and $\delta_i := \bar{R}_i - \bar{P}_i \geqslant 0$, we obtain that

$$y_i(x) = \frac{\beta_i e^{-\gamma_i x_i}}{\sum_{j=1}^{n} \beta_j e^{-\gamma_j x_j}}.$$

We link the QR Stackelberg security game with the generic notation as follows: The set of payoffs is $\{(V_v)_{v \in \mathcal{V} := \{1,\dots,2n\}}\} := \{(-\bar{P}_i)_{i \in \{1,\dots,n\}}, (-\bar{R}_i)_{i \in \{1,\dots,n\}}\}$, i.e., the set of all possible disutilities sorted in increasing order. Letting $i(v)$ being the target associated with outcome V_v—be it a penalty or a reward—the probabilities of having each outcome are as follows:

$$p_v(x) := \begin{cases} x_{i(v)}\beta_{i(v)}e^{-\gamma_{i(v)}x_{i(v)}} / \sum_{j=1}^{n} \beta_j e^{-\gamma_j x_j} & \text{If outcome } V_v \text{ is a reward} \\ \left(1 - x_{i(v)}\right) \beta_{i(v)}e^{-\gamma_{i(v)}x_{i(v)}} / \sum_{j=1}^{n} \beta_j e^{-\gamma_j x_j} & \text{If outcome } V_v \text{ is a penalty} \end{cases}$$

For convenience, let us define for each target $i \in \{1, \ldots, n\}$ the index $v^P(i)$ (respectively, $v^R(i)$) corresponding to the payoff of its penalty (respectively, reward).

5.2 *Efficient Solution*

We now see that any constraint of type (3) can be put in a tractable way in any optimization framework: In fact, they can be either piecewise linearly approximated or, in some reasonable cases, be equivalent to convex constraints. Because the adversary follows a QR, any type (3) constraint becomes

$$\sum_{i:v^P(i)\geq\bar{v}} \xi_{v^P(i)} \frac{\beta_i e^{-\gamma_i x_i}}{\sum_{j=1}^{n} \beta_j e^{-\gamma_j x_j}} (1 - x_i) + \sum_{i:v^R(i)\geq\bar{v}} \xi_{v^R(i)} \frac{\beta_i e^{-\gamma_i x_i}}{\sum_{j=1}^{n} \beta_j e^{-\gamma_j x_j}} x_i \leq \Xi \tag{16}$$

i.e., $\sum_{i:v^P(i)\geq\bar{v}} \xi_{v^P(i)}\beta_i e^{-\gamma_i x_i} (1 - x_i) + \sum_{i:v^R(i)\geq\bar{v}} \xi_{v^R(i)}\beta_i e^{-\gamma_i x_i} x_i \leq \Xi \sum_{i=1}^{n} \beta_i e^{-\gamma_i x_i}.$

Proposition 5 *The following statements hold:*

1. *The left-hand side of* (16) *is separable in the variables x_i and can be consequently piecewise linearly approximated via the use of integer variables (Vielma, 2015).*
2. *If the vector ξ is such that $\xi_1 \leq \xi_2 \leq \ldots \leq \xi_{|\mathcal{V}|}$,* (16) *can be cast as the following convex constraint:*

$$\sum_{i:v^P(i)\geq\bar{v}} \xi_{v^P(i)}\beta_i z_i - \Xi \sum_{i=1}^{n} \beta_i z_i + \sum_{i:v^P(i)\geq\bar{v}} \xi_{v^P(i)} \frac{\beta_i}{\gamma_i} z_i \ln z_i - \sum_{i:v^R(i)\geq\bar{v}} \xi_{v^R(i)} \frac{\beta_i}{\gamma_i} z_i \ln z_i \leq 0$$

after using the change of variables $x_i := -\ln(z_i)/\gamma_i$.

Proof The first part is immediate. For the second part, after using the change of variables $x_i := -\ln(z_i)/\gamma_i$, we obtain

$$\sum_{i:v^P(i)\geq\bar{v}} \xi_{v^P(i)}\beta_i z_i - \Xi\sum_{i=1}^{n}\beta_i z_i \\ + \sum_{i:v^P(i)\geq\bar{v}} \xi_{v^P(i)}\frac{\beta_i}{\gamma_i}z_i \ln z_i - \sum_{i:v^R(i)\geq\bar{v}} \xi_{v^R(i)}\frac{\beta_i}{\gamma_i}z_i \ln z_i \leq 0. \tag{17}$$

Because of the last term in the left-hand side, it is not obvious that constraints (17) define a convex set. However, if a term appears in the last sum of the left-hand side, it also appears in the penultimate term given that (1) $v^P(i) > v^R(i)$ and (2) the components ξ_v are in increasing order. Let us consider a single target i that appears in the complicating last term: its "contribution" wrt each x_i in constraint (17) is

$$\xi_{v^P(i)}\beta_i z_i - \Xi\beta_i z_i + \xi_{v^P(i)}\frac{\beta_i}{\gamma_i}z_i \ln z_i - \xi_{v^R(i)}\frac{\beta_i}{\gamma_i}z_i \ln z_i. \tag{18}$$

The first two terms in (18) do not cause any harm to the overall convexity because of their linearity, whereas the two last terms can be factored into

$$\left(\xi_{v^P(i)} - \xi_{v^R(i)}\right)\frac{\beta_i}{\gamma_i}z_i \ln z_i. \tag{19}$$

By hypothesis, we have $\xi_{v^P(i)} \geq \xi_{v^R(i)}$, making the term (19) convex.

The last proposition tells us that whenever the adversary follows a QR, using risk-aversion inducing constraints or objective functions is tractable in practice. More precisely, if $\mathcal{X}$ is defined by linear constraints:

1. The first result of Proposition 5 tells us that the full optimization problem can be cast as a mixed-integer linear optimization problem.
2. The second part tells us that if $\mathcal{X}$ is defined by r linear inequalities with nonnegative coefficients $(a^j)\top x \leq b_j, \forall j \in \{1,\ldots,r\}$, then the constraints defining $\mathcal{X}$ after the change of variables $x_i := -\ln(z_i)/\gamma_i$ translate into the r following convex constraints:

$$-\sum_{i=1}^{n}\frac{a_j^i}{\gamma_i}\ln(z_i) \leq b_i, \qquad \forall j \in \{1,\ldots,r\}$$

which is readily solvable by off-the-shelf interior point algorithms (e.g., Pirnay et al. (2012), Wachter and Biegler (2006))

6 Prospect Theory

Another case in which problem (1) can lead to a tractable solution problem has to do with approximate solutions of the SSG when both leader and follower consider a prospect theory decision model. As mentioned in Sect. 1, prospect theory assumes players deviate from the expected objective through distortion functions that modify the valuations and the probability of occurrence. Under this model, the disutility of the leader is

$$PT(D(x)) = \sum_{v \in \mathcal{V}} \pi(p_v)\nu(V_v) ,$$

where, given parameters $\lambda, \alpha, \beta \geq 0$, and $\delta \in [0, 1]$, the distortion functions are

$$\nu(z) = \begin{cases} (z - C)^\alpha & \text{if } z \geq C \\ -\lambda(-z + C)^\beta & \text{if } z < C \end{cases} \quad \text{and} \quad \pi(x) = \frac{x^\delta}{(x^\delta + (1 - x)^\delta)^{\frac{1}{\delta}}} .$$

where C represents the reference point for the valuation function. Similar expressions are used for the follower for its own distortion functions π' and ν'.

To propose a tractable model, we assume that the follower selects an optimal pure strategy (a target to attack), i.e., $y \in \{0, 1\}^n$—an assumption that holds for a linear objective of the subproblem. Since p_v is either $y_{i(v)}x_{i(v)}$ or $y_{i(v)}(1 - x_{i(v)})$, we have that $\pi(p_v)$ is either $y_{i(v)}\pi(x_{i(v)})$ or $y_{i(v)}\pi(1 - x_{i(v)})$. Therefore, we express (1) as

$$\begin{aligned} \max \ & \sum_{i=1}^{n} y_i \left[\pi(x_i)\nu(\bar{R}_i) + \pi(1 - x_i)\nu(\bar{P}_i)\right] \\ \text{s.t.} \ & x \in \mathcal{X} \\ & y = \operatorname{argmax} \sum_{i=1}^{n} y_i \left[\pi'(x_i)\nu'(P_i) + \pi'(1 - x_i)\nu'(R_i)\right] \\ & \qquad \text{s.t.} \ \sum_{i=1}^{n} y_i = 1, \quad y \in \{0, 1\}^n . \end{aligned}$$

This problem with nonlinear objectives in both problems can be solved approximately with piecewise linear approximations and integer variables. For this, consider that every x_i variable is partitioned into K segments with breakpoints $c_0, c_1, \ldots, c_K$, with $c_0 = 0$ and $c_K = 1$. The perturbation functions π and π' take values b_k and b'_k at breakpoints c_k for $k \in K$. To simplify the constraints, we will use a variable $z_{iK+1} = 0$.

$$
\begin{aligned}
\max\ & \gamma \\
\text{s.t.}\ & x \in \mathcal{X} \\
& \sum_{i=1}^{n} y_i = 1, \quad y \in \{0,1\}^n \\
& \sum_{k\in K} z_{ik} = 1, \quad z_i \in \{0,1\}^{|K|} && i \in I \\
& \sum_{k\in K} \hat{z}_{ik} = 1, \quad \hat{z}_i \in \{0,1\}^{|K|} && i \in I \\
& \sum_{k\in K} w_{ik} = 1, \quad w_i \in [0,1]^{|K|} && i \in I \\
& \sum_{k\in K} \hat{w}_{ik} = 1, \quad \hat{w}_i \in [0,1]^{|K|} && i \in I \\
& w_{ik} \le z_{ik} + z_{ik+1}, \quad \hat{w}_{ik} \le \hat{z}_{ik} + \hat{z}_{ik+1} && i \in I, k \in K \\
& x_i = \sum_{k\in K} c_k w_{ik}, \quad 1 - x_i = \sum_{k\in K} c_k \hat{w}_{ik} && i \in I \\
& q_i = \sum_{k\in K} b_k w_{ik}, \quad \hat{q}_i = \sum_{k\in K} b_k \hat{w}_{ik} && i \in I \\
& q_i' = \sum_{k\in K} b_k' w_{ik}, \quad \hat{q}_i' = \sum_{k\in K} b_k' \hat{w}_{ik} && i \in I \\
& 0 \le a - \left[q_i' v'(P_i) + \hat{q}_i' v'(R_i)\right] \le M(1 - y_i) && i \in I \\
& M(1 - y_i) + \left[q_i v(\bar{R}_i) + \hat{q}_i v(\bar{P}_i)\right] \ge \gamma && i \in I
\end{aligned}
$$

Here, variables z_{ik} and $\hat{z}_{ik}$ indicate which interval of the piecewise approximation is used for x_i and for $1 - x_i$, respectively. The value of the convex combination is given by variables w_{ik} and $\hat{w}_{ik}$, respectively. The values q_i, $\hat{q}_i$, q_i', and $\hat{q}_i'$ give the expressions of $\pi(x_i)$, $\pi(1 - x_i)$, $\pi'(x_i)$, and $\pi'(1 - x_i)$, respectively. We consider that this approximate mixed-integer optimization problem is a tractable model for the prospect theory approach.

7 Computational Results

7.1 *Expected Value and Entropy Minimization with QR Adversaries*

In Chicoisne and Ordóñez (2016), we studied risk-neutral and risk-averse objective models that minimize either the expected value $\mathbb{E}[D(x)]$ or an entropic risk measure $\alpha \ln \mathbb{E}[\exp(D(x)/\alpha)]$. The resulting models were able to solve instances within an hour with up to $n = 10{,}000$ targets for (1) a basic model where

$$
\mathcal{X}_0 := \left\{ x \in [0,1]^n : \sum_{i=1}^{n} x_i \le m \right\}
$$

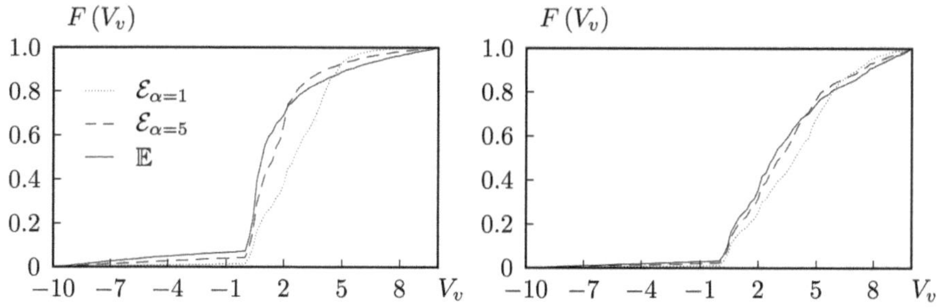

Fig. 1 Loss CDFs of the minimizers of $\mathcal{E}_\alpha$ and maximizers of $\mathbb{E}$ with $\mathcal{X}_0$ (left) and $\mathcal{X}_1$ (right)

Table 2 Difference of the optimal strategies in function of α as a % of the $\mathbb{E}$ solution's statistics ($\mathcal{X}_1$, $n = 1000$, $m = 100$)

	Objective minimized			
Statistic	$\mathcal{E}_{\alpha=1}$	$\mathcal{E}_{\alpha=2}$	$\mathcal{E}_{\alpha=5}$	$\mathcal{E}_{\alpha=7}$
$\mathbb{V}$	−32	−26	−17	−12
$\mathbb{E}$	−15	−5	−2	−1
Worst case $\mathbb{P}$	−68	−44	−14	−11
$\text{VaR}_{\epsilon=10\%}$	−9	−10	−6	−5
Exec. time (s)	6.324	5.131	4.085	3.862

and (2) a more concrete model with disjunctive and precedence constraints

$$\mathcal{X}_1 := \left\{ x \in \mathcal{X}_0 : \sum_{i \in \mathcal{D}_d} x_i \leq 1, \forall d \in \{1, \ldots, D\}, x_i \leq x_j, \forall (i, j) \in \mathcal{E} \right\}.$$

All payoffs R_i, $\bar{R}_i$, P_i, and $\bar{P}_i$ belong to $[-10, 10]$ in this subsection and the next.

As we can see in Fig. 1, the cumulative distributions corresponding to the risk-averse strategies (i.e., minimizing entropic risk measures of parameters $\alpha = 5$ and $\alpha = 10$) are stochastically dominating the risk-neutral strategies (i.e., minimizing the expected loss) in the tail of the distribution.

Further, we studied the influence of the entropic risk parameter α: in Table 2, we can see that as α increases—i.e., the defender becomes less risk-averse—the benefit in terms of variance, $\text{VaR}_{\epsilon=10\%}$, and the worst case probability reduction becomes less important but is significant for lower values of α. On another hand, these benefits come at the moderate cost of having an increased expected loss by about 1–15%.

7.2 *VaR_ϵ and $\mathbb{P}[D(x) \geqslant \widetilde{V}]$ Minimization*

Some preliminary experiments were conducted on minimizing VaR_ϵ and $\mathbb{P}[D(x) \geqslant \widetilde{V}]$ with mid-sized instances of $\mathcal{X}_0$ with $n = 400$ and $m = 60$. The thresholds $\widetilde{V}$

Table 3 Statistics of the optimal strategies for different objectives as a % of the $\mathbb{E}$ solution's ($\mathcal{X}_0$, $n = 400$, $m = 60$)

Statistic	Objective minimized						
	$\text{VaR}_{20\%}$	$\text{VaR}_{10\%}$	$\text{VaR}_{5\%}$	$\text{VaR}_{1\%}$	$\mathbb{P}_{100}$	$\mathbb{P}_{50}$	$\mathbb{P}_{20}$
$\mathbb{V}$	−9	−18	−26	−36	+9	+1	−34
$\mathbb{E}$	+6	+17	+29	+49	+10	+2	+45
$\text{VaR}_{\epsilon=2\%}$	−10	−6	−4	+5	+18	+5	+4
$\text{CVaR}_{\epsilon=2\%}$	−4	−10	−10	−8	+8	+3	−9
Exec. time (%)	+576	+595	+606	+600	−25	−36	−31

used when minimizing $\mathbb{P}[D(x) \geqslant \widetilde{V}]$ were chosen to be 100%, 50%, and 20% of the worst case disutility $-V_1$, noted, respectively, as $\mathbb{P}_{100}$, $\mathbb{P}_{50}$, and $\mathbb{P}_{20}$.

The results are summarized in Table 3 where we can see that the variance is consistently decreased by using risk measures instead of the expected value, and the more risk-averse the defender is, the greater the loss in expected outcome. Because minimizing VaR involves the solution of $O(\log_2 n)$ subproblems, it is significantly slower than minimizing the probability of being over a threshold.

7.3 Prospect Theory

Here, we present computational results evaluating the change in the solution of using and not using a prospect theory model over a small random instance with $n = 8$ targets. Payoffs are generated from $[-10, 10]$. We consider seven instances with this data, changing the number of security resources that the leader uses, with $m = \{1, 2, \ldots, 7\}$. We consider a piecewise linear approximation of the probability distortion function by partitioning $[0, 1]$ in five uniformly spaced breakpoints $K = 5$. We consider three different models, depending on which player considers a prospect theory or an expected utility objective. In particular, model *Neither* assumes both the leader and follower minimize the expected utility; model *Only Follower* has a follower with prospect theory and the leader with expected utility; and model *Both* assumes both players use a prospect theory objective.

In Table 4, we present the leader utility objective (expected utility for Neither and Only Follower) and a prospect theory objective in Both over the different instances. We observe, as instance number increases (and more security resources are used), the disutility decreases for all models. In addition, notice that changing the follower utility function does not cause significant change on the leader utility. Finally, the decrease in leader utility when the leader uses prospect theory is related to the diminishing returns of the utility perturbation because $0 \leq \alpha < 1$.

In Table 5, we present the change in leader expected utility as we modify the reference point C. The change is given as the difference between the leader expected utility of the Only Follower model minus the Neither model. As we change the

Table 4 Leader utility objective function. *Model* identifies if objective is prospect theory or expected utility

	Instances						
Model	1	2	3	4	5	6	7
Neither	4.4	5.8	3.8	5.5	1.1	1.6	0.5
Only follower	4.2	5.9	3.7	5.7	0.9	1.2	0.6
Both	1.1	1.9	1.1	1.8	0.2	0.1	0

Table 5 Expected leader utility difference (*Only Follower − Neither*) for different follower reference points

Reference	Instances						
point	1	2	3	4	5	6	7
−10	0	−0.1	1.5	4.3	3	1.8	0.7
−8	0	−0.1	1.5	3.3	2.9	1.8	0.7
−6	−0.1	−0.1	1.5	2.2	2.8	1.7	0.7
−4	−0.1	−0.1	1.5	2	2.1	1.6	0.7
−2	−0.1	0	1.9	2.9	2.7	1.8	0.7
0	−0.2	0.1	−0.1	0.2	−0.2	−0.4	0.1
2	0	0	0	0.4	0.5	0.3	0.1
4	−3	−0.7	1.4	1.8	1.3	0.8	0.2
6	−2.3	1.3	3	2.7	1.9	1.2	0.2
8	−1.9	2	3.5	3.2	2.1	1.2	0.2
10	0.4	3.2	3.8	3.4	2.1	1.2	0.2

follower reference point from −10 to 10 for all instances, the leader expected utility difference is U-shaped. This difference decreases and then increases. An explanation for this is because for a reference point close to 0, the distortion of the utility value of the follower is not so large and thus does not change much from the expected utility behavior. Largest changes are for extreme reference values in instances 3, 4, and 5, because in this situation, leader policy can be more different. In instance 1, most targets are not protected, while in instance 7, most targets are protected.

8 Conclusions

The Stackelberg security game considered has an uncertain leader utility whose outcome is a discrete random variable with a probability distribution that depends on the players' decisions. In addition, the payoffs for a given leader/follower action pair are invariant. The more common situation in optimization under uncertainty is that the decision variables influence the utility values, not the probability distribution.

We present several formulations for risk models of uncertainty that, for the leaders' utility, provide convex constraints or that can be approximated efficiently with a few integer variables. These are referred to as tractable models. We show that the difficulty of computing certain statistics changes depending on whether the decision variables determine the probability or the utility. In particular, VaR becomes tractable, while variance seems intractable for the leader utility, a situation that is reversed when the probabilities are given and utility values depend on decision variables.

Our computational results illustrate the tractability of the approach in two situations, when the follower uses a quantal response model and when the follower responds with pure strategies which enables the use of distortion functions (prospect theory) for the leader and follower. In the former, we show that we can compute different risk measures (entropic risk, VaR, and chance constraint), and in the latter, we compare the use or not of prospect theory for different players and the effect of the reference point. In Chicoisne and Ordóñez (2023), some particular cases of this model are studied. Specifically, that work considers explicitly the possibility of multiple adversaries. Extending this work for the multiple adversaries setting presented in Chicoisne and Ordóñez (2023) is straightforward, since the derivations in Sect. 3 can be used in every utility function and the leader utility is the weighted sum of the interaction between the leader with each follower. Further work is necessary to evaluate the tractability of a multiple follower SSG if the utilities depend in a more complicated nonlinear way of the multiple players' decisions. Another line of future research is exploring the use of these formulations in other stochastic optimization problems where the decision variables influence the probability distribution.

Acknowledgments We would like to thank Fernanda Jiménez for her help with the computational results of the prospect theory models. This work was partially funded by the Complex Engineering Systems Institute through grant CONICYT-PIA/PUENTE AFB220003 and CONICYT through grant FONDECYT 1201844.

References

An, B., Ordóñez, F., Tambe, M., Shieh, E., Yang, R., Baldwin, C., DiRenzo III, J., Moretti, K., Maule, B., & Meyer, G. (2013). A deployed quantal response-based patrol planning system for the U.S. Coast Guard. *Interfaces, 43*(5), 400–420.

Artzner, P., Delbaen, F., Eber, J.-M., & Heath, D. (1999). Coherent measures of risk. *Mathematical Finance, 9*(3), 203–228.

Balbás, A., Garrido, J., & Mayoral, S. (2009). Properties of distortion risk measures. *Methodology and Computing in Applied Probability, 11*(3), 385–399.

Ben-Akiva, M., & Lerman, S. R. (2018). *Discrete choice analysis: theory and application to travel demand.* Transportation Studies.

Bier, V. M. (2007). Choosing what to protect. *Risk Analysis, 27*(3), 607–620.

Brünner, T., Reiner, J., Natter, M., & Skiera, B. (2019). Prospect theory in a dynamic game: theory and evidence from online pay-per-bid auctions. *Journal of Economic Behavior & Organization, 164*, 215–234.

Brown, G., Carlyle, M., Salmerón, J., & Wood, K. (2006). Defending critical infrastructure. *Interfaces, 36*(6), 530–544.

Camerer, C. (1999). Behavioral economics: Reunifying psychology and economics. *Proceedings of the National Academy of Science, 96*, 10575–10577.

Charnes, A., & Cooper, W. W. (1959). Chance-constrained programming. *Management Science, 6*(1), 73–79.

Chicoisne, R., & Ordóñez, F. (2016). Risk averse Stackelberg security games with quantal response. In: *Proceedings of GameSec 2016, New York* (vol. LNCS 9996, pp. 83–100).
Chicoisne, R., & Ordóñez, F. (2023). Algorithms for a risk-averse Stackelberg game with multiple adversaries. *Computers & Operations Research*. Available online 4 August 2023, https://doi.org/10.1016/j.cor.2023.106367
Delle Fave, F. M., Jiang, A. X., Yin, Z., Zhang, C., Tambe, M., Kraus, S., & Sullivan, J. P. (2014). Game-theoretic security patrolling with dynamic execution uncertainty and a case study on a real transit system. *Journal of Artificial Intelligence Research, 50*, 321–367.
Dentcheva, D., & Ruszczynski, A. (2003). Optimization with stochastic dominance constraints. *SIAM Journal on Optimization, 14*(2), 548–566.
Dentcheva, D., & Ruszczyński, A. (2004). Semi-infinite probabilistic optimization: first-order stochastic dominance constrain. *Optimization, 53*(5–6), 583–601.
Freire, A. S., Moreno, E., & Yushimito, W. F. (2016). A branch-and-bound algorithm for the maximum capture problem with random utilities. *European Journal of Operational Research, 252*(1), 204–212.
Gensch, D. H., & Recker, W. W. (1979). The multinomial, multiattribute logit choice model. *Journal of Marketing Research, 16*(1), 124–132.
Haase, K., & Müller, S. (2014). A comparison of linear reformulations for multinomial logit choice probabilities in facility location models. *European Journal of Operational Research, 232*(3), 689–691.
Haile, P., Hortaçsu, A., & Kosenok, G. (2008). On the empirical content of quantal response equilibrium. *The American Economic Review, 98*(1), 180–200.
Jain, M., Tsai, J., Pita, J., Kiekintveld, C., Rathi, S., Tambe, M., & Ordóñez, F. (2010). Software assistants for randomized patrol planning for the LAX airport police and the federal air marshal service. *Interfaces, 40*(4), 276–290.
Kahneman, D., & Tversky, A. (1979). Prospect theory: An analysis of decision under risk. *Econometrica, 47*(2), 263–291.
Kar, D., Nguyen, T. H., Fang, F., Brown, M., Sinha, A., Tambe, M., & Jiang, A. X. (2017). Trends and applications in Stackelberg security games. *Handbook of dynamic game Theory* (pp. 1–47). Springer International Publishing.
Kiekintveld, C., Jain, M., Tsai, J., Pita, J., Ordóñez, F., & Tambe, M. (2009). Computing optimal randomized resource allocations for massive security games. In: *Proceedings of the 8th AAMAS Conference, Budapest, Hungary*, (pp. 689–696). International Foundation for AAMAS.
Ljubić, I., & Moreno, E. (2018). Outer approximation and submodular cuts for maximum capture facility location problems with random utilities. *European Journal of Operational Research, 266*(1), 46–56.
Markowitz, H. (1952). Portfolio selection. *The Journal of Finance, 7*(1), 77–91.
Mayer, J. (1992). Computational techniques for probabilistic constrained optimization problems. In *Stochastic optimization* (pp. 141–164). Springer.
McDermott, R. (2004). Prospect theory in political science: gains and losses from the first decade. *Political Psychology, 25*(2), 289–312.
McKelvey, R., & Palfrey, T. (1995). Quantal response equilibria for normal form games. *Games and Economic Behavior, 10*(1), 6–38.
Myerson, R. B. (2013). *Game theory*. Harvard University Press.
Paruchuri, P., Pearce, J., Marecki, J., Tambe, M., Ordóñez, F., & Kraus, S. (2008). Playing games for security: an efficient exact algorithm for solving Bayesian Stackelberg games. In *Proceedings of the 7th AAMAS Conference, Estoril, Portugal*. International Foundation for AAMAS.
Pirnay, H., Lopez-Negrete, R., & Biegler, L. (2012). Optimal sensitivity based on ipopt. *Mathematical Programming Computations, 4*(4), 307–331.
Pratt, J. W. (1964). Risk aversion in the small and in the large. *Econometrica: Journal of the Econometric Society, 32*(1/2), 122–136.
Rockafellar, R., & Uryasev, S. (2000). Optimization of conditional value-at-risk. *Journal of Risk, 2*, 21–42.

Shieh, E., An, B., Yang, R., Tambe, M., Baldwin, C., DiRenzo, J., Maule, B., & Meyer, G. (2012). PROTECT: a deployed game theoretic system to protect the ports of the United States. In *Proceedings of The 11th International Conference on Autonomous Agents and Multiagent Systems (AAMAS)*.

Stahl, D., II., & Wilson, P. (1994). Experimental evidence on players' models of other players. *Journal of Economic Behavior & Organization, 25*(3), 309–327.

Tversky, A., & Kahneman, D. (1986). Rational choice and the framing of decisions. *The Journal of Business, 59*(4), 251–278.

Vielma, J. (2015). Mixed integer linear programming formulation techniques. *SIAM Review, 57*, 3–57.

Von Stackelberg, H. (1952). *The theory of the market economy*. William Hodge.

Wachter, A., & Biegler, L. (2006). On the implementation of a primal-dual interior point filter line search algorithm for large-scale nonlinear programming. *Mathematical Programming, 106*(1), 25–57.

Wright, J., & Leyton-Brown, K. (2010). Beyond equilibrium: predicting human behavior in normal-form games. In *Proceedings of the 24th AAAI Conference on Artificial Intelligence, Atlanta, GA*.

Yang, R., Kiekintveld, C., Ordóñez, F., Tambe, M., & John, R. (2011). Improving resource allocation strategy against human adversaries in security games. In *22th IJCAI Proceedings, Barcelona, Spain* (vol. 22, pp. 458–464). AAAI Press.

Facility Location and Supply Chain Risk Analytics

Iris Heckmann, Stefan Nickel, and Francisco Saldanha-da-Gama

Abstract In this chapter, location analysis is put at the core of what is coined as supply chain risk analytics. This is accomplished by devising a hierarchy for the core elements underlying a new definition introduced for supply chain risk. The ultimate purpose of this analysis is to obtain risk-aware decisions for supply network design that adequately overcome the information gap. The above elements and concepts are operationalized through a capacitated facility location problem with an objective that has firm- and customer-oriented features and that includes demand uncertainty. This leads to a risk-aware optimization model, which is introduced for supply chain network design problems fulfilling gaps identified in contemporary research. Additionally, the value of considering risk-aware solutions is discussed in the context of supply chain management. Above all, this chapter closes a gap existing in the literature, namely, a missing clear objective and quantifiable definition of risk in supply chain management. Both researchers and practitioners can benefit from the contents of this chapter.

Keywords Supply chain risk · Risk-aware locations · Two-stage stochastic location model

---

I. Heckmann
CamelotITLab, Innovative Technologies Lab, Cologne, Germany
e-mail: ihec@camelot-itlab.com

S. Nickel (✉)
Karlsruhe Institute of Technology (KIT), Karlsruhe, Germany
e-mail: stefan.nickel@kit.edu

F. Saldanha-da-Gama
Departamento de Estatística e Investigação Operacional e Centro de Matemática, Aplicações Fundamentais e Investigação Operacional, Faculdade de Ciências, Universidade de Lisboa, Lisbon, Portugal
e-mail: fsgama@ciencias.ulisboa.pt

H. A. Eiselt, V. Marianov (eds.), *Uncertainty in Facility Location Problems*, International Series in Operations Research & Management Science 347, https://doi.org/10.1007/978-3-031-32338-6_7

1 Introduction

The role of facility location in logistics network design has been widely recognized (see, e.g., Klose & Drexl 2005, Melo et al. 2009, Heckmann & Nickel 2019). The impressive advances we have observed in stochastic programming in general and stochastic facility location problems in particular explain the increasing attention put on the role of facility location in supply chain planning under uncertainty (Correia & Saldanha-da-Gama, 2019; Dunke et al., 2016; Heckmann & Nickel, 2019).

It is commonly accepted that stochasticity leads to risk. However, the concept of risk is vague and strongly depends on the specific problem or area in which it is being adopted. In the context of supply chain planning, there has been a big debate on the topic with different risk concepts being introduced (Dunke et al., 2016; Heckmann et al., 2015).

In this chapter, we discuss how facility location can be put at the core of a comprehensive risk definition in supply chain network design. This, in turn, leads to a comprehensive stochastic facility location model. In this process, we highlight the relevant elements and their hierarchy for handling risk in supply chain management.

The remainder of the chapter is organized as follows. In Sect. 2, we discuss different aspects that lay the foundations for what is later coined as "supply chain risk analytics" and also for a new comprehensive definition for supply chain risk. In Sect. 3, we show how to operationalize the new risk definition by means of a stochastic facility location model. In Sect. 4, we assess the relevance of considering risk-aware solutions in the strategic context discussed. In Sect. 5, we illustrate all the concepts introduced in this chapter by using a small instance of the new model proposed. The chapter ends with several conclusions drawn from the contents and illustrations proposed.

2 Toward Supply Chain Risk Analytics

Over the past decades, the need for risk assessment and management in general and within the context of supply chain planning in particular has become increasingly relevant. This is attested by Aven (2016), Baryannis et al. (2019), Dunke et al. (2016), Heckmann et al. (2015), and Munir et al. (2020) as well as by the references therein. However, the concept of risk in the context of supply chain management—"supply chain risk"—has not been clearly defined. The reason for a heterogeneous and often ambiguous understanding of such concept has its roots in the long history and evolution of the notion of risk itself. Nevertheless, for practitioners and researchers, it would be important to have a standardized definition helping in the development of more comprehensive optimization models to better support decision-making.

In this section, we discuss recent attempts to formally define supply chain risk, and we introduce a new definition that ultimately accounts for all the aspects that, in our view, should be considered.

2.1 *Toward a Comprehensive Definition of Supply Chain Risk*

As pointed out by Heckmann et al. (2015), there is a vast literature dealing with risk in multiple fields. For the particular case of supply chain management, those authors identify a set of core characteristics underlying risk, namely, risk exposition, risk attitude, and risk objective. A central aspect indicating the need to cope with supply chain risk regards the objectives to achieve and their target values. In that same paper, the authors identify two types of goals in modern supply chains: functionality (effectiveness) and profitability (efficiency). The former refers, for instance, to the availability of resources or to the service level achieved; the latter demonstrates competitive advantage, which may depend on factors such as the logistics and supply chain costs (e.g., facility location or transportation of commodities). Whenever a supply chain is hindered to achieve both types of goals, supply chain risk may arise. This possibility, in turn, depends on the exposition of the supply chain to unexpected events—the risk exposition. Finally, risk exposition is determined by three elements: (i) disruptive triggers, (ii) time-dependent features, and (iii) the affected components of the supply chain.

A disrupting trigger is an event (e.g., a strike, a flood, a pandemic) that initiates unexpected changes with an unpredictable outcome. The second element—time—is important for assessing the status of a supply chain either before, while, or after a disruptive trigger occurs. For instance, if a labor strike occurs, then typically the more time the disruption lasts, the more severe are the consequences for the system as a whole. Finally, the affected supply chain components have to be identified to evaluate their capability to cope with changes that affect their inner processes.

Last but not the least, the relevance of not achieving the goals set for a supply chain—a prime indicator of the presence of risk—is further assessed by the risk attitude of the decision-maker. In particular, it depends on the subjective interpretation of the decision-maker as well as on how negative a deviation from the goal should be evaluated.

Considering the core characteristics just reviewed, Heckmann et al. (2015) offered the following definition: "Supply chain risk is the potential loss for a supply chain in terms of its target values of efficiency and effectiveness evoked by uncertain developments of supply chain characteristics whose changes were caused by the occurrence of triggering-events."

More recently, Dunke et al. (2016) argued that often supply chain risk is time-dependent, unlike suggested by the above definition. For instance, the target values set for the goals often vary in time; alternatively, a decision-maker may have a level of risk aversion that changes with the time (e.g., more experience/knowledge accumulated). In other words, time dependency cannot be neglected when considering the core elements underlying supply chain risk (risk objective, risk exposition, and risk attitude).

By gathering the insights provided by Heckmann et al. (2015) and Dunke et al. (2016), we offer an improved definition of supply chain risk:

> *Supply chain risk* is the *time-dependent potential loss* for a supply chain in terms of its *target values of profitability* and *functionality* evaluated by the decision maker's *nature* and evoked by *uncertain* changes of the *affected supply chain* and its processes whose changes were caused by the *occurrence* of *triggering-events*.

In the above definition, we highlight in italics the relevant aspects to consider. The concept of supply chain process was formally introduced by Dunke et al. (2016): it is any activity involved in procuring, producing, storing, or distributing goods or any service required to ensure the achievement of the goals set for the supply chain.

Next, we look into the different elements involved in the new definition just proposed.

2.2 *Hierarchy of the Core Characteristics of Supply Chain Risk*

In Fig. 1, we illustrate the *supply chain risk hierarchy*, embracing the main features of the new definition just proposed. By jointly considering these features, we can eventually operationalize supply chain risk.

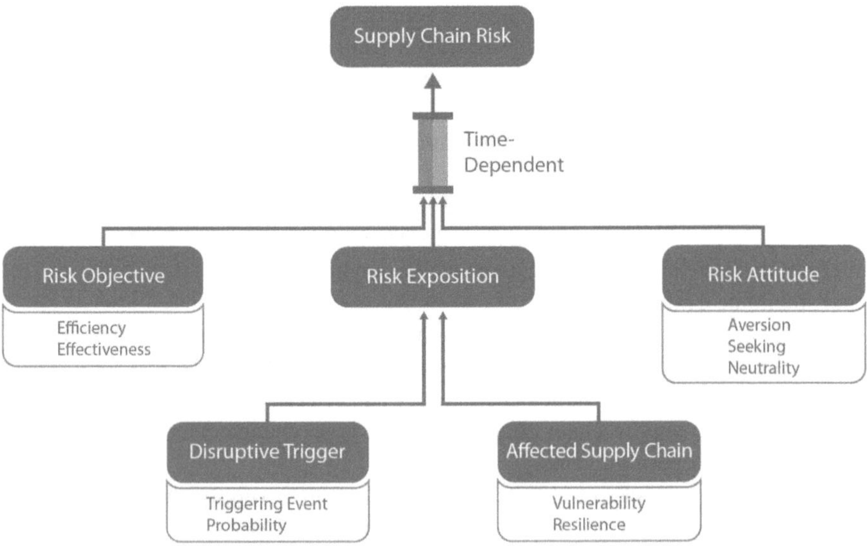

Fig. 1 The core characteristics of supply chain risk hierarchy

2.2.1 Time Dependency

The central role of time in supply chain risk has scarcely been considered in the literature. Nevertheless, some authors point out its importance when it comes to modeling supply chain risk (see, e.g., Hahn & Kuhn 2012, Lockamy & McCormack 2010, Manuj & Mentzer 2008, Sheffi 2005, and Wagner & Bode 2008).

A so-called disruption profile is a way to relate time with performance deterioration.[1] This was introduced by Sheffi (2005) and Sheffi & Rice (2005), and it was discussed by other authors such as Asbjørnslett (2009), Behdani (2013), Craighead et al. (2007), Lynch (2012), Melnyk et al. (2008), and Snyder et al. (2016). For example, a triggering event can lead to a huge capacity reduction (e.g., in terms of production) that recovers slowly over the planning horizon; it can also lead to a small reduction that can be quickly compensated.

Despite the distinct evolution over time usually observed for different parameters underlying a problem, the time-dependent integration of these parameters is of major relevance. For instance, in the case of the flooding in Thailand in 2011, an extreme capacity reduction for exporters of hard disk drives and solid-state drives came along with an increased demand for the new solid-state drives (Risk Response Network, 2011).

Accordingly, a quantitative approach for handling risk should account for preparedness with respect to possible disruptive triggers. Furthermore, capacity and demand shifts are more adequately represented as time-dependent parameters.

2.2.2 Risk Objective

Traditionally, risk is perceived as financial risk and assessed with metrics often used in finance like variance, mean-variance ratios, value-at-risk, or conditional value-at-risk (see, for instance, Sarykalin et al. 2008). In supply chain management, we also find works perceiving risk as financial risk (see, for instance, Nickel et al. 2012 and Osadchiy et al. 2016).

For the sake of competitiveness, supply chains need to accomplish a cost-efficient execution of supply chain processes. However, unlike the overall corporate business objective, the main purpose of a supply chain is to satisfy customers' demand. Therefore, it is inaccurate (not to say incorrect) to evaluate supply chain risk only in terms of financial risk, i.e., monetary loss.

While the availability of resources is captured by the concept of *effectiveness*, the competitive advantage is captured by the concept of *efficiency*. Therefore, as pointed out before, a supply chain should seek two main goals: efficiency and effectiveness. Both should be considered in a quantitative approach for handling supply chain risk.

---

[1] Performance deterioration is the difference (over a certain planning horizon) between the planned targeted supply chain performance value and the actual performance value—see Cui et al. (2010).

2.2.3 Decision-Maker's Nature

Like in most of the cases when dealing with uncertainty, the nature of the decision-maker when facing uncertainty is a core feature of supply chain risk. When a decision-maker is risk-seeking, she/he prefers a chance outcome than a certain return with the same expected utility (Kochenderfer et al., 2022). A risk-neutral decision-maker is indifferent to a chance outcome or a certain one with the same expected utility. This is a very common attitude assumed in the literature; the future assets are typically assessed by their expected value.

When managing a supply chain, two objectives should be considered: service level, i.e., proportion of the demand that is satisfied (an effectiveness-rents objective), and cost minimization (an efficiency-rents objective). One possibility for handling these objectives, which are often conflicting, is to weight them in a single objective function. By doing so, it is possible to give more strength to one of them or even to look for different trade-offs before making a final decision. Furthermore, by playing with the weights, we can actually capture different levels of financial risk aversion, and thus we can better adjust a model to the actual nature of the decision-maker. This is being highlighted by the stochastic facility location model to be introduced later in the chapter.

2.2.4 Risk Exposition

The exposition of a supply chain to risk is one of the core characteristics highlighted in Fig. 1 that is further specified by the disruptive triggers and the affected components of the supply chain.

Dunke et al. (2016) call potential trigger to an event that can negatively affect the efficiency and effectiveness of a supply chain process, resulting in a performance deterioration. A potential trigger becomes a disruptive trigger when its occurrence results in the actual deterioration of the supply chain performance.

The consequences of a disruptive trigger typically propagate through the entire supply network. For example, a negative event such as a strike or a wart can result in a capacity reduction or in a demand increase. It is the interplay of all supply chain processes and the actual state of the different supply chain features that define the resilience of the system. We recall the concept of supply chain resilience adopted by Heckmann (2015) and Ivanov (2018) among other authors: the ability of a supply chain to overcome vulnerability with the latter describing the extent to which a supply chain is susceptible to some event. Such interaction determines whether the first impact of the event on a process provokes the dis-functionality and/or non-profitability of consecutive processes, propagates through the entire network, and finally results in failing to achieve supply chain goals in terms of efficiency and/or effectiveness.

To endow supply chains with the ability to absorb (or to adjust to) the consequences of disruptive triggers, several additional decisions emerge. First, it is necessary to assess the need for increasing the supply chain resilience. If the supply chain is considered to be able to hedge against uncertain triggers, the installation of (further) risk countermeasures is not necessary; otherwise, the supply chain should be endowed with an occasional and temporary adjustment of its structures. For instance, a facility can be "protected" by options for temporary capacity expansion that can be activated if necessary. This is a way to ensure proper countermeasures for recovering production capacity, inventory levels, or handling throughput, just to mention some possibilities. Note that hedging against risk has costs and that is a decision problem by itself.

2.3 *Supply Chain Risk Analytics*

Most contemporary approaches for dealing with supply chain risk focus on reducing the financial consequences of uncertain and unexpected developments (Heckmann et al., 2015). They predominantly evaluate the impact of changes of monetary policies (prices) or fiscal policies (taxation) with measures developed for the quantification of financial risk. As discussed above, this is not the same as supply chain risk.

To this date, supply chain risk management suffers from the lack of clear and adequate quantitative measures respecting the characteristics of modern supply chains. Consequently, it is difficult to adequately quantify risk. Even if it is not possible to fully quantify supply chain risk through some measures, still supply chain risk and its related core characteristics need to be represented within supply chain models. Following these arguments, we introduce a new concept:

> *Supply chain risk analytics* is a bundle of mathematical methods and measurement techniques tailored for determining risk-aware solutions for supply chain design, planning and execution.

The remainder of this chapter materializes this new concept.

3 Supply Chain Risk Made Operational: A Stochastic Facility Location Model

A renewed definition of supply chain risk as presented in the previous section is interesting but is useless if not properly operationalized: it is important to look into ways for embedding such a definition into quantitative approaches to better support decision-making. As we show next, this can be accomplished by taking advantage from the strong role of location analysis in supply chain management, which allows putting location problems at the service of the above discussed concepts.

One fundamental facility location problem is the capacitated facility location problem (CFLP) also known as the fixed-charge facility location problem (Fernández & Landete, 2019). When it comes to strategic planning in supply chain management, it is often found at the core of comprehensive problems since it involves capacity constraints that can be related with production, inventory, or some type of handling.

3.1 Model Formulation: Embedding Time and Uncertainty

The CFLP consists of deciding where to locate a set of capacitated facilities and how to ship some commodity to a set of customers to minimize the total cost associated with the facilities and with the transportation of the commodity from the facilities to the customers.

As we have discussed in Sect. 2, when risk is to be accounted for, time and uncertainty cannot be neglected. We consider the planning horizon of interest divided into time periods. Different arguments support a discretized planning horizon (see, e.g., Nickel & Saldanha-da-Gama 2019). Regarding uncertainty, we consider its most common source in a facility location problem: demand. We assume that uncertainty can be captured by a finite (even if of large cardinality) set of scenarios; each scenario determines the demand of all customers in all periods. If we further assume that the probability associated with each scenario can be estimated, then we can resort to stochastic programming to derive an optimization model for the problem.

We introduce some notation to be used hereafter. I denotes the set of candidate locations for the facilities, J represents the set of customers or demand points, T stands for the set of time periods in the planning horizon, and S is the set of demand scenarios. Regarding the costs, we define f_i as the operation cost of a facility located at $i \in I$, c_{ij} is the unit transportation cost between facility $i \in I$ and customer $j \in J$, and r_j is the unit revenue provided by customer $j \in J$. Other parameters in our problem are q_i, which is the capacity of a facility operating at $i \in I$, d_{jts} representing the demand of customer $j \in J$ in period $t \in T$ under scenario $s \in S$, and π_s, which is the probability for scenario $s \in S$.

The above notation assumes a single commodity as well as a negligible monetary devaluation. In practice, this may not be the case but our analysis can be easily extended if necessary. Nevertheless, keeping the setting as simple as possible allows to better highlight several aspects of interest.

The decisions to make are represented by two sets of decision variables: binary variables y_i equal to 1 if and only if a facility is set operating at $i \in I$, and non-negative continuous variables x_{ijts} indicating the fraction of the demand of customer $j \in J$ in period $t \in T$ supplied from a facility operating at $i \in I$ under scenario $s \in S$. A multi-period stochastic CFLP can be formulated as follows:

$$\text{minimize} \quad \sum_{i \in I} f_i y_i + \sum_{s \in S} \pi_s \left(\sum_{t \in T} \sum_{i \in I} \sum_{j \in J} \left(c_{ij} - r_j\right) d_{jts} x_{ijts} \right), \tag{1}$$

$$\text{subject to} \quad \sum_{i \in I} x_{ijts} \leq 1, \qquad j \in J,\ t \in T,\ s \in S, \tag{2}$$

$$\sum_{j \in J} d_{jts} x_{ijts} \leq Q_i y_i, \quad i \in I,\ t \in T,\ s \in S, \tag{3}$$

$$x_{ijts} \geq 0, \qquad i \in I,\ j \in J,\ t \in T,\ s \in S, \tag{4}$$

$$y_i \in \{0, 1\}, \qquad i \in I. \tag{5}$$

The objective function (1) accounts for the total facilities' operation costs and the total expected cost for supplying the demand; Constraints (2) ensure that demand is not oversupplied; Inequalities (3) stand for the capacity of the facilities; finally, (4) and (5) state the domain of the decision variables. Due to the presence of revenues in the objective function, supplying all the demand of all customers in all periods and scenarios does not necessarily correspond to an optimal decision.

Having considered the above prototype model, we can now proceed by discussing how the different characteristics of supply chain risk identified in the previous section can be embedded in the model.

3.2 *Risk Objective: Efficiency and Effectiveness*

As discussed in Sect. 2.2, one relevant element in the supply chain risk hierarchy is the risk objective, which should gather both efficiency and effectiveness. Noticeably, the above model is already capturing supply chain efficiency in the objective function (1) by means of the total cost associated with the facilities (f_i, $i \in I$) and the transportation cost between facilities and customers, c_{ij} ($i \in I$, $j \in J$).

In turn, effectiveness is related with the ability of the supply chain to fulfill its function—to supply the customers according to their demand. This is not captured by the model. As we mentioned before, a way to account for effectiveness is to consider service level as a decision to make and then to use the objective function to penalize any shortage with respect to a minimum threshold defined by the decision-maker—the target service level.

To extend the model, we introduce some additional notation, namely, $\alpha^0 \in [0, 1]$, denoting the target service level, and h standing for the unit cost for service level shortage (w.r.t. target). The additional decision variables include α_s representing the service level achieved under scenario $s \in S$; Δ_s, the service level shortage (w.r.t. the target value) under scenario $s \in S$; and u_{jts}, the proportion of unsupplied demand for customer $j \in J$, in period $t \in T$ under scenario $s \in S$.

Using the above notation, for each $s \in S$, we have

$$\alpha_s = 1 - \frac{\sum_{j \in J} \sum_{t \in T} d_{jts} u_{jts}}{\sum_{j \in J} \sum_{t \in T} d_{jts}} \qquad \text{and} \qquad \Delta_s = \max\left\{0, \alpha^0 - \alpha_s\right\}.$$

Note that for each scenario $s \in S$, we have $\alpha_s \in [0, 1]$, with 0 indicating that no demand is supplied and 1 indicating that all demand is supplied.

Model (1)–(5) can now be extended as follows:

$$\text{minimize} \quad \sum_{i \in I} f_i y_i + \sum_{s \in S} \pi_s \left(h\,\Delta_s + \sum_{t \in T} \sum_{i \in I} \sum_{j \in J} (c_{ij} - r_j)\, d_{jts} x_{ijts} \right), \tag{6}$$

$$\text{subject to} \quad \sum_{i \in I} x_{ijts} + u_{jts} = 1, \qquad j \in J,\ t \in T,\ s \in S, \tag{7}$$

$$\Delta_s \geq \alpha^0 - \left(1 - \frac{\sum_{j \in J} \sum_{t \in T} d_{jts} u_{jts}}{\sum_{j \in J} \sum_{t \in T} d_{jts}} \right), \quad s \in S, \tag{8}$$

(3)–(5),

$$u_{jts} \geq 0, \qquad j \in J,\ t \in T,\ s \in S, \tag{9}$$

$$\Delta_s \geq 0, \qquad s \in S. \tag{10}$$

The possibility of having unsupplied demand has been captured within the context of supply chain network design problems in general and within facility location problems in particular (see, for instance, Cui et al. 2010, Miranda & Garrido 2009, and Nickel et al. 2012). In the case of facility location models, unsupplied demand can be easily embedded in a classical model since we can consider a dummy facility supplying all the missing demand. Nevertheless, explicitly considering the unsupplied demand as we are doing above helps to better illustrate the concepts we are discussing in this chapter.

When capacity becomes tight, it will be necessary to decide which customers to serve. In the above model, the unit revenues provided by the customers guide such decision. Furthermore, the model also allows each customer to be served by multiple facilities, which is of particular interest, when, for instance, disruptions prevent the supply from a major facility (Ang et al., 2017).

Finally, we note time and uncertainty being involved in both efficiency and effectiveness as it should be the case according to our preliminary discussion.

3.3 *The Attitude Toward Risk*

At a first glance, the extended model just presented seems to indicate that we are considering a risk-neutral decision-maker, because our objective function measures the expected value of the future outcome. However, a closer look shows that we can do more than that. A risk-aware optimization model for supply chain network design must seek a trade-off between customer satisfaction and cost. This suggests the use of a multicriteria optimization model. However, looking closely into our above model, we realize that this is being accomplished through scalarization. In fact, the above-mentioned trade-off is assured and "regulated" by parameter h. By playing with it, one can capture the extent to which the decision-maker is willing to invest for getting more operational capacity to increase the value of customer satisfaction. A decision-maker with a higher level of risk aversion with respect to investments in the supply chain will certainly choose smaller values for h. Accordingly, despite considering the expected value of the future assets, our objective function allows defining different risk profiles for decisions-makers when it comes to financial risk.

This analysis is still valid (and even better stressed) when other costs are included in the objective function as we do next.

3.4 Risk Exposition

In Sect. 2, we realized that risk exposition in a supply chain stems from disruptive triggers since it depends on their impact in the supply chain structures. The most common form of disruption in a supply chain concerns its capacity for fulfilling demand. In terms of our facility location problem, this corresponds to a capacity reduction, thus affecting constraints (3).

Let us assume that we can identify a set of scenarios in terms of one or several disruptive triggers and their impact in the operating capacity of the facilities. If we combine each of these scenarios with each scenario already in S, we get an extended set of scenarios, each determining all the uncertainty parameters. For this reason, w.l.o.g., we keep using the notation S for the enlarged set of scenarios. In particular, each scenario now specifies a capacity reduction (in [0, 1]) for each facility in each time period. We need one extra set of parameters, γ_{its}, standing for the proportion of the operational capacity of facility $i \in I$ in period $t \in T$ that is available under scenario $s \in S$. Thus, $\gamma_{its} \times Q_i$ becomes the actual capacity of facility $i \in I$ in period $t \in T$ under scenario $s \in S$.

A disruptive trigger can also affect the demand (either by increasing or decreasing it). Moreover, we may face a situation in which a combination of disruptive triggers leads to simultaneous changes in the operating capacity and in the demand. The notation adopted before for the demand can already capture this situation since scenario-indexed parameters are being used.

Also related with the risk exposition is the affected supply chain and its resilience. In this case, additional decisions may be required to ensure the resilience of the system, thus bringing it back to a status of efficiency and effectiveness. This can be incorporated in our modeling framework by considering temporary capacity expansions associated with extra capacity options (see, e.g., Xu & Nozick 2009 for supplier selection) and thus corresponding to a here-and-now decision. In fact, preparedness measures are adequate to react against a disruption but they call for a corporation to decide in advance about the possibilities that may need to be activated to mitigate sudden changes in the underlying setting. Such decisions are typically part of a contingency plan and thus are identified in advance. On the other hand, the specific decisions concerning when and at which level to activate an option are recourse decisions calling for a more detailed course of action that often can be made only after a disruption occurs, i.e., the actual future scenario becomes known.

Capacity expansion decisions have been considered by several authors as a way to react to changing demand although not explicitly in a risk-aware setting. The reader can refer to Aghezzaf (2005), Fleischmann et al. (2006), Hugo & Pistikopoulos (2005), Julka et al. (2007), Ko & Evans (2007), and Troncoso & Garrido (2005).

We define one extra set L, containing the capacity expansion levels available; each level determines a temporary increase in the operational capacity of a facility. Additional parameters are also required: g_i is the fixed cost associated with an option contracted for facility $i \in I$ to ensure a temporary capacity increase if necessary; b_ℓ stands for the unit capacity cost associated with capacity expansion level $\ell \in L$; and k_ℓ represents the amount of extra capacity associated with capacity expansion level $\ell \in L$.

The extended model makes use of the following two additional sets of binary variables: z_i is a binary variable equal to 1 if and only if a capacity expansion option is contracted for facility $i \in I$; $w_{it\ell s}$ is also binary: it is equal to 1 if and only if in scenario $s \in S$, expansion level $\ell \in L$ is installed at facility $i \in I$ in time period $t \in T$.

We can now formulate a risk-aware multi-period stochastic capacitated facility location problem that we denote by $\text{CFLP}_{\text{risk}}$:

$$\text{minimize} \quad \sum_{i \in I} (f_i y_i + g_i z_i) + \sum_{s \in S} \pi_s \Bigg[(h\,\Delta_s + \sum_{i \in I} \sum_{\ell \in L} \Big(b_\ell k_\ell \sum_{t \in T} w_{it\ell s} \Big) + \sum_{t \in T} \sum_{i \in I} \sum_{j \in J} (c_{ij} - r_j)\, d_{jts} x_{ijts} \Bigg], \tag{11}$$

$$\text{subject to} \quad \sum_{i \in I} x_{ijts} + u_{jts} = 1, \qquad j \in J,\ t \in T,\ s \in S, \tag{7}$$

$$\sum_{j \in J} d_{jts} x_{ijts} \le \gamma_{its}\, Q_i\, y_i + \sum_{\ell \in L} k_\ell w_{it\ell s}, \qquad i \in I,\ t \in T,\ s \in S, \tag{12}$$

$$z_i \le y_i, \qquad i \in I \tag{13}$$

$$\sum_{\ell \in L} w_{it\ell s} \le z_i, \qquad i \in I,\ t \in T,\ s \in S \tag{14}$$

$$\Delta_s \ge \alpha^0 - \left(1 - \frac{\sum_{j \in J} \sum_{t \in T} d_{jts} u_{jts}}{\sum_{j \in J} \sum_{t \in T} d_{jts}} \right), \qquad s \in S, \tag{8}$$

$$x_{ijts} \geq 0, \qquad i \in I,\ j \in J,\ t \in T,\ s \in S, \tag{4}$$

$$y_i \in \{0, 1\}, \qquad i \in I. \tag{5}$$

$$u_{jts} \geq 0, \qquad j \in J,\ t \in T,\ s \in S, \tag{9}$$

$$\Delta_s \geq 0, \qquad s \in S, \tag{10}$$

$$z_i \in \{0, 1\}, \qquad i \in I, \tag{15}$$

$$w_{it\ell s} \in \{0, 1\}, \qquad i \in I,\ t \in T,\ \ell \in L,\ s \in S. \tag{16}$$

The above model comprises a two-stage decision-making process: (i) a decision that is to be implemented now (facilities to operate and capacity expansion options to buy) and (ii) a recourse decision—thus defined for every possible future scenario. Therefore, we are facing a two-stage stochastic programming model aiming at minimizing the total cost in terms of operating facilities and contracted options plus the expected cost associated with the recourse actions (penalty for unsupplied demand, capacity expansion, and demand satisfaction).

In addition to the constraints introduced before, we consider now also constraints (12) ensuring that even with the temporary capacity expansions, we cannot satisfy more demand than the actual operating capacity (in every facility, period, and scenario). Constraints (13) ensure that options can only be contracted for operating facilities, whereas constraints (14) guarantee that operating capacity can only be expanded if an option was previously contracted. Finally, constraints (15) and (16) are the domain of the decision variables associated with capacity expansion.

4 The Value of a Risk-Aware Solution

The analysis presented so far led to a comprehensive stochastic facility location model that allows capturing different aspects underlying risk in supply chain network design. However, the model easily becomes a large-scale one and thus more difficult to tackle. Accordingly, it is important to check whether this increased difficulty is compensated by additional insights provided through the risk-aware solutions we obtain.

Two indicators are usually considered for looking into this aspects: the value of the stochastic solution (VSS) and the expected value of perfect information (EVPI) (Birge & Louveaux, 2011).

The VSS is the difference between the objective value of the stochastic problem evaluated using the optimal solution of the expected value problem (EEV)—single-scenario model resulting from replacing the random variables by their expectations—and the optimal value of the stochastic problem. Denote by $(\hat{y}, \hat{z})$ the optimal solution to the former, and let $\text{CFLP}_{\text{risk}}(\hat{y}, \hat{z})$ represent the $\text{CFLP}_{\text{risk}}$

when the first-stage decision is fixed according to $(\hat{y}, \hat{z})$. In our case, the VSS is formally defined as:

$$\text{VSS} = \mathscr{V}(\text{CFLP}_{\text{risk}}(\hat{y}, \hat{z})) - \mathscr{V}(\text{CFLP}_{\text{risk}}),$$

where $\mathscr{V}(\text{P})$ denotes the optimal objective value of model P.

The EVPI originates in decision theory and represents the value that the decision-maker is willing to pay to get perfect information about the future. It is determined by the difference between the optimal value of the stochastic problem and the wait-and-see-solution value. In turn, the latter is the expected value of the random variable that represents the optimal value of a single-scenario problem. Denoting by $\text{CFLP}_{\text{risk},s}$ the single-scenario problem induced by scenario $s \in S$, we have formally:

$$\text{EVPI} = \mathscr{V}(\text{CFLP}_{\text{risk}}) - \sum_{s \in S} \pi_s \mathscr{V}(\text{CFLP}_{\text{risk},s}).$$

As we have largely discussed, although stochasticity is a basic "ingredient" for risk, it is not the only one. Accordingly, we may ask whether the VSS and EVPI are appropriate measures for evaluating the relevance of a modeling framework capturing risk. In our opinion, the answer is a clear "no": we need some specific measure for that purpose.

The first aspect we should emphasize is that the need for a risk-aware supply chain design emerges if hedging against risk has some expected "positive value." Hence, it would be interesting to find a measure producing such a value. A second aspect we must point out is that the single-scenario problems induced by $\text{CFLP}_{\text{risk}}$ (i.e., models $\text{CFLP}_{\text{risk},s}$, $s \in S$) do not represent a *non-risk-aware* counterpart. Therefore, VSS and EVPI are not enough for evaluating the advantages of considering a risk-aware model and its solution.

Suppose we knew in advance the exact scenario, say s, we will be facing. In that case, neither $\text{CFLP}_{\text{risk}}$ nor $\text{CFLP}_{\text{risk},s}$ are appropriate models to consider since we know exactly the disruption profile that will occur. What model should be solved then? The answer is not straightforward.

At a first glance, the correct answer could be the deterministic model induced by scenario s in models (1)–(5). However, this is again not true since a solution obtained from that deterministic model may easily turn out to be infeasible for some scenarios (disruption profiles).

This discussion suggests that an adequate deterministic counterpart should be an extension of the single-scenario version of models (1)–(5) accounting for unsupplied demand. This way, it becomes possible to evaluate a solution induced by some scenario (rather than simply concluding that for some other scenarios, it is infeasible). Note also that under a single-scenario risk-free setting, it makes no sense to buy options for expanding the capacity since that feature stems from the uncertainty underlying the problem. Finally, in a deterministic setting, the service level is no longer uncertain; it directly results from the decision to make. Therefore, setting a target value for it does not make sense either.

The following multi-period stochastic CFLP (MPSCFLP$_s$) emerges as an adequate single-scenario model describing a risk-free setting:

$$\text{minimize} \quad \sum_{i\in I} f_i y_{is} + \sum_{t\in T}\sum_{j\in J} \hat{h} d_{jts} u_{jts} + \sum_{t\in T}\sum_{i\in I}\sum_{j\in J} (c_{ij} - r_j) d_{jts} x_{ijts}, \tag{17}$$

$$\text{subject to} \quad \sum_{i\in I} x_{ijts} + u_{jts} = 1, \qquad j \in J,\ t \in T, \tag{18}$$

$$\sum_{j\in J} d_{jts} x_{ijts} \le \gamma_{its}\, Q_i\, y_i, \qquad i \in I,\ t \in T, \tag{19}$$

$$x_{ijts} \ge 0, \qquad i \in I,\ j \in J,\ t \in T, \tag{20}$$

$$y_{is} \in \{0, 1\}, \qquad i \in I. \tag{21}$$

For each $s \in S$, we denote by $\mathscr{V}(\text{MPSCFLP}_s)$ the optimal value of the corresponding problem (17)–(21).

Definition 1.1 (Value of the Supply Chain Risk) The value of a risk-aware supply chain solution can now be defined as the difference between the optimal value of CFLP$_{\text{risk}}$ and the weighted value of the models induced by each and every scenario, i.e.,

$$\text{VSCR} = \mathscr{V}(\text{CFLP}_{\text{risk}}) - \sum_{s\in S} \pi_s\, \mathscr{V}(\text{MPSCFLP}_s) \tag{22}$$

The above definition bears resemblance with the EVPI. However, the differences are clear: (i) the way this value is computed is not the same as for the EVPI since the single-scenario problems do not result from considering one scenario in the original stochastic model and (ii) this is a measure that can in principle be computed within the context of any risk-aware supply chain network design model.

5 Illustration with a Simple Instance

In this section, a small instance of CFLP$_{\text{risk}}$ is used to highlight the relevance of the features we have been discussing in this chapter. We focus our attention on the following aspects: (i) plausibility of the solutions obtained by the model, (ii) value of supply chain risk consideration, and (iii) the effect of capturing uncertainty.

5.1 Data

For illustrative purposes, we limit the instance size to 2 facilities, 3 customers, and 12 periods in the planning horizon (e.g., weeks).

Noticeably, supply chain risk is not necessarily associated with catastrophic disruptions and big turmoils (such as a tsunami). They can stem from less smaller incidents like short circuits or machine failure at production facilities (see, e.g., Norrman & Jansson 2004). These minor incidents typically do not affect cost parameters. Hence, for the analysis presented next, we consider deterministic and time-invariant costs. In particular, for the potential facility 1, we take both the operational cost (f_{1t}) and the extra-capacity costs (g_{1t}) in each period equal to 100, doubling that value for facility 2 ($f_{2t} = g_{2t} = 200$). As for the capacities, we take the value 10 for both facilities. The unit revenue obtained from the customers (r_j) is set equal to 4 for the three customers. Additionally, we consider a service level adherence value $h = 30{,}000$. Finally, in Table 1, we introduce the other deterministic parameters that we consider in our illustration.

For evaluating model CFLP$_{\text{risk}}$, we generated several scenarios—disruption profiles—based on: (i) the start and end time of the disruption, (ii) the average speed of capacity decrease and recovery, and (iii) the absolute minimum for the operational capacity resulting from the disruption. We worked with three different scenarios that differ in terms of the minimum operational capacity. Scenario 1 reflects the situation of a major capacity reduction exposed to a minor (negligible) fall in demand. This scenario is assumed to have a probability of 0.1. The second scenario is defined by a less intense capacity reduction although it comes along with a higher demand level (that is assumed constant for the sake of the simplicity). A probability of 0.3 is assumed for this scenario. Finally, scenario 3 reflects a normal situation in which no disruption occurs and the operating capacity exceeds the demand. A probability of 0.6 is assumed for this case.

The values considered for the overall capacity and demand for the three scenarios are presented in Table 2. Customers are assumed to have the same demand, and thus the values presented for the demand in Table 2 must be divided by 3 to obtain the individual demand. In that table, we also present the overall capacity deficit in each period resulting from the scenario disruptions. A negative deficit indicates that there is a surplus of capacity w.r.t. demand. The data we are introducing is assuming a disruption lasting for 9 periods.

Table 1 Expansion levels and transportation costs for the illustrative example

Expansion level (ℓ)	1 (small)	2 (medium)	3 (large)
Expansion capacity k_ℓ	2	5	10
Expansion costs b_ℓ	5	8	10
Transportation costs	Customer		
c_{ij}	1	2	3
Facility 1	3	5	3
Facility 2	3	5	3

Table 2 Disruption scenarios for the prototype instance

	Time period											
	1	2	3	4	5	6	7	8	9	10	11	12
Scenario 1												
Operational capacity	20	11	2	4.25	6.5	8.75	11	13.25	15.5	17.75	20	20
Overall demand	12	12	12	12	12	12	12	12	12	12	12	12
Capacity deficit	−8	1	10	7.75	5.5	3.25	1	−1.25	−3.5	−5.75	−8	−8
Scenario 2												
Operational capacity	20	14	8	9.5	11	12.5	14	15.5	17	18.5	20	20
Overall demand	27	27	27	27	27	27	27	27	27	27	27	27
Capacity deficit	7	13	19	17.5	16	14.5	13	11.5	10	8.5	7	7
Scenario 3												
Operational capacity	20	20	20	20	20	20	20	20	20	20	20	20
Overall demand	18	18	18	18	18	18	18	18	18	18	18	18
Capacity deficit	−2	−2	−2	−2	−2	−2	−2	−2	−2	−2	−2	−2

All models were implemented using the Java optimization modeling library of the IBM ILOG Concert Technology. The experiments were solved with ILOG CPLEX 12.6, on an Intel Core i7-2640M PC with 2.8 GHz processors and 7.88 GB RAM.

5.2 *Solution Plausibility*

We solved model $\text{CFLP}_{\text{risk}}$ using the data above presented and setting a target service level $\alpha^0 = 0.95$. The optimal solution has value 3147 and calls for opening both facilities and for buying expansion options also for both locations ($y_i = z_i = 1$, $i = 1, 2$). The second-stage decision is more involved since it requires determining a course of action for each possible scenario. In Table 3, we present for each scenario (i) the expansion decisions for facility 1 ($w_{1t\ell s}$), (ii) the expansion decisions for facility 2 ($w_{2t\ell s}$), (iii) the unsupplied demand ($\sum_{j\in J} u_{jt1}$), and (iv) the service level (α_s).

The results are not surprising: capacity expansions are decided for scenarios 1 and 2. These capacity options are executed in the periods exhibiting capacity deficit, and the type of expansion is not over-dimensioned compared to the extra capacity required. It is also interesting to see how this solution differs from a solution to the simplified model $\text{CFLP}_{\text{risk}}(\hat{y}, \hat{z})$. The corresponding second-stage decision as well as the unsupplied demand and the scenario service levels are presented in Table 4 ("N/A" stands for "not applicable").

We realize that when capacity reduction is averaged, we observe no extra capacity being installed although it would have been necessary in scenario 2. This is not surprising since we know that averaging stochastic parameters leads often

Table 3 Detailed solution obtained when solving $\text{CFLP}_{\text{risk}}$

	Time period												Service level
	1	2	3	4	5	6	7	8	9	10	11	12	
Scenario 1													
$w_{1t\ell 1}$			2	2	2	2							
$w_{2t\ell 1}$			5	5	2	2	2						
$\sum_{j \in J} u_{jt1}$		1	2.0		0.5								
α_1													0.98
Scenario 2													
$w_{1t\ell 1}$	5	5	5	10	5	5	5	5	5	2	5	2	
$w_{2t\ell 1}$	2	5	10	5	10	10	5	5	5	5	2	5	
$\sum_{j \in J} u_{jt1}$		3	4	2.5	1		3	1.5		1.5			
α_2													0.95
Scenario 3													
$w_{1t\ell 1}$													
$w_{2t\ell 1}$													
$\sum_{j \in J} u_{jt1}$													
α_3													1.00

Table 4 Detailed solution obtained when solving $\text{CFLP}_{\text{risk}}(\hat{y}, \hat{z})$

	Time period												
	1	2	3	4	5	6	7	8	9	10	11	12	
Scenario 1													
$w_{1t\ell 1}$	N/A	N/A	N/A	N/A	N/A	N/A	N/A	N/A	N/A	N/A	N/A	N/A	
$w_{2t\ell 1}$			10	5	5	2							
$\sum_{j \in J} u_{jt1}$		1		2.75	0.5	1.25	1.0						
α_1													0.95
Scenario 2													
$w_{1t\ell 1}$	N/A	N/A	N/A	N/A	N/A	N/A	N/A	N/A	N/A	N/A	N/A	N/A	
$w_{2t\ell 1}$	10	10	10	10	10	10	10	10	10	10	10	10	
$\sum_{j \in J} u_{jt1}$		3	9	7.5	6	4.5	3	1.5					
α_2													0.95
Scenario 3													
$w_{1t\ell 1}$	N/A	N/A	N/A	N/A	N/A	N/A	N/A	N/A	N/A	N/A	N/A	N/A	
$w_{2t\ell 1}$													
$\sum_{j \in J} u_{jt1}$													
α_3													1.00

to an underestimation of uncertainty (Savage, 2012). Note also that the objective function we are considering strives for cost minimization and (only) punishes the non-achievement of the target service level. Therefore, solutions certainly contain unsatisfied demand.

Observing the solutions obtained, we conclude that the risk-aware model seems to better hedge against risk.

5.3 The Relevance of Capturing Uncertainty

We adopt the two measures discussed in Sect. 4: the VSS and the EVPI. Our analysis aims at understanding how "uncertainty" influences those measures. We designed an experiment consisting of 100 instances of the problem that differ in the 3 scenarios considered for demand and operating capacity. In particular, for each instance, the variability of the values associated with the demand and overall capacity change.

The base value for the overall demand is 6 (2 units per customer). As before, we assume that a disruption occurs at the end of period 2 which is fully recovered in period 11. Then, for each combination of the steps $\delta_d = 1, \ldots, 10$ and $\delta_c = 1, \ldots, 10$, we define three scenarios, each with probability $\frac{1}{3}$ as described next.

Scenario 1 corresponds to an increase in demand and a decrease in capacity:

total demand = $6 + \delta_d \times 0.5$,
total operating capacity after disruption (period 3) = $20 \times (0.5 - \delta_c \times 0.05)$.

Scenario 2 corresponds to the average scenario:

total demand = 6,
total operating capacity after disruption (period 3) = 20×0.5.

Scenario 3 corresponds to a decrease in the demand and an increase in the capacity:

total demand = $6 - \delta_d \times 0.5$,
total operating capacity after disruption (period 3) = $20 \times (0.5 + \delta_c \times 0.05)$.

In total, we have 100 combinations of the steps δ_d and δ_c. The larger the value of δ_d (δ_c), the larger the variability associated with the overall demand (initial capacity reduction). Now, we set the operating costs of the facilities equal to 1000 and 1500, respectively, for facilities 1 and 2. This led to the results depicted in Table 5a.

Observing this table, we conclude that the values greater than zero for the EVPI and VSS concentrate in the lower left corner of the table. We note that a darker cell indicates a larger value for EVPI. This indicates that the use of a stochastic programming model is especially adequate when we expect high demand variability. As the value of the stochastic solution reveals, considering all potential developments of uncertain parameters rather than averaging out the stochasticity leads to improved results.

Also in Table 5a, we observe positive values for the EVPI in the upper right corner. This indicates that the stochastic programming approach might also be suitable if variability of the capacity after disruption increases.

Table 5 Evolution of the EVPI and VSS

(a) Facility operating costs: (1000, 1500)

		Capacity steps																			
		1		2		3		4		5		6		7		8		9		10	
		EVPI	VSS	EVPI	VSS	EVPI	VSS	EVPI	VSS	EVPI	VSS	EVPI	VSS	EVPI	VSS	EVPI	VSS	EVPI	VSS	EVPI	VSS
Demand steps	1	0	0	0	0	0	0	0	0	0	0	0	0	0	0	0	0	34	0	103	0
	2	0	0	0	0	0	0	0	0	0	0	0	0	0	0	0	0	0	0	0	0
	3	0	0	0	0	0	0	0	0	0	0	0	0	0	0	0	0	0	0	0	0
	4	5	0	0	0	0	0	0	0	0	0	0	0	0	0	0	0	0	0	0	0
	5	160	0	67	0	47	0	27	0	7	0	0	0	0	0	0	0	0	0	0	0
	6	411	0	313	0	179	0	109	0	73	0	48	0	42	0	29	0	28	0	21	0
	7	576	146	510	108	496	0	362	0	196	0	130	0	110	0	89	0	70	0	51	0
	8	577	498	578	391	511	374	511	244	511	114	495	0	332	0	202	0	132	0	113	0
	9	580	894	580	800	579	694	513	683	513	560	513	437	513	313	476	193	473	73	422	0
	10	581	1240	581	1153	581	1058	581	968	515	963	515	845	515	728	515	610	478	496	475	382

(b) Facility operating costs: (500, 600)

		Capacity steps																			
		1		2		3		4		5		6		7		8		9		10	
		EVPI	VSS	EVPI	VSS	EVPI	VSS	EVPI	VSS	EVPI	VSS	EVPI	VSS	EVPI	VSS	EVPI	VSS	EVPI	VSS	EVPI	VSS
Demand steps	1	0	0	0	0	14	0	27	0	11	0	21	0	34	0	39	0	150	0	232	37
	2	38	0	19	0	0	0	0	0	2	0	0	0	0	0	0	0	8	0	33	0
	3	100	0	80	0	71	0	51	0	31	0	10	0	0	0	0	0	0	0	0	0
	4	131	41	124	28	114	18	101	11	91	1	78	5	54	0	42	0	22	0	9	0
	5	173	155	136	98	133	81	126	68	116	58	103	51	93	41	134	0	114	0	80	0
	6	177	401	174	305	174	171	138	138	131	109	121	95	118	91	114	82	196	0	188	0
	7	242	646	176	608	176	487	176	353	139	223	136	161	133	145	123	133	216	22	212	5
	8	245	997	244	891	178	874	178	744	178	614	178	484	141	357	134	234	128	171	228	51
	9	246	1394	246	1301	246	1194	179	1184	179	1060	179	937	179	813	143	693	139	573	212	376
	10	247	1740	247	1653	248	1558	247	1468	181	1463	181	1346	181	1228	181	1110	144	996	141	882

(continued)

Table 5 (continued)

(c) Facility operating costs: (100, 200)

		Capacity steps																			
		1		2		3		4		5		6		7		8		9		10	
		EVPI	VSS	EVPI	VSS	EVPI	VSS	EVPI	VSS	EVPI	VSS	EVPI	VSS	EVPI	VSS	EVPI	VSS	EVPI	VSS	EVPI	VSS
Demand steps	1	0	0	0	0	0	0	0	0	25	0	66	168	67	370	67	621	79	869	79	1119
	2	0	0	0	0	0	0	0	0	0	0	67	18	67	244	67	472	66	726	66	1001
	3	0	0	0	0	0	0	0	0	0	0	0	0	67	93	67	342	67	585	67	848
	4	0	0	0	0	0	0	0	0	0	0	0	0	19	0	66	179	67	456	67	728
	5	20	0	0	0	0	0	0	0	0	0	0	0	0	0	67	44	67	289	67	604
	6	24	0	21	0	21	0	0	0	0	0	0	0	0	0	0	0	67	167	67	436
	7	89	81	23	0	23	0	23	0	0	0	0	0	0	0	0	0	66	22	66	324
	8	92	284	91	84	25	0	25	0	25	0	25	0	0	0	0	0	0	0	75	185
	9	93	486	93	285	93	87	27	0	27	0	27	0	27	0	0	0	0	0	60	0
	10	95	676	95	486	95	288	95	95	28	0	28	0	28	0	28	0	0	0	0	0

We, therefore, conducted two additional experiments with modified facility operating costs: 500 and 600 for facilities 1 and 2, respectively, and 100 and 200, respectively, for facilities 1 and 2 as we had initially set. The results are presented in Tables 5b and c.

When decreasing the facility operating costs, the variability of the overall capacity after disruption becomes more relevant with respect to EVPI and VSS. Accordingly, for smaller facility operating costs, even if capacity is not expensive, the larger the variability observed, the more relevant the stochastic model. Once capacity is installed, the range of cost-minimizing actions in the presence of disruptions is limited, because the execution costs of expansion options are comparably small. In this situation, the application of stochastic programming compared to deterministic models seems to be less advantageous. If we consider the installation of warehouses, which is much cheaper than the opening of production facilities, demand uncertainty/variability, though it might be small, is intensified by capacity uncertainty/variability. Hence, the uncertainty/variability in both parameters motivates embedding stochasticity in the planning process. On the other hand, the installation of distribution centers or small regional warehouses is even cheaper. Thus, the allocation of capacity and capacity options has a wide range of possibilities.

Summing up, we conclude that the input data, especially the cost structure, has a clear influence in the VSS and EVPI.

5.4 *The Value of a Risk-Aware Solution*

We focus now on the relevance of capturing supply chain risk. With the purpose of making a clear distinction between operating a facility and expanding its capacity, we consider operating costs equal to 1000 and 1500 for facilities 1 and 2, respectively. Furthermore, we weigh the unsatisfied demand by parameter h, which is set equal to 50. Additionally, we set the target service level $\alpha^0 = 0.95$.

In Sect. 4, we derived an appropriate model to solve in case we know the occurring scenario, s: model MPSCFLP_s. We can solve this model for each scenario. Moreover, we can still look at the optimal solution obtained by model $\text{CFLP}_{\text{risk}}$. A comparison between solutions is provided in Fig. 2.

Model $\text{CFLP}_{\text{risk}}$ yields a solution that calls for opening both facilities and buying capacity options at both sites (this is highlighted by additional circles around the facility locations).

On the other hand, for scenario $s = 1$, model MPSCFLP_s suggests opening a single facility. The decisions determined by this non-risk-aware model reflect the case of a company that has not felt the need to consider supply chain risk. Risk features such as balancing efficiency- and effectiveness-related objectives are not applied for the model formulation. The deterministic solution for scenario 3, therefore, implies that all demands are fulfilled, because that is feasible.

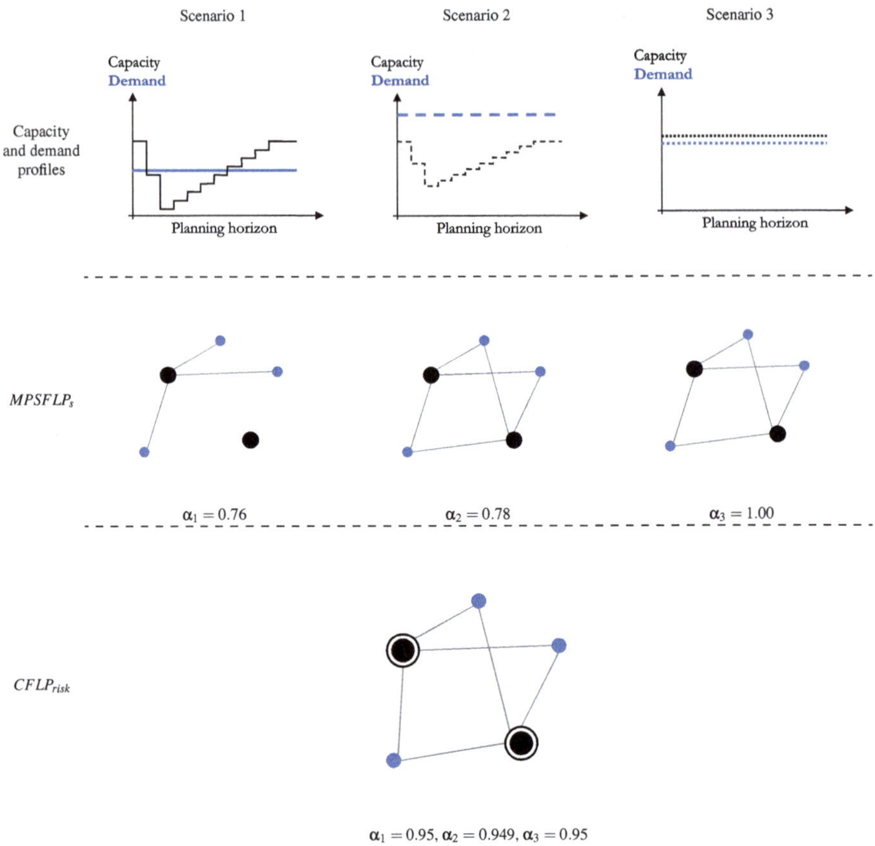

Fig. 2 Comparing the solutions obtained using $CFLP_{risk}$ and $MPSCFLP_s$ $(s = 1, 2, 3)$

We can now compute the value of a risk-aware solution according to (22) (the required information is depicted in Table 6): $\text{VSCR} = 0.1 \times 2725 + 0.3 \times 6100 + 0.6 \times 2500 - 3147 = 455.5$.

For the aforementioned business environment, the value of a risk-aware solution is positive and motivates the discussion presented in this chapter.

6 Conclusions

The analysis presented in this chapter aimed at finding risk-aware decisions for supply network design that adequately overcome the *information gap*. This was accomplished using the so-called prescriptive analytics.

Table 6 Detailed costs for the prototype instance with the optimal values necessary for computing the value of the risk-aware solution

$CFLP_{risk}$				
Facility costs	Scenario 1	Scenario 2	Scenario 3	
Operation	2500	2500	2500	
Expansion	450	1380	300	
Total	2950	3380	2800	
Optimal value				3147
$MPSCFLP_s$				
Facility costs	Scenario 1	Scenario 2	Scenario 3	
Operation	1000	2500	2500	
Expansion	0	0	0	
Total	1000	2500	2500	
Optimal value	2725	6100	2500	

We defined supply chain risk in a very precise way. By doing so, we are able to devise a hierarchy for the core elements underlying that new definition and eventually lay the foundations for what we can now call *supply chain risk analytics*.

The above discussion was complemented by showing that the new concepts can be operationalized. The development of an optimization model capturing the new risk definition also led to a new concept: value of supply chain risk consideration. Finally, we used a prototype example to show that the operationalization proposed is fully consistent with the new concepts developed.

The model presented in this chapter can (and should) be complemented by some follow-ups. It is true that the objective function considered in our optimization model can capture different attitudes toward financial risk. Nevertheless, they are all somehow risk-neutral since we are considering the expected value of the future assets. It would be interesting to specifically study other risk attitudes, namely, in terms of risk aversion.

The optimization model proposed is actually a mathematical structure that requires further study. In fact, the problem we consider is at the core of many supply chain network design problems, and thus having efficient tools for solving such a problem may be relevant especially if large instances need to be solved. In that case, we may have to resort to heuristic approaches.

Acknowledgments The authors would like to thank DAAD and National Funding from FCT-Fundação para a Ciência e Tecnologia, Portugal, under the project UIDB/04561/2020.

References

Aghezzaf, E. (2005). Capacity planning and warehouse location in supply chains with uncertain demands. *Journal of the Operational Research Society, 56*, 453–462.

Ang, E., Iancu, D. A., & Swinney, R. (2017). Disruption risk and optimal sourcing in multitier supply networks. *Management Science, 63*, 2397–2771.

Asbjørnslett, B. E. (2009). Assessing the vulnerability of supply chains. In: G. A. Zsidisin & B. Ritchie (Eds.), *Supply chain risk, International Series in Operations Research & Management Science* (vol. 124, pp. 15–33). Springer, Boston, MA

Aven, T. (2016). Risk assessment and risk management: review of recent advances on their foundation. *European Journal of Operational Research, 253*, 1–13.

Baryannis, G., Validi, S., Dani, S., & Antoniou, G. (2019). Supply chain risk management and artificial intelligence: state of the art and future research directions. *International Journal of Production Research, 57*, 2179–2202.

Behdani, B. (2013). Handling disruptions in supply chains: An integrated framework and an agent-based model. PhD thesis, Technische Universiteit Delft, The Netherlands

Birge, J., & Louveaux, F. (2011). *Introduction to Stochastic Programming* (2nd edn.). Springer New York.

Correia, I., & Saldanha-da-Gama, F. (2019). Facility location under uncertainty. In: G. Laporte, S. Nickel & F. Saldanha-da-Gama (Eds.), *Location Science* (2nd ed., pp. 185–213). Springer Cham.

Craighead, C. W., Blackhurst, J., Rungtusanatham, M. J., & Handfield, R. B. (2007). The severity of supply chain disruptions: design characteristics and mitigation capabilities. *Decision Sciences, 38*, 131–156.

Cui, T., Ouyang, Y., & Shen, Z. J. M. (2010). Reliable facility location design under the risk of disruptions. *Operations Research, 58*, 998–1011.

Dunke, F., Heckmann, I., Nickel, S., & Saldanha-da-Gama, F. (2016). Time traps in supply chains: Is optimal still good enough? *European Journal of Operational Research, 264*, 813–829.

Fernández, E., & Landete, M. (2019). Fixed-charge facility location problems. In: G. Laporte, S. Nickel & F. Saldanha-da-Gama (Eds.), *Location Science* (2nd ed., pp. 67–98). Springer Cham.

Fleischmann, B., Ferber, S., & Henrich, P. (2006). Strategic planning of BMW's global production network. *Interfaces, 36*, 194–208.

Hahn, G. J., & Kuhn, H. (2012). Value-based performance and risk management in supply chains: a robust optimization approach. *International Journal of Production Economics, 139*, 135–144.

Heckmann, I. (2015). Towards supply chain risk analytics: fundamentals, simulation, and optimization. PhD thesis, Karlsruhe Institute of Technology, KIT, Germany

Heckmann, I., & Nickel, S. (2019). Location logistics in supply chain management. In: G. Laporte S. Nickel & F. Saldanha-da-Gama (Eds.), *Location Science* (2nd ed., pp. 453–476). Springer Cham.

Heckmann, I., Comes, T., & Nickel, S. (2015). A critical review on supply chain risk – definition, measure and modeling. *Omega, 52*, 119–132.

Hugo, A., & Pistikopoulos, E. N. (2005). Environmentally conscious long-range planning and design of supply chain networks. *Journal of Cleaner Production, 13*, 1471–1491.

Ivanov, D. (2018). *Structural dynamics and resilience in supply chain risk management.* Springer Cham.

Julka, N., Baines, T., Tjahjono, B., Lendermann, P., & Vitanov, V. (2007). A review of multifactor capacity expansion models for manufacturing plants: searching for a holistic decision aid. *International Journal of Production Economics, 106*, 607–621.

Klose, A., & Drexl, A. (2005). Facility location models for distribution system design. *European Journal of Operational Research, 162*, 4–29.

Ko, H. J., & Evans, G. W. (2007). A genetic algorithm-based heuristic for the dynamic integrated forward/reverse logistics network for 3PLs. *Computers & Operations Research, 34*, 346–366.

Kochenderfer, M. J., Wheeler, T. A., & Wray, K. H. (2022). *Algorithms for decision making*. MIT Press.

Lockamy, A. I., & McCormack, K. (2010). Analysing risks in supply networks to facilitate outsourcing decisions. *International Journal of Production Research, 48*, 593–611.

Lynch, G. (2012). Supply chain risk management. In: H. Gurnani, A. Mehrotra, & S. Ray (Eds.), *Supply chain disruptions: theory and practice of managing risk* (pp. 319–337). Springer, London, UK.

Manuj, I., & Mentzer, J. T. (2008). Global supply chain risk management strategies. *International Journal of Physical Distribution & Logistics Management, 38*, 192–223.

Melnyk, S. A., Rodrigues, A., & Ragatz, G. L. (2008). Using simulation to investigate supply chain disruptions. In: G. A. Zsidisin & B. Ritchie (Eds.), *Supply chain risk: a handbook of assessment, management, and performance* (pp. 103–122). Springer, Boston, MD.

Melo, M. T., Nickel, S., & da Gama, F. S. (2009). Facility location and supply chain management—a review. *European Journal of Operational Research, 196*, 401–412.

Miranda, P. A., & Garrido, R. A. (2009). Inventory service-level optimization within distribution network design problem. *International Journal of Production Economics, 122*, 276–285.

Munir, M., Jajja, M. S. S., Chatha, K. A., & Farooq, S. (2020). Supply chain risk management and operational performance: The enabling role of supply chain integration. *International Journal of Production Economics, 227*, 107667.

Nickel, S., & Saldanha-da-Gama, F. (2019). Multi-period facility location. In: G. Laporte, S. Nickel & F. Saldanha-da-Gama (Eds.), *Location Science* (2nd ed., pp. 303–326). Springer Cham.

Nickel, S., Saldanha-da-Gama, F., & Ziegler, H. P. (2012). A multi-stage stochastic supply network design problem with financial decisions and risk management. *Omega, 40*, 511–524.

Norrman, A., & Jansson, U. (2004). Ericsson's proactive supply chain risk management approach after a serious sub-supplier accident. *International Journal of Physical Distribution & Logistics Management, 34*, 434–456.

Osadchiy, N., Gaur, V., & Seshadri, S. (2016). Systematic risk in supply chain networks. *Management Science, 62*, 1755–1777.

Risk Response Network. (2011). New models for addressing supply chain and transport risk. Tech. rep. The World Economic Forum.

Sarykalin, S., Serraino, G., & Uryasev, S. (2008). Value-at-risk vs. conditional value-at-risk in risk management and optimization. In: *Tutorials in operations research, INFORMS* (pp. 270–294).

Savage, S. L. (2012). *The flaw of averages: why we underestimate risk in the face of uncertainty*. John Wiley & Sons.

Sheffi, Y. (2005). *The resilient enterprise: overcoming vulnerability for competitive advantage*. Cambridge: The MIT Press.

Sheffi, Y., & Rice, B. (2005). A supply chain view of the resilient enterprise. Tech. rep. MIT Sloan Management review. http://sloanreview.mit.edu/article/.

Snyder, L. V., Atan, Z., Peng, P., Rong, Y., Schmitt, A. J., & Sinsoysal, B. (2016). OR/MS models for supply chain disruptions: a review. *IIE Transactions, 48*, 89–109.

Troncoso, J. J., & Garrido, R. A. (2005). Forestry production and logistics planning: an analysis using mixed-integer programming. *Forest Policy and Economics, 7*, 625–633.

Wagner, S., & Bode, C. (2008). An empirical examination of supply chain performance along several dimensions of risk. *Journal of Business Logistics, 29*, 307–325.

Xu, N., & Nozick, L. (2009). Modeling supplier selection and the use of option contracts for global supply chain design. *Computers & Operations Research, 36*, 2786–2800.

Designing for Resilience and Protection

Richard L. Church

Abstract Disasters can come in many forms ranging from natural events like floods, pandemics, and earthquakes to human-based disasters like terrorism and accidents like nuclear power plant failures. Because significant disasters do not happen very often, most system planners become complacent about risks, such as major floods and earthquakes. But when one does happen, such as in the case of the Tōhoku earthquake in Japan or the Metcalf Substation attack in California, it is easy to tell when the resilience of a system is inadequate. The value of advanced facility planning and modeling can help reduce the likelihood of incurring substantial losses when disaster strikes. Since the terrorist attack of 9/11, greater attention has been directed to studying facility system vulnerability as well as how a system might be enhanced to have a high degree of resilience. In this chapter, several different areas of facility system design and location are reviewed with the goal of demonstrating some major achievements in this fertile research area as well as outlining specific areas of research need.

Keywords Resilient design · Worst-case disruption · Bi-level and tri-level optimization · p-median location problem

1 Introduction

Location science has evolved since the 1960s with the development of relatively simple models, like the location set covering problem (Toregas et al., 1971) and the fixed charge plant location problem (Balinski, 1965), where the principal component was the selection of one or more sites for facility placement (ReVelle et al., 2008). Now problems may include many different components and decisions, like interacting hubs (Campbell & O'Kelly, 2012), integrated vehicle routing (Balakrishnan et

R. L. Church (✉)
University of California, Santa Barbara, CA, USA
e-mail: church@geog.ucsb.edu

H. A. Eiselt, V. Marianov (eds.), *Uncertainty in Facility Location Problems*, International Series in Operations Research & Management Science 347,
https://doi.org/10.1007/978-3-031-32338-6_8

al., 1987), and integrated transport system design (station locations, routes, rolling stock, and schedules) (Repolho et al., 2016). The notion of uncertainty has also been addressed in location design problems, starting with the stochastic nature of travel (Mirchandani & Odoni, 1979), different future scenarios (Daskin et al., 1997; Snyder & Daskin, 2006), and even the availability or reliability of facilities (Lim et al., 2010; Snyder & Daskin, 2005) and their servers (ReVelle & Hogan, 1989; Marianov & ReVelle, 1994, 1996). Other important features have included the simultaneous selection of locations and the allocation of different types of service (Schilling et al., 1979), systems of interrelated activities (Armour & Buffa, 1963; Moore & ReVelle, 1982; Weaver & Church, 1991), and competition (ReVelle, 1986; Eiselt et al., 1993, 2019). A number of these facility location problems have been designed to address uncertainty in facility availability, but these models are often cast to handle the natural variability in demands to be served, the probability of receiving service, and other stochastic elements that fall within the normal range of operation. The goal of this chapter is to describe some of the research that has addressed issues that fall outside of the normal spectrum of events, like an intentional strike to destroy a facility or a catastrophic flood that destroys one of more facilities in a system. The roots of this newer addition to the field of location science began after the terrorist strike of 9/11 when researchers started to address questions such as "What is critical?"; "Are there facilities that we cannot afford to lose in a system?"; and "Can we protect or fortify them?"

2 Background

The terrorist attacks of 9/11 on the World Trade Center in New York City and the Pentagon in Washington, DC, underscored that many important facilities may be quite easily disrupted and even destroyed. A case in point is the terrorist strike at the Metcalf Transmission Substation on April 16, 2013, near San Jose, California. Gunmen armed with AK-47 rifles shot at the substation and knocked out 17 large transformers in less than 20 min and escaped before police arrived. This substation is a key asset in the electrical grid supplying Silicon Valley. It took a concerted effort on the part of the electric company and the Independent Service Operator of the grid in California to avert a complete blackout. If that had occurred, it is estimated that restoring the grid in Silicon Valley might have taken months to accomplish. Subsequent work has shown that a simultaneous strike on 9 key substations across the USA could possibly sink the entire US electrical grid. To date none of the perpetrators have been caught or even identified. In response to this near disaster, the USA has focused on building a strategic stockpile of large transformers and requiring utilities to strengthen major substation facilities.

Another example of a disaster is the Tōhoku earthquake and tsunami of March 11, 2011, that killed more than 20,000 people and disrupted a significant portion of the northeast coast of Japan. Among the casualties was the Fukushima Daiichi nuclear plant. This power plant was protected by a sea wall that was too low to stop the Tsunami. The plant was flooded and operators lost control of the reactors due to the fact that the backup power generators were housed in the basement that was flooded which rendered the generators useless. Without power, the control system could not safely shut down the reactors, causing a meltdown. This particular event also underscores the need to fully assess risks and possible impacts to systems of facilities. The fact that advanced planning did not identify or quantify the risks of backup power placement underscores the need to use decision theory and other classical risk assessment techniques as well as to develop new models to quantify possible disruption.

Grubesic and Murray (2006) have discussed the fact that a failure of one system can lead to a failure of other systems, which they called a system of cascading failures. For example, the Tōhoku earthquake disrupted a number of auto part suppliers to Toyota. This disaster was so severe it took six months for Toyota to get global production back to normal levels (Shirouzu, 2021). Because of this major disruption Toyota initiated the development of a business continuity plan which was based upon knowing the source of every part and component in their supply chain. This included knowing the suppliers of every component of every product purchased by Toyota for their vehicles. The supply chain network database was so detailed that news of any disruption involving any company in their supply network could be quickly assessed as to the impact to Toyota or any of its prime suppliers. Microchips were one of the components that Toyota recognized as so critical to their production that the company forced their suppliers to keep a years' worth of inventory to meet Toyota's needs. This strategic policy has helped Toyota to maintain high levels of production during the current global shortage in microchips. Of course, such a strategic decision should be made on a cost-benefit basis, where the expected costs of inventory and advanced manufacturing of the microchips must be less than the estimated costs of delay in revenues from auto sales, the costs of disruption in manufacturing, and the impacts of delivery schedules and other supplied parts. This shortage is something that its competitors have been significantly hampered with during the COVID-19 pandemic of 2020–2022.

Another recent example of a natural disruption includes hurricane Ida of August, 2021, that disrupted major facilities like refineries and resin and plastics manufacturing in the Southeast USA. This disruption has cascaded through American industry, causing shortages of paint and other products based on these refined products. Disruptions can also be accidental like the 2005 chlorine spill in Graniteville, South Carolina, and the railroad bridge fire in Sacramento, California (Peterson & Church, 2008). Whatever the cause, it is important to address such possibilities in facility planning and location whether such events are intentional, natural, or accidental. Some issues are easily mitigated like keeping facilities out of flood plains, keeping backup generators in safe locations, and protecting facility perimeters. Other disruptions can be mitigated by providing backup facility capacity, holding larger inventories, and even fortifying buildings and roadways against earthquakes, attackers, and floods. That is, there are a great number of options in providing some degree of protection/fortification. There are also a number of options that can be used for the analysis of a system and location risk. For example, decision trees can be developed to represent possible disruptive events, their probabilities of occurrence, and even the costs and benefits of certain actions. Simulation can also be used to simulate a system operation under stress or potential disruption patterns given a hurricane scenario or a possible earthquake at a known location (e.g., along the San Andreas fault in California) as examples. The principal emphasis of this chapter is on identifying worst-case impacts to an existing system, optimizing protection resources to thwart as best as possible these maximally disruptive events, and even designing (by location) a system so that it is as resilient as possible to disruption. However, some attention is given to other disruptive events that are not necessarily maximally disruptive.

Before delving into the modeling of facility system disruption, it is important to underscore that this is part of a larger literature that has evolved in disaster/emergency management. Disaster management involves a number of possible options: (1) risk analysis of possible disruptions and costs; (2) strategic decision making on possible mitigation measures (measures that could reduce or even eliminate a potential major disruption); and (3) how to operate and respond after a major disruption with a damaged system or even a regional catastrophe like a famine or flood. Examples of advanced planning, protection, and response include humanitarian relief (Özdamar & Ertem, 2015), emergency warning systems (Murray et al., 2008; Mathews et al., 2017), evacuation modeling (Lindell, 2008; Cova & Church, 1997), shelter location (Jin et al., 2021; Dekle et al., 2005), designing and strengthening communication systems (Lei et al., 2019; Eiselt & Marianov, 2012; Nicholas & Alderson, 2015), and protecting electrical grids (Yuan & Zeng, 2020; Alguacil et al., 2014), among many others.

3 Initial Developments: Optimizing Disruption

Wollmer (1964) was the first to address disruption within the context of a military supply line. The objective was to strike arcs and render then incapable of being used while minimizing the resulting capacity of the network to transport material from an origin to a destination. Wollmer (1970) expanded this work on interdiction with several algorithms that targeted communication networks. Slater (1982) was one of the first to consider elements of a network to serve as a facility which was expanded on by Current et al. (1985). In essence, one can consider a facility to be represented as a point (Hakimi, 1964), a path (Current et al., 1985), a tree (Hutson & ReVelle, 1993), or some other connected portion of a network (e.g., a tour; see Current & Schilling, 1989). The original work in interdiction was oriented to flow paths and shortest paths on a network like the work of Wollmer. Church et al. (2004) were the first to consider the interdiction of point-based facilities. They started their work within the context of the p-median facility location problem.

The p-median problem is defined as follows (Hakimi, 1964, 1965):

> Find the p-positions on the network that minimize the total weighted distance of demand.

Consider a network of nodes and arcs. It is assumed that the demands are distributed among the nodes of the network and that facilities could serve all demand assigned to them. Since there are no capacity issues, then each demand is served by their closest facility. The weighted distance of serving a given demand is calculated as the demand weight multiplied by the distance that the demand is from its closest facility. Total weighted distance is the sum of all weighted distance assignments. Hakimi (1965) proved that at least one optimal solution to this problem consisted entirely of a subset of nodes of the network. Because of this groundbreaking theorem, researchers have concentrated on finding the best subset of p-nodal locations in solving this problem. But let us suppose that we have an existing set of operating facilities, where the service protocol is based upon the p-median problem, where facilities are not capacitated and demand can always be served by their closest operating facility. The natural question that arises when considering the possibility of interdiction is are some facilities more critical to system operation than others, and if so, which ones are? In an attempt to address this question, Church et al. (2004) posed the following problem:

> Of the p different locations of supply, find the subset of r facilities, which when removed, yields the highest level of weighted distance

They called this the r-interdiction median (RIM) problem. It represents the objective of an intelligent attacker, an agent that attempts to maximize harm or damage to a system. One can think that if an attack is made, then the system operator would respond by reallocating service to the remaining facilities so that we can formally call this an "attacker-operator" or "attacker-defender" model where after the attack and the loss of r-facilities, the operator reassigns those demands that have lost their service facility to the closest remaining facilities. This problem could be posed as a bi-level optimization problem, but the operator's demand reassignment can easily be incorporated into the same level as the attacker, thereby creating a simple one level optimization problem. To formulate this model, consider the following notation:

i=an index used to refer to demand locations, where I is the set of all demands
j=an index used to refer to demand locations, where F is the set of current facility locations
d_{ij}=the shortest distance between nodes i and j

$$s_j = \begin{cases} 1, \text{ if a facility located at } j \text{ is eliminated by interdiction} \\ 0, \text{ otherwise} \end{cases}$$

$$x_{ij} = \begin{cases} 1, \text{ if demand } i \text{ assigns to a facility at } j \text{ after interdiction} \\ 0, \text{ otherwise} \end{cases}$$

$T_{ij} = \{k \in F \mid k \neq j \text{ and } d_{ik} > d_{ij}\}$, the set of existing sites (not including j) that are as far or farther than j is from demand i

We can now formulate the r-interdiction median (RIM) problem as the following integer-programming problem (Church et al., 2004):

$$\text{RIM : Max } Z = \sum_{i \in I} \sum_{j \in F} a_i d_{ij} x_{ij} \tag{1}$$

Subject to:

$$\sum_{j \in F} x_{ij} = 1 \text{for each } i \in I \tag{2}$$

$$\sum_{j \in F} s_j = r \tag{3}$$

$$\sum_{k \in T_{ij}} x_{ik} \leq s_j \text{ for all } i \in I \text{ and for all } j \in F \tag{4}$$

$$s_j \in \{0, 1\} \text{ for all } j \in F \tag{5}$$

$$x_{ij} \in \{0, 1\} \text{ for all } i \in I \text{ and for all } j \in F \tag{6}$$

The objective (1) involves maximizing the weighted distance associated with assigning each demand to their closest open facility after interdiction. This means that the interdictor seeks the most harm to system operation, by attacking and destroying *r*-facilities out of *p*-facilities. Constraint (2) maintains that each demand must assign to a facility after interdiction. Constraint (3) establishes that exactly *r*-facilities are to be eliminated. Constraints (4) maintain that each demand must assign to their closest open facility after interdiction. This constraint restricts assignment of demand *i* to a site that is farther than facility *j* is to demand *i* unless facility *j* has been interdicted. Altogether, constraints (4) specify that demand *i* will assign to the closest remaining facility. The form of these constraints follows that of Hanjoul and Peeters (1987) and Church and Cohon (1976). Constraints (5) and (6) restrict the decision variables to be zero or one in value. However, the restrictions on the assignment variables are not necessary as they will be integer in value as long as the S_j values are integer. There are alternatives to formulating the closest assignment constraints (4), which have been explored in Scaparra and Church (2008).

Figure 1 presents results from applying the RIM model to an optimal 5 facility *p*-median solution involving the ReVelle–Swain (1970) dataset where the level of interdiction was $r = 2$ facilities. The weighted distance before interdiction is 2950. The weighted distance after interdiction is 6124, a substantial increase of weighted distance. The X's in the figure denote the facility locations that are interdicted.

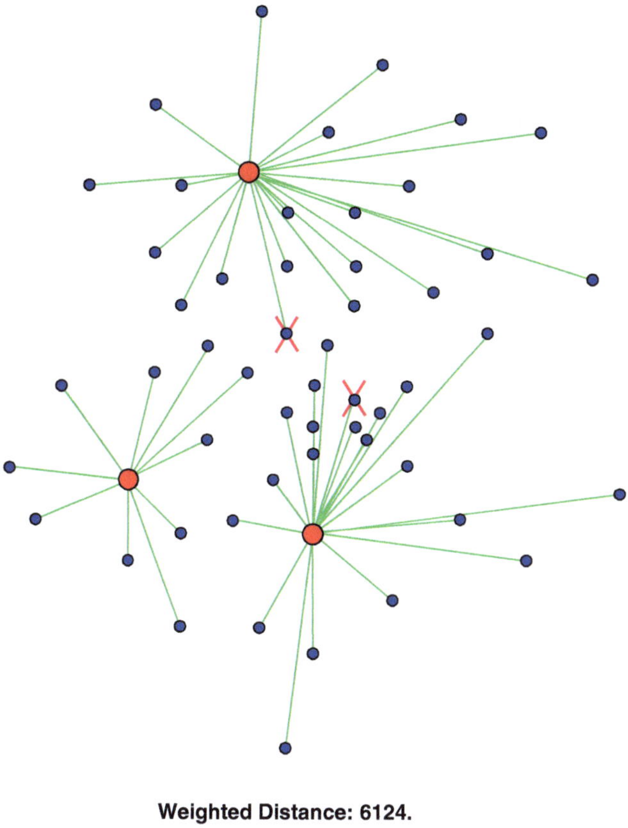

Fig. 1 An optimal 5-median solution with weighted distance of 2950 suffers worst-case loss of 2 facilities at nodes 1 and 3 resulting in a substantial increase of weighted distance to 6124. This solution was generated by the RIM model. The X's indicate the facility locations that are interdicted

4 Optimizing Protection

Church et al. (2004) demonstrated that even the elimination of one or two facilities can significantly impact service efficiency. A logical question to ask is what can be done to prevent such losses? That is, can we thwart interdiction? If facilities could be hardened to the extent that an interdictor would choose some other target, then the answer is a simple yes, especially if such fortification is very inexpensive. Fortification measures may be very simple, like building a fortified perimeter, or installing a security system. If resources are somewhat limited and the costs of fortification are high, it may be that only some of the facilities can be hardened or fortified. This is the central issue of the following problem:

Identify the set of q facilities to secure or harden, so that after interdiction, the remaining system operates as efficiently as possible.

The objective would be to use those resources to thwart an attack to the greatest extent possible. This is a form of what is called a "defender-attacker-defender" problem (Brown et al., 2005; Lazzaro, 2016). It represents a 3-level optimization problem. But since the RIM problem can be formulated as a single level optimization model, we can use the RIM model to build a bi-level optimization model to optimize fortification resources. Consider:

$$z_j = \begin{cases} 1, \text{if a facility located at } j \text{ is fortified} \\ 0, \text{otherwise} \end{cases}$$

If fortification or hardening will thwart an attack or will simply deter someone from attacking a facility, then we will assume that an intelligent attacker will choose to hit a different facility. That is, a fortified facility will not be attacked or if it was the attack would not be successful. This can be specified in the following constraint:

$$s_j \leq 1 - z_j \tag{7}$$

The bi-level optimization model is composed of the defender or systems planner deciding which facilities to fortify (represented by the z_j decision variables), followed by the attacker deciding which of the unfortified facilities to attack (represented by the s_j variables):

$$\text{RIMF minimize } H(z) \tag{8}$$

subject to:

$$\sum_{j \in F} z_j = q \tag{9}$$

$$z_j \in \{0, 1\} \text{ for all } j \in F \tag{10}$$

where:

$$H(z) = \max \sum_{i \in I} \sum_{j \in F} a_i d_{ij} x_{ij} \tag{11}$$

subject to:

$$\sum_{j \in F} x_{ij} = 1 \text{for all } i \in I \tag{12}$$

$$\sum_{j \in F} s_j = r \tag{13}$$

$$\sum_{k \in T_{ij}} x_{ik} \leq s_j \text{ for all } i \in I \text{ and for all } j \in F \tag{14}$$

$$s_j \leq 1 - z_j \text{ for all } j \in F \tag{15}$$

$$s_j \in \{0, 1\} \text{ for all } j \in F \tag{16}$$

$$x_{ij} \in \{0, 1\} \text{ for all } i \in I \text{ and for all } j \in F \tag{17}$$

The lower level of this bi-level problem is represented by conditions (11)–(17). This represents the interdictor's decisions to attack, which maximizes the weighted distance that results from the attack, given the decisions of the defender in terms of the facilities that have been fortified. This lower-level problem is the RIM model, with the extra condition that the attacker does not select a facility that has been fortified by the defender [constraint (15)]. Note that the integer restrictions on the x_{ij} variables are given for completeness, but as long as the s_j variables are integer in value, the x_{ij} will be integer as well and only need to be specified as non-negative variables when solving the problem.

The upper level of the model represents the defender's decision to fortify q of the existing facilities. The objective of the defender is to keep the value of $H(z)$ to be as small as possible. The value of $H(z)$ is based upon the response of the interdictor's decision (11) once the fortification plan is set by the defender. That is, any given fortification plan is met with an optimal interdiction plan by the attacker. The objective of the defender then represents finding the fortification plan which results in the lowest weighted distance after interdiction. This type of problem is called a Stackelberg game of a leader and follower, where the follower always responds in an optimal/competitive way to the leader's decisions. It is also important to note that when the number of facilities is small and the number of possible interdictions is also small, it is possible to formulate this problem as a single-level optimization problem (Church & Scaparra, 2007a). There are several strategies that have been developed to solve this specific or related problems (Scaparra & Church, 2008; Lozano & Smith, 2017) as well as solving such problems in general (Brown et al., 2008; Alderson et al., 2011).

The RIMF model was applied to the optimal 5 facility p-median solution that was generated using the ReVelle and Swain dataset (1970) and the solution is presented in Fig. 2. The solution was generated where the level of fortification was 2 facilities and the resulting interdiction was 2 facilities. The weighted distance of the RIM model where $r = 2$ (see Fig. 1) resulted in a weighted distance of 6124. If one optimally fortifies 2 facilities and is then subjected to a worst-case loss of 2 facilities, the resulting weighted distance is 4185. Thus, selective fortification can result in an improved ability to operate after interdiction.

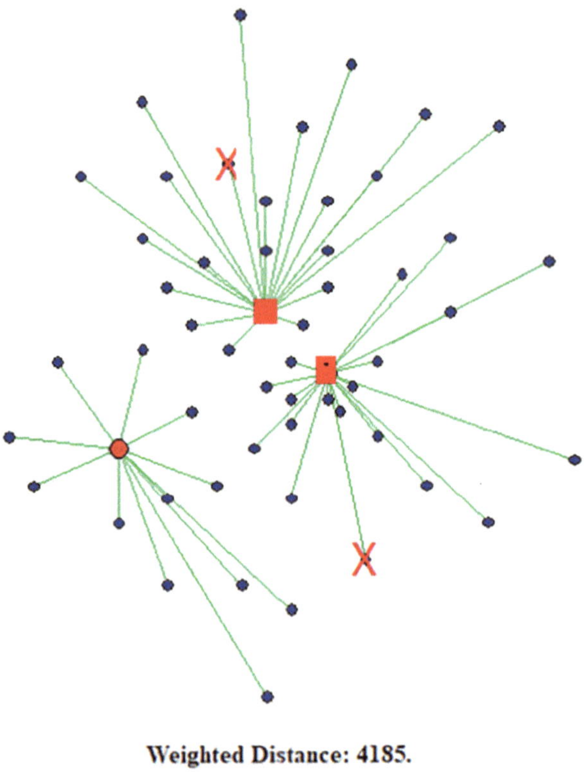

Fig. 2 An optimal fortification of 2 facilities in a five-facility configuration subject to a worst-case loss of 2 facilities results in a weighted distance of 4185. Thus, selected fortification of the facilities in a configuration reduces the impact of interdiction in a substantial way. The solution was generated by RIMF. The X's represent the facilities that are interdicted. The boxes depict the facility locations that are fortified

5 Adding Complexity

What makes the above model of fortification and interdiction somewhat easy to solve is that the problem can be posed as a bi-level optimization problem. Unfortunately, adding a relatively simple component to the problem, e.g., facility capacities, presents an additional level of complexity. The reason for this is that we can easily structure constraints that force assignment to the closest facility that has not been interdicted [see constraints (4) and (13)], but if a facility has a limited capacity to serve demands then a given demand allocation may not be to its closest open facility. Assignment may be to some facility that is farther away as capacity limits may force this to occur, or it may not be possible to serve this demand at all as not enough capacity remains in the system after interdiction. Thus, the problem of reallocating demand after interdiction when facilities have set capacities can only be handled in a three-level optimization problem. To formulate this model, consider the following additional or modified notation:

c_i= the capacity of facility j
t_{ij}= per unit cost for serving customer i from facility j
φ_i= the penalty for not serving customer i (per unit of demand)
u_i= units of demand i that cannot be served after interdiction

We can now formulate a capacitated version of the RIMF model as follows:

$$\text{CRIMF : Min } K(z) \tag{18}$$

subject to:

$$\sum_{j \in F} z_j = q \tag{19}$$

$$z_j \in \{0, 1\} \text{ for all } j \in F \tag{20}$$

where (ML 20–23):

$$K(z) = \max \ H(s) \tag{21}$$

subject to:

$$\sum_{j \in F} s_j \leq r \tag{22}$$

$$s_j \leq 1 - z_j \text{ for all } j \in F \tag{23}$$

$$s_j \in \{0, 1\} \text{ for all } j \in F \tag{24}$$

Where (LL: 24–28):

$$H(s) = \min \sum_{i \in I} \sum_{j \in F} t_{ij} x_{ij} + \sum_{i \in I} \varphi_i u_i \tag{25}$$

$$\sum_{j \in F} x_{ij} + u_i \geq a_i \tag{26}$$

$$\sum_{i \in I} x_{ij} \leq (1 - s_j)\, c_j \text{ for all } j \in F \tag{27}$$

$$x_{ij} \geq 0 \text{ for all } j \in F \text{ and for all } i \in I \quad (28)$$

$$u_i \geq 0 \text{ for all } i \in I \quad (29)$$

The upper level of the problem is represented by the objective (18) and conditions (19) and (20). Here the leader or defender is attempting to minimize the costs of suppling demand after interdiction plus any penalties associated with not meeting specific demands. We have included penalty values, φ_i, to reflect the penalty of not serving specific demands. The reason for this is that keeping the costs or weighted distance of assignment as low as possible would mean that mathematically we should not serve any demand and keep the costs at zero. We must include this penalty to ensure that demands can be met as long as there is capacity left somewhere in the system after interdiction and the costs of supplying a given demand i do not exceed the penalty φ_i . When facilities are not restricted by capacity issues, we assumed as in the RIM model that the closest remaining facility will serve each demand. But when each facility has an associated capacity, interdiction may reduce system capacity to the extent that not all demand can be served, so it is not possible to constrain that that each demand be served after interdiction in constraint (26).

The second or middle level of this 3-level optimization problem represents the attacker, where interdiction resources are allocated [constraint (22)] and interdiction involves only non-fortified facilities [constraint (23)]. The objective of the interdictor is the antithesis of the defender with the objective of maximizing the costs of supply and penalties incurred by the defender [Objective (21)]. However, the interdictor's objective is based upon the response of the defender in optimizing demand allocation in the lower or third level of the problem.

The lower level of bottom level represents the defender's allocation decisions after interdiction, given a fortification plan. This bottom level prevents facilities that have been interdicted to supply any demand [constraint (27)] and defines the amount of demand that has not been served [constraint (26)]. The defender's objective is to respond to the interdictions with the best distribution plan by minimizing distribution costs and penalty costs [objective (25)]. As stated before, solving bi-level and tri-level problems can be a complex task. One of the techniques developed to solve this problem can be found in Scaparra and Church (2012).

The three models described above represent some rather simple forms of facility system interdiction and fortification. They helped form the basis for a growing rich body of work in facility disruption and protection. Given an understanding of these three models, one can add other issues that can be important. For example, we could consider the interdiction of a system that has been designed to cover demands, that is an R-interdiction covering problem (see Church et al., 2004). Another important issue is the fact that a system operator/defender could rebuild or replace a facility. The real issue then becomes how long will an interdiction event continue to degrade a system operation or to what extent a facility has degraded service (Losada et al., 2012) as well as hedging against disruptions (Liberatore et al., 2012).

It is important to underscore the fact that there are many elements of disruptions and risk that should be addressed in planning, design, and operation of facility systems. The rest of this chapter is devoted to three very important research directions. Each of these directions will be described in greater detail starting with the need to compute risk/reliability envelopes of performance for existing facility systems. The second research area that is the development of simple models can be used to optimize/improve system fragility. The third area is that more work should be devoted to the development of facility location models that seek solutions that are inherently resilient, without specific attention to fortification or hardening. We will include a new prototype model for resilient design.

6 Beyond the Basics: Reliability Envelopes

Suppose that we have a system of p-operating facilities servicing a set of demands. Further suppose that this system was designed by the use of the p-median problem, using one of the well-known approaches for this (ReVelle & Swain, 1970; García et al., 2011; Elloumi, 2010; Church, 2008). Let us say that each demand is served periodically by a delivery vehicle. We can measure the overall efficiency of the system in terms of vehicle-miles of travel needed to supply all of the demand. Let us also say that some type of disruption could happen in which one or more facilities may be lost or damaged to the extent that they can no longer provide service without significant repair. This might occur for any number of possibilities ranging from natural disasters to an intentional strike due to a terrorist. If one or more facilities could be inoperable or destroyed it is only natural to ask: what is the resulting impact to system efficiency (i.e., the resulting increase in vehicle-miles of travel).

It makes sense to calculate possible losses of efficiency that may occur over a range of facility losses or impacts. Figure 3 depicts a hypothetical envelope of facility efficiency levels associated with possible facility losses. The x-axis depicts the level of facility losses or closures, which would naturally range from zero to p. The y-axis depicts the range of resulting system efficiencies associated with a given level of facility losses. If all p-facilities operate, then we will assume that the system operates at a level of efficiency of 100%. Any loss to this system will degrade operating efficiency. If all facilities are inoperable, we will define the resulting efficiency as zero percent (0%). However, if one facility is lost due to some reason, then some level of overall efficiency is lost and overall efficiency decreases. One can easily enumerate all possibilities of losing one facility and calculate the loss of efficiency for each of these p-instances. The best outcome is when the least important or critical facility is destroyed, and the worst outcome would happen if the most important or critical facility is lost. Other possible outcomes of lowered efficiency would occur between the best and worst outcomes.

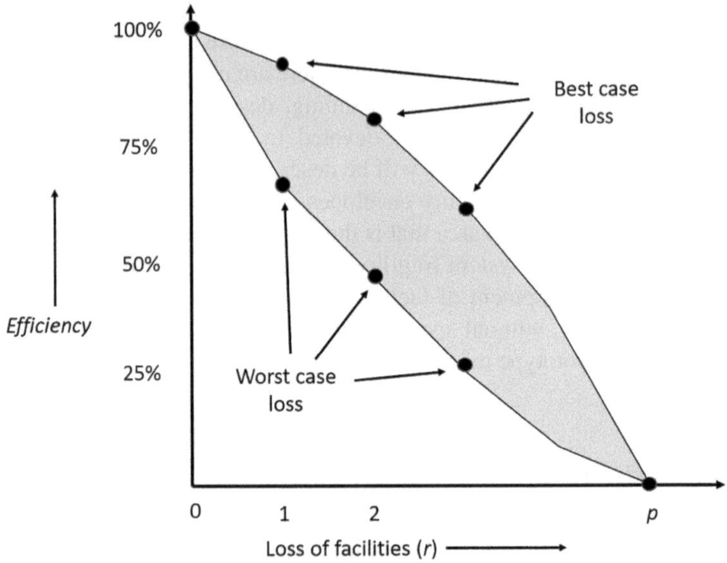

Fig. 3 A hypothetical operational efficiency envelope associated with a system of p-facilities, where r-facilities are rendered inoperable for some reason

In Fig. 3, the gray region is bordered by an upper curve of best-case losses of efficiency and a lower curve that depicts the worst-case losses of efficiency for each value of r. The gray region represents an envelope of possible operational efficiency levels due to possible system disruption and was originally developed by Kim and O'Kelly (2004) in analyzing potential impacts to a communication system. Church and Scaparra (2007b) used the concept to depict the operational region of a set of facilities which may be subject to losses.

The envelope defines the extent that facility interruptions could impact a system operation. There are many possible solutions which fall between the upper curve of best outcomes and the lower curve of worst possible outcomes. This is a simple but informative diagram, which gives managers a good picture of an operation. This is the type of information that could guide managers in making strategic decisions as to whether specific actions, like fortification, should be pursued. It should come as no surprise that the worst-case outcomes can be easily generated by the RIM model described in an earlier section of this chapter. The best-case outcomes must be generated in a different way. In a manner of speaking, the upper curve of best-case outcomes coincides with an intelligent facility closing strategy. That is, if you were to optimally close r facilities, which ones would they be? The answer of course is to close the facilities that have the least impact on system operation in terms of costs, or in this case the least possible impact in terms of an increase of weighted distance (vehicle miles for this example). Consider the following additional/modified notation:

$$x_{ij} = \begin{cases} 1, \text{ if demand } i \text{ assigns to an open facility at } j \\ 0, \text{ otherwise} \end{cases}$$

$$y_j = \begin{cases} 1, \text{ if the existing facility at } j \text{ is kept open} \\ 0, \text{ otherwise} \end{cases}$$

$$p = \text{the number of existing facilities, comprising set } F$$

$$r = \text{the number of facilities to be closed, ranging from 1 to } p - 1$$

Using the notation defined above, we can formulate an optimistic closing (OC) model as the following integer-linear programming problem:

$$\text{OC : Min } Z = \sum_{i \in I} \sum_{j \in F} a_i d_{ij} x_{ij} \tag{30}$$

subject to:

$$\sum_{j \in F} x_{ij} = 1 \text{ for each } i \in I \tag{31}$$

$$\sum_{j \in F} y_j = p - r \tag{32}$$

$$x_{ij} \leq y_j \text{ for each } i \in I \text{ and each } j \in F \tag{33}$$

$$x_{ij} \in \{0, 1\} \text{ for each } i \in I \text{ and each } j \in F \tag{34}$$

$$y_j \in \{0, 1\} \text{ for each } j \in F \tag{35}$$

The above model essentially closes r-facilities by selecting $p{-}r$ facilities to keep open. The choice of which r facilities is based upon minimizing the resulting weighted distance. Constraint (31) specifies that each demand must assign to a facility that remains open. Constraint (32) ensures that r facilities are closed, while keeping $p - r$ facilities open. Constraints (33) ensure that demand assignments are only made to facilities that are kept open. Finally, constraints (34) and (35) represent the integer restrictions on the variables. It can be easily shown that the integer restrictions in constraints (34) can be relaxed to non-negative conditions without impacting identifying optimal integer solutions. It is important to note that this model is a restricted form of the p-median model given in ReVelle and Swain (1970), but there are other ways in which this model can be formulated (see García et al., 2011).

Using the RIM and OC models the bounds of a reliability envelope can be determined. If one decides to fortify a specific set of facilities, then the exercise can be repeated to compute the range of outcomes that are possible for that fortification. Suppose that each facility is subject to a probability of disruption, e.g., the chances that it can be flooded. Then one can use Monte Carlo simulation to generate a frequency distribution of weighted distance values associated with a given probabilistic loss of r facilities. Another issue is that it is possible in that facilities after being impacted by a natural event of even an intentional strike could operate at a reduced level (e.g., operate at a capacity, or at increased cost). This means that not all impacts should be represented by discrete, complete facility closures. This means that the envelope is a cloud of possible outcomes, not just at discrete levels of facility loss. This is a prime area for future research.

7 Beyond the Basics: Simple Approaches to Address Fragility

Virtually all facilities operate within some type of network. Facility functions are contingent on such networks operating. Thus, to ensure lifelines and facility operations, the health of networks is a key issue. Modeling and designing fault-tolerant communication networks has a long history, beginning with Hakimi and others (Hakimi, 1969). Virtually all transport, communication systems, electrical transmission, and pipeline networks should be analyzed in order to identify the range of possible outcomes in terms of the loss of system operability as well as identify strategies in which to lessen those risks and potential damages. For example, a highway may have one bridge that is especially vulnerable to an earthquake or to flooding which might undermine the foundation. Whatever the risk is, it may be that this one component is especially at risk. What if the entire route is useless if that element is damaged? Then, it may be important to ensure that this one component is strengthened or protected so that the risk of losing an important route is substantially reduced. The overall strategy would be to identify the elements that if protected or reengineered could keep lifeline support systems in operation, e.g., water transportation, food, supplies, and communications. Each system needs to be analyzed within this perspective. The transportation system is a critical element in securing many of the lifeline systems (food, medications, personnel) in the event of an emergency so the transport infrastructure should be given a high priority for analysis as well as strengthening.

Designing a transportation system so that it can provide lifeline services, like food and emergency services, as well as support evacuation when needed was proposed by Viswanath and Peeta (2003). Suppose that there exists a region with an existing road network. The network represents roads or highways that connect towns or cities. The major cities represent the origins and destinations of specific services or commodities. The idea is that routes of commodities or services between all major towns should be supported if at all possible. Each town can be thought of as a place of demand or destination. Major supply locations can be represented as origins. Although a route between a given origin/destination pair should be efficient, the route cannot traverse along a given road unless that road has been seismically upgraded to withstand a major earthquake. The idea is to provide at least one hardened supply route for each city of town if at all possible. Viswanath and Peeta optimized road improvements subject to a budget constraint so that as many OD (origin-destination) pairs are supported by a seismically safe route. Each OD pair is represented as a unique commodity type k. They cast this as a two-objective problem. The first objective maximized the population served by access routes. The second objective minimized the transport cost of providing support for each OD provided access. The network is represented as an undirected graph and traffic flow can occur in either direction. Consider the following notation:

$$i, j, m = \text{indices used to represent towns and cities.}$$

$$k = \text{index of commodity or type of service that represents a specific OD pair}$$

$$a_m = \text{the population at center } m$$

$$A = \{(i, j) \mid \text{a road connects towns or cities } i \text{ and } j\}$$

$$E = \{(i, j) \mid \text{road link } (i, j) \text{ needs to be hardened if used}\}$$

$$i, j, m = \text{indices used to represent towns and cities.}$$

$$k = \text{index of commodity or type of service between a specific origin and destination}$$

$$a_m = \text{the population at center } m$$

$$O(k) = \text{the origin node } i \text{ for commodity route } k$$

$$D(k) = \text{the destination node } i \text{ for commodity route } k$$

$$c_{ij}^k = \text{the unit cost of routing commodity or service } k \text{ along link } (i, j) \in A$$

$$f_{ij} = \text{the cost of seismically upgrading road link } (i, j) \in A$$

$$B = \text{the budget for upgrading road links}$$

$$x_{ij}^k = \begin{cases} 1, \text{ if there is a unit of flow of commodity } k \text{ on link } (i, j) \\ 0, \text{ otherwise} \end{cases}$$

$$y_{ij} = \begin{cases} 1, if\ link\ (i,j) is\ used\ on\ a\ commodity\ flow\ path \\ 0, otherwise \end{cases}$$

$$Z_m^k = \begin{cases} 1, \text{if demand center } m \text{ is accessible from link on a commodity path } k \\ 0, \text{otherwise} \end{cases}$$

The formulation is as follows:

$$\text{Max} \sum_m \sum_k a_m z_m^k \quad (36)$$

$$where$$
$$m = D(k)$$

$$\text{Min} \sum_k \sum_{(i,j)\in A} \left(c_{ij}^k x_{ij}^k + c_{ji}^k x_{ji}^k\right) \quad (37)$$

subject to:

$$\sum_{(i,j)\in A} x_{ij}^k - \sum_{(j,i)\in A} x_{ji}^k = \begin{cases} z_m^k, if\ i = O(k) \\ z_m^k, if\ j = D(k) \quad \text{for each } i\&k \\ 0, \text{otherwise} \end{cases} \quad (38)$$

$$x_{ij}^k \le y_{ij} \text{ for all } k \text{ and } (i,j) \text{ and } (j,i) \in E \quad (39)$$

$$\sum_{(i,j)\in E} f_{ij} y_{ij} \le B \quad (40)$$

$$x_{ij}^k \in \{0,1\} \text{ and } x_{ji}^k \in \{0,1\} \text{ for all } k \text{ and } (i,j) \in E \quad (41)$$

$$y_{ij} \in \{0,1\} \text{ for all } (i,j) \in E \quad (42)$$

$$z_m \in \{0,1\} \text{ for all } m \quad (43)$$

The above model can be used to identify which routes should be made safe so that services can be transported or flow between as many communities as possible so that feasible evacuation and supply routes exist after an earthquake. The basic idea is to design the best "safe-routes" system within budget limitations and serve as many communities as possible as well as make the hardened routes as efficient as possible. Objective (36) maximizes the population that can be served by a hardened route between a given origin and destination pair. Objective (37) minimizes the cost of providing service along a hardened route for a given OD pair k. Constraint (38) represents that a flow path between a given OD pair exists or it does not. If it exists, then a complete safe path must connect that given OD pair k. Constraints (39) prevent a specific link from being used if its needs upgrading and has not been upgraded. Constraint (40) restricts the cost on seismically upgrading links to be less than a given budget B. Constraints 41 through 43 represent the restrictions on the variables. We know that the budget may prevent us from upgrading every weakness in a short period of time in a network, so one could envision using this model to prioritize the upgrading process. The real need is to develop relatively simple models that address fragility in networks, including water conveyance systems, communication networks, satellite systems, and critical supply facilities, among many others. Viswanath and Peeta (2003) give an application involving a form of this model in their paper.

This model is quite simple and conveys an important feature. By setting a standard for connection between towns and maximizing connections, one can structure a relatively simple model that optimizes the investment in strengthening. Such a simple model does not capture the possibility that it may be possible to reroute traffic along or around a destroyed element, to handle some of the traffic. Another issue is that this model is based upon the assumption that each city pair is connected by a single route or pathway, when indeed there can be several such route possibilities. That is the model could be extended by adding additional route options between a city pair where only one of the routes needs to be strengthened between a city pair in order for that city pair to be connected. Additionally, such a refined model could be defined where each city pair is represented by a specific commodity path that can be completed only when the arcs along a specific commodity path connecting a given city pair are fully strengthened [details for a related road investment problem for this approach are given in Scaparra and Church (2005)].

There are many types of problems where the issue is to keep routes or facilities available given the loss or damage to specific system components. For issues such as earthquakes and floods, or even a strike by a terrorist, lifelines of support need to be present for different needs like hospitals, food, police, etc. Making access possible for such lifelines is one way in which a system or a region can be made more resilient. Models such as the one given above can be simple yet powerful to aid in decision making.

8 Beyond the Basics: Resilient Design

Most of the models and discussion presented so far have addressed what could happen and developing plans to reduce that risk for systems in place, e.g., fortifying a subset of facilities and strengthening network segments. But what if we took possible disruption into account when we designed a facility system so that the resulting system was as resilient as possible without special efforts to fortify or harden any of the facilities. In this section, we present a new model to optimize resilience based upon the classic p-median problem. To start we might choose sites that are not close to fault lines or in flood plains, or low-lying areas. That is, we can screen out possible sites so that we reduce risk as much as possible before we actually solve a location problem. Consider the following facility location problem that is cast within the p-median framework:

> Locate a set of p-facilities in order to minimize weighted distance while at the same time minimize the resulting weighted distance when r of these facilities might be inoperable do to some natural or intentional event.

That is, when several facilities are lost to a system, we want the remaining configuration to be as resilient as possible, i.e., to be relatively efficient. Resilience is the capability of being able to bounce back from some disruptive event. The greater resilience, the faster and easier it is for a system to return to a fairly high level of efficiency. Consider the following additional or modified notation:

$J =$ the set of potential facility sites

$Z_0 =$ the weighted distance of the facility system when all p $-$ facilities are in operation

$Z_r =$ the weighted distance of a facility system when it is struck with a worst $-$ case loss of $r -$ facilities.

$$x_{ij}^{r} = \begin{cases} 1, \text{if demand } i \text{ asssigns to a facility at } j \text{ when the system} \\ \text{has a loss of } r \text{ facilities} \\ 0, \text{otherwise} \end{cases}$$

$$x_{ij}^{0} = \begin{cases} 1, \text{if demand } i \text{ asssigns to a facility at } j \text{ when the system} \\ \text{operates all } p \text{ facilities} \\ 0, \text{otherwise} \end{cases}$$

$$y_j = \begin{cases} 1, \text{if a facility is located at site } j \\ 0, \text{otherwise} \end{cases}$$

$$s_j = \begin{cases} 1, \text{if a facility located at site } j \text{ is interdicted} \\ 0, \text{otherwise} \end{cases}$$

$T_{ij} = \{k \in I \mid k \neq j \text{ and } d_{ik} > d_{ij}\}$, the set of potential facility sites (not including j) that are as far or farther than j is from demand i

We can now formulate the resilient design p-median (ReDe-PM) problem as a bi-level two-objective optimization problem as follows:

$$\text{RD} - \text{PMP} : \text{Min } Z_0 \tag{44}$$

$$\text{Min } Z_r \tag{45}$$

subject to:

$$Z_0 = \sum_{i \in I} \sum_{j \in J} a_i d_{ij} x_{ij}^{0} \tag{46}$$

$$\sum_{j \in J} x_{ij} = 1 \text{ for each } i \in I \tag{47}$$

$$\sum_{j \in J} y_j = p \tag{48}$$

$$x_{ij}^0 \leq y_j \text{ for all } i \in I \text{ and for all } j \in J \tag{49}$$

$$y_j \in \{0, 1\} \text{ for all } j \in J \tag{50}$$

$$x_{ij}^0 \in \{0, 1\} \text{ for all } i \in I \text{ and for all } j \in J \tag{51}$$

where:

$$Z_r(s) = \max \sum_{i \in I} \sum_{j \in F} a_i d_{ij} x_{ij}^r \tag{52}$$

subject to:

$$\sum_{j \in F} x_{ij}^r = 1 \text{ for all } i \in I \tag{53}$$

$$\sum_{j \in F} s_j = r \tag{54}$$

$$\sum_{k \in T_{ij}} x_{ij}^r \leq s_j \text{ for all } i \in I \text{ and for all } j \in J \tag{55}$$

$$s_j \leq y_j \text{for all } j \in F \tag{56}$$

$$x_{ij}^r \leq y_j - s_j \text{ for all } i \in I \text{ and for all } j \in J \tag{57}$$

$$s_j \in \{0, 1\} \text{ for all } j \in J \tag{58}$$

$$x_{ij}^r \in \{0, 1\} \text{ for all } i \in I \text{ and for all } j \in J \tag{59}$$

The ReDe-PM problem is composed of two levels. The top level comprises a classical p-median problem except that it has a modified objective. The first objective (44) is that of the p-median problem that involves minimizing the weighted distance associated with the location of p-facilities. The second objective (45) involves minimizing the weighted distance after a worst-case loss of r-facilities of the p-facilities that are being located. The selection of the r-facilities occurs in the lower level of the problem, where the intelligent interdictor takes the sites selected by the designer and identifies those sites that if removed increase the weighted distance the most. The idea is that the designer plans to optimize the system without interdiction (or loss) as well as its operation after possible losses of r-facilities. Since this is a two-objective model, one would presumably solve this model using a multi-objective approach in order to generate a trade-off of solutions between efficient but fragile design and less efficient but robust design.

Constraints (46)–(51) represent the classic *p*-median problem formulation of ReVelle and Swain (1970) and the lower-level problem [(objective (52)] and constraints (53) through (59) represent a form of the RIM model. Constraints (56) have been added in order to ensure that interdiction targets only those sites that have been selected for facilities in the upper level of the problem by the designer. Constraints (57) have been added to ensure that assignment after interdiction occurs only when a site has a facility that has not been interdicted. Constraints (55) ensure that a demand assigns to its closest open, non-interdicted facility.

One can consider the above a passive protection system model, where a degree of resilience against destructive events is built into the solution. To give an example of the type of solutions that might be generated from the use of this model, consider a solution that was generated when a zero weight is assigned to the first objective and a weight of one is specified for the second objective. That means the designer is interested in finding a solution that is as resilient as possible after interdiction. We

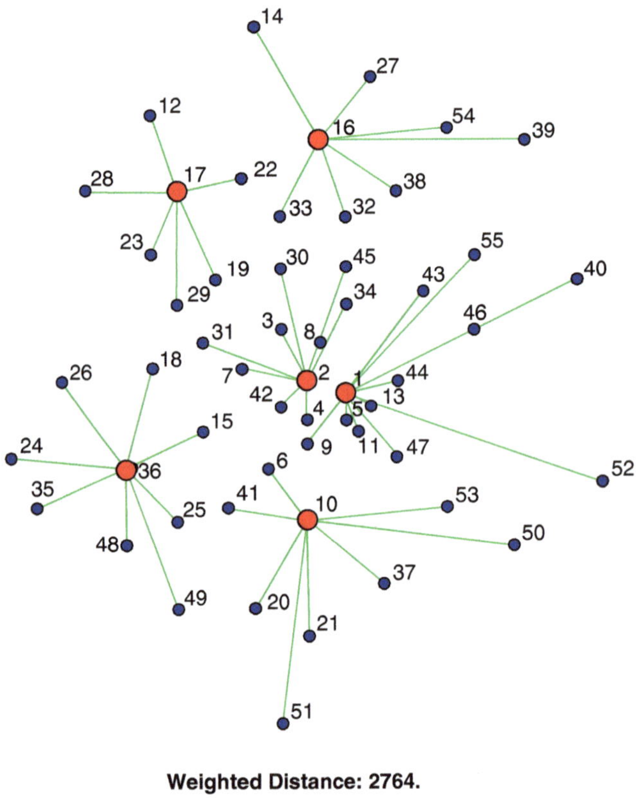

Fig. 4 A solution generated by the resilient design p-median model. This solution reduces the impact of a worst-case loss of one facility by 19% as compared to the worst-case loss of the optimal p-Median solution

used the classic 55-point dataset used by ReVelle and Swain (1970) and solved for the location of 6 facilities. The optimal solution to this problem is given in Figure 4. The weighted distance of this solution is 2764. The worst-case loss of a facility in this configuration occurs when site 2 is interdicted where weighted distance increases to 3252. This results in a weighted distance that is 19% better when the optimal 6 median solution is subjected to a loss of 1 facility. Thus, it is possible to design a system that is more resilient to facility disruption without fortification. It should be noted that for this particular problem the weighted distance of the optimal 6-median solution without interdiction is 4% less than the weighted distance of the optimal resilient pattern. Thus, there are trade-offs between seeking resilience and optimal efficiency without resilience. That is exactly what a systems planner should understand before making configuration decisions.

9 Summary and Conclusions

Disruptions to systems can occur due to human error or accident, by intentional due to sabotage by a terrorist, or due to a natural event like a devastating hurricane, earthquake, or flood. Perhaps the most notable events have occurred due to failures of a system after a natural. Small disruptions in facility operations can often be overcome in a reasonably short period of time, but in other disruptions facilities are rendered inoperable for months or even years. Events such as the Metcalf substation attack have raised concerns among government agencies, utilities, and private companies. Disruption of a facility or a system of facilities can occur due to loss of resources, the destruction of equipment, or even the collapse of a facility. There are numerous accounts of facility losses, ranging from explosions at grain storage centers, sugar refineries, and oil refineries to facilities collapsing in earthquakes or being flooded. Limits on resources can hamper a facility operation and reduce output. The auto industry is currently experiencing limits on computer chips that has degraded/reduced production. Despite the fact that there has been a surge in demand, chip production has been lost due to a drought in Taiwan and a chip plant destroyed by fire in Japan. Shortages and other events can cascade through a system causing damage many times greater than the initial event. In this chapter, we have presented a few models that have been developed to address risk of loss, ranging from using a model to identify worst-case losses within a facility system to optimizing fortification of facility assets in order to limit potential losses.

There are a number of research needs in this area of location science, ranging from the need to develop simple models that are designed to improve safety and reduce risk to more computer-intensive tasks of generating a range of possible outcomes as represented by reliability envelopes. In addition, models for resilient design need to be developed for a range of applications. Finally, specific applications in interdiction, fortification, and resilient design to electrical grids (Alguacil et al., 2014; Yuan & Zeng, 2020), communication systems and grids (Nicholas & Alderson, 2015; Lei et al., 2019), supply chains (Snyder et al., 2006; Snyder et al.,

2016), and critical manufacturing systems present major obstacles due to problem size, data needs, and lack of computational resources and algorithms. Bi-level and tri-level optimization problems present a major challenge in solving and in application due to possible problem sizes. The research frontier must also include the development of sophisticated AI techniques to detect weaknesses in systems such as hub networks and supply chains as well as the use of quantum computing.

References

Alderson, D. L., Brown, G. G., Carlyle, W. M., & Wood, R. K. (2011). *Solving defender-attacker-defender models for infrastructure defense*. Department of Operations Research, Naval Postgraduate School.

Alguacil, N., Delgadillo, A., & Arroyo, J. M. (2014). A trilevel programming approach for electric grid defense planning. *Computers and Operations Research, 41*, 282–290.

Armour, G. C., & Buffa, E. S. (1963). A heuristic algorithm and simulation approach to relative location of facilities. *Management Science, 9*(2), 294–309.

Balakrishnan, A., Ward, J. E., & Wong, R. T. (1987). Integrated facility location and vehicle routing models: Recent work and future prospects. *American Journal of Mathematical and Management Sciences, 7*(1–2), 35–61.

Balinski, M. L. (1965). Integer programming: Methods, uses, computations. *Management Science, 12*(3), 253–313.

Brown, G. G., Carlyle, W. M., Salmeron, J., & Wood, K. (2005). Analyzing the vulnerability of critical infrastructure to attack and planning defenses. In *Emerging theory, methods, and applications* (pp. 102–123). Informs.

Brown, G. G., Carlyle, W. M., & Wood, R. K. (2008). *Optimizing Department of Homeland Security Defense investments: Applying defender-attacker (−defender) optimization to terror risk assessment and mitigation*. Department of Operations Research, Naval Postgraduate School.

Campbell, J. F., & O'Kelly, M. E. (2012). Twenty-five years of hub location research. *Transportation Science, 46*(2), 153–169.

Church, R. L. (2008). BEAMR: An exact and approximate model for the p-median problem. *Computers & Operations Research, 35*(2), 417–426.

Church, R. L. & Cohon, J. L. (1976). *Multiobjective location analysis of regional energy facility siting problems*. Report prepared for the U.S. Energy Research and Development Administration (BNL 50567).

Church, R. L., & Scaparra, M. P. (2007a). Protecting critical assets: The r-interdiction median problem with fortification. *Geographical Analysis, 39*(2), 129–146.

Church, R., & Scaparra, M. P. (2007b). Analysis of facility systems' reliability when subject to attack or a natural disaster. In *Critical infrastructure* (pp. 221–241). Springer.

Church, R. L., Scaparra, M. P., & Middleton, R. S. (2004). Identifying critical infrastructure: The median and covering facility interdiction problems. *Annals of the Association of American Geographers, 94*(3), 491–502.

Cova, T. J., & Church, R. L. (1997). Modelling community evacuation vulnerability using GIS. *International Journal of Geographical Information Science, 11*(8), 763–784.

Current, J. R., & Schilling, D. A. (1989). The covering salesman problem. *Transportation Science, 23*(3), 208–213.

Current, J., ReVelle, C. R., & Cohon, J. L. (1985). The maximum covering/shortest path problem: A multiobjective network design and routing formulation. *European Journal of Operational Research, 21*(2), 189–199.

Daskin, M. S., Hesse, S. M., & Revelle, C. S. (1997). α-reliable p-minimax regret: A new model for strategic facility location modeling. *Location Science, 5*(4), 227–246.

Dekle, J., Lavieri, M. S., Martin, E., Emir-Farinas, H., & Francis, R. L. (2005). A Florida county locates disaster recovery centers. *Interfaces, 35*(2), 133–139.
Eiselt, H. A., & Marianov, V. (2012). Mobile phone tower location for survival after natural disasters. *European Journal of Operational Research, 216*(3), 563–572.
Eiselt, H. A., Laporte, G., & Thisse, J. F. (1993). Competitive location models: A framework and bibliography. *Transportation Science, 27*(1), 44–54.
Eiselt, H. A., Marianov, V., & Drezner, T. (2019). Competitive location models. In *Location science* (pp. 391–429). Springer.
Elloumi, S. (2010). A tighter formulation of the p-median problem. *Journal of Combinatorial Optimization, 19*(1), 69–83.
García, S., Labbé, M., & Marín, A. (2011). Solving large p-median problems with a radius formulation. *INFORMS Journal on Computing, 23*(4), 546–556.
Grubesic, T. H., & Murray, A. T. (2006). Vital nodes, interconnected infrastructures, and the geographies of network survivability. *Annals of the Association of American Geographers, 96*(1), 64–83.
Hakimi, S. L. (1964). Optimum locations of switching centers and the absolute centers and medians of a graph. *Operations Research, 12*(3), 450–459.
Hakimi, S. L. (1965). Optimum distribution of switching centers in a communication network and some related graph theoretic problems. *Operations Research, 13*(3), 462–475.
Hakimi, S. (1969). An algorithm for construction of the least vulnerable communication network or the graph with the maximum connectivity. *IEEE Transactions on Circuit Theory, 16*(2), 229–230.
Hanjoul, P., & Peeters, D. (1987). A facility location problem with clients' preference orderings. *Regional Science and Urban Economics, 17*, 451–473.
Hutson, V. A., & ReVelle, C. (1993). Indirect covering tree problems on spanning tree networks. *European Journal of Operational Research, 65*(1), 20–32.
Jin, J. G., Shen, Y., Hu, H., Fan, Y., & Yu, M. (2021). Optimizing underground shelter location and mass pedestrian evacuation in urban community areas: A case study of Shanghai. *Transportation Research Part A: Policy and Practice, 149*, 124–138.
Kim, H., & O'Kelly, M. (2004, November). Survivability of commercial backbones with peering: A case study of Korean networks. In *51st annual north American meetings of the regional science association international,* Seattle, WA.
Lazzaro, G. L. (2016). *Tri-level optimization algorithms for solving defender-attacker-defender network models*. Department of Operations Research, Naval Postgraduate School.
Lei, H., Huang, S., Liu, Y., & Zhang, T. (2019). Robust optimization for microgrid defense resource planning and allocation against multi-period attacks. *IEEE Transactions on Smart Grid, 10*(5), 5841–5850.
Liberatore, F., Scaparra, M. P., & Daskin, M. S. (2012). Hedging against disruptions with ripple effects in location analysis. *Omega, 40*(1), 21–30.
Lim, M., Daskin, M. S., Bassamboo, A., & Chopra, S. (2010). A facility reliability problem: Formulation, properties, and algorithm. *Naval Research Logistics (NRL), 57*(1), 58–70.
Lindell, M. K. (2008). EMBLEM2: An empirically based large scale evacuation time estimate model. *Transportation Research Part A: Policy and Practice, 42*(1), 140–154.
Losada, C., Scaparra, M. P., Church, R. L., & Daskin, M. S. (2012). The stochastic interdiction median problem with disruption intensity levels. *Annals of Operations Research, 201*(1), 345–365.
Lozano, L., & Smith, J. C. (2017). A backward sampling framework for interdiction problems with fortification. *INFORMS Journal on Computing, 29*(1), 123–139.
Marianov, V., & Revelle, C. (1994). The queuing probabilistic location set covering problem and some extensions. *Socio-Economic Planning Sciences, 28*(3), 167–178.
Marianov, V., & ReVelle, C. (1996). The queueing maximal availability location problem: A model for the siting of emergency vehicles. *European Journal of Operational Research, 93*(1), 110–120.

Mathews, A. J., Haffner, M., & Ellis, E. A. (2017). GIS-based modeling of tornado siren sound propagation: Refining spatial extent and coverage estimations. *International Journal of Disaster Risk Reduction, 23*, 36–44.

Mirchandani, P. B., & Odoni, A. R. (1979). Locations of medians on stochastic networks. *Transportation Science, 13*(2), 85–97.

Moore, G. C., & ReVelle, C. (1982). The hierarchical service location problem. *Management Science, 28*(7), 775–780.

Murray, A. T., O'Kelly, M. E., & Church, R. L. (2008). Regional service coverage modeling. *Computers and Operations Research, 35*(2), 339–355.

Nicholas, P. J., & Alderson, D. L. (2015). *Designing interference-robust wireless mesh networks using a defender-attacker-defender model*. Department of Operations Research, Naval Postgraduate School.

Özdamar, L., & Ertem, M. A. (2015). Models, solutions and enabling technologies in humanitarian logistics. *European Journal of Operational Research, 244*(1), 55–65.

Peterson, S. K., & Church, R. L. (2008). A framework for modeling rail transport vulnerability. *Growth and Change, 39*(4), 617–641.

Repolho, H. M., Church, R. L., & Antunes, A. P. (2016). Optimizing station location and fleet composition for a high-speed rail line. *Transportation Research Part E: Logistics and Transportation Review, 93*, 437–452.

ReVelle, C. (1986). The maximum capture or "sphere of influence" location problem: Hotelling revisited on a network. *Journal of Regional Science, 26*(2), 343–358.

ReVelle, C., & Hogan, K. (1989). The maximum availability location problem. *Transportation Science, 23*(3), 192–200.

ReVelle, C. S., & Swain, R. W. (1970). Central facilities location. *Geographical Analysis, 2*(1), 30–42.

Revelle, C. S., Eiselt, H. A., & Daskin, M. S. (2008). A bibliography for some fundamental problem categories in discrete location science. *European journal of operational research, 184*(3), 817–848.

Scaparra, M. P., & Church, R. L. (2005). A GRASP and path relinking heuristic for rural road network development. *Journal of Heuristics, 11*(1), 89–108.

Scaparra, M. P., & Church, R. L. (2008). A bilevel mixed-integer program for critical infrastructure protection planning. *Computers and Operations Research, 35*(6), 1905–1923.

Scaparra, M. P., & Church, R. (2012). Protecting supply systems to mitigate potential disaster: A model to fortify capacitated facilities. *International Regional Science Review, 35*(2), 188–210.

Schilling, D., Elzinga, D. J., Cohon, J., Church, R., & ReVelle, C. (1979). The TEAM/FLEET models for simultaneous facility and equipment siting. *Transportation Science, 13*(2), 163–175.

Shirouzu, N. (2021). *How Toyota thrives when the chips are down, Reuters* (https://www.reuters.com/article/us-japan-fukushima-anniversary-toyota-in/how-toyota-thrives-when-the-chips-are-down-idUSKBN2B1005).

Slater, P. J. (1982). Locating central paths in a graph. *Transportation Science, 16*(1), 1–18.

Snyder, L. V., & Daskin, M. S. (2005). Reliability models for facility location: The expected failure cost case. *Transportation Science, 39*(3), 400–416.

Snyder, L. V., & Daskin, M. S. (2006). Stochastic p-robust location problems. *IIE Transactions, 38*(11), 971–985.

Snyder, L. V., Scaparra, M. P., Daskin, M. S., & Church, R. L. (2006). Planning for disruptions in supply chain networks. In *Models, methods, and applications for innovative decision making* (pp. 234–257). INFORMS.

Snyder, L. V., Atan, Z., Peng, P., Rong, Y., Schmitt, A. J., & Sinsoysal, B. (2016). OR/MS models for supply chain disruptions: A review. *IIE Transactions, 48*(2), 89–109.

Toregas, C., Swain, R., ReVelle, C., & Bergman, L. (1971). The location of emergency service facilities. *Operations Research, 19*(6), 1363–1373.

Viswanath, K., & Peeta, S. (2003). Multicommodity maximal covering network design problem for planning critical routes for earthquake response. *Transportation Research Record, 1857*(1), 1–10.

Weaver, J. R., & Church, R. L. (1991). The nested hierarchical median facility location model. *INFOR: Information Systems and Operational Research, 29*(2), 100–102.

Wollmer, R. (1964). Removing arcs from a network. *Operations Research, 12*(6), 934–940.

Wollmer, R. D. (1970). Algorithms for targeting strikes in a lines-of-communication network. *Operations Research, 18*(3), 497–515.

Yuan, W., & Zeng, B. (2020). Cost-effective power grid protection through defender–attacker–defender model with corrective network topology control. *Energy Systems, 11*(4), 811–837.

Part III
Facility-Customer Response Time and Congested Facilities

Uncertainty in Facility Location Models for Emergency Medical Services

Eric G. Stratman, Justin J. Boutilier, and Laura A. Albert

Abstract Emergency medical service (EMS) systems aim to respond to emergency calls and provide life-saving care to patients. The location of EMS resources is critical to providing this care in a timely manner, and as a result, EMS facility location problems have received a tremendous amount of attention since the 1960s, and their advancement is directly tied to a wide range of facility location problems. This chapter reviews uncertainty in facility location problems applied to EMS systems and provides an intuition for and understanding of EMS problem settings. The chapter begins by explaining EMS response processes and the goals of the early deterministic models. Next, it introduces probabilistic formulations that account for uncertainty in ambulance availability, response time, and demand. Then, it highlights directions within the field and the role of uncertainty in these problem settings. This includes EMS systems with tiered units, systems that consider resource relocation, EMS systems in developing countries, and several other areas. Lastly, it concludes by providing insights into how these models are used in practice.

Keywords Facility location · Ambulance location · Emergency medicine · Public safety · Optimization under uncertainty

1 Introduction

Emergency medical service (EMS) systems aim to quickly respond to emergency calls and provide life-saving care to patients. The location of the responding EMS unit is critical to providing this care in a timely manner. As a result, EMS facility location problems have received a tremendous amount of attention since the 1960s, and their advancement is directly tied to a wide range of facility location problems. EMS facility location problems are similar to many other facility location problems:

E. G. Stratman · L. A. Albert (✉) · J. J. Boutilier
Industrial & Systems Engineering, University of Wisconsin-Madison, Madison, WI, USA
e-mail: egstratman@wisc.edu; jboutilier@wisc.edu; laura@engr.wisc.edu

H. A. Eiselt, V. Marianov (eds.), *Uncertainty in Facility Location Problems*, International Series in Operations Research & Management Science 347,
https://doi.org/10.1007/978-3-031-32338-6_9

requests for emergency aid (EMS calls) are demand points at geographic locations that must be serviced by EMS units (ambulances, aircraft, etc.) positioned at strategically located facilities. Optimization models determine the optimal location of EMS stations and/or the optimal allocation of vehicles to a set of candidate locations to best meet this demand.

The early EMS facility location models were adapted from simple deterministic location models such as the Maximal Coverage Location Problem (MCLP) and P-Median Problem (PMP). These models provided high-level insight; however, they did not account for the inherent uncertainty of EMS systems and created a need for more advanced techniques. To address this need, probabilistic models were created specifically for EMS response. Initially, these models were simple extensions of deterministic models and addressed a single stochastic element associated with EMS response, such as uncertainty in ambulance availability, response time, or demand. Over time, improved computing power, new algorithms, and increased data availability allowed more advanced models to emerge. Today, the problem of locating EMS facilities with uncertainty is still a growing area of research, where researchers continue to refine existing models and investigate settings with new sources of uncertainty.

This chapter reviews uncertainty in facility location problems applied to EMS systems. The goal is to provide the reader with an intuition for and understanding of the EMS problem setting. This chapter does not provide formulations for the discussed models and is not a comprehensive literature review. Instead, it focuses on the main themes and approaches found in facility location research applied to EMS. The remainder of the chapter is structured as follows. Section 2 reviews the EMS response process in greater detail. Section 3 briefly presents deterministic EMS facility location models that serve as the basis for more advanced models. Section 4 introduces probabilistic formulations that account for uncertainty in ambulance availability, response time, and demand. Section 5 highlights interesting directions and applications of EMS facility location problems. Section 6 discusses common themes in the successful implementations of EMS facility location models. Lastly, Sect. 7 provides references to formal literature reviews that discuss EMS facility location problems.

2 EMS Background

2.1 The EMS Response Process

Most EMS systems emphasize the importance of timely care; however, there are two primary classifications of modern EMS systems. The *scoop and run method* is practiced in countries such as the United States, Canada, the United Kingdom, New Zealand, and Australia. Under this model, the EMS system seeks to quickly reach a patient, provide minimal pre-hospital care, and then deliver the patient to a care facility for further treatment (e.g., hospital emergency department). The alternative approach is the *stay and stabilize method*, which is practiced in countries such as Germany, France, Greece, Malta, and Austria. In these systems, fewer patients are delivered to care facilities. Although many studies have compared the outcomes and cost-effectiveness of both methods, differences in operational standards and context make it nearly impossible to determine if one approach is better than the other (Al-Shaqsi, 2010a). However, these differences may influence the way one would model the system and the response time threshold chosen. Furthermore, EMS agencies may be a public service, operated or funded by a government, or a private for-profit business. Once again, there is no clear answer to which approach is better in general; rather, this discrepancy is primarily driven by national and cultural approaches to healthcare (Narad & Gillespie, 1998). With this in mind, no two EMS systems are exactly the same, and every EMS system must adapt to differences in available resources, geographic challenges, EMS infrastructure, legal requirements, and cultural dynamics. We initially present models for an EMS system with a centralized dispatching system under the scoop and run method with a single type of ambulance. This distinction is made since these are the assumptions that many of the early EMS facility location models used. In Sect. 5, we explore settings where these assumptions are relaxed.

The EMS response process under the scoop and run model is as follows. (1) A medical emergency occurs and someone calls an emergency telephone line. On the phone, an EMS call taker asks the caller a series of questions to determine where the patient is located and estimate their condition. Typically, these questions are scripted by a computer system. (2) One or more EMS vehicles are dispatched to the patient's location. In many areas, police or fire vehicles may also be dispatched to provide basic care, if they can reach the patient sooner. (3) The EMS vehicle arrives at the scene of the patient. (4) The EMS personnel treat the patient. (5) The patient is loaded into the vehicle and transported to a care facility. (6) Finally, the EMS vehicle reaches the care facility and transfers the patient. (7) After serving the patient, the EMS vehicle prepares for the next patient by returning to a station or another location. Figure 1 summarizes this process using the standard names for each time interval.

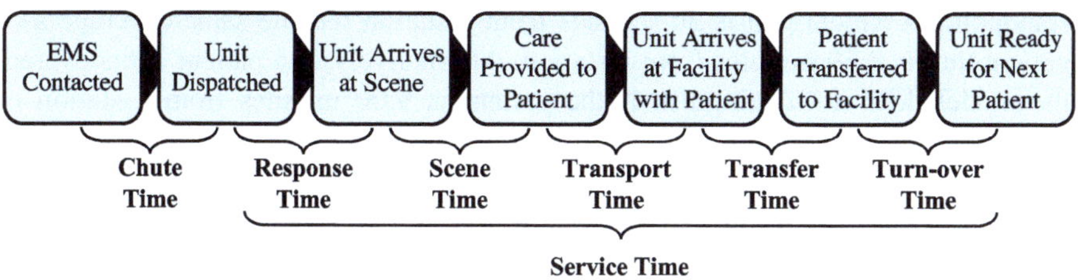

Fig. 1 The EMS process under the scoop and run model. The intervals beneath the figure indicate various time components of the EMS process

2.2 Response Time, Response Time Threshold, and Coverage

Improving patient outcomes following an emergency is the primary objective of any EMS system. Despite this simple goal, patient outcomes are difficult to quantify due to their qualitative nature. *Response time* is defined as the time interval between the moment an EMS vehicle is dispatched and an EMS vehicle arrives at the scene of a patient (see Fig. 1). In cardiac arrest patients, a 1-minute reduction in response time increases the odds of survival by 1–10% in a nonlinear, decaying, relationship (Stoesser et al., 2021; Holmén et al., 2020). Therefore, most EMS agencies use response time as the primary performance metric and as a proxy for outcomes since it is easy to measure and understand. Consequentially, nearly all EMS facility location problems have an objective function that evaluates response time.

The *response time threshold* (RTT) is a response time standard that many EMS agencies are held to. An EMS response time within the RTT is usually considered acceptable, and a response time above this threshold is usually considered to be too slow. In North America, the most widely used RTT in urban areas is 9 minutes for 90% of EMS responses (Fitch, 2005).

This standard is often traced back to several studies from the late 1970s and 1980s that concluded that a patient's odds of survival following an out-of-hospital cardiac arrest (OHCA) decrease rapidly after this window (Mullie et al., 1989). In non-urban areas, this standard may be extended to account for longer travel distances and hard-to-reach areas. For example, the US state of California recommends that EMS agencies should respond within 20 minutes to patients in rural areas 90% of the time (Narad & Driesbock, 1999). Figure 2 presents the response time distribution by urbanicity in the United States. Globally, RTT standards vary due to the available resources, the type of EMS system used, and the national and cultural approaches to healthcare (see Table 1). In any setting, the RTT is extensively used in EMS facility location problems because it provides a simple classification for coverage. A request for EMS service that can be reached by an ambulance within the RTT is considered *covered*, whereas one that is beyond the RTT is not. Coverage is then implemented as a constraint or objective in EMS facility location problems.

We note that although intuitive and easy to measure, RTT coverage is a binary metric. Consider an urban system with an RTT of 9 minutes. From a modeling standpoint, a patient that is 30 seconds from a station has the same coverage as a patient that is 8:59 minutes from a station. Alternatively, if a patient is uncovered, the model does not distinguish if the patient is 9:00 minutes from a station or 30 minutes from a station. Therefore, throughout this chapter, we also highlight several approaches to encourage timely response without using the RTT. This includes minimizing average response time (Sect. 3), acknowledging the inherent uncertainty in response time (Sect. 4.3.1), and using the objective function to encourage favorable patient outcomes (Sect. 4.3.3).

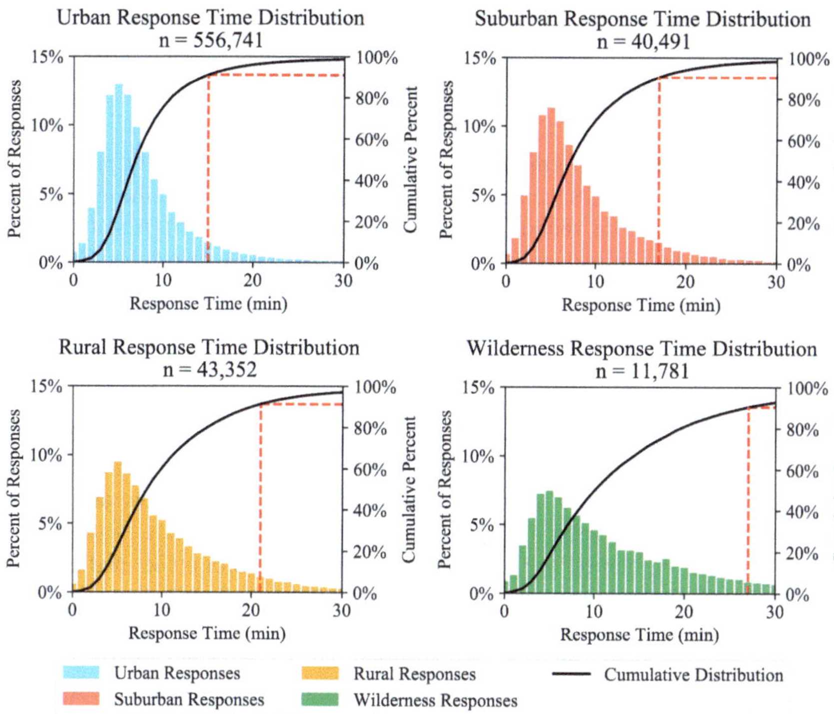

Fig. 2 Response time distribution by urban and rural areas in the United States (Jan-Feb 2020). The 90th percentile of response times is indicated by the dashed red line. As shown, the 90th percentile for urban and rural patients is outside the RTT standard of 9 and 20 minutes, respectively (data provided by NEMSIS)

Table 1 RTT standards may be driven by the resources available to the country, the type of EMS system used, and the national and cultural approaches to healthcare. For example, Hong Kong uses the same RTT standard for urgent and non-urgent patients (Fitch, 2005; Krafft et al., 2003)

Location	RTT for life-threatening emergencies (min)	Compliance goal	Type of system
Richmond, VA, USA	8:59	90%	Scoop and run
West Midlands, UK	8:00	75%	
Hong Kong	12:00[a]	92%	
Bonn, Germany	7:59	90%	Stay and stabilize
Genoa, Italy	8:00	No data	
Ulleval, Norway	9:39	No data	

[a] For life-threatening and non-life-threatening emergencies

3 Deterministic EMS Facility Location

3.1 Deterministic Single Coverage Models

We start by introducing deterministic models, inspired by or directly pulled from pioneering research on EMS facility location problems. We note that the term *station* is used loosely within this section (and throughout EMS research). A station could be any location where ambulances are positioned before being dispatched. Common locations include EMS bases, hospitals, and public parking lots.

- The Location Set Covering Model (LSCM) positions the fewest number of stations at a set of predefined locations, such that all demand nodes are within the RTT from at least one station (Toregas et al., 1971).
- The Maximal Coverage Location Problem (MCLP) weighs each demand node by its generated demand and then positions a limited number of stations at a set of predefined locations to maximize demand within the RTT from at least one station (Church & ReVelle, 1974).
- The P-Median Problem (PMP) positions p stations at a set of predefined locations to minimize the average response time to all demand nodes, weighted by generated demand (Hakimi, 1964).
- The P-Center Problem (PCP) positions p stations at a set of predefined locations to minimize the maximum response time to a demand node (Hakimi, 1965).

These models are the building blocks for the models with uncertainty and demonstrate the varying goals of an EMS system. For example, the LSCM requires RTT coverage for all demand nodes; however, the resulting solutions may not be achievable in an EMS system with limited resources. The MCLP maximizes RTT coverage using a limited number of stations; however, it ignores the response time of demand nodes located outside the RTT, which may lead to inequity in response time. The PMP minimizes the average response time and considers the effect of station location on all demand nodes. However, it may reach fewer patients within the RTT, which may lead to worse patient outcomes. Lastly, while the MCLP and PMP are inherently biased to serve areas with higher demand, the PCP minimizes the longest response time of all the demand nodes to provide near-homogeneous and more equitable service. However, the resulting solution may be an inefficient use of EMS resources.

To illustrate the importance of these modeling decisions, Fig. 3 provides a comparison of the MCLP and PMP applied to a fictional problem instance with five stations. The MCLP maximizes the demand nodes located within the RTT of a station, depicted by the gray dashed line, by locating several stations near the area with the highest density of demand to achieve a maximal coverage of 85.7%. As shown, the solution provides excellent coverage to the areas of denser demand (this may be a more populated area, like a city). Given the importance of a 9-minute response time for survival following cardiac arrest, one could argue that the MCLP is preferred since it can serve the most patients within this threshold. However,

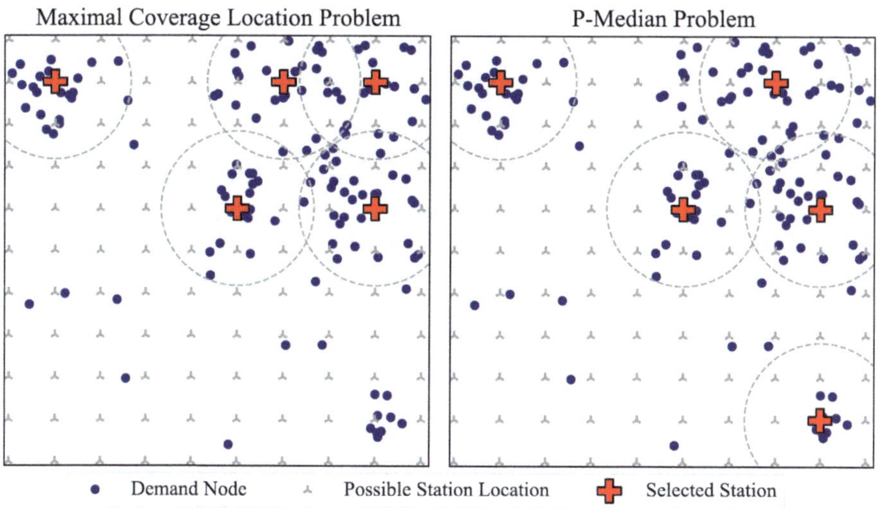

Model	% in RTT (< 9 min)	Average Response Time	90th-Percentile Response Time	Maximum Response Time
MCLP	85.7%	7.1 min	17.7 min	27.7 min
PMP	82.7%	6.1 min	11.0 min	26.3 min

Fig. 3 The deterministic single coverage models represent the various goals of EMS systems

the MCLP provides poor coverage to outlier demand nodes. As shown, the 90th percentile response time is 17.7 minutes, well beyond the RTT and the threshold for resuscitation following cardiac arrest.

Alternately, the PMP minimizes the average response time of all demand nodes. While the PMP must provide quick response to areas with many demand nodes, the response times of the outlier demand nodes are also included within the objective function. The stations in the PMP solution are slightly more dispersed to attain a minimum average response time of 6.1 minutes and a 90th percentile response time of 11.0 minutes. This solution is preferable for the areas with fewer demand nodes (these may be the more rural areas). However, only 82.7% of demand points are covered within 9 minutes, which could lead to overall worse outcomes. These are the trade-offs researchers must be aware of. The formulation of an EMS facility location model must align with the objectives of the EMS system, and the researcher should communicate potential unforeseen implications of their model with practitioners.

3.2 Deterministic Multi-coverage Models

The models mentioned above assume that a station is always available to serve demand. In practice, all the vehicles at a particular station may be busy serving other patients when a request for aid is received. To hedge against this likelihood, *deterministic multi-coverage models* aim to increase the number of stations that can cover each demand node. This recognition of uncertainty in station availability is a precursor to the probabilistic availability models discussed in Sect. 4.1. There are two primary approaches to model multiple coverage: models that encourage multiple coverage through the objective function and models that enforce multiple coverage through constraints. Models that encourage multiple coverage reward demand nodes covered multiple times within their objective function. For example, the Hierarchical Objective Set Covering (HOSC) model is an extension of the LSCM that positions the fewest number of stations to cover each demand node once and maximizes the demand nodes covered multiple times (Daskin & Stern, 1981). Similarly, the Double Standard Model (DSM) combines the MCLP and LSCM to introduce the idea of multiple coverage radii. The model requires that a specified proportion of the demand be located within a distance r_1 of a station, all demand be located within a distance r_2 such that $r_2 > r_1$, and maximizes the demand covered twice within r_1 (Gendreau et al., 1997).

Models that use the second approach, enforcing multiple coverage via constraints, require a demand node to be located within the RTT from a predefined number of stations to be considered covered. This is the approach used in the Tandem Equipment Allocation Model (TEAM), a model that was initially constructed for fire systems with multiple types of vehicles (Schilling et al., 1979). Although this approach may make sense for fire systems, this method is unrealistic in EMS as covering more demand nodes with a single station is often preferable to covering just a few demand nodes with multiple stations. The Backup Coverage Problem II (BACOP II) is an MCLP-type model that combines both approaches. BACOP II rewards demand nodes that are covered by a single station with weight w and demand nodes that are covered twice with weight $(1 - w)$, allowing the decision-maker to control the trade-off between single and double coverage by adjusting $w \in [0, 1]$ (Hogan & Revelle, 1986). This idea of encouraged and enforced reliability is similar to the trade-off of expected coverage models and chance-constrained models, which we discuss in Sects. 4.1.1 and 4.1.2, respectively.

4 Probabilistic EMS Facility Location

The models presented in Sect. 3 are the fundamental facility location models as applied to EMS systems. However, they rely on several unrealistic assumptions. The single coverage models assume that a station is always available to serve incoming demand within a known response time. The multi-coverage models hedge against station unavailability, but do not quantify this reliability. In this section, we present EMS facility location models that address the uncertainty in EMS systems to provide more accurate and dependable results. Section 4.1 reviews approaches to model the uncertainty in ambulance availability. Section 4.2 focuses specifically on the uncertainty in the arrival of EMS requests. Finally, Sect. 4.3 reviews methods to model the uncertainty in EMS response time.

4.1 Uncertainty in Vehicle Availability

Ambulance availability is (typically) defined as the probability that an ambulance will be available when one is needed to respond to an EMS request. In practice, availability is a high-level metric that depends on congestion in the system and is influenced by demand and service time. However, from a modeling perspective, it provides a convenient way to assess how well EMS requests are fulfilled. In Sect. 4.1.1, we review expected coverage models, which account for this uncertainty within the objective function, and in Sect. 4.1.2, we present models that guarantee a level of reliability through chance constraints.

4.1.1 Expected Coverage Facility Location Models

An *expected coverage model* optimizes the long-run probability that a request for EMS service is appropriately covered. These models often embed probability formulations within their objective functions as an extension to the MCLP. The Maximum Expected Covering Location Problem (MEXCLP) is often credited as the first EMS facility location problem to incorporate uncertainty in its formulation and is the seminal contributor of expected coverage models applied to EMS (Daskin, 1983). MEXCLP uses an objective function similar to the MCLP; however, it adjusts the value of covering a demand node by the long-run probability that an ambulance is available within its RTT. This probability is derived using a binomial probability distribution and a system-wide busy fraction, an input of the model that represents the long-run fraction of time an ambulance is unavailable to be assigned to arriving calls. We note that this busy fraction is believed to be the same for all ambulances in the system. Figure 4 shows how the objective function of MEXCLP is derived.

Several other expected coverage models are direct extensions of MEXCLP. The Multiple-coverage One-unit FLEET (MOFLEET) model extends this idea for a set number of ambulances and stations (Bianchi et al., 1988). The Generalized Maximum Expected Coverage (GMEXC) model provides a framework with varying time standards for each additional coverage of a node (Daskin et al., 1988). The Maximum Expected Covering Location with Time Variation (TIMEXCLP) is an extension that considers demand patterns over time (Repede & Bernardo, 1994). Although MEXCLP and its direct extensions are a tremendous step forward for the field, they rely on several limiting assumptions:

1. All ambulances share the same system-wide busy fraction.
2. Ambulances operate independently.

Deriving the MEXCLP Objective Function:
Let $m'_j \in \{0, ..., M\}$ be the number of ambulances that cover demand node $j \in J$. Let x_{jm} be a binary decision variable such that $x_{jm} = 1$ if $m \leq m'_j$, else $x_{jm} = 0$. Let p be the system busy fraction and let d_j represent the weight (generated demand) of demand node $j \in J$.

$$\max \sum_{j \in J} d_j * Pr(\text{An ambulance is available to serve demand node } j \in J)$$

$$\max \sum_{j \in J} d_j * \left(1 - Pr(\text{All } m'_j \text{ ambulances that cover } j \in J \text{ are busy})\right)$$

$$\max \sum_{j \in J} d_j * (1 - p^{m'_j})$$

$$\max \sum_{j \in J} d_j \sum_{m=1}^{m'_j} (p^{m-1} - p^m)$$

$$\max_x \sum_{j \in J} d_j \sum_{m=1}^{M} (1-p)p^{m-1} x_{jm}$$

Fig. 4 Deriving the MEXCLP objective function

3. The ambulance busy fraction is invariant to the locations and assignments of the ambulances and patients.

These assumptions are not true in practice. In any EMS system, certain ambulances may be utilized more than others depending on how many patients require care in a given area. Moreover, how these patients are assigned to a given ambulance also impacts the utilization of other vehicles. Lastly, the distance between an ambulance and its assigned patient impacts how long an ambulance must travel and consequentially its busy fraction. Therefore, it is unsurprising that researchers soon found methods to overcome these limitations.

The Hypercube Queuing Model (HQM) is a method that uses queuing theory to determine the steady-state behavior of servers in a multi-server system (Larson, 1974), can be approximated using correction factors for computational simplicity (Larson, 1975), and can distinguish service times dependent on unique server-patient assignments (Jarvis, 1985). This stream of research led to the Approximate MEXCLP (AMEXCLP), a direct extension of MEXCLP that uses correction factors to account for ambulance interdependencies. In this work, the authors provide an application of the HQM to waive all three assumptions noted above (Batta et al., 1989). Other models to directly use the HQM approximation in expected coverage models include an extension of MEXCLP to allow for two vehicle types (McLay, 2009) and a model to locate facilities in a system with a cut-off priority queue (Yoon & Albert, 2018). We note that these HQM approaches assume that EMS requests arrive according to a Poisson process and ambulance service times follow an exponential service time. The Poisson arrival assumption is consistent with real-world data (Kim & Whitt, 2014), and we elaborate on this in Sect. 4.2. The exponential service time assumption may deviate from reality, but several papers have shown that the performance of models does not critically depend on the choice of service time distribution (Ansari et al., 2017; Jagtenberg et al., 2017). Another expected coverage approach addresses the assumptions of MEXCLP without the HQM approximation using two models. The first model assumes that there is no interaction between stations to screen many solutions, and then this assumption is lifted in the second model for a given ambulance allocation. These models utilize an Erlang loss queue, which does not assume that EMS service follows an exponential distribution; however, they assume that all calls that arrive while all servers are busy are unable to be served (Restrepo et al., 2009).

4.1.2 Chance-Constrained Facility Location Models

All of the models mentioned in the previous section maximize expected coverage. A shortcoming of this approach is that it does not provide a guarantee on the reliability of coverage (probability a call arrives and can be served by an ambulance within its RTT) for a particular demand node. Presumably, an EMS agency would like to cover demand nodes with a given level of reliability. An alternative method is to use a *chance-constrained model* that requires a demand node to be covered with a specified level of reliability. For example, in the Maximum Availability Location Problem (MALP I and MALP II), a method is presented to determine how many ambulances must be positioned within the RTT of the demand nodes for a given reliability level. The authors present an MCLP variant in which a demand node is considered covered only if there are enough ambulances within its RTT to meet the reliability level (ReVelle & Hogan, 1989). The Probabilistic Location Set Covering Problem (PLSCP) is a similar approach applied to the LSCM (ReVelle & Hogan, 1988). However, these models are limited by the same three assumptions described above.

To address these assumptions, queuing theory is used to determine station-specific busy fractions, and refinements of these models include the QPLSCP and QMALP (Marianov & ReVelle, 1994). The use of queuing theory is then extended in the Probabilistic Location-Allocation Set Covering Model with a variable number of servers (PLASCη). PLASCη positions the minimum number of servers and allocates the demand nodes such that a patient will be served within a given time limit with pre-specified reliability (Marianov & Serra, 2002). The Extended Maximum Availability Location Problem (EMALP) uses the hypercube correction factor to adjust MALP to account for server interdependence (Galvão et al., 2005). Once again, using queuing theory, these models assume that ambulances arrive according to a Poisson process and assume an exponential service time or general service time with a loss (zero-length) queue.

4.2 *Uncertainty in Arrival Rate*

The models presented in Sect. 4.1 explore the likelihood that ambulances are busy and unable to serve a request for help. In an EMS system, the arrival rate (the

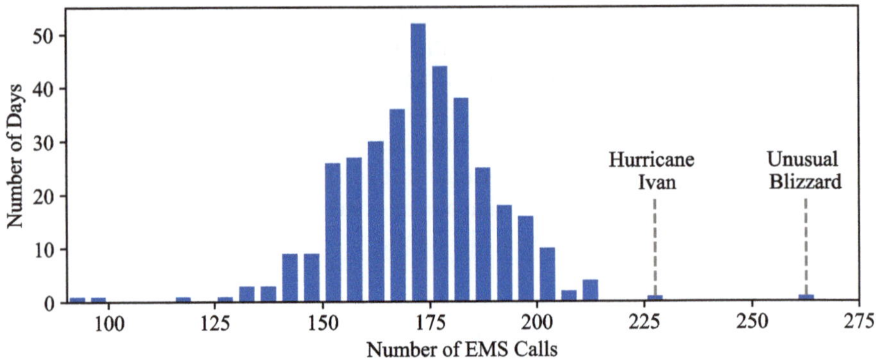

Fig. 5 Histogram of daily EMS demand in Mecklenburg County, North Carolina (2004), which includes the city of Charlotte and the surrounding area. The 10th percentile in demand is 151 requests per day and the 90th percentile is 194 requests per day. An unusual blizzard caused demand to spike to 264 requests in a single day (data provided by MEDIC)

number and distribution of requests for EMS service) is a factor that significantly affects availability. As shown in Fig. 5, demand for EMS requests is highly variable. Additionally, disasters and other events can cause EMS demand to spike well-beyond standard levels and prevent an EMS system from providing timely and reliable care to all patients. In this section, we explore models that explicitly model arrival rates and are better equipped to examine how this variability in demand affects the optimal locations of EMS resources.

4.2.1 Facility Location Models with Probabilistic Arrivals

A method to capture uncertainty in the arrival of EMS requests is to incorporate the underlying arrival distribution into the constraints of EMS facility location problems. One of the first models to do so is Rel-P, a model that considers the arrival distribution in a reliability constraint, under the assumption that a station serves all demand within its RTT as an upper bound (Ball et al., 1993). In many ways, Rel-P is a variation of the chance-constrained models presented in Sect. 4.1.2, since it constrains the likelihood of a request arriving while all servers are busy. However, this likelihood is derived assuming calls arrive according to a Poisson process. Due to the discrete nature, independence, and time invariance (at a high level) of EMS requests, using the Poisson distribution is a safe assumption used in most models and is consistent with real-world data (Kim & Whitt, 2014). A more explicit formulation was later proposed using a stochastic integer program in which the marginal probability distribution for the number of arrivals within each region is captured within a constraint (Beraldi et al., 2004). To adjust for server dependence, this idea is extended to a two-stage model, where the first level locates the EMS facilities and the second level determines how the demand is allocated between them (Beraldi & Bruni, 2009).

Robust and scenario optimization are other methods to model uncertainty in call arrivals. Rather than embedding the arrival distribution into the model, these approaches consider a finite set of demand realizations, called an *uncertainty set*. An element of this set may represent the number of EMS calls in each area of a city on a particular day, and the set contains different historical or expected demand patterns across multiple days. In robust optimization, the model finds the best EMS allocation that satisfies the given constraints across all demand realizations in the uncertainty set (Zhang, 2014; Boutilier & Chan, 2020). Scenario optimization is similar; however, each element in the uncertainty set is associated with a probability. These probabilities are used within the model to constrain the likelihood that some condition is violated across all scenarios (Noyan, 2010; Nickel et al., 2016). The validity of robust and scenario optimization models depends heavily on the variation captured within the uncertainty set. However, robust and scenario optimization offer several advantages; these approaches avoid using complicated constraints derived from probability distributions, they provide more tractable approaches for larger problem instances, and they rely on fewer assumptions regarding the underlying probability distributions than stochastic modeling approaches (Zhang, 2014).

4.2.2 Predicting Arrival Rates

Traditionally, demand is estimated for EMS facility location models by summarizing historical data, under the assumption that future demand will behave similarly. However, there is a growing stream of research that uses advanced machine learning and analytical models to predict demand. These methods are especially useful when planning for population growth or in areas without reliable historic data.

Early approaches that explore population, demographic, and spatial information associated with EMS demand found that the demand for ambulances is highly predictable using socioeconomic and land-use factors (Kamenetzky et al., 1982). For example, ambulance use is often higher in low-income, non-white, and elderly populations. Other approaches have focused on the daily, weekly, or seasonal cycles of EMS demand using a variety of methods (Channouf et al., 2007; Matteson et al., 2011). Lastly, there are even more granular models that combine spatial and temporal aspects in their forecasts (Zhou et al., 2015; Sariyer et al., 2017). Figure 6 demonstrates the daily temporal trend of EMS requests. Emergency demand is known to follow a circadian rhythm and is highest midday (Bagai et al., 2013). We discuss models that alter their deployment throughout the day in Sect. 5.2.1. We note these prediction models were developed using methods that require accurate historical data. In settings without historical EMS records, such as low- and middle-income countries (LMICs), methods have been developed to heavily rely on population estimates and other high-level features (Fujiwara et al., 1987; Boutilier & Chan, 2020).

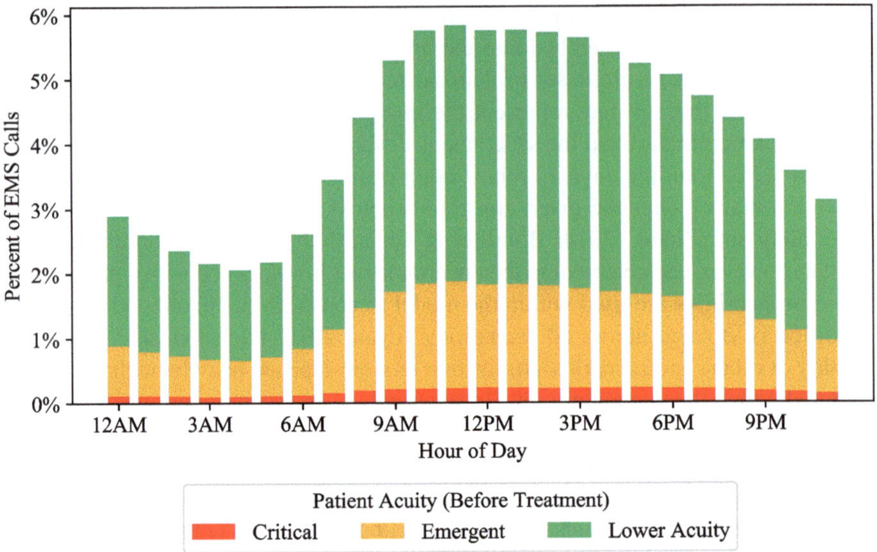

Fig. 6 Daily cycle of EMS demand in the United States (Jan-Feb 2020). EMS demand is highest between 10 AM and 3 PM. Demand variability is more prominent in patients with lower acuity and emergent needs than for patients with critical needs (data provided by NEMSIS)

4.3 Uncertainty in Response Time

As described in Sect. 2, response time is defined as the time interval between the moment an EMS vehicle is dispatched and an EMS vehicle arrives at the scene of a patient. The majority of the models presented so far assume that response time and an ambulance's ability to respond within the RTT is known with certainty. In practice, traffic, weather, and other delays can cause uncertainty in the response time of EMS services and can severely affect patient outcomes (Kunkel & McLay, 2013). In this section, we review models that explicitly consider uncertainty in response time.

4.3.1 Facility Location Models with Probabilistic Response Time

The MCLP with Probabilistic Response (MCLP+PR) and MEXCLP with Probabilistic Response (MEXCLP + PR) are simple extensions of the MCLP and MEXCLP, respectively, which use a parameter in the objective function that represents the likelihood a station can reach a demand node within the time limit (Daskin, 1987; Erkut et al., 2009). A similar idea uses gradual covering to reflect coverage uncertainty (Berman et al., 2010; Eiselt & Marianov, 2009). Probabilistic response times are also considered in other models that account for server interdependencies (Goldberg & Paz, 1991), service level constraints (Alsalloum & Rand, 2006), and pre-trip (chute time) delays (Ingolfsson et al., 2008). Robust optimization (as described in Sect. 4.2.1) has also been used to model uncertainty in response times (Boutilier & Chan, 2020). Lastly, while the models discussed thus far focus on the RTT or the average response time, some researchers specifically constrain the tail of EMS response times, such as the 90th percentile of response time, using a value at risk (VaR) approach (Krishnan et al., 2016; Chan et al, 2017; Boutilier & Chan, 2022).

A closely related source of uncertainty is uncertainty in the total service time: the time interval from the moment an EMS unit is dispatched until it is available to respond to another call (see Fig. 1). While uncertainty in response time affects an ambulance's ability to respond to a call within the RTT, service time affects an ambulance's availability to serve future calls. For this reason, many of the queuing methods presented in Sect. 4.1 may be used to represent the distribution of service time.

4.3.2 Predicting Response Time

Many EMS models use distance from an EMS unit to a demand node as an approximation for response time. However, it is important to recognize that distance and response time have a nonlinear relationship (Budge et al., 2010). The earliest to explore this relationship modeled EMS response time using a variety of factors such as distance, acceleration (Ingolfsson et al., 2003), road type (Goldberg et al., 1990), and time of day (Hausner, 1975). Modern approaches continue to use similar inputs with more refined techniques (Do et al., 2013; Fleischman et al., 2013; Westgate et al., 2016). In the last decade, data availability has allowed for more accurate traffic pattern predictions and routing (Kok et al., 2012), which, in tandem with machine learning techniques, continues to allow for more precise response time predictions (Vlahogianni et al., 2014; Woodard et al., 2017; Boutilier & Chan, 2020).

4.3.3 Response Time and Patient Outcomes

The models explored so far emphasize timely response. However, improving patient outcomes is the ultimate goal of an EMS system. Some researchers in this area question whether EMS systems overvalue response time and the response time threshold at the expense of other factors important to effective care (Al-Shaqsi, 2010b). One method to ensure optimal outcomes is to replace the objective function based on response time with one that directly measures the survival likelihood of cardiac arrest patients (Erkut et al., 2008) or multiple patient types (Knight et al., 2012). Another approach is to carefully select response time thresholds that maximize survival (McLay & Mayorga, 2010). However, these approaches do not transfer well to patients with conditions that are non-life-threatening. Furthermore, predicting how EMS response affects outcomes is difficult; patient outcomes and response time do not have a linear relationship (Holmén et al., 2020), and there are many other factors that influence outcomes (Boutilier & Chan, 2022). For example, the odds of survival for a cardiac arrest incident observed by a bystander are 2.31 times higher than for an unobserved incident (Stoesser et al., 2021).

5 Notable Directions in EMS Facility Location

The methods and models described in Sect. 4 are a broad sampling of the techniques used to solve EMS facility location problems with uncertainty. These techniques account for many of the unknown elements of EMS systems and more accurately represent the underlying dynamics of a traditional EMS system than deterministic models. However, as mentioned in Sect. 2, it is difficult to generalize EMS systems. The methods presented in Sect. 4 may fail to capture the critical dynamics of more complicated or less studied settings. In this section, we explore a few of the many directions of EMS facility location problems and the advanced modeling techniques developed to capture their unique features. Section 5.1 reviews methods to model EMS systems with multiple vehicle types. Section 5.2 explores EMS facility location models that allow ambulance relocation. Section 5.3 discusses the unique challenges of EMS in low- and middle-income countries. Lastly, Sect. 5.4 briefly highlights a few other new applications and directions of EMS facility location research.

5.1 EMS Vehicle Types and Tiered EMS Systems

In the United States, it is estimated that over 65% of EMS responses are for patients with conditions that are non-life-threatening and unlikely to progress (data provided by NEMSIS). To avoid unnecessary utilization of expensive and resource-intensive ambulances for these non-urgent patients, many EMS systems implement a *tiered EMS system* in which EMS vehicles with different levels of training and resources are matched to the needs of a patient. Some advantages of a tiered EMS system are that lower-acuity ambulances are less resource-intensive, they typically allow for larger EMS fleets, and they can enable shorter response times for critical patients (Stout et al., 2000). However, patient misclassification and under-triage are risks that must be carefully managed in such a system (Wilson et al., 1992). In Sect. 5.1.1, we review the types of vehicles that may be available in a tiered EMS system, and in Sect. 5.1.2, we describe the facility location models that explore the unique dynamics and uncertainties of tiered EMS systems.

5.1.1 Tiered EMS Vehicle Types

In North America, transport-equipped EMS vehicles are traditionally categorized by the level of care they may provide a patient. An *advanced life support* (ALS) unit is a larger and more resource-intensive ambulance, usually with a truck chassis-cab and a large box-like rear compartment (see Fig. 7a). An ALS ambulance is staffed by highly trained paramedics equipped to provide advanced care such as administering intravenous fluids, providing controlled medication, using advanced airway techniques, and monitoring cardiac conditions in addition to providing basic noninvasive care. Regions that do not use a tiered EMS system often use an all-ALS ambulance system. Alternatively, a *basic life support* (BLS) unit is a less resource-intensive ambulance usually staffed by emergency medical technicians who are only equipped to provide noninvasive care such as performing cardiopulmonary resuscitation (CPR), immobilizing broken bones, administering oral medications, and providing oxygen (see Fig. 7b) (Al-Shaqsi, 2010a). Studies have found that 85% of calls in the United States are within the scope of BLS ambulances (Pozner et al., 2004), but despite their ability to serve the majority of patients, 5 of the 50 US

Fig. 7 Different types of vehicles in a tiered EMS system. (**a**) ALS ambulances are equipped to perform invasive care, and are preferred for time-sensitive patients in a tiered system. Photo by Eric Stratman. (**b**) BLS ambulances are equipped to perform non-invasive care, and are preferred for lower acuity patients in a tiered system. https://unsplash.com/photos/T5TojXFNnjk (See: https://unsplash.com/license). (**c**) EMS NTVs may be the sole respondent under the stay and stabilize EMS model or they may support transport vehicles under the scoop and run model. https://pixabay.com/photos/car-city-medicine-automobile-4368213/ (See: https://pixabay.com/service/license/). (**d**) Fire and Police units, which may act as NTVs, are playing an increasingly important role in many EMS systems. https://pixabay.com/photos/fire-in-houston-houston-texas-texas-3252193/ (See: https://pixabay.com/service/license/)

states (Georgia, Hawaii, Missouri, South Dakota, Washington) do not license BLS agencies as of 2011 (National Highway Traffic Safety Administration, 2014).

A *non-transport vehicle* (NTV) is another common component of an EMS system and is defined as any vehicle equipped to respond to an EMS request and provide on-scene care, but not intended to transport the patient (Crawford & Wilson, 2019). NTV is a very broad definition. A traditional NTV is often a sports utility vehicle (SUV) staffed by medical personnel (see Fig. 7c). This is the type of vehicle used in European-style stay and stabilize systems where an NTV staffed by a medical doctor is the sole respondent to the majority of patients. In a North American scoop and run system, an NTV may also be a medically trained fire or police unit. From 1977 to 2015, the number of fire incidents has decreased by 59% (Ahrens, 2017). Due to this extra capacity and the strategic positioning of fire companies, researchers and practitioners are exploring ways to use these resources in EMS systems (Swersey et al., 1993; McLay & Moore, 2012).

5.1.2 Facility Location with Tiered EMS

Tiered EMS systems bring new dynamics and modeling challenges to EMS facility location problems. In addition to providing reliable and timely EMS response, a tiered EMS system must ensure that the types of resources responding to a patient are appropriate for their need. Modeling this dynamic is often the central challenge of tiered EMS facility location models. There are three general approaches to tiered EMS models: (1) models that assume all patients are of equal priority, (2) models that differentiate between patient priority and assume priority is known with certainty at the time of dispatch, and (3) models that differentiate between patient priority and acknowledge the inherent uncertainty in patient classification. We explore all three of these approaches within this section.

Tiered EMS Without Patient Prioritization Many of the early tiered EMS facility location models are extensions of the models reviewed in Sect. 4. They provide insights into EMS strategy; however, they do not explicitly consider how a tiered EMS unit will respond to individual patients or patient priority classes. Rather, the tiered units are used to redefine coverage. For example, the simplest approach to model a tiered system is to require a covered demand node to be within the RTT of all vehicle types. This is the approach of the early models that position resources for *fire systems* with multiple vehicle types (Marianov & ReVelle, 1992). However, unlike fire emergencies that often require multiple types of vehicles, EMS emergencies offer more flexibility because they usually only need one appropriate ambulance. Another approach is to use lower-acuity vehicle types to fill the coverage gaps of the higher-acuity ambulances. For example, a covered demand node must be within an RTT from an ALS ambulance; however, this RTT is more lenient (longer) for demand nodes that can be quickly reached by a BLS ambulance (Mandell, 1998). A slightly different tiered approach allows BLS ambulances to be the sole respondent to patients believed to meet their criteria for response; however, there is a chance that these patients need advanced care and trigger an ALS unit to also be dispatched (Marianov & Serra, 2001). While these methods utilize ALS and BLS units in a variety of ways, none of these approaches prioritize the need of patients with truly life-threatening emergencies.

Tiered EMS with Known Patient Prioritization The manner in which tiered vehicles are used for patients with varying conditions and levels of urgency is an important dynamic of tiered EMS systems with many practical implications. One approach is to use higher-acuity ambulances for high-priority patients and use lower-acuity ambulances as the sole respondent to lower-acuity patients. Only when all higher-acuity ambulances are busy will a lower-acuity ambulance be used as the sole respondent for a high-priority patient. The previously mentioned MEXCLP2 uses such an approach (McLay, 2009) as do recent papers that explore more complicated vehicle substitutions (Nelas & Dias, 2021). However, in practice, these types of vehicle substitutions are likely unfavorable, and some models adjust the reward of responding to a call depending on if the patient-vehicle type match. In other words, they provide a full reward when the responding ambulance meets or exceeds the patient need and a partial reward when the assigned vehicle is lower than a patient's need (Chong et al., 2016; Yoon et al., 2021) or do not allow any coverage when the vehicle does not meet a patient's need (Boujemaa et al., 2018). While these models may account for the uncertainty in demand, service time, and ambulance availability, they assume that the information used to prioritize patient response and inform ambulance type decisions is known with certainty. In practice, EMS dispatchers must predict this information using the limited information provided during the initial EMS request, which may be inaccurate.

Tiered EMS with Patient Prioritization Under Uncertainty Following a request for emergency care, an EMS dispatcher must use the limited information provided during the phone screening to decide which EMS unit should respond to the patient. Often EMS personnel act conservatively and assume a higher level of urgency (over-triage) to avoid delaying the patient from potentially life-saving care. Although there is no standard for appropriate rates of over- and under-triage, previous studies have estimated that over- and under-triage rates are as high as 78% and 4%, respectively (Dami et al., 2015). Optimal dispatching models have demonstrated that high uncertainty in patient classification leads to more urgent responses to mid-priority patients (McLay & Mayorga, 2013), and many researchers continue to investigate methods to assess and improve EMS triage accuracy (Bohm & Kurland, 2018; Alotaibi et al., 2021). However, it is unlikely that an EMS system will ever be perfectly accurate, and one method to hedge against this risk is through *multiple response*: the EMS practice in which multiple units are sent to the same patient. This approach ensures that the vehicle that best meets a patient's need responds to the patient while allowing the unneeded vehicle to serve another patient sooner and, hence, influences the optimal location of EMS resources (Yoon & Albert, 2020). Surprisingly, misclassification in patient priority and the subsequent mitigating actions is not well explored in EMS facility location model literature and is certainly an area of opportunity for the field.

5.2 *EMS Systems with Relocation*

Thus far, we have presented the location of EMS units as static and implicitly assumed that whenever an EMS unit is not in use, it is waiting at a specific location. However, given the cyclic trends in EMS demand and the mobility of EMS units, *ambulance relocation*, defined as strategically moving EMS vehicles to improve EMS coverage, is a topic that has received much attention in EMS research. This is also commonly called System Status Management (SSM) in practice. There are two primary classifications for EMS relocation models: *multi-period relocation* and *dynamic relocation*. A multi-period relocation model relocates EMS units to account for the spatial-temporal trends in EMS demand, such as those described in Sect. 4.2.2. A dynamic relocation model relocates EMS units to account for gaps in coverage that form as EMS units become busy. In this section, we review both approaches.

5.2.1 Multi-period Relocation Models

Multi-period relocation models account for the cyclic spatial-temporal changes in demand for EMS service. For example, during the day, many people are at an office, school, or another commercial location. A strategic EMS system may increase the number of units located near commercial areas during these times. Then, during the evening, most people return home; therefore, it may be advantageous to shift EMS units to residential areas during these times (see Fig. 8). This is the basic idea of multi-period relocation models.

The Maximal Expected Coverage Location Model with Time Variation (TIMEXCLP) is often credited as the first EMS relocation model (Repede & Bernardo, 1994). TIMEXCLP is an adaptation of MEXCLP developed in partnership with an EMS system in Louisville, Kentucky, and accounts for changes in the EMS fleet size and spatial-temporal demand patterns. This application demonstrates the importance of relocation by showing that 95% coverage, which required nine ambulances under the MEXCLP solution, is achievable with eight ambulances if relocation is used. This model does not utilize the advancements of the HQM approach and is therefore subject to the limiting assumptions described in Sect. 4.1.1. Nearly 15 years later, the Dynamic Available Coverage Location Model (DACL), a multi-period LSCM model, was suggested as an extension of the QPLCP (mentioned in Sect. 4.1.2) with a correction factor to account for server dependence (Rajagopalan et al., 2008). However, a possible shortcoming of both of these models is they do not limit ambulance relocations. In practice, a system that relocates ambulances too often could be difficult to implement, increase the workload for EMS personnel, and cause fatigue (Bledsoe, 2003). To minimize relocations, TIMEXCLP was later extended to include relocation and facility costs (van den Berg & Aardal, 2015), and DACL was extended to minimize relocations (Saydam et al., 2013). Some additional multi-period relocation models include extensions that account for time-dependent travel

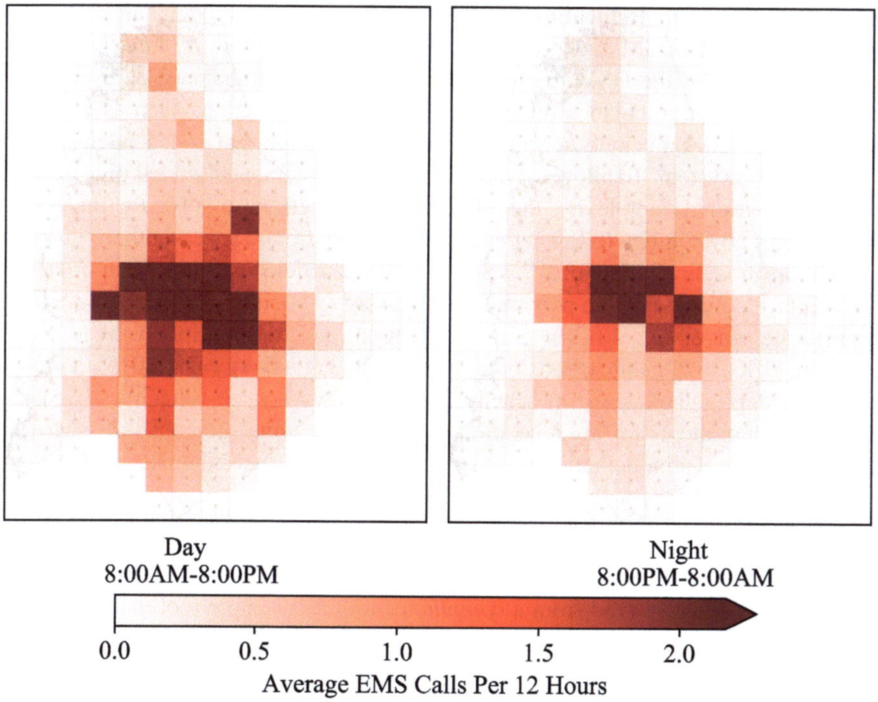

Fig. 8 Heat map of EMS demand in Mecklenburg County, North Carolina (2004), which includes the city of Charlotte and the surrounding area. The time of day influences the volume and distribution of EMS demand (data provided by MEDIC)

times (Schmid & Doerner, 2010; Degel et al., 2015), a multi-period extension of the DSM (Başsar et al., 2011), a model that considers personnel workloads (Enayati et al., 2018), and a model tailored for tiered EMS systems (Boujemaa et al., 2020).

5.2.2 Dynamic Relocation Models

Whereas multi-period relocation models account for predictable changes in EMS demand, dynamic relocation models actively adjust ambulance deployment based on the state of the system. As ambulances become busy, dynamic relocation models shift other EMS units to fill coverage gaps. The first models in this area were *real-time* dynamic models, which were intended to be solved every time an ambulance was dispatched. However, due to computational challenges and complexity, implementing these models in practice is difficult. Alternatively, *offline* dynamic models provide pre-computed strategies that can be implemented by following a table or other tools. Lastly, we separate relocation models that use *approximate dynamic programming* to generate admissible solutions to larger problems while accounting for additional levels of information (e.g., ongoing trip duration, travel destination, attributes of queued demand, etc.).

Real-Time Dynamic Relocation Models The first dynamic model, named RP^t, is an extension of the DSM (Gendreau et al., 2001). At its simplest, RP^t can be viewed as an extension of the DSM that is solved whenever an ambulance is dispatched. However, to limit ambulance relocation and undesirable actions such as a round-trip, RP^t includes a penalty term in its objective function that is dependent on the history of the system. To allow for the model to be implemented in real time despite the computational complexity, the authors offer a method to solve for the future dispatching possibilities following each decision. There are several other real-time relocation models similar to RP^t. Some of these extensions include approaches that replace coverage with a preparedness measure (Andersson & Väarbrand, 2007; van Barneveld et al., 2017a), a model that allows for multiple vehicle types (Mason, 2013), a dynamic extension of MEXCLP (Jagtenberg et al., 2015), a model that considers ambulance shift schedules (Naoum-Sawaya & Elhedhli, 2013), an extension that allows for the relaxation of the double-coverage constraint (Moeini et al., 2015), and a model tailored for rural areas (van Barneveld et al., 2017a).

Offline Dynamic Relocation Models Real-time relocation models are difficult to implement in practice due to computational complexity, technology limitations, and workflow burdens. EMS relocation models that focus on developing strategies that can be used offline or read from a *compliance table*—a simple tool that tells EMS managers where they should locate ambulances given high-level metrics (number of available ambulances, time of day)—are preferred in many settings. For example, as mentioned previously, the RP^t computes future dispatching possibilities following each EMS unit dispatch. However, in the event of two EMS requests in a short period, the model may not have been solved to completion. To address this, the authors later proposed the Maximal Expected Coverage Relocation Problem (MECRP), one of the earliest offline models (Gendreau et al., 2006). MECRP pre-computes the optimal ambulance location given the number of ambulances while limiting the number of ambulance moves. Other models similar to MECRP that generate offline compliance tables include a Markov-chain model that can quickly assess the performance of many compliance tables (Alanis et al., 2013), an extension that allows for two types of vehicles (van Barneveld et al., 2017b), an approach to consider the variation of demand patterns over time and changes in response times (Nair & Miller-Hooks, 2009), and a model that considers call priorities (Sudtachat et al., 2014).

ADP Dynamic Relocation Models Dynamic Programming (DP) is well-suited to optimize processes with sequential decisions, such as ambulance relocation. However, due to the high level of detail needed to accurately represent an EMS system, DP models quickly become intractable (due to large state spaces). Approximate Dynamic Programming (ADP) is a powerful tool that can be used to overcome this limitation and generate admissible solutions to complex stochastic and dynamic problems. For ADP applied to EMS relocation, the research by Maxwell et al. (2010) is often cited as the seminal work in this field. In their paper, the authors determine where an EMS unit should be positioned after it transfers a patient to the hospital. To inform these decisions, the ADP model uses a *basis function*, a linear function that encodes many details about the state of the EMS system to predict the cost of a relocation decision. The inputs of this basis function are inspired by queuing theory and an EMS location model. The weights of these inputs are tuned using least squares regression to minimize the error between the basis function and a cost found via simulation. It is important to note that in any ADP model, the final relocation strategy may not be the optimal solution because model convergence is highly dependent on the basis function and how the parameters are tuned. However, an admissible solution generated by an ADP can inform relocation decisions much faster than other real-time models, offers greater flexibility than the offline compliance tables since the user can deviate from the suggested solution, and allows for solving higher-dimensional problems. Two other refinements to EMS relocation using ADP include an approach that considers time-dependent travel times and demand (Schmid, 2012) and an approach that allows for the relocation of any idle EMS unit (Nasrollahzadeh et al., 2018). We note that since ADPs do not guarantee an optimal solution, methods can be used to develop bounds on the optimal solution (Maxwell et al., 2014; Nasrollahzadeh et al., 2018).

5.3 *EMS in Low- and Middle-Income Countries*

Time-sensitive medical emergencies, like cardiac arrests, motor vehicle accidents, and child or maternal health issues, are a major health concern in low- and middle-income countries (LMICs), comprising over 50% of all deaths (Moresky et al., 2019). Globally, LMICs are home to approximately 85% of the world's population and 90% of all healthcare emergencies (Lecky et al., 2020; Prydz & Wadhwa, 2019). As a consequence, researchers and international organizations have stressed the need for increased focus on emergency medical care in LMICs with several calls to action (United Nations, 2010). Despite these calls and the widespread evidence that emergency medical care in LMICs saves lives, poor access and availability continues to be a major problem, with the lack of EMS as one of the main barriers (Moresky et al., 2019; Lecky et al., 2020). Many of the high-level challenges associated with EMS in LMICs are well-suited to operations research and management science tools. For example, recent surveys indicate that poor performance is a major barrier for patients attempting to access EMS services, with 20–30% of those surveyed indicating they wanted to take an ambulance, but response times were too slow or that an ambulance was not available (i.e., busy at another call) (Boutilier & Chan, 2020). Moreover, most LMICs are resource-constrained, presenting an opportunity to optimize the use of EMS resources to improve effectiveness and efficiency.

Early research on EMS in LMICs was primarily focused on rural areas, including a variant of the MLCP to determine the optimal location of rural health clinics in Colombia (Bennett et al., 1982). A few years later, motivated by the lack of an existing EMS system in the Dominican Republic, the HOSC formulation (see Sect. 3.2) was extended to include two objectives: one to minimize the number of resources and one to maximize demand covered multiple times (Eaton et al., 1986). At the time, existing models that accounted for ambulance availability required a pre-determined "busy fraction" for each ambulance base, which did not exist for the problem of designing a new EMS system, and motivated the authors to develop a multiple coverage formulation. More recently, researchers have shifted their focus to urban areas in LMICs. The first paper to explore EMS response optimization in a developing urban center leveraged MEXCLP to determine ambulance locations and conduct a case study in Bangkok, Thailand (Fujiwara et al., 1987). Since then, there have been several papers that optimize EMS response in urban centers around the world (Boutilier & Chan, 2020).

In general, EMS network design and response optimization in LMICs presents several unique challenges that are not typically found in high-income settings, which we highlight below:

- Traffic can be extremely heavy (especially in urban areas), and it is often not the cultural norm to yield for ambulances, implying that ambulances must face the same traffic as regular road users. In contrast, ambulances are typically able to drive "fast" in high-income countries where it is the cultural norm to yield and where infrastructure is designed with EMS in mind (e.g., extra wide shoulders on freeways). The difference in road speed and consequential uncertainty in response time must be accounted for in models designing EMS systems for LMICs (Boutilier & Chan, 2020).
- Many LMICs do not have a centralized EMS system, instead relying on decentralized ambulance services comprised of private and hospital-owned fleets. The lack of a centralized EMS system (or even a coordinated public health system) presents significant challenges associated with data collection, including data that may be required for many facility location models (Eaton et al., 1986). In addition, the decentralized nature of these services can lead to a phenomenon known as "ambulance abandonment," where patients simultaneously request service from multiple ambulance providers and take the first one to arrive (Marla et al., 2021). Future work is needed to explore the competitive nature of these decentralized EMS systems.
- Road infrastructure in LMICs is typically not designed for modern EMS resources, especially in dense urban areas and "old towns." For example, many areas in urban slums cannot be accessed by four-wheeled vehicles (like ALS or BLS ambulances) and require non-traditional ambulance designs, such as motorcycles and three-wheeled vehicles (see Fig. 9a). Moreover, current ambulance designs and staff certifications in many LMICs are akin to the early EMS systems in high-income settings, where ambulances were staffed by "ambulance drivers" with limited first-aid training (rather than certified paramedics or physicians) and focused on patient transport, rather than treatment. These differences require innovative solutions and models designed to account for multiple vehicle types, restricted road access, limited transportation options, and lack of treatment in place.

Fig. 9 EMS facility location models can be adapted for settings that present unique modeling challenges. (**a**) Non-traditional ambulance designs, such a motorcycles, are common in LMIC settings. Photo by Justin Boutilier. (**b**) EMS helicopters offer increased speed and maneuverability for remote patients. https://pixabay.com/photos/helicopter-rescue-6392253/ (https://pixabay.com/service/license/). (**c**) UH-60M Blackhawk helicopters provide urgent care and evacuation to military casualties. Photo by Nolan Donahue, Given to and edited by Eric Stratman. (**d**) Search and rescue applications utilize sea and air units with various capabilities. https://pixabay.com/photos/coast-guard-sea-maritime-rescue-6718306/ (https://pixabay.com/service/license/)

5.4 *Other Unique Settings in EMS Facility Location*

The directions discussed in this section are a sampling of the unique areas in EMS facility location research. There are numerous directions for applications of EMS facility location (aircraft, military, following natural disasters, wilderness applications, etc.), and in this section, we provide a summary of a couple of these areas to demonstrate how EMS facility location models may be adapted to unique settings.

AED Location Models Automatic electronic defibrillators (AEDs) are portable devices designed to deliver life-saving electric shocks to a patient experiencing an out-of-hospital cardiac arrest (OHCA). AEDs are safe and easy to use by untrained bystanders and significantly improve an OHCA patient's chances of survival. An OHCA patient that receives an AED shock administered by a bystander is 2.36 times more likely to survive than a patient without AED shock, and the impact of EMS delays is significantly reduced following AED treatment (Stoesser et al., 2021). To improve accessibility, location models similar to those in EMS are used to located AEDs and have performed better than the traditional population-based AED location methods (Chan et al., 2013). In these models, predicting the location and timing of OHCA events is especially important since AEDs are often retrieved by bystanders that only travel a short distance from the patient (approximately 100 meters) and may face barriers to AED access (Sun et al., 2016). AED location research builds upon the EMS facility location models and explores dynamics unique to the AED application including uncertainty in the time before an OHCA is detected (Dao et al., 2012), the dimensionality of locating AEDs in multi-story buildings (Dao et al., 2012; Chan, 2017), and emphasise on the spatial probability distribution of demand (Chan et al, 2017).

Aircraft Location Models in EMS Systems Aircraft are an important feature of some EMS systems. Although more expensive and dangerous than ground units, transportation aircraft may be advantageous in some settings due to their increased speed and maneuverability (Steenhoff et al., 2022). Generally, it is only recommended that helicopters be used for urgent patients that cannot be reached by ground-based units in 30 minutes (Godfrey & Loyd, 2021), and airplanes are only used for missions that require over 120 miles (193 km) of travel (Urdaneta et al., 1987). Historical records indicate that 64% of all patients served by aircraft suffered a traumatic injury (data provided by NEMSIS). Given the number of trauma patients served by EMS aircraft, many of the aircraft location models consider the optimal allocation of both trauma centers and EMS aircraft (Branas et al., 2001); however, this leads to added complexity when computing system busy fractions (Cho et al., 2014). Another approach is to exploit the added range of the aircraft for backup coverage in a tiered system with ground-based units (Erdemir et al., 2010).

In addition to airplanes and helicopters, unmanned aerial vehicles such as drones are becoming more common in healthcare applications (Scott & Scott, 2017). Several studies explore how drones can be used to transport medical supplies to rural areas (Kim et al., 2017; Knoblauch et al., 2019). In EMS systems, drones perform a variety of functions including delivering AEDs, critical blood products, and medication before EMS arrives to enable timely care to patients (Johnson et al., 2021). This is especially true in areas with limited EMS coverage. In a drone AED delivery system in Toronto, the optimal deployment of drone bases allows for a nearly 7-minute reduction in the 90th percentile of urban response time (Boutilier et al., 2017). In Sweden, a similar system using drones has already saved the life of a cardiac arrest patient (Schierbeck et al., 2021; BBC, 2022). This demonstrates the value of emerging technology in EMS and how EMS facility location research can be used to influence the deployment of these advanced technologies.

Military Medical Evacuation Location Models Military medical evacuation (MEDEVAC) systems provide urgent care and evacuation to military casualties. Unlike EMS systems in which the primary time-sensitive medical emergency is often cardiac arrest, loss of blood is the direct cause of 85% of soldiers killed in action (Garrett, 2013). To account for the treatment of battlefield injuries,

MEDEVAC coverage models generally allow for longer RTT thresholds of 60 minutes for high-priority patients (Bastian et al., 2012; Grannan et al., 2015) or prioritize patients by their condition (Fulton et al., 2010). Other works specify that MEDEVAC systems need to focus on the time interval from when a soldier is injured until the time that they are delivered to a facility equipped to safely perform blood transfusion, with a focus on the inherent stochasticity of this process (Lejeune & Margot, 2018). Another key difference is in the arrivals; while traditional EMS setting requests can be viewed as independent events, battlefield injuries generally occur in batches following an attack. Therefore, some of the MEDEVAC models try to avoid queuing assumptions that rely on a Poisson process. These models generally account for non-Poisson arrivals through either empirical or simulated data with some form of robust or scenario optimization (Fulton et al., 2010; Bastian, 2010; Bastian et al., 2012). We also note that the objectives of MEDEVAC facility location may balance facility vulnerability to adverse attacks (Bastian, 2010) and the required capabilities of MEDEVAC units (Bastian et al., 2012). Lastly, models that inform MEDEVAC dispatching policies are also influential within this stream of research (Rettke et al., 2016; Keneally et al., 2016).

Search and Rescue Location Models Search and rescue (SAR) resource allocation is a field with many connections to EMS facility location problems. SAR systems, such as those deployed by national coast guard units, provide potentially life-saving aid to people in danger, and the location of resources is critical to providing this care in a timely manner. However, unlike EMS, the exact location of the person in need may be unknown in SAR missions, and conditions (weather, time of day) may dramatically impact the search effort, which is reflected in modeling approaches (Başsdemir & Melih, 2000; Abi-Zeid & Frost, 2005). Furthermore, unlike EMS systems where there is ample data to predict where an EMS request might be generated, the vastness of the areas patrolled by SAR makes it difficult to predict where future requests will occur, especially in maritime settings. Models have used a variety of methods to predict future demand (Azofra et al., 2007; Akbari et al., 2018). Additionally, SAR units must balance scheduled duties, operational and political rules, and fleet capabilities for a variety of watercraft and aircraft (Wagner et al., 2012; Karatas et al., 2017). We refer the reader to the chapter of Karatas et al. (2019) for a detailed review of SAR models.

6 Implementation of EMS Facility Location Models

The goal of EMS facility location problems is to develop strategies to better serve patients. Despite all of the academic contributions cited within this chapter, not all papers *directly* contribute to the development of an actual EMS implementation. As stated by Chaiken (1978), "careful studies of the actual use of models by decision-makers have drawn sobering conclusions about the chances that such models will actually be applied as intended." Although many models indirectly bolster the field by developing newer and more elegant approaches, this field needs to keep the ultimate goal in mind and avoid turning into an academic exercise. From our review, we summarize the lessons learned and the important themes from several implementation-based papers.

Design for the User Successful implementations of EMS facility location models need to be designed with the decision-maker in mind. While we found numerous implementations of deterministic and simple probabilistic models, we found very few implementations of the more complicated probabilistic models. This is likely because these simpler models are easier to explain and be accepted by an EMS agency. This is the same rationale that made offline relocation models more preferable in our discussion in Sect. 5.2.2. Similarly, in an implementation of the probabilistic MEXCLP in Lexington, Kentucky, that project was largely successful because the authors worked closely with the decision-makers, utilized a graphical interface to explain their ideas (Repede et al., 1993), and gained the EMS agency's trust by modifying their model with their feedback (Repede & Bernardo, 1994). Similar collaborations and validations are noted in other implementations (Brandeau & Larson, 1986; Goldberg et al., 1990).

Guide EMS Practice EMS facility location models must be timely and ahead of EMS practice. Following the implementation of an EMS facility location design in Morgantown, West Virginia, the researchers suggested that sophisticated models are only appropriate when given sufficient time for careful model formulation (Baker et al., 1980). This is perhaps why many of the most documented EMS facility location model implementations are in LMICs with emerging EMS infrastructures, as discussed in Sect. 5.3. Unlike areas with well-established EMS operations that are unlikely to change their existing strategy or facility locations, these emerging infrastructures are less committed and often open to change. The recent implementations of EMS facility location models to drone AED delivery systems is another example of research being timely and accessible to strategic planning efforts (Boutilier et al., 2017; Schierbeck et al., 2021; Boutilier & Chan, 2022).

Acknowledge EMS Legal Limitations Any implementation will be subject to many legislative barriers. According to the General Data Protection Regulation (GDPR) law of the European Union, human intervention can be requested, and an adequate explanation must be provided for any automated decision (Olhede & Wolfe, 2018). Therefore, EMS location models must be robust enough that they still perform well should they be overridden by a human decision-maker and intuitive enough should an explanation be requested. Once again, these requirements limit the complexity of EMS location models. Similarly, EMS agencies are required to respond to calls in a reasonable manner. An EMS agency may prefer to always send the closest or most highly trained EMS unit to patients to avoid potential legal pitfalls. Within the next decade, we hope that EMS researchers expand the visibility of their work to lawmakers to demonstrate the benefit of more progressive EMS strategies and allow for more cost-effective, efficient, and patient-centered EMS care.

Focus on Accessibility Although not every EMS facility location model is designed with a specific implementation in mind, there are many ways in which EMS facility location research can make an impact. Many EMS implementations use commercial tools, such as ESRI's Geographic Information System (GIS) software, to inform their location problem (Foo et al., 2010). Perhaps the best way for EMS facility location modelers to have a direct impact in the field is through closer collaboration with these accessible tools. As another example, the relocation model developed by Mason (2013) has been implemented by multiple ambulance operators through an associated software package.

We conclude by acknowledging that it was difficult to find clear summaries of EMS systems that implement the results of facility location models. As a field dedicated to continuous improvement, we believe EMS models in action should receive greater attention (even when unsuccessful) to guide future directions and modeling efforts.

7 Additional Resources

In this chapter, we reviewed the models and methods of EMS facility location problems under uncertainty. We focused on the overall themes within this field and aimed to provide the reader with an understanding of these models and their evolution. In this section, we summarize resources that present more in-depth technical reviews of this field for a comprehensive literature review. Since there is no shortage of facility location review papers, we limit the cited works to those that focus on EMS or address another focus of this chapter.

- The work of ReVelle et al. (1977) is the earliest literature review of EMS facility location problems. They present the deterministic LSCM, MCLP, and PMP models context-free and discuss the application of facility location problems to EMS and other emergency services.
- The review by Brotcorne et al. (2003) summarizes the formulation of deterministic single and multi-coverage models as well as the early probabilistic models. They conclude with a summary of an early dynamic model and predicted their continued rise within the field.
- Goldberg (2004) provides a review of the methods that support EMS facility location models, such as travel time modeling, demand prediction, testing model validity, and queuing. They classify existing models using a different method than the previous reviews and cite models that were directly implemented.
- The chapter by Henderson (2011) provides a specific focus on EMS models with relocation, such as those discussed in Sect. 5.2, and provides additional discussion of this practice.
- The summary by Başsar et al. (2012) provides a helpful taxonomy of facility location problems as applied to emergency services. They classify models based

on their solution methodology, application, objective function, and parameters in a concise table.

- Aringhieri et al. (2017) offer a slightly different perspective than the other review papers. In the first part of their paper, they address EMS facility location problems with sections related to probability, stochastic, relocation, and equity. Then, they discuss EMS dispatching and routing policies and the important connection with other components of the healthcare system.
- The chapter by Karatas et al. (2019) provides a review of military facility location problems and provides greater details about the MEDEVAC and SAR settings discussed in Sect. 5.4.
- The review paper by Bélanger et al. (2019) provides an in-depth review of deterministic and probabilistic models, providing more details on the model formulations. They also offer a thorough discussion on equity, relocation, and dispatching decisions in EMS systems.

Acknowledgments The authors acknowledge the Mecklenburg EMS Agency and NEMSIS for data used in this chapter. The content reproduced from the NEMSIS Database remains the property of the National Highway Traffic Safety Administration. The National Highway Traffic Safety Administration is not responsible for any claims arising from works based on the original data, text, tables, or figures. The third author was in part supported by the National Science Foundation Award 2000986. Any opinions, findings, and conclusions or recommendations expressed in this material are those of the authors and do not necessarily reflect the views of the US Government or the National Science Foundation.

References

Abi-Zeid, I., & Frost, J. R. (2005). SARPlan: a decision support system for Canadian Search and Rescue operations. *European Journal of Operational Research, 162*(3), 630–653.

Ahrens, M. (2017). *Trends and patterns of U.S. fire loss*. Quincy, MA: National Fire Protection Association.

Akbari, A., Pelot, R., & Eiselt, H. A. (2018). A modular capacitated multiobjective model for locating maritime search and rescue vessels. *Annals of Operations Research, 267*(1), 3–28.

Al-Shaqsi, S. (2010a). Models of international emergency medical service (EMS) systems. *Oman Medical Journal, 25*(4), 320–323.

Al-Shaqsi, S. (2010b). Response time as a sole performance indicator in EMS: Pitfalls and solutions. *Open Access Emergency Medicine, 2*, 1–6.

Alanis, R., Ingolfsson, A., & Kolfal, B. (2013). A Markov chain model for an EMS system with repositioning. *Production and Operations Management, 22*(1), 216–231.

Alotaibi, A., Alghamdi, A., Reynard, C., & Body, R. (2021). Accuracy of emergency medical services (EMS) telephone triage in identifying acute coronary syndrome (ACS) for patients with chest pain: a systematic literature review. *BMJ Open, 11*(8), e045815.

Alsalloum, O. I., & Rand, G. K. (2006). Extensions to emergency vehicle location models. *Computers & Operations Research, 33*(9), 2725–2743.

Andersson, T., & Väarbrand, P. (2007). Decision support tools for ambulance dispatch and relocation. *The Journal of the Operational Research Society, 58*(2), 195–201.

Ansari, S., Yoon, S., & Albert, L. A. (2017). An approximate hypercube model for public service systems with co-located servers and multiple response. *Transportation Research Part E: Logistics and Transportation Review, 103*, 143–157.

Aringhieri, R., Bruni, M. E., Khodaparasti, S., & van Essen, J. T. (2017). Emergency medical services and beyond: addressing new challenges through a wide literature review. *Computers & Operations Research, 78*, 349–368.

Azofra, M., Péerez-Labajos, C. A., Blanco, B., & Achúutegui, J. J. (2007). Optimum placement of sea rescue resources. *Safety Science, 45*(9), 941–951.

Bélanger, V., Ruiz, A., & Soriano, P. (2019). Recent optimization models and trends in location, relocation, and dispatching of emergency medical vehicles. *European Journal of Operational Research, 272*(1), 1–23.

Başsar, A., Çatay, B., & Ünlüyurt, T. (2011). A multi-period double coverage approach for locating the emergency medical service stations in Istanbul. *The Journal of the Operational Research Society, 62*(4), 627–637.

Başsar, A., Çatay, B., & Ünlüyurt, T. (2012). A taxonomy for emergency service station location problem. *Optimization Letters, 6*(6), 1147–1160.

Başsdemir, M. M. (2000). Locating search and rescue stations in the Aegean and western Mediterranean regions of Turkey. *Journal of Aeronautics and Space Technologies, 1*(3), 63–76.

Bagai, A., McNally, B. F., Al-Khatib, S. M., Brent Myers, J., Kim, S., Karlsson, L., Torp-Pedersen, C., Wissenberg, M., van Diepen, S., Fosbol, E. L., Monk, L., Abella, B. S., Granger, C. B., & Jollis, J. G. (2013). Temporal differences in out-of-hospital cardiac arrest incidence and survival. *Circulation, 128*(24), 2595–2602.

Baker, D. W. & Byrd, J. (1980). A lesson in timing: a nonemergency solution to an emergency service decision. *Interfaces, 10*(3), 30–33.

Ball, M. O. & Lin, F. L. (1993). A reliability model applied to emergency service vehicle location. *Operations Research, 41*(1), 18–36.

Bastian, N. D. (2010). A robust, multi-criteria modeling approach for optimizing aeromedical evacuation asset emplacement. *The Journal of Defense Modeling and Simulation, 7*(1), 5–23.

Bastian, N. D., Fulton, L. V., Mitchell, R., Pollard, W., Wierschem, D., & Wilson, R. (2012). The future of vertical lift: initial insights for aircraft capability and medical planning. *Military Medicine, 177*(7), 863–869.

Batta, R., Dolan, J. M., & Krishnamuthy, N. N. (1989). The maximal expected covering location problem: revisited. *Transportation Science, 23*(4), 277–287.

BBC. (2022). Drone helps save cardiac arrest patient in Sweden. *BBC News*.

Bennett, V. L., Eaton, D. J., & Church, R. L. (1982). Selecting sites for rural health workers. *Social Science & Medicine, 16*(1), 63–72.

Beraldi, P., & Bruni, M. E. (2009). A probabilistic model applied to emergency service vehicle location. *European Journal of Operational Research, 196*(1), 323–331.

Beraldi, P., Bruni, M. E., & Conforti, D. (2004). Designing robust emergency medical service via stochastic programming. *European Journal of Operational Research, 158*(1), 183–193.

Berman, O., Drezner, Z., & Krass, D. (2010). Generalized coverage: new developments in covering location models. *Computers & Operations Research, 37*(10), 1675–1687.

Bianchi, G., & Church, R. L. (1988). A hybrid fleet model for emergency medical service system design. *Social Science & Medicine, 26*(1), 163–171.

Bledsoe, B. (2003). EMS myth #7: system status management lowers response times and enhances patient care. *EMSWorld, 32*(9), 158–159.

Bohm, K., & Kurland, L. (2018). The accuracy of medical dispatch - a systematic review. *Scandinavian Journal of Trauma, Resuscitation and Emergency Medicine, 26*, 94.

Boujemaa, R., Jebali, A., Hammami, S., Ruiz, A., & Bouchriha, H. (2018). A stochastic approach for designing two-tiered emergency medical service systems. *Flexible Services and Manufacturing Journal, 30*(1), 123–152.

Boujemaa, R., Jebali, A., Hammami, S., & Ruiz, A. (2020). Multi-period stochastic programming models for two-tiered emergency medical service system. *Computers & Operations Research, 123*, 104974.

Boutilier, J. J., & Chan, T. C. Y. (2020). Ambulance emergency response optimization in developing countries. *Operations Research, 68*(5), 1315–1334.

Boutilier, J. J., & Chan, T. C. Y. (2022). Drone network design for cardiac arrest response. *Manufacturing & Service Operations Management, 24*(5), 2387–2796.

Boutilier, J. J., Brooks, S. C., Janmohamed, A., Byers, A., Buick, J. E., Zhan, C., Schoellig, A. P., Cheskes, S., Morrison, L. J., Chan, T. C. Y., & Rescu Epistry Investigators. (2017). Optimizing a drone network to deliver automated external defibrillators. *Circulation, 135*(25), 2454–2465.

Branas, C. C., & Revelle, C. S. (2001). An iterative switching heuristic to locate hospitals and helicopters. *Socio-Economic Planning Sciences, 35*(1), 11–30.

Brandeau, M., & Larson, R. C. (1986). Extending and applying the hypercube queueing model to deploy ambulances in Boston. *National Emergency Training Center, 22*, 121–153.

Brotcorne, L., Laporte, G., & Semet, F. (2003). Ambulance location and relocation models. *European Journal of Operational Research, 147*(3), 451–463.

Budge, S., Ingolfsson, A., & Zerom, D. (2010). Empirical analysis of ambulance travel times: the case of Calgary emergency medical services. *Management Science, 56*(4), 716–723.

Chaiken, J. M. (1978). Transfer of emergency service deployment models to operating agencies. *Management Science, 24*(7), 719–731.

Chan, T. C. Y. (2017). Rise and shock: optimal defibrillator placement in a high-rise building. *Prehospital Emergency Care, 21*(3), 309–314.

Chan, T. C. Y., Shen, Z.-J. M., & Siddiq, A. (2017). Robust defibrillator deployment under cardiac arrest location uncertainty via row-and-column generation. *Operations Research, 66*(2), 358–379.

Chan, T. C. Y., Li, H., Lebovic, G., Tang, S. K., Chan, J. Y. T., Cheng, H. C. K., Morrison, L. J., & Brooks, S. C. (2013). Identifying locations for public access defibrillators using mathematical optimization. *Circulation, 127*(17), 1801–1809.

Channouf, N., L'Ecuyer, P., Ingolfsson, A., & Avramidis, A. (2007). The application of forecasting techniques to modeling emergency medical system calls in Calgary, Alberta. *Health Care Management Science, 10*, 25–45.

Cho, S.-H., Jang, H., Lee, T., & Turner, J. (2014). Simultaneous location of trauma centers and helicopters for emergency medical service planning. *Operations Research, 62*(4), 751–771.

Chong, K. C., Henderson, S. G., & Lewis, M. E. (2016). The vehicle mix decision in emergency medical service systems. *Manufacturing & Service Operations Management, 18*(3), 347–360.

Church, R., & ReVelle, C. (1974). The maximal covering location problem. *Papers of the Regional Science Association, 32*(1), 101–118.

Crawford, W., & Wilson, S. (2019). Emergency medical services. *Alabama Department of Public Health Administrative Code* (p. 6). Alabama State Board of Health.

Dami, F., Golay, C., Pasquier, M., Fuchs, V., Carron, P.-N., & Hugli, O. (2015). Prehospital triage accuracy in a criteria based dispatch centre. *BMC Emergency Medicine, 15*, 32.

Dao, T. H. D., Zhou, Y., Thill, J.-C., & Delmelle, E. (2012). Spatiotemporal location modeling in a 3D indoor environment: the case of AEDs as emergency medical devices. *International Journal of Geographical Information Science, 26*(3), 469–494.

Daskin, M. S. (1983). A maximum expected covering location model: formulation, properties and heuristic solution. *Transportation Science, 17*(1), 48–70.

Daskin, M. S. (1987). Location, dispatching and routing models for emergency services with stochastic travel times. In A. Ghosh, & G. Rushton (Eds.), *Spatial analysis and location-allocation models* (pp. 224–265). Van Nostrand Reinhold Co.

Daskin, M. S. & Stern, E. H. (1981). A hierarchical objective set covering model for emergency medical service vehicle deployment. *Transportation Science, 15*(2), 137.

Daskin, M. S., Hogan, K., & ReVelle, C. (1988). Integration of multiple, excess, backup, and expected covering models. *Environment and Planning B: Planning and Design, 15*, 15–35.

Degel, D., Wiesche, L., Rachuba, S., & Werners, B. (2015). Time-dependent ambulance allocation considering data-driven empirically required coverage. *Health Care Management Science, 18*(4), 444–458.

Do, Y. K., Foo, K., Ng, Y. Y., & Ong, M. E. H. (2013). A quantile regression analysis of ambulance response time. *Prehospital Emergency Care, 17*(2), 170–176.

Eaton, D. J., Sánchez, U. H. MI., Lantigua, R. R., & Morgan, J. (1986). Determining ambulance deployment in Santo Domingo, Dominican Republic. *The Journal of the Operational Research Society, 37*(2), 113–126.

Eiselt, H. A. & Marianov, V. (2009). Gradual location set covering with service quality. *Socio-Economic Planning Sciences, 43*(2), 121–130.

Enayati, S., Mayorga, M. E., Rajagopalan, H. K., & Saydam, C. (2018). Real-time ambulance redeployment approach to improve service coverage with fair and restricted workload for EMS providers. *Omega, 79*, 67–80.

Erdemir, E. T., Batta, R., Rogerson, P. A., Blatt, A., & Flanigan, M. (2010). Joint ground and air emergency medical services coverage models: A greedy heuristic solution approach. *European Journal of Operational Research, 207*(2), 736–749.

Erkut, E., Ingolfsson, A., & Erdoğan, G. (2008). Ambulance location for maximum survival. *Naval Research Logistics, 55*(1), 42–58.

Erkut, E., Ingolfsson, A., Sim, T., & Erdoğan, G. (2009). Computational comparison of five maximal covering models for locating ambulances. *Geographical Analysis, 41*(1), 43–65.

Fitch, J. (2005). Response times: myths, measurement and management. *The Journal of Emergency Medicine, 30*(9), 47.

Fleischman, R. J., Lundquist, M., Jui, J., Newgard, C. D., & Warden, C. (2013). Predicting ambulance time of arrival to the emergency department using global positioning system and Google Maps. *Prehospital Emergency Care: Official Journal of the National Association of EMS Physicians and the National Association of State EMS Directors, 17*(4), 458–465.

Foo, C. P. Z., Ahghari, M., & MacDonald, R. D. (2010). Use of Geographic Information Systems to determine new helipad locations and improve timely response while mitigating risk of helicopter emergency medical services operations. *Prehospital Emergency Care, 14*(4), 461–468.

Fujiwara, O., Makjamroen, T., & Gupta, K. K. (1987). Ambulance deployment analysis: a case study of Bangkok. *European Journal of Operational Research, 31*(1), 9–18.

Fulton, L. V., Lasdon, L. S., McDaniel, R. R., & Nicholas Coppola, M. (2010). Two-stage stochastic optimization for the allocation of medical assets in steady-state combat operations. *The Journal of Defense Modeling and Simulation, 7*(2), 89–102.

Galvão, R. D., Chiyoshi, F. Y., & Morabito, R. (2005). Towards unified formulations and extensions of two classical probabilistic location models. *Computers & Operations Research, 32*(1), 15–33.

Garrett, M. X. (2013). *USCENTCOM review of MEDEVAC procedures in Afghanistan*. Technical Report. United States Central Command.

Gendreau, M., Laporte, G., & Semet, F. (1997). Solving an ambulance location model by Tabu search. *Location Science, 5*(2), 75–88.

Gendreau, M., Laporte, G., & Semet, F. (2001). A dynamic model and parallel Tabu search heuristic for real-time ambulance relocation. *Parallel Computing, 27*(12), 1641–1653.

Gendreau, M., Laporte, G., & Semet, F. (2006). The maximal expected coverage relocation problem for emergency vehicles. *The Journal of the Operational Research Society, 57*(1), 22–28.

Godfrey, A., & Loyd, J. W. (2021). EMS helicopter activation. *StatPearls*. Treasure Island (FL): StatPearls Publishing.

Goldberg, J. (2004). Operations research models for the deployment of emergency services vehicles. *EMS Management Journal, 1*(1), 20.

Goldberg, J., & Paz, L. (1991). Locating emergency vehicle bases when service time depends on call location. *Transportation Science, 25*(4), 264–280.

Goldberg, J., Dietrich, R., Chen, J. M., George Mitwasi, M., Valenzuela, T., & Criss, E. (1990). Validating and applying a model for locating emergency medical vehicles in Tuczon, AZ. *European Journal of Operational Research, 49*(3), 308–324.
Grannan, B. C., Bastian, N. D., & McLay, L. A. (2015). A maximum expected covering problem for locating and dispatching two classes of military medical evacuation air assets. *Optimization Letters, 9*(8), 1511–1531.
Hakimi, S. L. (1964). Optimum locations of switching centers and the absolute centers and medians of a graph. *Operations Research, 12*(3), 450–459.
Hakimi, S. L. (1965). Optimum distribution of switching centers in a communication network and some related graph theoretic problems. *Operations Research, 13*(3), 462–475.
Hausner, J. (1975). *Determining the travel characteristics of emergency service vehicles*. New York: RAND Corporation.
Henderson, S. G. (2011). Operations research tools for addressing current challenges in emergency medical services. In J. J. Cochran, L. A. Cox, P. Keskinocak, J. P. Kharoufeh, & J. Cole Smith (Eds.), *Wiley encyclopedia of operations research and management science*. Hoboken, NJ, USA: John Wiley & Sons, Inc. ISBN: 978-0-470-40053-1.
Hogan, K., & Revelle, C. (1986). Concepts and applications of backup coverage. *Management Science, 32*(11), 1434–1444.
Holmén, J., Herlitz, J., Ricksten, S.-E., Strömsöe, A., Hagberg, E., Axelsson, C., & Rawshani, A. (2020). Shortening ambulance response time increases survival in out-of-hospital cardiac arrest. *Journal of the American Heart Association, 9*(21), e017048.
Ingolfsson, A., Erkut, E., & Budge, S. (2003). Simulation of single start station for Edmonton EMS. *Journal of the Operational Research Society, 54*(7), 736–746.
Ingolfsson, A., Budge, S., & Erkut, E. (2008). Optimal ambulance location with random delays and travel times. *Health Care Management Science, 11*(3), 262–274.
Jagtenberg, C. J., Bhulai, S., & van der Mei, R. D. (2015). An efficient heuristic for real-time ambulance redeployment. *Operations Research for Health Care, 4*, 27–35.
Jagtenberg, C. J., Bhulai, S., & van der Mei, R. D. (2017). Dynamic ambulance dispatching: is the closest-idle policy always optimal? *Health Care Management Science, 20*(4), 517–531.
Jarvis, J. P. (1985). Approximating the equilibrium behavior of multi-server loss systems. *Management Science, 31*(2), 235–239.
Johnson, A. M., Cunningham, C. J., Arnold, E., Rosamond, W. D., & Zégre-Hemsey, J. K. (2021). Impact of using drones in emergency medicine: what does the future hold? *Open Access Emergency Medicine, 13*, 487–498.
Kamenetzky, R. D., Shuman, L. J., & Wolfe, H. (1982). Estimating need and demand for prehospital care. *Operations Research, 30*(6), 1148–1167.
Karatas, M., Razi, N., & Gunal, M. M. (2017). An ILP and simulation model to optimize search and rescue helicopter operations. *Journal of the Operational Research Society, 68*(11), 1335–1351.
Karatas, M., Yakıcı, E., & Razi, N. (2019). Military facility location problems: a brief survey. *Operations Research for Military Organizations*, 1–27. ISBN: 978-1-5225-5513-1.
Keneally, S. K., Robbins, M. J., & Lunday, B. J. (2016). A Markov decision process model for the optimal dispatch of military medical evacuation assets. *Health Care Management Science, 19*(2), 111–129.
Kim, S.-H., & Whitt, W. (2014). Are call center and hospital arrivals well modeled by nonhomogeneous Poisson processes? *Manufacturing & Service Operations Management, 16*(3), 464–480.
Kim, S. J., Lim, G. J., Cho, J., & Côté, M. J. (2017). Drone-aided healthcare services for patients with chronic diseases in rural areas. *Journal of Intelligent & Robotic Systems, 88*(1), 163–180.
Knight, V. A., Harper, P. R., & Smith, L. (2012). Ambulance allocation for maximal survival with heterogeneous outcome measures. *Omega, 40*(6), 918–926. Special Issue on Forecasting in Management Science.
Knoblauch, A. M., de la Rosa, S., Sherman, J., Blauvelt, C., Matemba, C., Maxim, L., Defawe, O. D., Gueye, A., Robertson, J., McKinney, J., Brew, J., Paz, E., Small, P. M., Tanner, M., Rakotosamimanana, N., & Lapierre, S. G. (2019). Bi-directional drones to strengthen

healthcare provision: experiences and lessons from Madagascar, Malawi and Senegal. *BMJ Global Health, 4*(4), e001541.

Kok, A. L., Hans, E. W., & Schutten, J. M. J. (2012). Vehicle routing under time-dependent travel times: the impact of congestion avoidance. *Computers & Operations Research, 39*(5), 910–918.

Krafft, T., García-Castrillo Riesgo, L., Fischer, M., Lippert, F., Overton, J., & Robertson-Steel, I. (2003). *EMS data-based health surveillance system project report* (p. 79). Munich: European Emergency Data Project.

Krishnan, K., Marla, L., & Yue, Y. (2016). Robust ambulance allocation using risk-based metrics. In *2016 8th International Conference on Communication Systems and Networks*, 1–6.

Kunkel, A., & McLay, L. A. (2013). Determining minimum staffing levels during snowstorms using an integrated simulation, regression, and reliability model. *Health Care Management Science, 16*(1), 14–26.

Larson, R. C. (1974). A hypercube queuing model for facility location and redistricting in urban emergency services. *Computers & Operations Research, 1*(1), 67–95.

Larson, R. C. (1975). Approximating the performance of urban emergency service systems. *Operations Research, 23*, 845–868.

Lecky, F. E., Reynolds, T., Otesile, O., Hollis, S., Turner, J., Fuller, G., Sammy, I., Williams-Johnson, J., Geduld, H., Tenner, A. G., French, S., Govia, I., Balen, J., Goodacre, S., Marahatta, S. B., De-Vries, S., Sawe, H. R., El-Shinawi, M., Mfinanga, J., Rubiano, A. M., Chebbi, H., Do Shin, S., Ferrer, J. M. E., Haddadi, M., Firew, T., Taubert, K., Lee, A., Convocar, P., Jamaluddin, S., Kotecha, S., Abu Yaqeen, E., Wells, K., & Wallis, L. (2020). Harnessing interdisciplinary collaboration to improve emergency care in low- and middle-income countries (LMICs): results of research prioritisation setting exercise. *BMC Emergency Medicine, 20*(1), 68.

Lejeune, M. A., & Margot, F. (2018). Aeromedical battlefield evacuation under endogenous uncertainty in casualty delivery times. *Management Science, 64*(12), 5481–5496.

Mandell, M. B. (1998). Covering models for two-tiered emergency medical services systems. *Location Science, 6*(1), 355–368.

Marianov, V., & ReVelle, C. (1992). A probabilistic fire-protection siting model with joint vehicle reliability requirements. *Papers in Regional Science, 71*(3), 217–241.

Marianov, V., & Revelle, C. (1994). The queuing probabilistic location set covering problem and some extensions. *Socio-Economic Planning Sciences, 28*(3), 167–178.

Marianov, V., & Serra, D. (2001). Hierarchical location-allocation models for congested systems. *European Journal of Operational Research, 135*(1), 195–208.

Marianov, V., & Serra, D. (2002). Location–allocation of multiple-server service centers with constrained queues or waiting times. *Annals of Operations Research, 111*(1), 35–50.

Marla, L., Krishnan, K., & Deo, S. (2021). Managing EMS systems with user abandonment in emerging economies. *IISE Transactions, 53*(4), 389–406.

Mason, A. J. (2013). Simulation and real-time optimised relocation for improving ambulance operations. In B. T. Denton (Ed.), *Handbook of healthcare operations management* (vol. 184, pp. 289–317). New York, NY: Springer New York. ISBN: 978-1-4614-5885-2.

Matteson, D. S., McLean, M. W., Woodard, D. B., & Henderson, S. G. (2011). Forecasting emergency medical service call arrival rates. *The Annals of Applied Statistics, 5*(2), 1379–1406.

Maxwell, M. S., Restrepo, M., Henderson, S. G., & Topaloglu, H. (2010). Approximate dynamic programming for ambulance redeployment. *INFORMS Journal on Computing, 22*(2), 266–281.

Maxwell, M. S., Cao Ni, E., Tong, C., Henderson, S. G., Topaloglu, H., & Hunter, S. R. (2014). A bound on the performance of an optimal ambulance redeployment policy. *Operations Research, 62*(5), 1014–1027.

McLay, L. A. (2009). A maximum expected covering location model with two types of servers. *IIE Transactions, 41*(8), 730–741.

McLay, L. A. & Mayorga, M. E. (2010). Evaluating emergency medical service performance measures. *Health Care Management Science, 13*(2), 124–136.

McLay, L. A. & Mayorga, M. E. (2013). A model for optimally dispatching ambulances to emergency calls with classification errors in patient priorities. *IIE Transactions, 45*(1), 1–24.

McLay, L. A. & Moore, H. (2012). Hanover county improves its response to emergency medical 911 patients. *Interfaces, 42*(4), 380–394.

Moeini, M., Jemai, Z., & Sahin, E. (2015). Location and relocation problems in the context of the emergency medical service systems: a case study. *Central European Journal of Operations Research, 23*(3), 641–658.

Moresky, R. T., Razzak, J., Reynolds, T., Wallis, L. A., Wachira, B. W., Nyirenda, M., Carlo, W. A., Lin, J., Patel, S., Bhoi, S., Risko, N., Wendle, L. A., & Calvello Hynes, E. J. (2019). Advancing research on emergency care systems in low-income and middle-income countries: ensuring high-quality care delivery systems. *BMJ Global Health, 4*(Suppl 6), e001265.

Mullie, A., Van Hoeyweghen, R., & Quets, A. (1989). Influence of time intervals on outcome of CPR. *Resuscitation, 17*, S23–S33.

Nair, R., & Miller-Hooks, E. (2009). Evaluation of relocation strategies for emergency medical service vehicles. *Transportation Research Record, 2137*(1), 63–73.

Naoum-Sawaya, J., & Elhedhli, S. (2013). A stochastic optimization model for real-time ambulance redeployment. *Computers & Operations Research, 40*(8), 1972–1978.

Narad, R. A., & Driesbock, K. R. (1999). Regulation of ambulance response times in California. *Prehospital Emergency Care, 3*(2), 131–135.

Narad, R. A., & Gillespie, W. (1998). The public vs private debate: Separating facts from values. *Prehospital Emergency Care, 2*(3), 196–202.

Nasrollahzadeh, A. A., Khademi, A., & Mayorga, M. E. (2018). Real-time ambulance dispatching and relocation. *Manufacturing & Service Operations Management, 20*(3), 467–480.

National Highway Traffic Safety Administration. (2014). *Traffic safety facts: EMS research note: EMS system demographics*. Department of Transportation Highway Safety 812 041; DOT HS 812 041. United States National Highway Traffic Safety Administration.

Nelas, J., & Dias, J. (2021). Locating emergency vehicles: modelling the substitutability of resources and the impact of delays in the arrival of assistance. *Operations Research Perspectives, 8*, 100202.

Nickel, S., Reuter-Oppermann, M., & Saldanha-da-Gama, F. (2016). Ambulance location under stochastic demand: a sampling approach. *Operations Research for Health Care, 8*, 24–32.

Noyan, N. (2010). Alternate risk measures for emergency medical service system design. *Annals of Operations Research, 181*(1), 559–589.

Olhede, S. C. & Wolfe, P. J. (2018). The growing ubiquity of algorithms in society: implications, impacts and innovations. *Philosophical Transactions of the Royal Society A: Mathematical, Physical and Engineering Sciences, 376*(2128), 20170364.

Pozner, C. N., Zane, R., Nelson, S. J., & Levine, M. (2004). International EMS systems: The United States: past, present, and future. *Resuscitation, 60*(3), 239–244.

Prydz, E. B., & Wadhwa, D. (2019). *Classifying countries by income*. The World Bank.

Rajagopalan, H. K., Saydam, C., & Xiao, J. (2008). A multiperiod set covering location model for dynamic redeployment of ambulances. *Computers & Operations Research, 35*(3), 814–826.

Repede, J. F., & Bernardo, J. J. (1994). Developing and validating a decision support system for locating emergency medical vehicles in Louisville, Kentucky. *European Journal of Operational Research, 75*(3), 567–581.

Repede, J. F., Jeffries, C. J., & Hubbard, E. (1993). ALIAS: a graphical user interface for an ambulance location model. *International Journal of Operations & Production Management, 13*(12), 36–46.

Restrepo, M., Henderson, S. G., & Topaloglu, H. (2009). Erlang loss models for the static deployment of ambulances. *Health Care Management Science, 12*(1), 67–79.

Rettke, A., Robbins, M., & Lunday, B. (2016). Approximate dynamic programming for the dispatch of military medical evacuation assets. *European Journal of Operational Research, 254*, 824–839.

ReVelle, C., & Hogan, K. (1988). A reliability-constrained siting model with local estimates of busy fractions. *Environment and Planning B: Planning and Design, 15*(2), 143–152.

ReVelle, C., & Hogan, K. (1989). The maximum availability location problem. *Transportation Science, 23*(3), 192–200.

ReVelle, C., Bigman, D., Schilling, D., Cohon, J., & Church, R. (1977). Facility location: a review of context-free and EMS models. *Health Services Research, 12*(2), 129–146.

Sariyer, G., Ataman, M. G., Akay, S., Sofuoglu, T., & Sofuoglu, Z. (2017). An analysis of emergency medical services demand: Time of day, day of the week, and location in the city. *Turkish Journal of Emergency Medicine, 17*(2), 42–47.

Saydam, C., Rajagopalan, H. K., Sharer, E., & Lawrimore-Belanger, K. (2013). The dynamic redeployment coverage location model. *Health Systems, 2*(2), 103–119.

Schierbeck, S., Hollenberg, J., Nord, A., Svensson, L., Nordberg, P., Ringh, M., Forsberg, S., Lundgren, P., Axelsson, C., & Claesson, A. (2021). Automated external defibrillators delivered by drones to patients with suspected out-of-hospital cardiac arrest. *European Heart Journal, 42*, 1478–1487.

Schilling, D., Jack Elzinga, D., Cohon, J., Church, R. L., & ReVelle, C. (1979). The TEAM/FLEET models for simultaneous facility and equipment siting. *Transportation Science, 13*(2), 163–175.

Schmid, V. (2012). Solving the dynamic ambulance relocation and dispatching problem using approximate dynamic programming. *European Journal of Operational Research, 219*(3), 611–621.

Schmid, V., & Doerner, K. F. (2010). Ambulance location and relocation problems with time-dependent travel times. *European Journal of Operational Research, 207*(3), 1293–1303.

Scott, J., & Scott, C. (2017). Drone delivery models for healthcare. In: *Hawaii International Conference on System Sciences.*

Steenhoff, T. C., Siddiqui, D. I., & Zohn, S. F. (2022). EMS air medical transport. *StatPearls.* Treasure Island (FL): StatPearls Publishing.

Stoesser, C. E., Boutilier, J. J., Sun, C. L. F., Brooks, S. C., Cheskes, S., Dainty, K. N., Feldman, M., Ko, D. T., Lin, S., Morrison, L. J., Scales, D. C., & Chan, T. C. Y. (2021). Moderating effects of out-of-hospital cardiac arrest characteristics on the association between EMS response time and survival. *Resuscitation, 169*, 31–38.

Stout, J., Pepe, P. E., & Mosesso, V. N. (2000). All–advanced life support vs tiered-response ambulance systems. *Prehospital Emergency Care, 4*(1), 1–6.

Sudtachat, K., Mayorga, M. E., & McLay, L. A. (2014). Recommendations for dispatching emergency vehicles under multitiered response via simulation. *International Transactions in Operational Research, 21*(4), 581–617.

Sun, C. L. F., Demirtas, D., Brooks, S. C., Morrison, L. J., & Chan, T. C. Y. (2016). Overcoming spatial and temporal barriers to public access defibrillators via optimization. *Journal of the American College of Cardiology, 68*(8), 836–845.

Swersey, A. J., Goldring, L., & Geyer, E. D. (1993). Improving fire department productivity: merging fire and emergency medical units in New Haven. *Interfaces, 23*(1), 109–129.

Toregas, C., Swain, R., ReVelle, C., & Bergman, L. (1971). The location of emergency service facilities. *Operations Research, 19*(6), 1363–1373.

United Nations. (2010). The millennium development goals report. Technical Report. New York, NY.

Urdaneta, L. F., Miller, B. K., Ringenberg, B. J., Cram, A. E., & Scott, D. H. (1987). Role of an emergency helicopter transport service in rural trauma. *Archives of Surgery, 122*(9), 992–996.

van Barneveld, T. C., Bhulai, S., & van der Mei, R. D. (2017a). A dynamic ambulance management model for rural areas: computing redeployment actions for relevant performance measures. *Health Care Management Science, 20*(2), 165–186.

van Barneveld, T. C., van der Mei, R. D., & Bhulai, S. (2017b). Compliance tables for an EMS system with two types of medical response units. *Computers & Operations Research, 80*, 68—81.

van den Berg, P. L. & Aardal, K. (2015). Time-dependent MEXCLP with startup and relocation cost. *European Journal of Operational Research, 242*(2), 383–389.

Vlahogianni, E. I., Karlaftis, M. G., & Golias, J. C. (2014). Short-term traffic forecasting: where we are and where we're going. *Transportation Research Part C: Emerging Technologies, 43*, 3–19.

Wagner, M. R., & Radovilsky, Z. (2012). Optimizing boat resources at the U.S. Coast Guard: deterministic and stochastic models. *Operations Research, 60*(5), 1035–1049.

Westgate, B. S., Woodard, D. B., Matteson, D. S., & Henderson, S. G. (2016). Large-network travel time distribution estimation for ambulances. *European Journal of Operational Research, 252*(1), 322–333.

Wilson, B., Gratton, M. C., Overton, J., & Watson, W. A. (1992). Unexpected ALS procedures on non-emergency ambulance calls: The value of a single-tier system. *Prehospital and Disaster Medicine, 7*(4), 380–382.

Woodard, D., Nogin, G., Koch, P., Racz, D., Goldszmidt, M., & Horvitz, E. (2017). Predicting travel time reliability using mobile phone GPS data. *Transportation Research Part C: Emerging Technologies, 75*, 30–44.

Yoon, S., & Albert, L. A. (2018). An expected coverage model with a cutoff priority queue. *Health Care Management Science, 21*(4), 517–533.

Yoon, S., & Albert, L. A. (2020). A dynamic ambulance routing model with multiple response. *Transportation Research Part E: Logistics and Transportation Review, 133*, 101807.

Yoon, S., Albert, L. A., & White, V. M. (2021). A stochastic programming approach for locating and dispatching two types of ambulances. *Transportation Science, 55*(2), 275–296.

Zhang, Z.-H., & Jiang, H. (2014). A robust counterpart approach to the bi-objective emergency medical service design problem. *Applied Mathematical Modelling, 38*(3), 1033–1040.

Zhou, Z., Matteson, D. S., Woodard, D. B., Henderson, S. G., & Micheas, A. C. (2015). A spatio-temporal point process model for ambulance demand. *Journal of the American Statistical Association, 110*(509), 6–15.

Location of Public Facilities Under Congestion

Robert Aboolian and Majid Karimi

Abstract In this chapter, we present location models for congested public facilities. In contrast with classical location models in which the demand and services are deterministic, we consider settings where consumers generate streams of stochastic demand for service, and service times are stochastic, which leads to congestion in facilities. Because of the congestion, consumers will either wait to receive services or leave the facility without being served. Location-allocation decisions in congested facilities are particularly important in applications of public service systems, with applications ranging from the design of preventive healthcare networks to welfare service systems. We particularly focus on congestion models in public location theory. After a brief review of congestion models and their impact on our understanding of public facilities, we detail state-of-the-art research in the operations research literature. To organize our view of the current literature, we present a unifying classification of public facility location models with congestion and present relevant models, solution approaches, and their strengths and limitations. We conclude this chapter by discussing the current research opportunities for location scientists in public location theory from the lens of stochastic modeling and congestion.

Keywords Location of public facilities · Congested facilities · Service system design

R. Aboolian (✉) · M. Karimi
College of Business Administration, California State University San Marcos, San Marcos, San Marcos, CA, USA
e-mail: raboolia@csusm.edu; mkarimi@csusm.edu

H. A. Eiselt, V. Marianov (eds.), *Uncertainty in Facility Location Problems*, International Series in Operations Research & Management Science 347, https://doi.org/10.1007/978-3-031-32338-6_10

1 Location Analysis in Public Sector

From the early work that started from the seminal work of Weber (Alfred, 1929), considering the location of a factory between two resources and a single market, to the influential work by Cooper in locating warehouses, distribution centers, communication centers, or production facilities (Cooper, 1967), much of the early developments in location science have been attributed to the location-allocation problems concerning the private sector.

Until the 1960s, facility location problems (FLPs) were mainly involved in determining the optimal placement of private facilities. Before the seminal work of Teitz's "Toward a theory of urban public facility location" (Teitz, 1968), no clear distinction was made between public and private facility location models. Motivated by the role of public facilities in urban planning, Teitz (1968) introduced a framework for locating urban public facilities. Taking a systematic approach, Teitz describes public facility systems as:

> components of the city whose primary function is to facilitate the provision of goods and services declared to be wholly or partly within the domain of government. (Teitz, 1968, p. 38)

In the operations research literature, Revelle et al. (1970) were one of the earliest to distinguish location problems into two categories of private and public. Revelle et al. (1970) introduced public FLPs as those involving private sector FLPs with the extra challenge that objectives and constraints may not be easily defined or quantifiable.

Armed with the modeling and methodologies of operations research, location scientists have tackled public FLPs since the 1970s. In its most general structure, public FLPs' lack of a market-driven environment and their emphasis on welfare–as opposed to private FLPs' profit-oriented decisions—differentiates public facility location theory from private sector location theory. Unlike the private FLP's clarity of objective in minimizing cost or maximizing profit, the public FLP's goal of maximizing welfare is less straightforward and more debated.

The difficulties of modeling the concept of welfare led to the introduction of welfare proxies. The earlier literature uses three proxies of *accessibility*, *maximum distance*, and *participation/coverage* in an optimization framework, all of which are subject to budget (investment) constraints. The other optimization framework is to minimize the planner's cost to achieve a certain level of a welfare proxy. Accessibility can be measured with respect to the (weighted) average of consumers' travel distance or travel time between facilities and consumers. If a planner's objective is to increase a systems' accessibility, then the FLP can be defined as minimizing the total average (weighted or maximum) travel distance (or travel time). Participation/coverage is based on the assumption that a system's characteristics determine the amount of covered (served) potential demand. If a planner's objective is to increase the participation/coverage, then the FLP can be defined as maximizing the overall demand capture or realization. Maximum distance can also be used as a proxy, in that a planners' goal is to minimize the maximum distance (or time) a consumer must travel to receive services. A convenient categorization of the welfare proxies (accessibility, maximum distance, and participation/coverage) is through their mathematical models of p median, p center, and covering problem, respectively. See Laporte et al. (2019, Chapters 1–4) for extended coverage of these problems and their related literature.

The simple and elegant formulation of public FLPs as optimization problems was significantly influential in many real-world applications. For example, the City of Baltimore used the *central facility location models* of ReVelle and Swain (1970) to determine the closure of surplus fire stations (Schilling et al., 1979).

Despite their advantages, however, the early framework of optimizing welfare proxies subject to constraints on investment gave rise to the conflicts of *efficiency* (in financial terms) and (social) *equity* (McAllister, 1976). In particular, in the absence of a competitive market, the earlier normative approaches to solve location-allocation decisions in public FLPs often lead to configurations with few large facilities that may not be equitable (fair) in maximizing the served demand (see DeVerteuil (2000) for a comprehensive review). At its core, the efficiency-equity conflicts in location decisions exemplify the broader efficiency-equity trade-off in economics and policy (Okun, 2015). Additionally, and from a modeling perspective, difficulties in modeling the concept of fairness further complicate the issues of equity and efficiency.

To address the *equity-efficiency conflicts*, the location science literature focuses on two paradigms of research: behavioral studies of location science and normative studies of multi-objective optimization. The behavioral studies seek to understand the location-allocation problems in a broader context of *"human agency"* by studying the impact of location decisions on behaviors, experiences, and choices of those involved. In multi-objective optimization approaches, researchers have developed normative frameworks to incorporate efficiency and welfare simultaneously. It is important to note that, even though the incorporation of multi-objective optimization problems presents a main distinction between public FLPs and private FLPs, some instances of public FLPs—which do not involve equity-efficiency constraints—can be applied to private FLPs, which may or may not lead to higher efficiency in private settings. For example, the p median problem is an applicable model for public FLPs since it minimizes transportation costs for consumers, but it could also be applied to private FLPs since it maximizes consumer access by minimizing the average customer-facility distance. However, we believe that given the different patterns of purchasing behavior from consumers, to model the location of private facilities, a model with an implicit objective to maximize revenue or profit is more appropriate.

A domain where the aforementioned paradigms of research—behavioral and operational—naturally meet is the design of *service systems*. Unlike the traditional normative models, viewing public facilities as systems providing services to individuals (or groups) opens the door for incorporating features and parameters beyond the measurement of welfare. When viewed as a service system, the public (and private) FLP models allow us to study individuals' behavior alongside the broader system goals, thus expanding the context of location-allocation decisions. Service systems are also more suited to multi-objective optimization, as service provision often involves balancing conflicting objectives.

A key feature of the service system view is its stochastic nature. Unlike the static and deterministic frameworks, service systems are complex, and frequency involves tactical and strategic decisions intended for a considerable time.

To deal with the uncertainties of service systems, the location-allocation decisions are often formulated as *stochastic location models* (SLMs) (Berman & Krass, 2019). In contrast with classical location models in which the demand and services are deterministic, SLMs consider settings where consumers generate streams of stochastic demands for service, and service times are stochastic. Stochastic demand and service time lead to congestion in facilities, resulting in waiting to receive services or lost demand. SLMs are particularly important in the applications of public service systems, with applications ranging from the design of preventive healthcare networks to welfare service systems.

The first paper to include uncertainty in public FLP is by Carbone (1974), in which the number of users at each node is assumed to be a random variable, but since it does not include stochastic service times, there is no congestion modeled in the facilities. Congestion models in public location theory are the main focus of this chapter. Section 2 briefly reviews congestion models and their impact on our understanding of public FLPs, detailing state-of-the-art research in the operations research literature. In Sect. 3, we further delve into current research opportunities for location scientists in public location theory from the lens of stochastic modeling and congestion.

2 Congestion and Its Impact on Public Sector Location Decisions

Nearly five decades ago, Revelle et al. (1970) offered an apt description of the application of *analytical location models*: *These methods of analysis are no panacea for pouring out "optimal" solutions since the real world with its immense complexity tends to defy exact analogs. The results of analyzing these models may be optimal and exact in reference to the models, but they are not necessarily the optimal results for the real world. Rather, the results are regarded as an aid to the analyst's intuition and not as a replacement for it.*

Since Revelle et al. (1970) made those remarks, analytical models have remained a crucial technique for supporting location analysts in their decision-making. As we discussed in Sect. 1, a key contributor to the location models' success—especially in public sector applications—is the consideration of uncertainty that results in congestion in the facilities. In this section, we review Public Facility Location Problem with Congestion (PFLPC) in greater detail and discuss state-of-the-art models. Given the fact that most services offered by public facilities are offered at the facility, we focus our attention on public facilities with immobile servers, where consumers travel to the facilities to receive their service. In Sect. 2.1, we provide an overview of the PFLPC with immobile servers and offer a general categorization based on modeling assumptions. In Sect. 2.2, we review the exact optimization problems considered along with their solution methodologies.

2.1 Public Facility Location Problem with Congestion

In its broadest sense, the uncertainty of PFLPC with immobile servers (from now on simply referred to as PFLPC) is stemmed from the stochastic supply—in the form of service times—and demand. As such, a service provider's decision must involve the capacity of service, as well as its location. One of the most challenging aspects of such decisions is *congestion*. A byproduct of stochastic service times and stochastic demand, congestion occurs because demand cannot be served in its entirety, and consumers must either wait in queues or leave service systems. To facilitate our review of the PFLPC, their strengths, and their limitations, we categorize PFLPC based on their modeling attributes.

One of the broadest categorizations of PFLPC is based on the consideration of elastic vs. inelastic demand. Here, elasticity can be defined as the changes in demand with respect to changes to consumers' utility (or disutility) toward a facility offering those services. For instance, in situations where the demand decreases when the waiting time to receive the services increases, the demand is said to be elastic to waiting time. In PFLPC, demand may or may not be elastic. If a public service provider is the only source of an essential service, the demand for such service is often inelastic. In other words, all consumers are willing to travel any distance and wait any time to receive the service (e.g., visiting a specialist in a public healthcare system). On the other hand, demand elasticity is more common in situations where the public and private providers co-exist or service offered by the public facility might not be considered essential to all consumers (e.g., receiving a COVID-19 vaccination).

PFLPC also have varying objectives depending on the perspective taken to formulate an optimization problem. The three most common views when determining the PFLPC's goals are:

- *Consumer Perspective*: From consumers' perspective, the goal is to optimize the public system for consumers, usually given a limited budget or service capacity. The most common goals here are to:
  - Maximize consumer participation.
  - Maximize the coverage of the service.
- *Service Provider Perspective*: From providers' perspective, the goal is to optimize the public system for the service provider, usually by ensuring to maintain a minimum service quality for consumers. The most common goals here are to:
  - Minimize investment in service capacity.
  - Minimize the overall operating cost of the service provider.

- *Balanced Perspective*: From a society's perspective, the goal is to optimize the overall societal benefit by balancing the cost incurred by both consumers and the service provider and allowing the objective function to find the optimal capacity. FLP models with a balanced perspective are also known as *socially optimal models*. The most common goals here are to:
  - Minimize the sum of operating cost to service provider and accessibility cost to consumers.
  - Maximize the overall benefit to the public.

When considering the constraints of PFLPC, the main restrictions revolve around the assignment of consumers to service facilities. A service provider could determine the assignment, or the choice could be given to consumers to use the service facility they prefer. Such models are often referred to as *directed choice* and *consumer choice*, respectively.

For consumer choice models, in particular, a more detailed comparison classifies PFLPC by the factor used for consumers' choice (or their utility). The three most frequent factors used for consumers' choice are the following:

- *Proximity*: Some models use distance as the main factor for consumers' choice. A standard assumption in such models is that consumers travel to the closest open facility to receive services.
- *Access Time*: When access time is used as the factor, consumers are assumed to be choosing the facility with the least access time—the sum of travel, waiting, and service times.
- *Disutility*: When disutility is used as the factor, consumers are assumed to be choosing the facility with the least disutility—which could include access time as well as other attraction attributes.

PFLPC could also be classified into planned vs. unplanned congestion. In planned congestion, we would like to limit the waiting time for consumers. This is achieved either by planning the capacity at each facility when the capacity is a decision variable or by directing the consumers to the facilities, especially in directed choice models. In unplanned congestion, we allow the system to decide the optimal capacity and waiting time in facilities with respect to its specific objective. In this case, especially when the capacity at each facility is given, we could end up with facilities with a low utilization rate.

Another classification of PFLPC is the consideration for demand coverage. In environments where demand cannot be satisfied in its entirety, some PFLPC consider partial demand allocation, which could be probabilistic (when there is a chance of demand loss) or deterministic (when a certain percentage of demand must be satisfied).

To gain predictive insight into the congestion attributes of service systems, location researchers often use *queuing theory*. From a queuing perspective, the most common categorization of PFLPC is to consider Kendall's notation of their queuing model, among which the most common are M/M/1, M/M/k, and M/G/1 models. These queuing systems are often laid out in *network* structures. In location analysis, a network is an abstract model of a spatial environment which often considers two sets of demand and facility nodes, with edges representing the distance or travel time.

In this chapter, we use $N = \{1, 2, \cdots, n\}$ to refer to the set of demand nodes and $M = \{1, 2, \cdots, m\}$ to refer to the set of facilities. We reserve the index i to refer to a demand node and j to indicate a facility throughout the chapter. In the queuing systems, we use the symbol μ for service capacity and λ for the mean demand rate. For M/M/k models, in particular, we use the parameter k to refer to the number of servers. We often discuss the system waiting and travel time, which we denote with W and t, respectively.

When discussing the optimization formulations, we reserve x_j to denote the binary decision variable that is 1 if *facility* j—the facility at node j—is open and 0 otherwise. To indicate the assignment of demand to facilities, we use the binary variable y_{ij}, which is 1 if the demand at node i is served at facility j. In some cases, demand may be partially served at a facility. In such situations, we denote the assignment of demand to facilities with a continuous variable $0 \leq y_{ij} \leq 1$ representing the fraction of node i's demand served at facility j. When a continuous variable y_{ij} is used, we often use the binary variable z_{ij} that is 1 if any fraction of the demand at node i is served at facility j and 0 otherwise.

2.2 *Models and Solutions Methods*

As discussed in Sect. 2.1, models in PFLPC can be classified into two major categories of elastic and inelastic demand. Most elastic demand models are used for consumer choice models, yet models with inelastic demand have been used for both directed and consumer choice models.

2.2.1 PFLPC Models with Inelastic Demand

Models with inelastic demand can be classified into three groups. The first group of models considers a service provider's perspective to optimize the provider's performance measures while maintaining a minimum service quality for consumers. The second group of models considers the consumers' perspective to optimize service quality while maintaining a minimum service performance. The third group of models, also known as *socially optimal models*, considers a more balanced perspective and optimizes performance and quality simultaneously.

2.2.1.1 PFLPC Models with Service Provider Perspective

Given the restrictions on operating budgets, service providers often aim to minimize their operating costs while maintaining a minimum service quality. In PFLPC models, service qualities are usually measured by congestion metrics such as queue lengths or waiting times. Consequently, maintaining service quality can be achieved by setting an upper bound on waiting time or the number of people waiting to receive services. Alternatively, certain levels of service quality can be maintained by assuring a lower bound on the probability of waiting or queue length not exceeding a certain level. In such settings, consumers can be directed to facilities. A simple formulation of PFLPC from a service provider's perspective is presented in (1). In the following formulation, the capacity at each facility is assumed to be known, and the parameter f_j represents the sum of fixed and capacity costs at facility j.

$$\min \sum_j f_j x_j \tag{1a}$$

Subject to

$$\sum_j y_{ij} = 1, \ \forall i, \tag{1b}$$

$$y_{ij} \leq x_j, \ \forall i, j, \tag{1c}$$

$$\mathbb{E}(W_j) \leq W^{\max} \text{ or } \mathbb{E}(L_j) \leq L^{\max} \text{ or}$$

$$\mathbb{P}(W_j \leq W^{\max}) \geq \beta \text{ or } \mathbb{P}(L_j \leq L^{\max}) \geq \beta, \ \forall j, \tag{1d}$$

$$y_{ij}, x_j \in \{0, 1\}, \ \forall i, j, \tag{1e}$$

in which $\mathbb{P}(\cdot)$ and $\mathbb{E}(\cdot)$ denote the probability and the expected values, respectively.

As discussed earlier, the objective in (1) is to minimize the provider's total cost. Constraints (1b) ensure that each demand node is covered by exactly one facility, and Constraints (1c) prevent an assignment of a demand node to a closed facility. Constraints (1d) ensure a minimum level of service quality, in which $W^{\max}$ and $L^{\max}$ represent a maximum allowed waiting time and a maximum length of queuing line, respectively.

This model can be used for both M/M/1 and M/M/k queuing systems where the service rate μ and number of parallel servers k (for M/M/K) are assumed to be known. Similar to Wang et al. (2004), the above model can also be modified for the case where the capacity at each location is a decision variable, and f_j is no longer a fixed value but a function of the capacity at facility j.

The main complexity in solving models such as (1) is the nonlinear constraints of (1d). An efficient solution, however, can be obtained via a linearization of (1d), which results in a *mixed-integer linear program* (MILP), for which many efficient solution approaches exist (Marianov & Serra, 1998).

2.2.1.2 PFLPC Models with Consumer Perspective

Marianov and Serra (1998) describe the first PFLPC model with inelastic demand to study the service design problem from consumers' perspective. Marianov and Serra (1998) consider a *probabilistic maximal covering location problem*, in which the goal is to find a location-allocation for p facilities with known capacities to maximize the consumers' participation while maintaining a minimum service quality.

A general formulation of PFLPC from consumers' perspective is presented in (2).

$$\max \sum_j \sum_i \lambda_i y_{ij} \tag{2a}$$

Subject to

$$\sum_j y_{ij} \leq 1, \ \forall i, \tag{2b}$$

$$y_{ij} \leq x_j, \ \forall i, j, \tag{2c}$$

$$\sum_j x_j = p, \tag{2d}$$

$$\mathbb{E}(W_j) \leq W^{\max} \text{ or } \mathbb{E}(L_j) \leq L^{\max} \text{ or}$$

$$\mathbb{P}(W_j \leq W^{\max}) \geq \beta \text{ or } \mathbb{P}(L_j \leq L^{\max}) \geq \beta, \ \forall j, \tag{2e}$$

$$y_{ij}, x_j \in \{0, 1\}, \ \forall i, j. \tag{2f}$$

The objective in (2a) is to maximize consumer coverage or the total served demand. Similar to the formulation from a provider's perspective, Constraints (2b) ensure that each demand node is covered by at most one facility, Constraints (2c) prevent an assignment of a demand node to a closed facility, and Constraints (2e) ensure a minimum level of service quality.

Marianov and Serra (1998) show that the nonlinearity of service quality of Constraints (2e) can be linearized by converting them to equivalent constraints that limit the arrival rate in each facility. Consequently, the problem in (2) becomes an MILP, which can be solved with any off-the-shelf linear optimization software. In cases where the capacity is not predetermined, Marianov and Serra (2002) also consider a generalization of the problem in (2), in which capacities are modeled as decision variables at each facility.

Maximizing participation is not the only way to incorporate consumers' perspectives; a service system can also be designed to optimize service quality. Consumer perspective models that attempt to optimize service quality are often constrained by a limited budget or an available service capacity. These models could consider a fixed capacity at each facility or consider capacity as a decision variable.

Wang et al. (2002) consider total system waiting time—the sum of travel time and the time spent waiting to receive services—as the primary measure of service quality. The authors formulate an optimization problem to minimize service waiting time given a set number of facilities with a fixed capacity. The authors consider the following formulation with a system under which each facility operates an M/M/1 queues:

$$\min \sum_j \sum_i \lambda_i t_{ij} y_{ij} + \sum_j \sum_i \frac{\lambda_i y_{ij}}{\mu - \sum_k \lambda_k y_{kj}} \tag{3a}$$

Subject to

(2b)–(2c)

$$\sum_k t_{ik} y_{ij} \leq \left(t_{ij} - M\right) x_j + M, \ \forall i, j, \tag{3b}$$

$$y_{ij}, x_j \in \{0, 1\}, \ \forall i, j. \tag{3c}$$

The first term of (3a) represents the total travel time, and the second term measures the total time consumers spend to receive services. To optimize quality, the objective is to minimize the (expected) system waiting time. Here, Constraints (3b) ensure that consumers are assigned to a facility that is closest to them.

To solve (3), Wang et al. (2002) suggest a *Lagrangian relaxation*. As the main complexity of (3) stems from its nonlinear (and concave) objective, we argue that ϵ-approximation solutions—such as the *tangent linear approximation* (TLA) incorporated in Aboolian et al. (2007)—may provide a more efficient solution.

Multiple generalizations of (3) have been considered in the literature. Berman and Drezner (2007) consider a similar setting to Wang et al. (2002), while considering decision variables for service capacities and allowing the optimization problem to determine the optimal capacities.

For an M/M/k queue, let $W_j\left(\sum_i \lambda_i y_{ij}, \mu, k_j\right)$ be the waiting time at facility j with k_j servers (each with a service rate of μ) and a demand rate of $\sum_i \lambda_i y_{ij}$.

Berman and Drezner (2007) consider the following formulation by incorporating M/M/k queues:

$$\min \sum_j \sum_i \lambda_i t_{ij} y_{ij} + \sum_j \sum_i \lambda_i y_{ij} W_j \left(\sum_i \lambda_i y_{ij}, \mu, k_j\right) \tag{4a}$$

Subject to

(2b)–(2c)

$$\sum_j k_j = p, \tag{4b}$$

$$k_j \leq p x_j, \ \forall j, \tag{4c}$$

$$\sum_{k} t_{ik} y_{ij} \leq \left(t_{ij} - M\right) x_j + M, \ \forall i, j, \tag{4d}$$

$$y_{ij}, x_j \in \{0, 1\}, k_j \in \{0, 1, \ldots, p\}, \ \forall i, j. \tag{4e}$$

The objective in (4a) is to minimize consumers' total traveling and waiting time. Constraint (4b) is the limited capacity constraint, and it ensures that only a total of p discrete servers will be assigned to open facilities. Constraints (4c) prevent an assignment of server(s) to a closed facility, and Constraints (4d) assign consumers to the closest facility.

Due to nonlinearity of the objective function, Berman and Drezner (2007) did not provide an exact approach but instead used tabu search as a heuristic to solve the problem.

Similar to Berman and Drezner (2007), Aboolian et al. (2009) consider capacity at each facility as a decision variable while attempting to minimize the maximum consumers' access time (travel + average waiting time). This problem is modeled for M/M/k queues with the following formulation:

$$\min Z \tag{5a}$$

Subject to

(2b) – (2c)

(4b) – (4d)

$$Z \geq t_{ij} y_{ij} + W_j \left(\sum_{i} \lambda_i y_{ij}, \mu, k_j \right), \ \forall i, j, \tag{5b}$$

$$y_{ij}, x_j \in \{0, 1\}, k_j \in \{0, 1, \ldots, p\}, \ \forall i, j. \tag{5c}$$

The objective in (5a) is to minimize the maximum consumer traveling and waiting time. This is enforced using Constraints (5b).

Due to nonlinearity of Constraints (5b), Aboolian et al. (2009) did not offer any exact approach but used a genetic algorithm to solve the problem.

It is interesting to note that Aboolian et al. (2009) is a p server version of the p center location problem, and Berman and Drezner (2007) is the p server version of the p median location problem.

2.2.1.3 PFLPC Models with Socially Optimal Perspective

Designing service systems from the perspective of providers or consumers may provide an inefficient solution due to missing one of the parties' outlooks. A broader approach to designing service systems is to take a societal approach by balancing providers' objectives with consumer needs.

Given the societal scope of PFLPC, socially optimal models are particularly interesting in designing service systems in the public sector.

Designing socially optimal service systems has been the focus of the attention of many researchers for the past 20 years.

One of the most frequent approaches to do so is to minimize the overall cost to consumers (cost incurred for traveling, waiting, and service time) plus the service providers' operating costs (fixed facility and variable capacity costs). Amiri (1997), Aboolian et al. (2008), Elhedhli (2006), Castillo et al. (2009), Vidyarthi and Jayaswal (2014), Elhedhli et al. (2018), and Aboolian and Karimi (2023a) are examples in which a socially optimal service system design problem is considered. Except for Aboolian et al. (2008) and Aboolian and Karimi (2023a), all papers in socially optimal models are directed choice models. The majority of the models in the literature also consider an M/M/1 (or M/G/1) queue for each facility. The general problem formulation for these models is as follows:

$$\min \sum_j f_j x_j + \sum_j \omega\left(\mu_j\right) + \sum_i \sum_j c_{ij} \lambda_i y_{ij}$$

$$+\alpha \sum_j \sum_i \lambda_i y_{ij} W_j \left(\sum_i \lambda_i y_{ij}, \mu_j, \sigma_j \right), \tag{6a}$$

Subject to

$$\sum_j y_{ij} = 1, \ \forall i, \tag{6b}$$

$$y_{ij} \leq x_j, \ \forall i, j, \tag{6c}$$

$$0 \leq \mu_j \leq x_j, \ \forall j, \tag{6d}$$

$$\sum_i \lambda_i y_{ij} \leq \mu_j, \ \forall j, \tag{6e}$$

$$y_{ij}, x_j \in \{0, 1\}, \ \forall i, j. \tag{6f}$$

The objective function of (6a) represents the total cost to the provider and consumers. The first two terms of (6a) are the total fixed and capacity costs, in which f_j denotes the fixed cost of operating facility j and $\omega(\cdot)$ is the capacity cost function. It is often assumed that capacity cost is a concave function of service capacity. As such, the function $\omega(\cdot)$ is usually modeled as a linear or concave function. The last two terms of (6a) are the consumers' traveling and waiting costs, in which c_{ij} models the traveling cost of consumers who travel from node i to facility j to receive services and α is a scaling factor determining how much consumers dislike waiting time.

Constraints (6b) ensure that each demand node is covered by exactly one facility, and Constraints (6c) prevent an assignment of a demand node to a closed facility. To ensure no capacity is assigned to closed facilities, we enforce Constraints (6d).

We also need to make certain there are enough capacities at each facility, a condition ensured by Constraints (6e).

Even though it is not explicitly mentioned, the current formulation in (6) assumes a directed choice, in which the service provider assigns facilities to consumers. In cases where distance can be used as the proxy for consumer choice, the following constraint can modify the above formulation to ensure consumer choices are considered:

$$\sum_{j'} c_{ij'} y_{ij'} \leq \left(c_{ij} - L\right) x_j + L,\ \forall j \tag{7}$$

The formulation in (6) is also for different queuing systems.

For M/G/1 queues, we can calculate waiting time—the function W_j—using Pollaczek-Khinchine formula (Pollaczek, 1930) as follows:

$$W_j\left(\sum_i \lambda_i y_{ij}, \mu_j, \sigma_j\right) = \frac{\sum_i \lambda_i y_{ij}\left(1 + \sigma_j^2 \mu_j^2\right)}{2\mu_j\left(\mu_j - \sum_i \lambda_i y_{ij}\right)} + \frac{1}{\mu_j};\ \forall j \tag{8}$$

In M/M/1 queues, $\sigma_j = 1/\mu_j$; thus, the value for $W_j\left(\sum_i \lambda_i y_{ij}, \mu_j, \sigma_j\right)$ takes the following simplified form:

$$W_j\left(\sum_i \lambda_i y_{ij}, \mu_j, \sigma_j\right) = \frac{1}{\mu_j - \sum_i \lambda_i y_{ij}};\ \forall j \tag{9}$$

Amiri (1997), Elhedhli (2006), Castillo et al. (2009) solved the model with M/M/1 queues and discrete capacity options for a directed choice model. Amiri (1997) used Lagrangian relaxation and Elhedhli (2006) used Bender's decomposition to solve the problem. Aboolian et al. (2022) used Search and Cut algorithm to solve the model with M/M/1 queues and discrete capacity options for a consumer choice model. Aboolian et al. (2008) used a branch and bound method to solve the model with M/M/k queues for a consumer choice model. Vidyarthi and Jayaswal (2014) used piecewise linear approximation to solve a model with M/G/1 queues, with discrete capacity options for a directed choice model. Elhedhli et al. (2018) solved the model with M/G/1 queues, with continuous capacity options and concave capacity cost functions and for a directed choice model. To solve this problem, they used Lagrangian relaxation to decompose the problem and reformulate the subproblems as second-order cone programs that are solved at multiple utilization levels. Aboolian and Karimi (2023a) used an advanced version of Search and Cut algorithm to solve a model with M/G/1 queues, with continuous capacity options and concave capacity cost functions both for a directed and consumer choice models.

2.2.2 PFLPC Models with Elastic Demand

In this section, we will review the PFLPC models with elastic demand. PFLPC models often consider a decay function of demand for changes in consumers' disutility to capture demand elasticity. Consequently, the decision variable y_{ij} takes a continuous form—$y_{ij} \in [0, 1]$—to represent the fraction of demand at node i who are served at facility j.

Formally, PFLPC models with elastic demand consider the following indirect relationship between the consumer disutility and the realized demand at node i:

$$\mathcal{Y}_i = f(\mathfrak{D}_i), \tag{10}$$

in which the function $f(\cdot)$ is a *decay function*, $\mathfrak{D}_i$ represents the disutility of consumers at node i, and $\mathcal{Y}_i$ is the realized demand at node i, i.e., $\mathcal{Y}_i \triangleq \sum_j y_{ij}$. We note that the disutility of consumers at node i, $\mathfrak{D}_i$, is usually defined as an increasing function of the factors which could negatively influence consumers' use of services (e.g., travel time, waiting time, etc.).

Inversely, one can also derive consumers' disutility as a function of the realized demand at node i.

$$\mathfrak{D}_i(\mathcal{Y}_i) = f^{-1}(\mathcal{Y}_i), \tag{11}$$

This inverse relationship, in particular, has an intuitive interpretation: for a given value of $\mathcal{Y}_i$, $\mathfrak{D}_i(\mathcal{Y}_i)$ can be interpreted as the largest disutility the consumers at node i would incur to receive services.

One of the earliest works considering such settings is Verter and Lapierre (2002). As an implementation of partial coverage, Verter and Lapierre (2002) used a linear decay function in modeling participation in preventive healthcare programs. Berman and Krass (2002) presented the gradual coverage decay function using a step function. Berman et al. (2003) presented the gradual coverage decay model with two prespecified threshold distances, where a consumer is considered fully covered within the first threshold, partially covered between the two thresholds, and "not covered" otherwise. A common theme among the works mentioned above is considering the consumer's access to the facilities, not necessarily to the offered service. This is because these studies do not incorporate the congestion at the facilities.

As noted in (10), PFLPC models with elastic demand assume the consumer demand for a facility is a function of consumer disutility. Consequently, PFLPC models with elastic demand often consider a consumer choice setting, assuming consumers choose the facility (or facilities) that minimizes their disutility. Consumers' disutility impacts their choice and, in turn, the realized demand at each facility. Thus, PFLPC models with elastic demand often consider *equilibrium conditions* to determine consumers' choices in a service system environment.

Since consumers incur most of their disutility from travel and waiting time, we can model the disutility of node i's consumers who chose facility j as the following weighted sum:

$$\mathfrak{d}_{ij} = \alpha t_{ij} + \beta W_j, \tag{12}$$

in which $\alpha, \beta \in \mathbb{R}$ represent the unequal sensitivity to the disutility incurred by waiting and travel time. In particular, the value of α and β capture the variety of situations under which consumers incur different disutility from different sources of time spent to receive services. For instance, when $\alpha > \beta$, consumers dislike the time spent traveling to a facility more than the time spent at the facility to receive services.

It is important to note that models that do not include waiting time in the consumers' disutility are less realistic but are generally easier to solve since they do not require any equilibrium conditions.

Given the location of open facilities and allocation of service rates, when selecting which facility to use to obtain services, consumers are assumed to choose those facilities that minimize their disutility. This view of consumer choice is closely related to the concept of *user-optimized* models in transportation networks (Nagurney, 1998).

In a user-optimized model of a transportation network, rational users select their routes with their own self-interest in mind, and the *equilibrium pattern* satisfies the following principle called Wardrop's first principle—named after Wardrop (1952):

> *The journey times of all routes* [of a transportation network that are] *actually used are equal and less than those which would be experienced by a single vehicle on any unused route.*

Put differently, in a user-optimized equilibrium of a transportation network, only those paths that have minimal *user costs* (a more general measure for "journey time") are used, and their costs are equal to the travel disutility associated with traveling between locations (Dafermos, 1982).

PFLPC models with elastic demand use a similar analogy to define a user equilibrium for a service system design problem in a location-allocation framework. In particular, given a set of open facilities $S \subseteq M$ and service rates $\mu_j, j \in S$, a user equilibrium is defined as follows:

$$\mathfrak{d}_{ij} = \left(\alpha t_{ij} + \beta W_j\left(\mathbf{y}^\star\right)\right) \begin{cases} = \mathfrak{D}_i^\star\left(\mathcal{Y}_i^\star\right) & \text{if } y_{ij}^\star > 0 \\ > \mathfrak{D}_i^\star\left(\mathcal{Y}_i^\star\right) & \text{if } y_{ij}^\star = 0, \end{cases} \quad \text{for } i \in N,\ j \in S, \tag{13}$$

where $y_{ij}^\star$ is the allocation of consumer demand at equilibrium and $\mathcal{Y}_i^\star = \sum_j y_{ij}^\star$.

The equilibrium conditions in (13) state that if any fraction of consumers at location i use facility j to receive services, then the consumers' disutility from node i to facility j must be equal to the largest disutility the consumers would accept to travel to facility j to receive services. Alternatively, if no consumer from location i is traveling to facility j to receive services, their disutility (if they would have chosen facility j) is more than the maximum allowed disutility. As such, under (13), consumers cannot reduce their disutility by unilaterally deviating from the assigned allocations.

Models with elastic demand can be classified into three similar groups as brought in inelastic demand.

2.2.2.1 PFLPC Models with Elastic Demand and Consumer Perspective

The overall objective of these models is to maximize consumer participation given a limited server capacity (budget).

Marianov et al. (2005), Zhang et al. (2010), Drezner and Drezner (2011) all consider M/M/k queuing systems and determine how to allocate p available servers among facilities and represent the consumers' sensitivity to the waiting time at a facility.

Marianov et al. (2005) used heuristic concentration to solve small-scale hypothetical problem instances. Drezner and Drezner (2011) considered a similar model where the demand at each facility is a function of waiting time, so inherently, it becomes a function of itself and requires equilibrium conditions to solve the problem. They modeled the problem with the following problem formulation:

$$\min \sum_i \sum_j \lambda_i t_{ij} y_{ij} + \sum_j W_j \left(\sum_i \lambda_i y_{ij}, \mu, k_j \right), \tag{14a}$$

Subject to

(2b) – (2c)

(4b) – (4c)

$$y_{ij} = \frac{e^{-\theta(d_{ij} + W_j(\sum_i \lambda_i y_{ij}, \mu, k_j))x_j}}{\sum_r e^{-\theta(d_{ir} + W_r(\sum_i \lambda_i y_{ir}, \mu, k_r))x_r}}, \ \forall i, j, \tag{14b}$$

$$y_{ij}, x_j \in \{0, 1\}, k_j \in \mathbb{N}, \ \forall i, j.$$

Please note that the formulation is similar to the consumer perspective model introduced above. The only difference is that the equilibrium condition (14b), which assigns demand in a consumer node to all open facilities, is replaced with (4d), which assigns all the demand in a consumer node to its closest facility. Drezner and Drezner (2011) used tabu search heuristic to solve the problem without any guarantee to find a solution for equilibrium conditions (14b).

Zhang et al. (2010) consider *access time*—the sum of travel and waiting time—as the main factor for consumers' disutility, i.e., $\eth_{ij} = t_{ij} + W_j$ and $\alpha = \beta = 1$ in (12). Given the equilibrium conditions, Zhang et al. (2010) formulate the problem as a bilevel problem where the allocation of consumers to facilities is determined in the lower level and the location of facilities and their capacity level are determined in the upper level. Their approach is most efficient when the number of capacity options is relatively small. However, bilevel programming would be remarkably inefficient when capacity is modeled as a continuous variable (as in M/M/1 queuing systems).

Similarly, Aboolian et al. (2016) also consider access time as the main proxy for consumers' disutility. Unlike Zhang et al. (2010), however, Aboolian et al. (2016) consider an M/M/1 queuing systems where service rate (service capacity) at each facility is a decision variable and determine how to allocate C service capacity among facilities and represent the consumers' sensitivity to the waiting time at a facility. Even with equilibrium conditions that are required for these models, they were able to formulate and solve the problem as an exact (single-level) mixed-integer problem (MIP).

Here is the original nonlinear formulation of the model:

$$\max \quad \sum_i \sum_j \lambda_i^{\max} y_{ij} \tag{15a}$$

subject to :

(2b)–(2c)

$$\mu_j - \sum_i \lambda_i^{\max} y_{ij} - \frac{x_j}{W^{\max}} \geq 0 \, , \; j \in M \, , \tag{15b}$$

$$x_j \mu^{\min} \leq \mu_j \leq x_j \mu^{\max} \, , \; j \in M \, , \tag{15c}$$

$$\sum_j \mu_j = C^{\max} \, , \tag{15d}$$

$$t_{ij} + \frac{1}{\epsilon\left(1 - x_j\right) + \mu_j - \sum_{r \in N} \lambda_r^{\max} y_{rj}} - \frac{f_i^{\max} - \sum_{k \in M} y_{ik}}{\alpha} \geq 0 \, , \quad i \in N, j \in M \, , \tag{15e}$$

$$y_{ij} \left(t_{ij} + \frac{1}{\epsilon\left(1 - x_j\right) + \mu_j - \sum_{r \in N} \lambda_r^{\max} y_{rj}} - \frac{f_i^{\max} - \sum_{k \in M} y_{ik}}{\alpha} \right) = 0 \, , \quad i \in N, j \in M \, , \tag{15f}$$

$$\mu_j, y_{ij} \geq 0, x_j \in \{0, 1\} \, , \; i \in N, j \in M \, .$$

Constraints (15b) ensure the minimum capacity required at each facility to ensure system stability, and Constraints (15c) limit the capacity at each facility to what is attainable. To ensure the overall capacity assigned to open facilities does not exceed the available level, we enforce Constraints (15d). Constraints (15e) and (15f) are the consumer traffic equilibrium conditions, which when met consumers have no incentive to change their choice of facilities.

To solve this model, Aboolian et al. (2016) introduced the linear epsilon-optimal MIP equivalent of the model, which could be solved efficiently.

2.2.2.2 PFLPC Models with Elastic Demand and Service Provider Perspective

The overall objective of these models is to minimize service providers' operating cost while maintaining a minimum level of service coverage and/or consumer participation.

To our knowledge, Aboolian et al. (2022) is the only work introducing such a perspective when demand is elastic. Compared to the previous models used in the consumer perspective with elastic demand, they introduced and used a more advanced disutility function with different consumer sensitivities to waiting and travel time, which aligns more with reality given that in two separate studies by Newman (1984) and Jan et al. (2000), survey respondents identified travel time, although important, was not as significant a factor as waiting time.

Aboolian et al. (2022) incorporate consumer choice by considering settings under which consumers would like to minimize their disutility from travel and waiting times when choosing which facility to patronize. In their setting, consumers' disutility is defined as $\mathfrak{d}_{ij} = t_{ij} + \beta W_j$, i.e., $\alpha = 1$ in (12).

Aboolian et al. (2022) model the eventual choice of facilities as a user equilibrium problem, where at equilibrium, consumers do not have any incentive to change their choices. The original nonlinear formulation of the model is as follows:

$$\min \quad \sum_{j\in M} h_j x_j + \sum_{j\in M} c_j \mu_j$$

subject to :

$$z_{ij} \le x_j\,,\ i \in N, j \in M\,, \tag{16a}$$

$$y_{ij} \le z_{ij}\,,\ i \in N, j \in M\,, \tag{16b}$$

$$\sum_{j\in M} z_{ij} \ge 1\,,\ i \in N\,, \tag{16c}$$

$$\sum_{i\in N} \lambda_i^{\max} y_{ij} \ge \lambda^{\min}\, x_j, j \in M\,, \tag{16d}$$

$$\left(t_{ij} + \frac{\beta}{\mu_j - \sum_{r\in N} \lambda_r^{\max} y_{rj}} + \frac{\ln\left(\sum_{k\in S} \lambda_i^{\max} y_{ik}\right)}{\alpha}\right) y_{ij} = 0\,, \quad i \in N, j \in M\,, \tag{16e}$$

$$t_{ij} + \frac{\beta}{\mu_j - \sum_{r\in N} \lambda_r^{\max} y_{rj}} + \frac{\ln\left(\sum_{k\in S} \lambda_i^{\max} y_{ik}\right)}{\alpha} \ge 0\,, \quad i \in N, j \in M\,, \tag{16f}$$

$$t_{ij} + \frac{\beta x_j}{\mu_j - \sum_{k\in N} \lambda_i^{\max} y_{kj} + (1-x_j)\epsilon} \leq \tau^{\max} + (1-z_{ij})L\,,$$
$$i \in N, j \in M\,, \tag{16g}$$

$$\mu_j \geq \sum_{i\in N} \lambda_i^{\max} y_{ij}\,,\ j \in M\,, \tag{16h}$$

$$\mu_j, y_{ij} \geq 0, x_j, z_{ij} \in \{0,1\}\,,\ i \in N, j \in M$$

The objective function terms of this problem represent the total fixed and the capacity costs. Constraints (16a) ensure that no consumer is assigned to a closed facility. Constraints (16b) guarantee that if node i is not assigned to facility j, the demand fraction of a node i to facility j is zero and prohibit an assignment of demand fractions above 100%, otherwise. Constraints (16c) ensure that each demand node is covered by at least one facility. Constraints (16d) will make sure that each open facility is assigned at least the minimum demand. Constraints (16e) and (16f) make certain the demand allocation equilibrium conditions are met. Constraints (16g) make sure that consumers' disutility is capped at $\tau^{\max}$, and L is a large enough number that ensures (16g) is enforced only when $z_{ij} = 1$. Finally, Constraints (16h) ensure that the service rate at each facility is greater than or equal to the demand rate at that facility—a stability condition for Markovian queuing systems. We note that Constraints (16e) and (16f) make this problem highly nonlinear.

Although this model belongs to the class of models with equilibrium constraints that mostly offer heuristic approaches, and only a few offer efficient optimization schemes, Aboolian et al. (2022) exploit the specific structure of the problem, allowing to reformulate a highly nonlinear problem as a MILP without any approximation, and as a result, they could find exact optimal solutions for fairly large instances. The MILP reformulation of the model is as follows:

$$\min \quad \sum_{j\in M} h_j x_j + \sum_{j\in M}\sum_{i\in N} c_j \lambda_i^{\max} y_{ij} + \sum_{j\in M} c_j P_j$$

subject to (16a)–(16c),

$$z_{jj} \geq x_j\,,\ j \in M\,, \tag{17a}$$

$$\lambda_j^{\max} y_{jj} \geq \lambda^{\min} x_j, j \in M\,, \tag{17b}$$

$$P_j \geq \left(\frac{\beta}{\tau^{\max} - t_{ij}}\right) z_{ij}\,,\ i \in N, j \in M\,, \tag{17c}$$

$$\sum_{k\in M} y_{jk} \geq e^{-\alpha(\tau^{\max} - t_{ij})} z_{ij}\,,\ i \in N, j \in M\,, \tag{17d}$$

$$t_{ij} z_{ij} \leq \tau^{\max}\,,\ i \in N, j \in M\,, \tag{17e}$$

$$\sum_{k \in M} y_{ik} \geq q_{ij} \sum_{k \in M} y_{jk} - 1 + x_j \,, \; i \in N, j \in M \,, \tag{17f}$$

$$\sum_{k \in M} y_{ik} \leq q_{ij} \sum_{k \in M} y_{jk} + 1 - z_{ij} \,, \; i \in N, j \in M \,, \tag{17g}$$

$$P_j, y_{ij} \geq 0, x_j, z_{ij} \in \{0, 1\} \,, \; i \in N, j \in M \,,$$

in which decision variables $P_j, j \in M$ are the inverse of W_j such that $P_j = \mu_j - \sum_{j \in M} \lambda_i^{\max} y_{ij}, j \in M$. Here Constraints (17a) and (17b) are based on Lemma 1 of Aboolian et al. (2022), which states that if a facility is opened at $j \in M$, then at least some of the consumers at j will use its services. Note that Constraints (17b) replace (16d). Constraints (17c), (17d), and (17e) ensure that the consumers' disutility is capped at $\tau^{\max}$. Given the objective function, Constraints (17c) more specifically determine the values for P_js. Constraints (17f) and (17g) like (16e) and (16f) make certain that the allocation of consumer demand to facilities is a user equilibrium, in which consumers have no incentive to change their selected facilities.

The MILP formulation is particularly beneficial as there are many out-of-the-box optimization implementations of MILPs that are highly efficient.

2.2.2.3 PFLPC Models with Elastic Demand and Socially Optimal Perspective

Governments around the globe are actively involved in providing essential services, such as healthcare, transportation, education, and utilities. In contrast with the private sectors' mission to maximize profit, governments' mandate is to maximize societal benefit by acting as public agents, and to have a more realistic model, the consumers' behavioral preferences should be considered in the utility (disutility) functions. When designing service systems, many models in the public sector focus on maximizing accessibility to public services to increase societal benefit. The idea is to (re)design the public service to maximize the number of people who will benefit from the program given a limited budget, thus using accessibility as a proxy for benefit. Such models fail to capture the marginal benefits—savings in costs to taxpayers by adding an extra unit of service capacity.

In a recent work, Aboolian and Karimi (2023c) introduce a more balanced approach to model PFLPC with elastic demand. The objective is to find the right balance for both the consumer and the service provider (public government). To do this, Aboolian and Karimi (2023c) study the problem of determining the optimal number, locations, and capacities of a network of facilities to maximize the public's overall benefit. They define the overall benefit as the difference in savings for the public by participating in services and the cost of the provided service capacity.

Like other PFLPC models with elastic demand, the consumers would like to maximize their utility (minimize their disutility) when choosing which facility to patronize. In Aboolian and Karimi (2023c), consumers' disutility takes the general form of (12). The authors consider a user equilibrium problem, where, at equilibrium, consumers have no incentive to change their choices.

The mathematical formulation of this benefit maximization problem is the following nonlinear mixed-integer program:

$$\max \quad v \sum_{i \in N} \sum_{j \in M} \lambda_i y_{ij} - c \sum_{j \in M} \mu_j$$

subject to :

$$y_{ij} \leq x_j \,, \; i \in N, j \in M \,, \tag{18a}$$

$$\left(\alpha\, t_{ij} + \frac{\beta}{\mu_j - \sum_{r \in N} \lambda_r y_{rj}} + \frac{\ln\left(\sum_{k \in M} y_{ik}\right)}{\gamma} \right) y_{ij} = 0 \,,$$

$$i \in N, j \in M \,, \tag{18b}$$

$$\alpha\, t_{ij} + \frac{\beta}{\mu_j - \sum_{r \in N} \lambda_r y_{rj}} + \frac{\ln\left(\sum_{k \in M} y_{ik}\right)}{\gamma} \geq 0 \,, \; i \in N, j \in M \,, \tag{18c}$$

$$\sum_{i \in N} \lambda_i y_{ij} \leq \mu_j \,, \; j \in M \,, \tag{18d}$$

$$\mu^{\min} x_j \leq \mu_j \leq \mu^{\max} x_j \,, \; j \in M \,, \tag{18e}$$

$$\frac{1}{\mu_j - \sum_{i \in N} \lambda_i y_{ij}} \leq W^{\max} \,, \; j \in M \,, \tag{18f}$$

$$\sum_{i \in N} \lambda_i y_{ij} \geq \lambda^{\min} \,, \; j \in M \,, \tag{18g}$$

$$\mu_j, y_{ij} \geq 0, x_j \in \{0, 1\} \,, \; i \in N, j \in M \,.$$

The objective function terms of this problem represent the total benefit of serving demand as the total value added minus the cost of the service system. Constraints (18a) ensure that demand is only assigned to open facilities while capping the demand fractions at 1. Constraints (18b) and (18c) make certain the demand allocation equilibrium conditions are met. Constraints (18d) ensure stability by enforcing steady-state service capacities. Constraints (18e) ensure that the service rates are bounded within the desired range. Constraints (18f) guarantee a maximum waiting time of $W^{\max}$ across the entire system. Lastly, Constraints (18g) guarantee every open facility serves a demand rate of at least $\lambda^{\min}$.

The main sources of complexity when solving BNDP are the nonlinear Constraints (18b) and (18c). To solve this problem, Aboolian and Karimi (2023c) offer a reformulation of it with linear constraints and a nonlinear objective. The reformulation is amenable to a *tangent-line piecewise linear approximation* (TLA)

(Aboolian et al., 2007) technique, which allows them to ϵ-optimally solve this problem.

When it comes to solution approaches used to solve the models for these three different perspectives, there is not much of a difference, with the exception that in the service sector perspective, the model can be transformed into a linear model without any approximation, while the other two perspectives use a linear approximation to ϵ-optimally solve their respective model. However, the results of the facility location and capacity allocation for different perspectives could be far from similar. In our experience, the consumer or service provider perspective solutions most often than not result in a not socially optimal solutions.

3 Current Challenges and Research Opportunities in Location Analysis for Public Sector

3.1 Current Congestion Models in Public Location Theory: Limitations and Extensions

Since the early facility location models, location science has come a long way in supporting location analysts and policymakers. As discussed in Sect. 1, the consumer-centric view of service system design and the incorporation of decision-making under uncertainty have played a big part in the recent advancements of public FLPs. In Sect. 2, we also argued that congestion models play a significant role in public FLPs as they consider service system design problems in stochastic environments. There are situations, however, where current congestion models may not provide helpful insight. Following the human-centric view of contemporary location applications, we identify two particular settings in which the current congestion model can, and arguably should, be extended to provide further insight.

3.1.1 Incorporating Incentive Initiatives in PFLPC Models

To overcome the various challenges caused by the COVID-19 pandemic, many branches of science and technology have come up with novel solutions and artifacts. Even with the grant of emergency permits, arguably, one of the most influential breakthroughs was the development of COVID-19 vaccines in record time—a process that normally requires 10 to 15 years. However, healthcare providers and policymakers soon found a new operational challenge when distributing vaccines: *vaccine hesitancy*. To increase the population of vaccinated individuals, many local and federal governments started to offer incentives in monetary prizes, raffles, gift cards, etc. (Pennings & Symons, 2021). Clearly, this is a government view of the societal benefits which should include consumers' behavioral preferences in the utility (disutility) functions.

From an operational perspective, the distribution of vaccines is an operational location-allocation problem, and location analysis can provide valuable support to decision-makers in deciding the number and capacity of vaccination centers and where to locate them optimally. PFLPC models, in particular, can be helpful due to their stochastic nature.

In its Quality Chasm report, the Institute of Medicine (IOM) concludes that to effectively prevent and manage chronic disease, the US healthcare system requires a major realignment of incentives (Plsek, 2001). The need for such realignment has led to *"pay for prevention"* initiatives to reduce the high cost that preventable disease could bring to health providers and consumers (Casalino et al., 2003). There are many papers that have studied the benefits of incentives for preventive healthcare practices such as immunization (Kerpelman et al., 2000), cancer screening (Mayer & Kellogg, 1989), and prenatal care (Laken & Ager, 1995) to name a few.

Current PFLPC models do not incorporate incentives for receiving services. In fact, in most congestion models, users are assumed to incur disutility—in the form of waiting or travel time—when receiving services.

In the PFLPC context, the novel idea could be to find the right balance of location, service capacity, and incentive level for facilities that will maximize societal benefit.

The addition of incentives, although costly to the system, reduces consumers' disutility, which in turn increases their participation and adds to the overall benefit of the participation. Overall, introducing the incentive level as a decision variable to PFLPC makes it an interesting problem to examine.

Aboolian and Karimi (2023b) introduce such a model assuming a linear relationship between the consumer participation and savings to system due to participation (e.g., cost reduction due to preventing a disease). This model while appropriate for preventive services such as non-contagious diseases is not suitable for preventive healthcare for contagious diseases (e.g., vaccination for COVID-19). This is due to the nonlinear relationship between the vaccination and system savings due to vaccination.

3.1.2 Individual Preferences and Behavioral Decision-Making

When modeling the individual decision-making process, a key assumption in congestion models is that individuals are *"rational entities"* and make optimal decisions. There is now mounting evidence in behavioral decision-making that illustrates various types of biases affecting individuals' decisions. Bounded rationality, for instance, considers a broader setting for individual decision-makers and addresses the differences between perfect rationality and observed human behavior.

Given the people-centric agenda in public service system design—including the congestion models discussed in this chapter—understanding how individuals (consumers, service providers, or central decision-makers) behave and incorporating micro-level individual behavior can lead to better designs and improved processes.

In particular, two behavioral elements are directly related to the public service system design: *behavioral decision-making* and *behavioral queues*.

3.1.2.1 Behavioral Decision-Making

In contemporary service system design problems, individuals are often assumed to behave rationally: choose the utility-maximizing (disutility-minimizing) alternative, choose the closest facility, do not stray from the equilibrium allocation, and other similar assumptions. There are, however, systematic cognitive biases that can be incorporated when considering individuals' choices.

For instance, in the *anchoring effect* (Kahneman et al., 1982), individuals often use an initial piece of information (or experience) to make subsequent judgments. If consumers use a particular facility to receive services, for example, they may choose the said facility in the future, even though they may incur a higher waiting time in others.

Decisions made by human decision-makers in congestion models—made by individual consumers or servers at public facilities—are often subject to uncertainty. Congestion models often assume that individual decision-makers use probabilities to determine expected outcomes and choose the best course of action. Behavioral studies, however, often present various behavioral biases when it comes to the interpretation of probabilities. When considering *probability weighting functions*, for instance, individuals often substitute outcome weights for probabilities and assign a higher probability to unlikely outcomes or lower probabilities to almost certain outcomes (Kahneman & Tversky, 2013).

Incorporating human biases when modeling consumers' or servers' decision-making can broaden the application of congestion models in public FLPs and is in line with the human agency research agenda discussed in Sect. 1. To this end, we identify the applications of prospect theory in location analysis as one of the main research opportunities in location analysis for the public sector.

3.1.2.2 Behavioral Queues

Queuing theory is an invaluable tool that allows location analysts to gain predictive insight into the system behavior of service systems and predict operational metrics such as waiting time, service time, capacity, and utilization. Given the importance of such measures on the quality of location-allocation decisions, any assumption regarding the formation of queues is subject to validation. Behavioral queuing departs from many such assumptions by considering micro-level decision-making that often involves behavioral biases in human judgment and decision-making.

For example, queuing models often assume that consumers' disutility from waiting is linearly decreasing in waiting time. Many empirical studies have challenged the idea and shown that the level of individuals' disutility from waiting may decline with time or steadily increase frustration (Kocas, 2015).

Allon and Kremer (2018) offer the following conceptualization of behavioral queues to identify the impact of micro-level individual behavior on the macro-level service outputs.

$$\text{System Waiting} = \text{Net Utility} \times \text{Throughput} \tag{19}$$

in which *net utility* is defined as the gross value of receiving services minus the weight disutility. In particular, net utility is defined as follows:

$$\text{Net Utility} = v - c_W T_W - c_S T_S, \tag{20}$$

for which v is the gross "service quality"; T_W and T_S represent waiting and service times, respectively; and c_W and c_S denote the cost of waiting in queues and in services, respectively. Throughput can also be viewed as the multiplication of the arrival rate by the probability of individuals waiting to receive services.

$$\text{Throughput} = \sum_i \lambda_i \mathbb{P}\left(v \geq \theta_i\right), \tag{21}$$

in which θ_i denotes a *patience* threshold of consumer i.

The above framework allows location analysts to consider various aspects of behavioral queues. For example, one of the main findings in the psychology of waiting lines is the perception of individuals regarding the *occupied* vs. *unoccupied* time, in which the latter *"feels"* longer than the former, i.e., $c_W > c_S$ (Allon & Kremer, 2018).

Considering the behavioral aspects of queues and the psychology of waiting lines can further bridge the gap between the theory and application of service system design and improve location-allocation decisions in the public sector, presenting another research opportunity for location researchers and practitioners.

3.2 Public-Private Relationships

The study of location problems from a service system design perspective integrates the strategic and tactical location-allocation problems with the individual micro-decisions. As discussed in Sects. 1 and 2, this integration allows us to analyze realistic situations influenced by individual behavior, such as social optimum and user equilibrium environments. Additionally, the service system design perspective facilitates the investigation of cross-sector collaborations.

Similar to how firms achieve better efficiency when integrating decisions in cross-functional collaborations (like Sales and Operations Planning), cross-sector collaborations present similar opportunities. The literature often refers to such cross-sector collaborations as *public-private partnership* (PPP)—commonly defined as *cooperative institutional arrangements between public and private sectors.* (Wang et al., 2018). Given their expansive advantages, PPP projects have been applied in many public sector domains such as infrastructure, transportation, water, energy, environment protection, humanitarian aids, health, and more (see Wang et al. (2018) for a detailed review of PPP adoptions).

Governments have employed PPPs to offer a long-term and sustainable service provision that provides a more efficient expenditure of taxpayer money. Some of the cited benefits of employing PPPs are accelerated infrastructure development, increased value for money, and improved service quality (see, e.g., Yong (2010)[Ch. 3]). In healthcare, in particular, PPPs are linked with an improvement in access, quality, and efficiency (Sekhri et al., 2011).

There are few studies of PPPs in service systems in the operations research and management science domain, primarily contemporary works carried out in the past decade, most of which have been conducted in healthcare applications. This is perhaps unsurprising given the many instances of public-private healthcare systems—often referred to as two-tier healthcare—coexisting worldwide.

Andritsos and Tang (2014), one of the earliest operations research works to study PPPs in healthcare service systems, consider the impact of private healthcare providers on operations of public healthcare systems and the increased patient choice. The authors adopt a game-theoretical queuing model to investigate the effects of PPPs on welfare requirements, cost, and patient waiting time. The authors find that the public healthcare systems can reduce costs given welfare requirements without increasing the patients' waiting time. Andritsos and Aflaki (2015) consider a competitive PPP setting under which a public hospital and a private hospital choose their service capacity. Their setting is particularly applicable in situations where the government uses (and subsidizes) the private healthcare system to satisfy demand when the public healthcare system is overly congested. Andritsos and Aflaki (2015) also adopt a game-theoretical queuing model and show that providing unconditional subsidies to the private hospital leads to a decrease in public hospital capacity, which in turn causes an increase in the public hospital's waiting time. This strand of research has been further generalized to consumer choice (Qian et al., 2017), coordination (Hua et al., 2016), sustainability (Zhang & Yin, 2021), and more.

Some of the main characteristics of the current studies of PPPs in service systems are the focus on strategic interaction of the public and private entities, the emphasis on user equilibrium behavior of consumers, and the focus on service quality in conjunction with the cost and benefit of service provision. Researchers often model the strategic interaction between the public and private partners as competition (duopoly games) or cooperation (coordination games). Game theory has been a particularly suitable tool to deal with the inherent complexity of PPPs' strategic interaction and the parties' possible conflicting interests. In particular, when engaged in a PPP, a government's role may change from supervision to cooperation or competition, and oversight affects both partners' objectives and restrictions. When considered, consumer choice has been related to the system waiting time—often modeled as the waiting time in an M/M/1 queue—and user equilibriums have been formulated to determine consumer demands. Equilibrium concepts are especially useful for studying the impact of consumers' micro-behavior on the macro-patterns of demand. When it comes to assessing a particular PPP setting, the literature often contrasts the cost of operations with the quality of service, which is often measured via consumers' waiting time.

In all studies mentioned above, however, location decisions have been overlooked. In particular, the operational decisions of capacity allocation or tactical decisions of public procurement for short- to medium-lived services is the main focus of the PPP literature in service system design. Yet, one of the most important benefits of PPPs is associated with their *"long-term contract"* property (Yong, 2010, Section 3.1). Given their strategic nature and long-term commitment, location decisions can play a significant role in the success of PPP initiatives, as incorporating the exact detail of location decisions when analyzing PPPs may provide further insights into system parameters such as cost, benefit, and quality.

Location decisions also allow (public or private) analysts to assess the allocation of risk in a broader sense of risk-benefit analysis. Broadly speaking, each PPP initiative can be viewed on a continuum of risk-sharing between two extremes. On the one end, governments may outsource the provision of services to private entities while bearing the entirety of the risk involved. For instance, governments often outsource waste management services on a short-to-medium-term basis. On the other end, the government can utilize *privatization* by transferring service provision—and its cost and revenue—entirely to private sectors who bear almost all the risk involved. The privatization of telephone landlines is perhaps the most prominent example of this extreme.

An arrangement under which the risk of service provision is shared between partners is more likely to be successful and sustainable in the long term (Liu et al., 2015). Determining the number, location, and ownership of facility locations allows for a more accurate risk assessment and brings further transparency in the level of *shared risk*. Furthermore, from an optimization point of view, determining the optimal number, location, and capacities of service facilities without consideration for public-private interactions may lead to an inefficient allocation of resources.

Location decisions are also important to the partnership agreement nature of PPPs. In particular, PPP environments often exhibit information asymmetry, in which the public partner may know more about the consumer population, and the private partner may know more about the cost of service provision (Aben et al., 2021). In the absence of shared information, for example, partners can identify and target more profitable segments of the market, leading to a less efficient service provision compared to no-partnership benchmarks. To mitigate adverse outcomes caused by information asymmetry, partners can rely on long-term location decisions as signaling mechanisms to communicate credible actions and establish trust.

References

Aben, T. A., van der Valk, W., Roehrich, J. K., & Selviaridis, K. (2021). Managing information asymmetry in public–private relationships undergoing a digital transformation: the role of contractual and relational governance. *International Journal of Operations & Production Management, 41*(7), 1145–1191.

Aboolian, R., Berman, O., & Drezner, Z. (2008). Location and allocation of service units on a congested network. *IIE Transactions, 40*(4), 422–433.

Aboolian, R., Berman, O., & Drezner, Z. (2009). The multiple server center location problem. *Annals of Operations Research, 167*(1), 337–352.

Aboolian, R., Berman, O., & Karimi, M. (2022). Probabilistic set covering location problem in congested networks. *Transportation Science, 56*(2), 528–542.

Aboolian, R., Berman, O., & Krass, D. (2007). Competitive facility location model with concave demand. *European Journal of Operational Research, 181*(2), 598–619.

Aboolian, R., Berman, O., & Verter, V. (2016). Maximal accessibility network design in the public sector. *Transportation Science, 50*(1), 336–347.

Aboolian, R., Elhedhli, S., & Karimi, M. (2022). An efficient approach for service system design with immobile servers, stochastic demand, congestion, and consumer choice. *Journal of Supply Chain and Operations Management, 20*(1), 1.

Aboolian, R., & Karimi, M. (2023a). An Efficient Approach for Service System Design with Continuous Service Rate and Concave Capacity Cost. (Working Paper)

Aboolian, R., & Karimi, M. (2023b). Benefit Maximizing Network Design for Preventive Health Care System Using Incentives. (Working Paper)

Aboolian, R., & Karimi, M. (2023c). Benefit Maximizing Network Design in the Public Sector. (Working Paper)

Alfred, W. (1929). *Theory of the location of industries.* The University of Chicago Press, Chicago, Ill.

Allon, G., & Kremer, M. (2018). Behavioral foundations of queueing systems. *The Handbook of Behavioral Operations, 9325*, 325–366.

Amiri, A. (1997). Solution procedures for the service system design problem. *Computers & Operations Research, 24*(1), 49–60.

Andritsos, D. A., & Aflaki, S. (2015). Competition and the operational performance of hospitals: the role of hospital objectives. *Production and Operations Management, 24*(11), 1812–1832.

Andritsos, D. A., & Tang, C. S. (2014). Introducing competition in healthcare services: the role of private care and increased patient mobility. *European Journal of Operational Research, 234*(3), 898–909.

Berman, O., & Drezner, Z. (2007). The multiple server location problem. *Journal of the Operational Research Society, 58*(1), 91–99.

Berman, O., & Krass, D. (2002). The generalized maximal covering location problem. *Computers & Operations Research, 29*(6), 563–581.

Berman, O., & Krass, D. (2019). Stochastic location models with congestion. In: *Location science* (pp. 477–535). Springer.

Berman, O., Krass, D., & Drezner, Z. (2003). The gradual covering decay location problem on a network. *European Journal of Operational Research, 151*(3), 474–480.

Carbone, R. (1974). Public facilities location under stochastic demand. *INFOR: Information Systems and Operational Research, 12*(3), 261–270.

Casalino, L., Gillies, R. R., Shortell, S. M., Schmittdiel, J. A., Bodenheimer, T., Robinson, J. C., Rundall, T., Oswald, N., Schauffler, H., & Wang, M. C. (2003). External incentives, information technology, and organized processes to improve health care quality for patients with chronic diseases. *JAMA, 289*(4), 434–441.

Castillo, I., Ingolfsson, A., & Sim, T. (2009). Social optimal location of facilities with fixed servers, stochastic demand, and congestion. *Production and Operations Management, 18*(6), 721–736.

Cooper, L. (1967). Solutions of generalized locational equilibrium models. *Journal of Regional Science, 7*(1), 1–18.
Dafermos, S. (1982). The general multimodal network equilibrium problem with elastic demand. *Networks, 12*(1), 57–72.
DeVerteuil, G. (2000). Reconsidering the legacy of urban public facility location theory in human geography. *Progress in Human Geography, 24*(1), 47–69.
Drezner, T., & Drezner, Z. (2011). The gravity multiple server location problem. *Computers & Operations Research, 38*(3), 694–701.
Elhedhli, S. (2006). Service system design with immobile servers, stochastic demand, and congestion. *Manufacturing & Service Operations Management, 8*(1), 92–97.
Elhedhli, S., Wang, Y., & Saif, A. (2018). Service system design with immobile servers, stochastic demand and concave-cost capacity selection. *Computers & Operations Research, 94*, 65–75.
Hua, Z., Chen, W., & Zhang, Z. G. (2016). Competition and coordination in two-tier public service systems under government fiscal policy. *Production and Operations Management, 25*(8), 1430–1448.
Jan, S., Mooney, G., Ryan, M., Bruggemann, K., & Alexander, K. (2000). The use of conjoint analysis to elicit community preferences in public health research: a case study of hospital services in South Australia. *Australian and New Zealand Journal of Public Health, 24*(1), 64–70.
Kahneman, D., Slovic, S. P., Slovic, P., & Tversky, A. (1982). *Judgment under uncertainty: heuristics and biases*. Cambridge University Press.
Kahneman, D., & Tversky, A. (2013). Prospect theory: an analysis of decision under risk. In *Handbook of the fundamentals of financial decision making: Part I* (pp. 99–127). World Scientific.
Kerpelman, L. C., Connell, D. B., & Gunn, W. J. (2000). Effect of a monetary sanction on immunization rates of recipients of aid to families with dependent children. *JAMA, 284*(1), 53–59.
Kocas, C. (2015). An extension of Osuna's model to observable queues. *Journal of Mathematical Psychology, 66*, 53–58.
Laken, M. P., & Ager, J. (1995). Using incentives to increase participation in prenatal care. *Obstetrics & Gynecology, 85*(3), 326–329.
Laporte, G., Nickel, S., & Saldanha-da Gama, F. (2019). Introduction to location science. In: *Location science* (pp. 1–21). Springer.
Liu, J., Love, P. E., Smith, J., Regan, M., & Davis, P. R. (2015). Life cycle critical success factors for public-private partnership infrastructure projects. *Journal of Management in Engineering, 31*(5), 04014073.
Marianov, V., Rios, M., & Barros, F. J. (2005). Allocating servers to facilities, when demand is elastic to travel and waiting times. *RAIRO-Operations Research, 39*(3), 143–162.
Marianov, V., & Serra, D. (1998). Probabilistic, maximal covering location–allocation models for congested systems. *Journal of Regional Science, 38*(3), 401–424.
Marianov, V., & Serra, D. (2002). Location–allocation of multiple-server service centers with constrained queues or waiting times. *Annals of Operations Research, 111*(1), 35–50.
Mayer, J. A., & Kellogg, M. C. (1989). Promoting mammography appointment making. *Journal of Behavioral Medicine, 12*(6), 605–611.
McAllister, D. M. (1976). Equity and efficiency in public facility location. *Geographical Analysis, 8*(1), 47–63.
Nagurney, A. (1998). *Network economics: a variational inequality approach* (vol. 10). Springer Science & Business Media.
Newman, R. G. (1984). A conjoint analysis in outpatient clinic preferences. *Journal of Health Care Marketing, 4*(1), 41–49.
Okun, A. M. (2015). *Equality and efficiency: the big tradeoff*. Brookings Institution Press.
Pennings, S., & Symons, X. (2021). Persuasion, not coercion or incentivisation, is the best means of promoting covid-19 vaccination. *Journal of Medical Ethics, 47*(10), 709–711.

Plsek, P. (2001). Institute of medicine. *Crossing the quality chasm: a new health system for the 21st century*. Washington, DC: National Academies Press.

Pollaczek, F. (1930). Über eine aufgabe der wahrscheinlichkeitstheorie. i. *Mathematische Zeitschrift, 32*(1), 64–100.

Qian, Q., Guo, P., & Lindsey, R. (2017). Comparison of subsidy schemes for reducing waiting times in healthcare systems. *Production and Operations Management, 26*(11), 2033–2049.

Revelle, C., Marks, D., & Liebman, J. C. (1970). An analysis of private and public sector location models. *Management Science, 16*(11), 692–707.

ReVelle, C. S., & Swain, R. W. (1970). Central facilities location. *Geographical Analysis, 2*(1), 30–42.

Schilling, D., Elzinga, D. J., Cohon, J., Church, R., & ReVelle, C. (1979). The team/fleet models for simultaneous facility and equipment siting. *Transportation Science, 13*(2), 163–175.

Sekhri, N., Feachem, R., & Ni, A. (2011). Public-private integrated partnerships demonstrate the potential to improve health care access, quality, and efficiency. *Health Affairs, 30*(8), 1498–1507.

Teitz, M. B. (1968). Toward a theory of urban public facility location. In: *Papers of the Regional Science Association* (vol. 21, pp. 35–51). Springer.

Verter, V., & Lapierre, S. D. (2002). Location of preventive health care facilities. *Annals of Operations Research, 110*(1-4), 123–132.

Vidyarthi, N., & Jayaswal, S. (2014). Efficient solution of a class of location–allocation problems with stochastic demand and congestion. *Computers & Operations Research, 48*, 20–30.

Wang, H., Xiong, W., Wu, G., & Zhu, D. (2018). Public–private partnership in public administration discipline: a literature review. *Public Management Review, 20*(2), 293–316.

Wang, Q., Batta, R., & Rump, C. M. (2002). Algorithms for a facility location problem with stochastic customer demand and immobile servers. *Annals of Operations Research, 111*(1), 17–34.

Wang, Q., Batta, R., & Rump, C. M. (2004). Facility location models for immobile servers with stochastic demand. *Naval Research Logistics (NRL), 51*(1), 137–152.

Wardrop, J. G. (1952). Road paper. some theoretical aspects of road traffic research. *Proceedings of the Institution of Civil Engineers, 1*(3), 325–362.

Yong, H. K. (2010). *Public-private partnerships policy and practice: a reference guide*. Commonwealth Secretariat.

Zhang, Y., Berman, O., Marcotte, P., & Verter, V. (2010). A bilevel model for preventive healthcare facility network design with congestion. *IIE Transactions, 42*(12), 865–880.

Zhang, Z. G., & Yin, X. (2021). Designing a sustainable two-tier service system with customer's asymmetric preference for servers. *Production and Operations Management, 30*(11), 3856–3880.

Stochastic Gradual Covering Location Models

Zvi Drezner

Abstract Most location models assume that the parameters are given and fixed. Demand for services is known, and the distance to the facility is given. Real-world parameters are not fixed but follow a probability distribution such as a normal distribution. Therefore, stochastic models estimate the results (cost, profit, cover) more accurately.

In cover models, facilities need to be located in an area to provide service to a set of demand points. Demand points that are within a given distance are covered. Two main objectives are investigated in the literature: provide as much cover as possible with a given number of facilities and minimize the number of facilities required to provide full cover. In gradual cover models, up to a certain distance, the demand point is fully covered, and beyond a greater distance, it is not covered at all. Between these two extreme distances, the demand point is partially covered.

In this chapter, we summarize gradual cover models emphasizing on models that have stochastic parameters. We also propose a new model analyzing a stochastic version of the directional graduate cover.

Keywords Location analysis · Cover models · Gradual cover models · Stochastic analysis

1 Introduction

Most location models assume that the parameters are given and fixed. Demand for services is "known," and the distance to the facility is given. If an ambulance or a fire truck needs to get to a customer within 10 min, the time is translated to a distance, for example, 3 miles, even though the travel speed may depend on traffic conditions

Z. Drezner (✉)
College of Business and Economics, California State University-Fullerton, Fullerton, CA, USA
e-mail: zdrezner@fullerton.edu

H. A. Eiselt, V. Marianov (eds.), *Uncertainty in Facility Location Problems*,
International Series in Operations Research & Management Science 347,
https://doi.org/10.1007/978-3-031-32338-6_11

and is not a constant. A customer is considered covered within 3 miles even though only a proportion of the customers are "covered."

Customers are assumed to be located at "demand points" even though in most applications customers reside in neighborhoods that are defined by regions. Not all customers residing in a neighborhood have the same distance to the facility. Francis et al. (2009, 2000) analyzed the selection of a point that represents a set of demand points or an area. Drezner and Drezner (1997) showed that the squared distance between a demand point located at a center of a circular area and a facility should be increased proportionally to the circle's area.

It is probably easier to formulate and solve models with known parameters, but in reality, stochastic models estimate the results (cost, profit, cover) more accurately. Real-world parameters are not fixed but follow a probability distribution such as a normal distribution.

2 Cover Models

Facilities need to be located in an area to provide service to a set of demand points. Demand points that are within a given distance are covered, meaning that they are getting the services under consideration (Church & ReVelle, 1974; ReVelle et al., 1976). Two main objectives are investigated in the literature: (i) provide as much cover as possible with a given number of facilities and (ii) minimize the number of facilities required to provide full cover. Such models are used for cover provided by emergency facilities such as ambulances, police cars, or fire trucks. They are also used to model cover by transmission towers such as cell phone towers, TV or radio transmission towers, and radar coverage, among others. For a review of cover models, see Plastria (2002), García and Marín (2015), Snyder (2011), Church and Murray (2018). Drezner et al. (2011, 2012) applied the cover concept to competing facilities. Each competing facility has a "sphere of influence" (Launhardt, 1885; Fetter, 1924; Lösch, 1954; Christaller, 1966; ReVelle, 1986), and customers patronize a facility up to a certain distance.

A different covering model where facilities "cooperate" in providing cover was proposed in Berman et al. (2010). Each facility emits a signal (such as light posts in a parking lot, warning sirens) whose strength declines according to a distance decay function. A point is covered if the combined signal from all facilities exceeds a certain threshold. For example, a parking spot is covered if the total light received at that spot exceeds a given threshold. Recent papers on the cooperative cover are Morohosi and Furuta (2017), Karatas (2017), Wang and Chen (2017), Bagherinejad et al. (2018).

3 Gradual Cover Models

In the gradual cover models, up to a certain distance R_1, the demand point is fully covered and beyond a greater distance R_2, it is not covered at all. Between these two extreme distances, the demand point is partially covered. Suppose that the cover distance in traditional cover models is 3 miles. At a distance of 2.99 miles, the demand point is fully covered while at a distance of 3.01 miles, it is not covered at all. This assumption may be convenient for analyzing and solving covering problems. However, in reality, cover does not drop abruptly but the decline in cover is gradual.

Various notations are defined in gradual cover models. To be consistent throughout this chapter, we define the following variables:

Notation

D Cover distance by non-gradual cover models.
d Distance between a facility and a demand point.
R_1 Full coverage for distance $d \leq R_1$.
R_2 No coverage for distance $d \geq R_2$.
R $= \frac{R_1+R_2}{2}$.
ΔR $= R_2 - R_1$.
r radius of a circle centered at the demand point.

Church and Roberts (1984) were the first to propose the gradual cover model (also referred to as partial cover). The facilities must be located within a finite set of potential locations. Drezner et al. (1998) investigated the gradual cover model in the plane for locating competing facilities. They model the partial cover by a logit function. The network version with a step-wise cover function is discussed in Berman and Krass (2002). The network and discrete models with a general non-increasing cover function were analyzed in Berman et al. (2003b). The single-facility planar model with a linearly decreasing cover function between R_1 and R_2 was optimally solved in Drezner et al. (2004) by the Big Triangle Small Triangle (BTST) optimization method (Drezner & Suzuki, 2004). It can also be solved by the Big Square Small Square (BSSS) method (Hansen et al., 1981). Location of several facilities can be solved optimally by the Big Cube Small Cube method (Schöbel & Scholz, 2010). Reasonable run time can be achieved for locating up to three facilities. Additional references include Karasakal and Karasakal (2004), Eiselt and Marianov (2009), Drezner and Drezner (2014), Berman et al. (2019).

3.1 Estimating Partial Cover of a Demand Point Covered by Several Facilities

An important issue in gradual cover models is the estimation of the total cover when a demand point is covered by several facilities. In traditional non-gradual cover models where a demand point is either fully covered by a facility, or not covered at all, the rule is straightforward. A demand point is covered if and only if it is covered by at least one facility.

This issue is discussed in Berman et al. (2019). They proposed several "axioms" and observations that we term properties, and we added Property 7:

Property 1: The total cover is between 0 and 1.
Property 2: If the partial coverage from a facility increases unilaterally, the joint coverage cannot decrease.
Property 3: Adding facilities that provide no coverage cannot change the joint coverage received by a demand point.
Property 4: The joint coverage is not lower than the partial coverage received from any one facility.
Property 5: If a demand point receives positive coverage from only one facility, then the joint coverage equals to the individual coverage.
Property 6: If a demand point is covered fully from any one facility, then the joint coverage is full as well.
Property 7: If all the distances between the demand point and the facilities do not increase, the total cover of the demand point cannot decrease.

We prove the following theorem based on Property 7:

Theorem 1 *The optimal locations of the facilities that maximize the total cover are in the convex hull of the demand points.*

Proof By a theorem in Wendell and Hurter (1973), for any location outside the convex hull of a set of points, there is a location in the convex hull that is closer to each of the points generating the convex hull. Therefore, for any location outside the convex hull of the demand points, there is a location in the convex hull with a better value of the objective function because all distances are shorter. If there is a facility outside the convex hull, a better location for that facility exists in the convex hull. The theorem follows by mathematical induction. □

Let c_j be the partial cover of a demand point by facility j for $j = 1, \ldots, p$. Eiselt and Marianov (2009) proposed a total partial cover of $\min\left\{\sum_{j=1}^{p} c_j, 1\right\}$. Partial cover can be interpreted as the probability of cover. Assuming that the partial covers are not correlated, the total partial cover is: $1 - \prod_{j=1}^{p}(1 - c_j)$ (Berman et al., 2003a; Drezner & Wesolowsky, 1997; Drezner & Drezner, 2008). The directional gradual cover discussed in Sect. 3.4 leads to a different rule for the total cover of several facilities.

3.2 Step-Wise Gradual Cover

Church and Roberts (1984) and Berman and Krass (2002) proposed a step-wise decline in cover. A sequence of $k > 1$ distances $R_1 < R_2 < \ldots, R_k$ is defined with associated partial covers $p_1 = 1 > p_2 > \ldots > p_k = 0$. Up to a distance R_1 cover is full ($p_1 = 1$). For distances $R_i < d \leq R_{i+1}$ for $1 \leq i \leq k-1$, the cover is p_{i+1}, and for $d > R_k$ cover is zero.

3.3 Linear Decline Gradual Cover

The simplest model for gradual cover is a linear decline in cover between R_1 and R_2 as suggested in Drezner et al. (2004). For $d \leq R_1$ cover is full (cover of one), and for $d \geq R_2$ cover is zero. For $R_1 \leq d \leq R_2$, the partial cover is $\frac{R_2-d}{\Delta R}$. When $R_2 \to R_1(\Delta R \to 0)$, the linear decline model converges to the non-gradual cover model.

3.4 The Directional Gradual Cover

Drezner et al. (2019a) proposed a different approach to estimate partial cover defined as "directional gradual cover." This model is distinguished from the others based on the assumption that each customer point is not a point, but an area. As in gradual cover models, up to distance R_1 a point is fully covered and beyond a distance R_2 it is not covered at all. Each demand point is replaced by a circle of radius $\frac{\Delta R}{2} = \frac{R_2-R_1}{2}$ and a facility covers points within a distance $R = \frac{R_1+R_2}{2}$ that can be different for different facilities. The intersection area between the disk centered at the demand point and the disk of the coverage radius $\frac{R_1+R_2}{2}$ centered at the facility is calculated. The ratio between the intersection area and the area of the disk centered at the demand point is the partial cover of that demand point.

The proportion of cover (for complete details, see Drezner et al. (2019a)) is:

$$c(d) = \begin{cases} 1 & \Big| d \leq R_1 \\ \frac{1}{2\pi}\left[2\theta - \sin 2\theta\right] + \frac{1}{2\pi}\frac{(R_1+R_2)^2}{(R_2-R_1)^2}\left[2\phi - \sin 2\phi\right] & \Big| R_1 \leq d \leq R_2 \\ 0 & \Big| d \geq R_2 \end{cases}$$

where $\theta = \arccos \frac{d^2-R_1R_2}{d(R_2-R_1)}$; $\phi = \arccos \frac{d^2+R_1R_2}{d(R_1+R_2)}$. For $d = R_1$: $\theta = \arccos(-1) = \pi$, and $\phi = \arccos(1) = 0$, and therefore $c(d) = 1$. For $d = R_2$: $\theta = \phi = \arccos(1) = 0$, and $c(d) = 0$.

Drezner et al. (2019a) tested discrete problems where there is a given set of potential locations for the facilities. Drezner et al. (2020a) investigated the objective of maximizing the minimum cover among the demand points rather than the total cover. Drezner et al. (2021) investigated maximizing the total cover when the facilities can be located anywhere in the plane.

In gradual cover models, it is not obvious how to estimate the total cover if a demand point is partially covered by several facilities as discussed in Sect. 3.1. In the directional gradual cover (Drezner et al., 2019a), if a demand point is partially covered by two or more facilities, the total cover (area) depends on the distances between the facilities and the demand point, and on the *directions* of the facilities from the demand point.

Demand points are usually not mathematical points but represent communities that occupy an area and not all the residents at the demand "point" are located at the same point. Therefore, facilities at different directions cover different parts of the area represented by the demand point. For example, consider one demand point and three facilities depicted in Fig. 1. Facilities 1 and 2 cover some of the northern part of the community and facility 3 covers part of the southern part of the community. Suppose that only facilities 1 and 2 exist in the area. The facilities are located to the north of the demand point, and there is an overlap between the covered areas. Therefore, the total cover is the area covered by facility 2 and facility 1 does not contribute to the total cover. If one facility (either 1 or 2) is located to the north and facility 3 to the south, there is usually a smaller overlap if at all. Since all facilities' disks in Fig. 1 do not cover the demand point itself, the total area is the sum of the areas because there is no overlap. By any other gradual cover model, the total cover is calculated by the partial covers, and the total cover is the same regardless of the directions of the facilities.

When the radius of the demand point is zero, the directional gradual cover function is a discontinuous curve and the model is equivalent to the traditional non-gradual cover. The demand point is either fully covered or not covered at all.

Fig. 1 Three facilities and one demand point

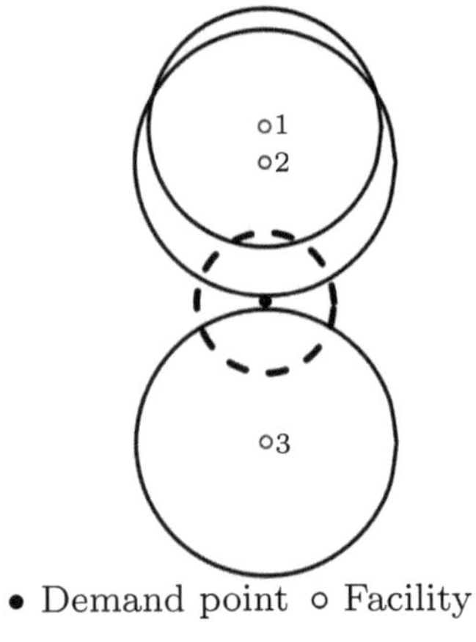

3.5 Random Limits of Gradual Cover

Drezner et al. (2010) modified the linear decline model by a model where R_1, the lowest distance when partial cover starts to decline from full cover, and R_2, the upper limit of the distance beyond which there is no cover at all are random variables. They assumed that cover declines linearly between the random values of R_1 and R_2. Other decline functions can also be investigated in a similar fashion. We summarize the formulations reported in that paper.

Let the cover radius used in the non-gradual covering model be D. Let $\phi_1(d)$ and $\phi_2(d)$ be the density function of the probability that R_1 and R_2, respectively, are at distance d. Each demand point may have different values for R_1 and R_2. Let $c(d)$ be the expected cover at distance d. If $d \leq R_1$, the cover is one. If $d \geq R_2$, the cover is zero. For $R_1 \leq d \leq R_2$ the cover is $\frac{R_2-d}{R_2-R_1}$. Therefore, the expected cover at distance d, $c(d)$, is

$$c(d) = Pr(d \leq R_1) + \int_0^d \int_d^\infty \frac{z-d}{z-y} \phi_1(y)\phi_2(z)dzdy \qquad (1)$$

Note that in (1) it is assumed that $\phi_1(d)$ and $\phi_2(d)$ are independent distributions. If they are correlated, then $\phi_1(y)\phi_2(z)$ should be replaced with a bi-variate distribution. The expected cover $c(d)$ can be calculated by numerical integration.

They analyzed the case where both distributions are uniform on both sides of a radius D and obtained an explicit formula for $c(d)$. Consider the following uniform distributions for a given $D > 0$ (the traditional non-gradual cover radius) and a range $\sigma \leq D$ for each one. Consequently, $R_1 = D - \sigma$, and $R_2 = D + \sigma$.

$$\phi_1(d) = \begin{cases} \frac{1}{\sigma} \mid R_1 \leq d \leq D \\ 0 \mid \text{Otherwise} \end{cases} \quad ; \quad \phi_2(d) = \begin{cases} \frac{1}{\sigma} \mid D \leq d \leq R_2 \\ 0 \mid \text{Otherwise} \end{cases}$$

The function $c(d)$ is (for complete details, see Drezner et al. (2010)):

$$c(d) = \begin{cases} 1 & \Big| d \leq R_1 \\ \frac{1+u}{2} + 2u \ln 2 - \frac{1}{2}\left\{(u+1)^2 \ln(1+u) - u^2 \ln u\right\} & \Big| R_1 \leq d \leq D \\ \frac{1-w}{2} - 2w \ln 2 + \frac{1}{2}\left\{(w+1)^2 \ln(1+w) - w^2 \ln w\right\} & \Big| D \leq d \leq R_2 \\ 0 & \Big| d \geq R_2 \end{cases} \qquad (2)$$

where $u = \frac{D-d}{\sigma}$ and $w = \frac{d-D}{\sigma}$. Note that $0 \leq u, w \leq 1$.

In Fig. 2, we depict the expression for the expected partial cover $c(d)$ by Eq. (2) for $R_1 = 1$ and $R_2 = 3$. By the linear decline gradual cover proposed in Drezner et al. (2004) with fixed values of R_1 and R_2, the graph has a line connecting $d = 1$ and $c(d) = 1$ with $d = 3$ and $c(d) = 0$ rather than the depicted curve. When $\sigma \to 0$, the random limit model converges to the non-gradual cover model.

3.6 The Logit Gradual Cover Function

Drezner et al. (1998) suggested a logit function

$$\frac{1}{1 + e^{\alpha+\beta d+\gamma d^2}}$$

for the partial cover. Drezner et al. (2020b) applied a simpler version of the logit function with only one parameter α:

$$\frac{1 + e^{\alpha}}{e^{\alpha} + e^{\alpha \frac{d}{R}}}. \tag{3}$$

These logit functions do not restrict the cover to be partial only between R_1 and R_2. A relatively large value of α is required as depicted in Figure 1 in Drezner et al. (2020b). In Fig. 2 below, a value $\alpha = 10$ was used so that the cover up to R_1 is very close to 1 and the partial cover for a distance greater than R_2 is very small.

When $\alpha \to \infty$, the model converges to the traditional non-gradual cover model. For $d < R$, $e^{\alpha \frac{d}{R}} << e^{\alpha}$ and becomes negligible compared to e^{α}. Consequently, the ratio is close to 1. For $d > R$, $e^{\alpha \frac{d}{R}} >> e^{\alpha}$ and e^{α} becomes negligible compared to $e^{\alpha \frac{d}{R}}$. Consequently, the ratio is close to 0.

3.7 An Inverse Cumulative Normal Distribution

Berman et al. (2019) considered the situation that an ambulance, police car, and fire truck needs to reach a demand point within a given time threshold. The time it takes to reach a demand point at distance d has a probability distribution that can be assumed normal by the central limit theorem. The mean of the distribution is μ at which the probability of reaching the demand point in time is 0.5. The standard deviation of the normal distribution, σ, reflects the variability of the travel time. When $\sigma \to 0$, the inverse cumulative normal model converges to the non-gradual cover model. There is a likely time to reach the demand point within the given threshold. Therefore, the probability of not reaching a demand point within the time threshold is the cumulative normal distribution and the probability of reaching it is the inverse cumulative normal.

Budge et al. (2010) performed an empirical study of over 7000 ambulance trips in the city of Calgary in Alberta, Canada, and developed a graph of the probability, which is a measure of coverage, that an ambulance will reach a patient within a given time as a function of the distance (the "Golden Half Hour"). The probability graph developed in their study is almost identical to the inverse cumulative normal curve.

3.8 Correlated Binomial

The distribution of ambulance trips in Budge et al. (2010) can be interpreted as a binomial distribution of events. Success is when the ambulance arrived on time and failure if it did not. The limit of a binomial distribution is a normal distribution. The underlying assumption of a binomial distribution is that the events are not correlated. What if the events are correlated? Drezner and Farnum (1993) developed a "generalized binomial distribution" (GBD) for correlated Bernoulli processes. See also Drezner (2019).

An initial probability of success p is given. An association factor θ, which is similar to the correlation coefficient, is given. Suppose that in the first k events, the number of successes is s. The probability of success in the next event is $(1-\theta)p+\theta\frac{s}{k}$. $\theta=0$ yields the "standard" binomial distribution where the probability of success is p regardless of the number of successes so far. On the other extreme, for $\theta=1$, if the first event is a success, all subsequent events are successes regardless of the value of p. The probability distribution when $\theta=1$ consists of two values success with probability of p and failure with probability $1-p$. It is not a bell shape distribution as is obtained by uncorrelated binomial.

When $\theta>0$, if the proportion of successes so far is greater than p, the probability of success in the next event is greater than p. If the rate of successes is below p, the probability of success is less than p. For example, in sport events, a "good" team that has a good record of successes so far in the season is more likely to succeed in the next game. Drezner and Farnum (1993) showed that the mean of the distribution is np, the same as the binomial distribution, but the variance is $p(1-p)\frac{n-\frac{1}{B(n,2\theta)}}{1-2\theta}$. They found that in baseball games $\theta=0.397$. For complete details, see Drezner and Farnum (1993).

Drezner (2006) further investigated the limit of the GBD. It is proven that for $\theta\leq 0.5$ the limit of the GBD, as the number of trials increases to infinity, is the normal distribution. For $\theta>0.5$, it can be bi-modal. It was also found, by analyzing real data, that the grade distribution of 1023 multiple choice exams yielded $\theta=0.5921$ and the number of wins of NBA teams at the end of the season yielded $\theta=0.5765$; both are not a normal distribution. The percentage of "wins" for both exam scores and NBA teams are not random but depend on the skill of the individual. Bhootra et al. (2015) investigated the performance of mutual funds by the GBD and found that the performance of mutual funds is not random, but the skill of the managers plays an important role in their performance.

An interesting gradual decline function is the inverse of the limit of the cumulative GBD. For $\theta \leq 0.5$, the function is the inverse cumulative normal distribution, but for $\theta > 0.5$, the distribution can be bi-modal. For $\theta = 1$, the distribution is either success or failure, which is actually the traditional non-gradual cover function. There is no gradual cover; the cover drops abruptly from full cover to no cover. In Fig. 2, the partial cover function is depicted for $\theta = 0.9$.

3.9 Comparing Gradual Cover Functions

In Fig. 2, gradual decline in cover functions are depicted for $R_1 = 1$ and $R_2 = 3$. In the original non-gradual cover models, there is an abrupt decline in cover at a certain distance (distance of $R = 2$ in the figure) from full cover to no cover. Church and Roberts (1984) and Berman and Krass (2002) proposed a step-wise decline in cover discussed in Sect. 3.2. Such approach still has discontinuities in the cover as a function of the distance. Drezner et al. (2004) proposed a linear decline in cover between R_1 and R_2, discussed in Sect. 3.3. This model is continuous but has a discontinuous derivative at $d = R_1$ and at $d = R_2$. Drezner et al. (2010) proposed that R_1 and R_2 are random variables rather than fixed values. Their model is discussed in Sect. 3.5. This partial cover function is continuous with a continuous derivative. By the directional cover, described in Sect. 3.4, it is close to linear decline and is the only function in Fig. 2 that is not equal to 0.5 at $d = 2$. This is because the intersection area between the circles when the circle of radius R passes through the demand point is not half of the circle's area. The logit-based gradual cover, discussed in Sect. 3.6, is based on Eq. (3). For $\alpha = 10$, which was used in the figure, the shape of the partial cover function resembles the random function shape. This shape also resembles the inverse normal distribution function discussed in Sect. 3.7 for a standard deviation $\sigma = \frac{R-r}{6}$. The correlated binomial model, discussed in Sect. 3.8, is the inverse cumulative distribution of a possibly bi-modal distribution for $\theta > 0.5$. The graph depicted in the figure is calculated by a simulation using $\theta = 0.9$. The density function is bi-modal and the derivative of the curve has a sharper decline near the two modes and a shallow decline near $d = 2$, which is the low point of the density function between the two modes.

3.10 Summary and Discussion of Gradual Cover Models

In the original gradual cover model, there is an abrupt decline from full cover to partial cover. In reality, cover does not drop abruptly. Earlier models of gradual cover attempted to rectify it by defining a decline of coverage by a step-wise or a linear function (Sects. 3.2, 3.3). More recently (see Sect. 3.4), it is assumed that every demand "point" is actually a neighborhood and not all customers are at the same distance from a facility.

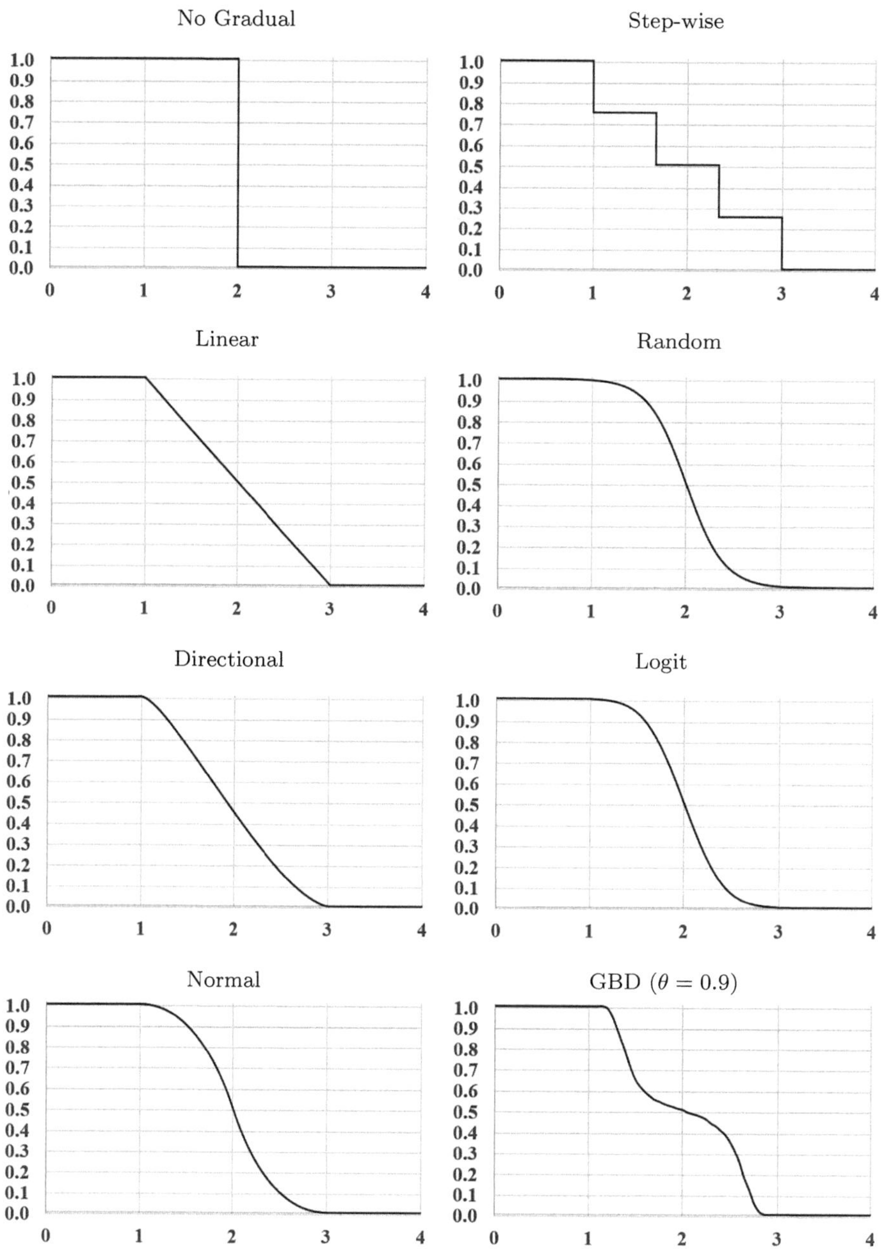

Fig. 2 Different gradual cover functions for $R_1 = 1$ and $R_2 = 3$

Subsequent models discussed in Sects. 3.5–3.8 assume that the parameters of the gradual cover models are random rather than having fixed values. Such an assumption is closer to reality and provides more flexibility. For example, in Sect. 3.5, it is assumed that the start of partial cover R_1 and the start of no cover R_2 are random variables. In Sect. 4, we propose and test a new model assuming that the parameters of the directional gradual cover (Sect. 3.4) are random variables, which makes the model yet closer to reality.

4 The Stochastic Directional Gradual Cover Model

In this section, we propose a stochastic formulation for the directional gradual cover model. We incorporate standard gradual cover approaches into the directional cover model. As in the directional gradual cover, the demand point is defined by a circle of radius r. The facility does not cover a point in the plane by a disk of radius D, but there are two radii $R_1 \leq D \leq R_2$ so that a point is fully covered within the circle of radius R_1, and not covered at all outside the circle of radius R_2. A point in the ring between R_1 and R_2 is partially covered. Each point in the circle of radius r centered at the demand point is covered at a proportion between 0 and 1. The cover of a demand point is the integral over the circle centered at the demand point, where the integrand at any point in the circle is its partial cover.

In Fig. 3, a typical cover of one demand point by one facility is depicted. The intersection area within a radius R_1 is fully covered. The area beyond R_2 is not covered. The intersection area with the ring between R_1 and R_2 is partially covered. In the original directional gradual cover (Drezner et al., 2019a), $R_1 = R_2$, the ring is a circle, and there is no area with partial cover.

The partial cover between R_1 and R_2 can be defined in many ways. For example, the gradual cover can be defined by a reverse cumulative of a distribution: (i) a normal distribution centered at D with $R = D + 3\sigma$ and $r = D - 3\sigma$ discussed in Sect. 3.7, (ii) a beta distribution, and (iii) a logit distribution (Drezner et al., 2020b) discussed in Sect. 3.6. We opted in the computational experiments to define it as declining linearly between R_1 and R_2, which is a reverse cumulative of the uniform distribution, as proposed in Drezner et al. (2004) and discussed in Sect. 3.3. It is as easy to implement it by any gradual cover function as long as an explicit formula for the gradual decline is available.

Suppose that k facilities are located in the area. Each point in the plane may be partially covered by several facilities. Let the proportions of cover of a point (not necessarily the demand point) by facility $1 \leq j \leq k$ be $0 \leq p_j \leq 1$. This proportion p_j can be calculated by a linear decline between R_1 and R_2, or any other rule. Interpreting these proportions as uncorrelated probabilities leads to a total cover of the point, P:

$$P = 1 - \prod_{j=1}^{k}(1 - p_j) \tag{4}$$

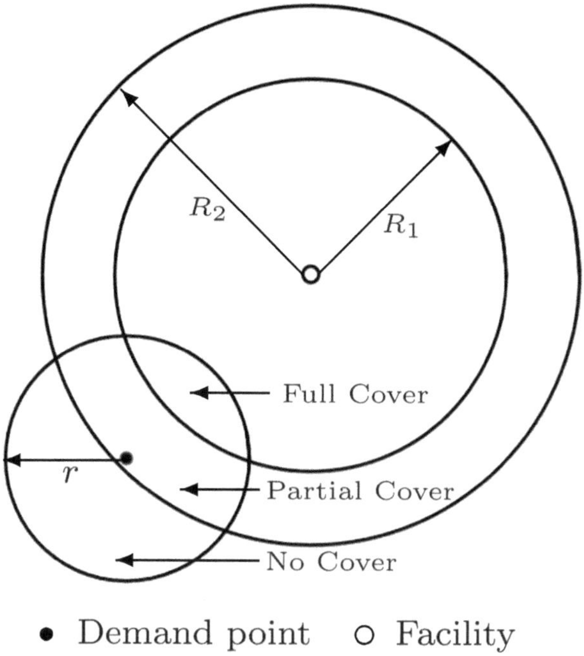

Fig. 3 Stochastic directional gradual cover

Note that if $p_j = 0$, facility j does not affect the total cover, and if $p_j = 1$ for some j, the total cover is full at 100% regardless of the other proportions.

4.1 Calculating the Total Cover

The total cover of a demand point is calculated by a two-dimensional integral in the disk centered at the demand point. The partial cover at each point (the integrand) in the disk is calculated by Eq. (4).

In the original directional cover model (Drezner et al., 2019a), the cover area of a demand point is the union of intersection areas between the circles centered at the facilities and the circle centered at the demand point. If at least one facility provides full cover, the total cover is full. Facilities that do not provide any cover can be removed from consideration. If, for example, there are five facilities that provide partial cover, it seems intractable to develop an explicit formula for the union of the five areas. It is possible to calculate the proportion of the circumference of a circle c with a radius $0 \le \rho \le r$, centered at the demand point that is covered. Each circle centered at a facility covers part of the circumference between two angles θ_1 and θ_2, which are the intersection points between the circle centered at the facility and circle c. The proportion of cover is the union of these parts of the circumference.

The total cover area of a demand point of radius r can be found by integration. Consider a circle of radius ρ for $0 \le \rho \le r$ centered at the demand point. Let $\gamma(\rho)$ be the proportion of the circumference of the circle of radius ρ that is covered. The total cover area A is

$$A = \int_0^r 2\pi\rho\gamma(\rho)d\rho \ , \qquad (5)$$

and the joint cover of a demand point of radius r is

$$Cover = \frac{A}{\pi r^2} = \frac{1}{\pi r^2}\int_0^r 2\pi\rho\gamma(\rho)d\rho \ . \qquad (6)$$

Note that if the circumference of every circle of radius ρ is covered, $\gamma(\rho) = 1$, then $Cover = 1$ by Eq. (6).

Drezner et al. (2019a) applied this calculation and evaluated the total covered area by Gaussian numerical quadrature based on Legendre polynomials. (Abramowitz & Stegun, 1972). For complete details, see Drezner et al. (2019a).

In the stochastic directional gradual cover, it is not simple to calculate the proportion of a circumference of a circle that is covered. The "circle" centered at the facility is actually a ring with various proportions of covers in the ring. There is no clear way to evaluate the intersection between the ring and the circumference of the circle. It is calculated by an integral for every facility and the union is not straightforward to calculate.

In the stochastic directional cover, as depicted in Fig. 3, it is not sufficient to calculate the union of the covered areas because some of the areas are partially covered and Eq. (4) need to be applied to each individual point. We therefore propose to evaluate the total cover numerically by the hexagonal pattern in the circle centered at the demand point as detailed in Drezner et al. (2021, 2019b). The points in the hexagonal pattern are defined by two sequences (all the combinations of the two lists for x and y):
$x = 0, \pm 1, \pm 2, \ldots;\quad y = 0, \pm\sqrt{3}, \pm 2\sqrt{3}, \ldots$, and
$x = \pm\frac{1}{2}, \pm\frac{3}{2}, \pm\frac{5}{2}, \ldots;\quad y = \pm\frac{\sqrt{3}}{2}, \pm\frac{3\sqrt{3}}{2} \pm \frac{5\sqrt{3}}{2}, \ldots$.

These points cover the plane with hexagons centered at each point with sides as perpendicular bisectors to six adjacent points. The area of each hexagon is $\frac{\sqrt{3}}{2}$. All the points satisfying $x^2 + y^2 \le M$ for some M are selected. For example, $M = 220$ results in $N = 805$ points. To get N hexagons that cover a disk of radius r, we multiply the coordinates by a factor K so that $\frac{\sqrt{3}}{2}NK^2 = \pi r^2$. Leading to a factor of $K = r\sqrt{\frac{2\pi}{N\sqrt{3}}}$ to get an hexagonal pattern in a circle of radius r centered at the origin (0, 0). Note that few hexagons have parts outside the circle and a small part of the area of the disk is not covered, but the total area of the hexagons is equal to the circle's area. The perimeter of the covered area is a little ragged.

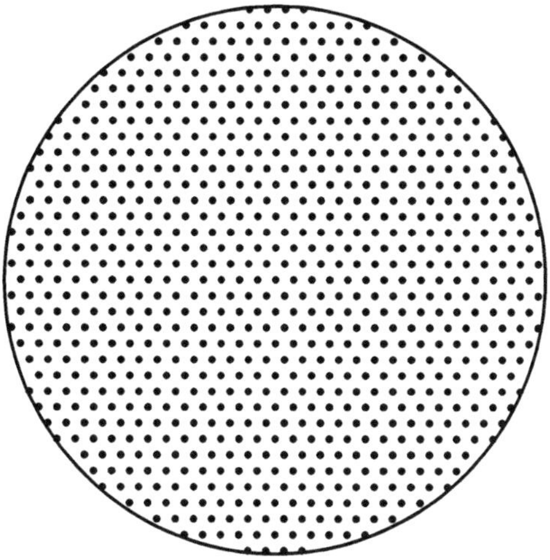

Fig. 4 Hexagonal pattern of 805 points

Drezner et al. (2019b) investigated applying different areas for hexagons that are close to the circle's perimeter and are either trimmed by the perimeter or have extra area bordered by the perimeter, and the results hardly changed. Therefore, such a refinement is not suggested.

There are many values of the number of hexagonal points that can be selected. We propose to select 805 points in the hexagonal pattern (see Fig. 4) that lead to a good estimate of the integral (Drezner et al., 2021). The partial cover for each point is calculated by Eq. (4), the sum S of the partial covers for all 805 points is calculated, and the partial cover of the demand point is $\frac{S}{805}$.

For example, consider a disk of radius 1 centered at the origin (demand point). The disk is partially covered by a facility located at (2,0). For a radius $D \leq 1$, there is zero cover. As D increases, partial cover increases up to $D = 3$. For $D \geq 3$, there is full cover. The area can be calculated exactly by Eq. (1). In Table 1, the exact area is compared with the hexagonal numerical integration for $N = 805$ points for various values of $1 \leq D \leq 3$. The average difference is 0.005. In one case, the difference exceeds 0.01 and in all other cases, it is below 0.01.

4.2 Investigating the Stochastic Directional Gradual Cover

Any solution method that was applied for (heuristically) solving directional gradual cover models can be applied for the stochastic model. Rather than calculating the value of the objective function numerically by the directional objective, it is calculated by the stochastic objective. The "black box" providing the total partial cover by the directional model is replaced by a black box providing total partial cover by the stochastic objective. For example, Drezner et al. (2019a) applied the ascent algorithm, Tabu search (Glover & Laguna, 1997), and simulated annealing

Table 1 Comparing exact proportion of the area to the hexagonal result

D	(1)	(2)	(3)	D	(1)	(2)	(3)
1.0	0.000	0.000	0.000	2.1	0.509	0.512	0.003
1.1	0.015	0.014	0.001	2.2	0.572	0.573	0.001
1.2	0.043	0.039	0.004	2.3	0.636	0.645	0.009
1.3	0.079	0.075	0.004	2.4	0.699	0.708	0.009
1.4	0.120	0.114	0.006	2.5	0.762	0.770	0.008
1.5	0.166	0.155	0.011	2.6	0.822	0.829	0.007
1.6	0.217	0.210	0.007	2.7	0.879	0.888	0.009
1.7	0.270	0.266	0.004	2.8	0.931	0.937	0.006
1.8	0.327	0.323	0.004	2.9	0.974	0.976	0.002
1.9	0.386	0.379	0.007	3.0	1.000	1.000	0.000
2.0	0.447	0.446	0.001				

(1) The exact proportion of the area
(2) The hexagonal pattern proportion
(3) Absolute value of the difference

(Kirkpatrick et al., 1983). Drezner et al. (2020a) applied the same heuristics but generated good starting solutions. Drezner et al. (2021) constructed a genetic algorithm (Holland, 1975; Goldberg, 2006) and solved the continuous case by SNOPT (Gill et al., 2005) and Nelder-Mead (Nelder & Mead, 1965; Dennis & Woods, 1987).

The justification for using gradual decline in cover rather than abrupt drop in cover is that it provides better estimates for the actual cover observed in real applications. The question is not whether one model provides greater coverage than another but which one estimates the cover more accurately. We believe that in reality, cover is stochastic in nature and does not drop abruptly. Therefore, stochastic gradual cover estimates the total cover more accurately because it imitates reality better. There are examples that total cover by one approach is greater than the total cover by another approach for facilities located at the same location. However, this does not mean that one model provides more "actual" cover than the other.

Consider locating one facility to cover four demand points, each with a weight of 1, located on the vertices of a square of side length of 1. By the non-gradual cover objective, if $D < 0.5$, a maximum of one demand point is covered for a total cover of 1. The four circles of radius D centered at the demand points do not intersect. For $0.5 \le D < \frac{\sqrt{2}}{2}$, the total cover is 2 because the only four intersections are of two circles of radius D. For $D \ge \frac{\sqrt{2}}{2}$, all four circles intersect and cover the center of the square, and locating a facility there covers all 4 demand points for a total cover of 4.

For the gradual cover model with the linear decline, it is fair to compare the non-gradual results with D being at the center of the segment connecting R_1 and R_2. For $D = 0.499$, we investigate the range of $R_1 = 0.499 - \sigma$ to $R_2 = 0.499 + \sigma$. Since $R_1 \geq 0$ by definition, then $\sigma \leq 0.499$. If we locate the facility at the center of the square, all four distances are equal to $\frac{\sqrt{2}}{2}$. The partial cover of each demand point is $\frac{0.499+\sigma-\frac{\sqrt{2}}{2}}{2\sigma}$ and the total cover is $4\frac{0.499+\sigma-\frac{\sqrt{2}}{2}}{2\sigma} = 2 + \frac{0.998-\sqrt{2}}{\sigma}$. This total cover is greater than 1 for $\sigma > \sqrt{2} - 0.998 \approx 0.416$, which is better than the non-gradual optimal solution. However, for $\sigma < 0.416$, the non-gradual solution is better.

A different question is whether the gradual cover objective for a given location of the facility is higher or lower than the non-gradual cover. It is easy to construct examples both ways. Consider the example of four demand points on the vertices of a square of side 1 and a facility located at the center of the square. For $D = 0.499$ with $R_1 = 0.499 - \sigma$ and $R_2 = 0.499 + \sigma$ discussed above, for $\sigma > 0.416$, the gradual cover is higher than the non-gradual cover (which is 0). However, for $D > \frac{\sqrt{2}}{2}$, non-gradual cover is 4 while partial cover is less than 4 when $R_1 < \frac{\sqrt{2}}{2}$.

We compared the combined cover by the stochastic directional gradual cover model calculated for given locations of facilities to the directional cover and non-gradual cover model. For the comparison, we generated problems with $n = 100$ demand points and up to 100 facilities by a pseudo-random number generator.

In order to allow for future comparisons, the problems were generated by the pseudo-random number generator described in Drezner et al. (2019c). It is based on the pseudo-random number generator proposed in Law and Kelton (1991). A sequence r_k of integer numbers in the open range (0, 100,000) is generated. A starting seed r_1, which is the first number in the sequence, and a multiplier λ, which is an odd number not divisible by 5, are selected. We used $\lambda = 12{,}219$. The sequence is generated by the following rule for $k \geq 1$:

$$r_{k+1} = \lambda r_k - \left\lfloor \frac{\lambda r_k}{100{,}000} \right\rfloor \times 100{,}000.$$

The random number between 0 and 10 is $\frac{r_k}{10{,}000}$.

For demand points (with coordinates between 0 and 10), the x coordinates were generated by $r_1 = 97$, and for the y-coordinates, we used $r_1 = 367$. For the weights, we used $r_1 = 12{,}347$ and $w_i = 1 + \frac{r_i}{100{,}000}$ so $1 < w_i < 2$. Facilities were generated by $r_1 = 23{,}431$ for the x-coordinates and $r_1 = 56{,}407$ for the y-coordinates.

The points are depicted in Fig. 5, and the first 10 points are listed in Table 2. For the non-gradual cover model, a facility covers a demand point within a distance of 3. For directional cover models, each demand point is defined by a circle of radius $r = 1$. For the directional cover (Drezner et al., 2019a), the facility covers points within a distance of 3. For the stochastic directional model, the facility covers a point in a range between 2 and 4. At a distance of 2, the cover is full and at a distance of 4, there is no cover. Cover declines linearly between 2 and 4.

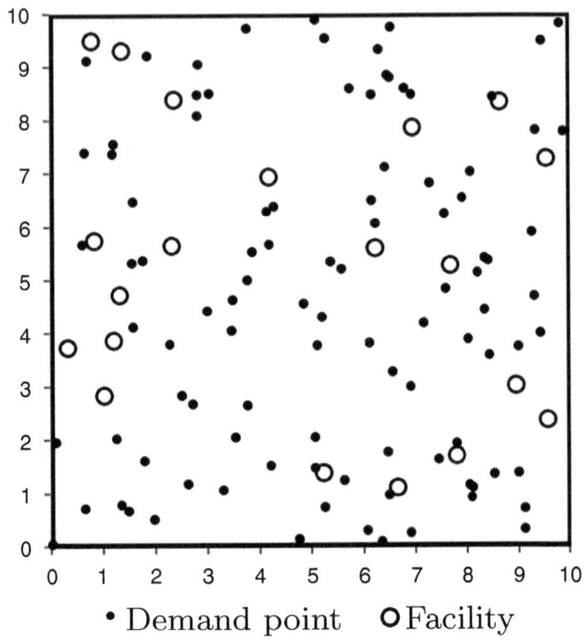

Fig. 5 The 100 demand points and 20 facilities

Table 2 The first 10 pseudo-randomly generated points

	Demand points			Facilities	
i	x	y	w	x	y
1	0.0097	0.0367	1.12347	2.3431	5.6407
2	8.5243	8.4373	1.67993	0.3389	3.7133
3	8.4217	5.3687	1.06467	1.0191	2.8127
4	4.7523	0.1453	1.20273	2.3829	8.3813
5	8.3537	5.4207	1.15787	6.6551	1.1047
6	3.8603	5.5333	1.01353	8.6669	8.3293
7	9.0057	1.3927	1.32307	0.8511	5.7167
8	0.6483	7.4013	1.59233	9.5909	2.3573
9	1.5777	6.4847	1.68027	1.2071	3.8487
10	7.9163	6.5493	1.21913	9.5549	7.2653

The comparison of the proportion cover of all 100 demand points by p facilities for the traditional cover model that is termed non-gradual, the directional gradual cover (Drezner et al., 2019a), the stochastic directional gradual cover model proposed in this chapter, are reported in Table 3. The best covers for each p are marked in boldface. Note that these are not the optimal solutions but the values of the objective function at facilities locations that were randomly generated and depicted in Table 2. The procedures were coded in FORTRAN using double precision

Table 3 Comparing non-gradual, directional, and stochastic directional covers

p	(1)	(2)	(3)
1	**0.21401**	0.21033	0.21345
2	0.24379	0.26227	**0.26877**
3	**0.33405**	0.32161	0.32185
4	0.38197	0.39176	**0.39670**
5	0.62032	**0.64340**	0.63884
6	0.82693	**0.83375**	0.82034
7	0.82693	**0.83375**	0.82120
8	0.88800	**0.90097**	0.88162
9	0.88800	**0.90102**	0.88621
10	0.89564	**0.91078**	0.90567
11	0.89564	**0.91078**	0.90746
12	**0.95949**	0.95661	0.94921
13	**0.95949**	0.95661	0.95054
14	**0.96974**	0.96901	0.96306
15	**0.98192**	0.98098	0.97823
16	**1.00000**	0.99028	0.98590
17	**1.00000**	0.99028	0.98633
18	**1.00000**	0.99031	0.98675
19	**1.00000**	0.99031	0.98726
20	**1.00000**	0.99031	0.98732
Average	0.79430	0.79676	0.79184

(1) Non-gradual proportion cover
(2) Directional proportion cover
(3) Stochastic proportion cover

arithmetic and were compiled by an Intel 11.1 FORTRAN compiler using one thread with no parallel processing. The programs were run on a desktop with the Intel i7-6700 3.4GHz CPU processor and 16GB RAM. We do not report the run times because they are mostly less than a millisecond.

Since the demand points are basically located randomly and uniformly in the square, the proportion of cover does not vary by much. The majority of the best proportion of cover was found by the non-gradual cover approach, especially for $p \geq 12$. The average was the highest (not by much) for the directional model.

We found the proportions for $p = 1, 2, \ldots, 100$. The non-gradual model provided full cover for $p \geq 16$, the directional cover model yielded full cover for $p \geq 54$, and the stochastic directional cover model for $p \geq 75$.

References

Abramowitz, M., & Stegun, I. (1972). *Handbook of mathematical functions*. New York, NY: Dover Publications Inc.

Bagherinejad, J., Bashiri, M., & Nikzad, H. (2018). General form of a cooperative gradual maximal covering location problem. *Journal of Industrial Engineering International, 14*, 241–253.

Berman, O., Drezner, Z., & Krass, D. (2010). Cooperative cover location problems: the planar case. *IIE Transactions, 42*, 232–246.

Berman, O., Drezner, Z., & Krass, D. (2019). The multiple gradual cover location problem. *Jornal of the Operational Research Society, 70*, 931–940.

Berman, O., Drezner, Z., & Wesolowsky, G. O. (2003a). The expropriation location problem. *Journal of the Operational Research Society, 54*, 769–776.

Berman, O., & Krass, D. (2002). The generalized maximal covering location problem. *Computers & Operations Research, 29*, 563–591.

Berman, O., Krass, D., & Drezner, Z. (2003b). The gradual covering decay location problem on a network. *European Journal of Operational Research, 151*, 474–480.

Bhootra, A., Drezner, Z., Schwarz, C., & Stohs, M. H. (2015). Mutual fund performance: luck or skill? *International Journal of Business, 20*, 52–63.

Budge, S., Ingolfsson, A., & Zerom, D. (2010). Empirical analysis of ambulance travel times: the case of Calgary emergency medical services. *Management Science, 56*, 716–723.

Christaller, W. (1966). *Central places in Southern Germany*. Englewood Cliffs, NJ: Prentice-Hall.

Church, R. L., & Murray, A. (2018). Location covering models: history, applications, and advancements. *Advances in Spatial Science*. https://doi.org/10.1080/13658816.2019.1634271

Church, R. L., & ReVelle, C. S. (1974). The maximal covering location problem. *Papers of the Regional Science Association, 32*, 101–118.

Church, R. L., & Roberts, K. L. (1984). Generalized coverage models and public facility location. *Papers of the Regional Science Association, 53*, 117–135.

Dennis, J., & Woods, D. J. (1987). Optimization on microcomputers: the Nelder-Mead simplex algorithm. In: A. Wouk (Ed.), *New computing environments: microcomputers in large-scale computing* (pp. 116–122). Philadelphia: SIAM Publications.

Drezner, T., & Drezner, Z. (1997). Replacing discrete demand with continuous demand in a competitive facility location problem. *Naval Research Logistics, 44*, 81–95.

Drezner, T., & Drezner, Z. (2008). Lost demand in a competitive environment. *Journal of the Operational Research Society, 59*, 362–371.

Drezner, T., & Drezner, Z. (2014). The maximin gradual cover location problem. *OR Spectrum, 36*, 903–921.

Drezner, T., Drezner, Z., & Goldstein, Z. (2010). A stochastic gradual cover location problem. *Naval Research Logistics, 57*, 367–372.

Drezner, T., Drezner, Z., & Kalczynski, P. (2011). A cover-based competitive location model. *Journal of the Operational Research Society, 62*, 100–113.

Drezner, T., Drezner, Z., & Kalczynski, P. (2012). Strategic competitive location: Improving existing and establishing new facilities. *Journal of the Operational Research Society, 63*, 1720–1730.

Drezner, T., Drezner, Z., & Kalczynski, P. (2019a). A directional approach to gradual cover. *TOP, 27*, 70–93.

Drezner, T., Drezner, Z., & Kalczynski, P. (2020a). Directional approach to gradual cover: a maximin objective. *Computational Management Science, 17*, 121–139.

Drezner, T., Drezner, Z., & Kalczynski, P. (2020b). A gradual cover competitive facility location model. *OR Spectrum, 42*, 333–354.

Drezner, T., Drezner, Z., & Kalczynski, P. (2021). Directional approach to gradual cover: the continuous case. *Computational Management Science, 18*, 25–47.

Drezner, T., Drezner, Z., & Suzuki, A. (2019b). A cover based competitive facility location model with continuous demand. *Naval Research Logistics, 66*, 565–581.

Drezner, Z. (2006). On the limit of the generalized binomial distribution. *Communications in Statistics: Theory and Methods, 35*, 209–221.

Drezner, Z. (2019). My career and contributions. In: H.A. Eiselt & V. Marianov (Eds.), *Contributions to location analysis - in honor of Zvi Drezner's 75th birthday* (pp. 1–67). Switzerland: Springer Nature.

Drezner, Z., & Farnum, N. (1993). A generalized binomial distribution. *Communications in Statistics-Theory and Methods, 22*, 3051–3063.

Drezner, Z., Kalczynski, P., & Salhi, S. (2019c). The multiple obnoxious facilities location problem on the plane: A Voronoi based heuristic. *OMEGA: The International Journal of Management Science, 87*, 105–116.

Drezner, Z., & Suzuki, A. (2004). The big triangle small triangle method for the solution of non-convex facility location problems. *Operations Research, 52*, 128–135.

Drezner, Z., & Wesolowsky, G. O. (1997). On the best location of signal detectors. *IIE Transactions, 29*, 1007–1015.

Drezner, Z., Wesolowsky, G. O., & Drezner, T. (1998). On the logit approach to competitive facility location. *Journal of Regional Science, 38*, 313–327.

Drezner, Z., Wesolowsky, G. O., & Drezner, T. (2004). The gradual covering problem. *Naval Research Logistics, 51*, 841–855.

Eiselt, H. A., & Marianov, V. (2009). Gradual location set covering with service quality. *Socio-Economic Planning Sciences, 43*, 121–130.

Fetter, F. A. (1924). The economic law of market areas. *The Quarterly Journal of Economics, 38*, 520–529.

Francis, R. L., Lowe, T. J., Rayco, M. B., & Tamir, A. (2009). Aggregation error for location models: survey and analysis. *Annals of Operations Research, 167*, 171–208.

Francis, R. L., Lowe, T. J., & Tamir, A. (2000). Aggregation error bounds for a class of location models. *Operations Reserach, 48*, 294–307.

García, S., & Marín, A. (2015). Covering location problems. In: G. Laporte, S. Nickel & F. S. da Gama (Eds.), *Location science* (pp. 93–114). Heidelberg: Springer.

Gill, P. E., Murray, W., & Saunders, M. A. (2005). SNOPT: an SQP algorithm for large-scale constrained optimization. *SIAM Review, 47*, 99–131.

Glover, F., & Laguna, M. (1997). *Tabu search*. Boston: Kluwer Academic Publishers.

Goldberg, D. E. (2006). *Genetic algorithms*. Delhi, India: Pearson Education.

Hansen, P., Peeters, D., & Thisse, J.-F. (1981). On the location of an obnoxious facility. *Sistemi Urbani, 3*, 299–317.

Holland, J. H. (1975). *Adaptation in natural and artificial systems*. Ann Arbor, MI: University of Michigan Press.

Karasakal, O., & Karasakal, E. (2004). A maximal covering location model in the presence of partial coverage. *Computers & Operations Research, 31*, 15–26.

Karatas, M. (2017). A multi-objective facility location problem in the presence of variable gradual coverage performance and cooperative cover. *European Journal of Operational Research, 262*, 1040–1051.

Kirkpatrick, S., Gelat, C. D., & Vecchi, M. P. (1983). Optimization by simulated annealing. *Science, 220*, 671–680.

Launhardt, W. (1885). *Mathematische Begründung der Volkswirthschaftslehre*. W. Engelmann.

Law, A. M., & Kelton, W. D. (1991). *Simulation modeling and analysis* (2nd ed.). New York: McGraw-Hill.

Lösch, A. (1954). *The economics of location*. New Haven, CT: Yale University Press.

Morohosi, H., & Furuta, T. (2017). Two approaches to cooperative covering location problem and their application to ambulance deployment. In: *Operations Research Proceedings 2015* (pp. 361–366). Cham, Switzerland: Springer.

Nelder, J. A., & Mead, R. (1965). A simplex method for function minimization. *The Computer Journal, 7*, 308–313.

Plastria, F. (2002). Continuous covering location problems. In: Z. Drezner & H. W. Hamacher (Eds.), *Facility location: applications and theory* (pp. 39–83). Springer.

ReVelle, C. (1986). The maximum capture or sphere of influence problem: Hotelling revisited on a network. *Journal of Regional Science, 26*, 343–357.
ReVelle, C., Toregas, C., & Falkson, L. (1976). Applications of the location set covering problem. *Geographical Analysis, 8*, 65–76.
Schöbel, A., & Scholz, D. (2010). The big cube small cube solution method for multidimensional facility location problems. *Computers & Operations Research, 37*, 115–122.
Snyder, L. V. (2011). Covering problems. In: H. A. Eiselt & V. Marianov (Eds.), *Foundations of location analysis* (pp. 109–135). New York: Springer.
Wang, S.-C., & Chen, T.-C. (2017). Multi-objective competitive location problem with distance-based attractiveness and its best non-dominated solution. *Applied Mathematical Modelling, 47*, 785–795.
Wendell, R. E., & Hurter, A. P. (1973). Location theory, dominance and convexity. *Operations Research, 21*, 314–320.

Equity in Stochastic Healthcare Facility Location

Karmel S. Shehadeh and Lawrence V. Snyder

Abstract We consider issues of equity in stochastic facility location models for healthcare applications. We explore how uncertainty exacerbates inequity and examine several equity measures that can be used for stochastic healthcare location modeling. We analyze the limited literature on this subject and highlight areas of opportunity for developing tractable, reliable, and data-driven approaches that might be applicable within and outside healthcare operations. Our primary focus is on exploring various ways to model uncertainty, equity, and facility location, including modeling aspects (e.g., tractability and accuracy) and outcomes (e.g., equity/fairness/access performance metrics vs. traditional metrics like cost and service levels).

Keywords Equity · Healthcare · Facility location · Uncertainty · Inequity-averse optimization · Stochastic optimization

1 Introduction

Equity and *uncertainty* concerns arise naturally in many real-life applications (e.g., healthcare scheduling, facility location, disaster response operations, air traffic control, etc.). Thus, incorporating equity and uncertainty in optimization contexts is necessary in order to make accurate, equitable, and robust decisions. Unfortunately, however, accounting for equity is a complex task, primarily because there is no unique notion of equity that is universally accepted; "equity" is generally understood to refer to the fair allocation of resources, but a precise definition often depends on the context. Moreover, uncertainty, an intrinsic property of real-life applications, interacts with equity in complicated and poorly understood ways. In particular, uncertainty complicates the quantification of efficiency loss (e.g., minimizing the

K. S. Shehadeh (✉) · L. V. Snyder
Department of Industrial and Systems Engineering, Lehigh University, Bethlehem, PA, USA
e-mail: kas720@lehigh.edu; kshehadeh@lehigh.edu; larry.snyder@lehigh.edu

H. A. Eiselt, V. Marianov (eds.), *Uncertainty in Facility Location Problems*, International Series in Operations Research & Management Science 347,
https://doi.org/10.1007/978-3-031-32338-6_12

fixed cost associated with establishing and locating outpatient clinics) incurred in the pursuit of equity (e.g., equitably allocating outpatient clinics across geographical areas). For example, we may need to open a larger number of specialty clinics (and thus incur higher fixed and operational costs) to ensure equitable allocation of outpatient clinics across geographical areas. However, the number of specialty clinics needed and the associated costs are not easy to quantify under random demands for specialty care, which depend on other random factors such time, location, and outspread of chronic and infectious diseases. In addition, deterministic models of most real-world problems are often challenging. Thus, incorporating both uncertainty and equity metrics or constraints may increase the complexity of these problems.

Historically, equity has been mainly considered in the public sector, with considerably less attention in the private sector. In this chapter, we focus on the issue of equity in stochastic facility location models for healthcare applications. The main motivation behind the attempt to establish equity in healthcare in general—and the need for inequity-averse models for healthcare facility (HCF) location in particular—is, of course, an ethical one: *humans have equal rights, and therefore, nobody should be discriminated against by inequitable access to healthcare services or distribution of healthcare services.* Another more pragmatic motivation for striving for equitable HCF location–allocation solutions is *the need to avoid adverse health outcomes of vulnerable populations who often do not have proper or equal access to HCF* (Gutjahr & Fischer, 2018). Finally, extending classical models with an emphasis on equity and equity–uncertainty interaction is practically relevant and technically interesting.

We first analyze different aspects and measures of equity in the literature. Then, we analyze recent static and mobile HCF location models to explore various ways to model uncertainty (demand, service time, travel time, etc.), equity, and facility location, including modeling aspects (e.g., tractability and accuracy) and outcomes (e.g., equity-, fairness-, or access-based performance metrics vs. traditional metrics like cost and service levels). Our goal is not to provide a comprehensive survey; instead, our goal is to highlight the issue of equity and access and the need for data-driven and tractable models to address emerging stochastic HCF problems.

The remainder of this chapter is organized as follows: In Sect. 2, we analyze existing definitions and metrics of equity as well as common methods to model these metrics. In Sect. 3, we briefly discuss the challenges of incorporating uncertainty and equity and demonstrate through a simple example that uncertainty and equity interact in ways that should not be ignored. In Sect. 4, we provide a high-level analysis of recent stochastic HCF location literature with a particular focus on studies that proposed and analyzed inequity-averse approaches. Maybe not surprisingly, and sadly, this analysis reveals that there is limited literature considering equity and equity–uncertainty interaction. Nevertheless, there are many opportunities to use the powerful tools of operations research to address equity concerns in emerging HCF location problems and derive inequity-averse stochastic HCF location approaches. In Sect. 5, we present some future research opportunities and open questions.

2 What Is Equity, Anyway?

To account for *equity* in facility location or any other decisions, one first needs to provide an exact meaning of equity. Despite the importance of the subject, there is no unique notion (definition) of equity that is generally accepted. Instead, there is a wide variety of notions of equity and fairness in the economics and decision theory literature that depend on the context. The primary concern for equity in resource allocation is treating entities fairly, such that everyone receives the same level of service and no one is at a disadvantage. The allocated resource(s) can be a particular good, or bad, or a chance of good or bad. The entities can be a population, group of people at some location or belonging to some social classes or organizations, etc. In general, most equity literature aims at equal distribution of benefits or disutilities between entities (Mostajabdaveh et al., 2019). Although there is no single equity concept that we can use to design inequity-averse HCF location models, there are four key areas of equity research for health systems (Cardoso et al., 2015, 2016). These are:

- *Equity of access*: Informally, accessibility is the relative ease by which patients can reach a healthcare facility from a given location (Hawthorne & Kwan, 2013; Jin et al., 2015; Wang, 2012). Thus, patients should receive the care they need as close as possible to their place of residence or employment. In the case of emergency services, accessibility is the ability of a healthcare provider to reach the patients. Accessibility measures include both spatial and nonspatial factors (Wang, 2012). Spatial factors include the spatial separation between supply (e.g., surgical centers) and demand (e.g., patient population needing surgical care) and how they are connected in space. Thus, it is a classical aspect in location analysis. Nonspatial factors include demographic (e.g., age, gender, sex, etc.) and socioeconomic (e.g., income, poverty, female-headed households, etc.) variables, which also vary across geographical areas.
- *Equity of utilization*: Utilization refers to the satisfied demand for different services. Ensuring equity of utilization means providing roughly equal service levels across services. Note that this diverges from the concept of "*deliver the cheapest service*" often observed in location models that seek cost minimization.
- *Socioeconomic equity*: Socioeconomic equity stipulates that the unsatisfied demand for population groups with lower income should not be greater than that of groups with higher income. Decision-makers may also want to ensure that unmet demand for lower-income or vulnerable population groups is sufficiently low.
- *Geographical equity*: Geographical equity refers to the ability of the system to provide relatively equal levels of unsatisfied (satisfied) demand across geographical areas. Decision-makers often want to ensure that unsatisfied demand is not vastly different across geographical areas, or that some geographical areas do not lack healthcare service entirely.

As with other performance measures in any optimization problem, one can account for equity in these areas by putting the related metrics in the objective function and/or in the constraints.

Equity research in other application domains additionally considers *social equity* and *diversity*. The *social equity* concept quantifies equity based on how any good received is proportional to the need (Levinson, 2010). For example, the volume of the demand for a particular health service may differ among demand nodes in rural and urban areas. If only a fraction of the demand can be satisfied, measures such as the proportion of the satisfied demand can be used to measure equity and service quality (Karsu & Morton, 2015). *Diversity* is another concept that is indirectly related to equity. Suppose, for example, that we want to select a set of locations to open vaccination centers. The decision-maker may have concerns about diversity because they want certain population groups to have a certain degree of coverage or access to vaccination by the chosen location. One way to achieve this is to use *quotas*, that is, ensuring that a certain proportion of the vaccination centers will be located to cover the groups of concern (Karsu & Morton, 2015). This approach treats people with different characteristics differently, such that the selected locations are diversified enough in the sense that they cover diverse groups of concerns.

In contrast to most of the HCF literature (see Sect. 4 and Ahmadi-Javid et al. (2017)), inequity measurement has found explicit and extensive consideration in the economic and decision theory literature and a few discussions in the humanitarian logistics literature. The commonly accepted theme is that *there is no one-size-fits-all solution to ensure equity, and customized methods are needed to measure and handle application-specific equity concerns*. Using transparent and explicit criteria that determine what is equitable and what is not is useful in ensuring that the decisions are acceptable, equitable, and implementable in practice.

There are also different operations research (OR) methods and metrics for incorporating equity in the decision process. The precise interpretation of each depends on both the structure of the problem at hand and the decision-maker's understanding of equity (Karsu & Morton, 2015). Karsu and Morton (2015) give a comprehensive and deep survey on the use of equity concepts connected with optimization models. In the following subsections, we provide a high-level overview and analysis of the most common equity measures detailed in Karsu and Morton (2015).

2.1 The Rawlsian Approach

The Rawlsian approach (Rawls, 1999) is one of the oldest, most common, and simplistic approaches used in OR to incorporate equity in optimization models. This approach represents equity preference by focusing on the worst-off entity, that is, the minimum outcome level in a distribution. One can enforce a constraint that ensures that the minimum outcome is larger than a predefined level or seek to maximize the minimum outcome. For example, in the *p*-center problem, a Rawlsian approach would seek to locate p facilities to minimize the maximum distance between any demand point and its nearest facility. One common criticism of the Rawlsian approach is that it focuses exclusively on the worst case and ignores the performance for all other entities. Some studies extend the Rawlsian approach to a maximum lexicographic approach. That is, the welfare of the worst-off is first maximized subject to resource and other constraints, then the second worst-off, then the third worst-off, and so on (Kostreva et al., 2004). As pointed out by Karsu and Morton (2015), the lexicographic maximin approach is a regularization of the Rawlsian maximin approach and is inequity averse.

2.2 Approaches Based on Inequity Indices

Various studies that involve equity incorporate an inequity index into the model, which often assigns a scalar value to any given distribution showing the degree of inequity. Inequity indices are often used to assess the disparity in distribution, and so they are related to several mathematical concepts of dispersion and variance. Inequity indices respect the anonymity property (Chakravarty, 1999) and often equal 0 when perfect equity occurs. *Anonymity property* indicates that an inequity measure does not depend on the labeling of individuals. As pointed out by Panzera and Postiglione (2020), the anonymity property implies that an inequity measure is permutation invariant, which means that very different spatial patterns can give rise to the same measure. Suppose we have a set of $i \in I$ individuals (or groups). Let x_i denote the outcome value at node i. Below we use x to briefly summarize the most commonly used inequity indices in the literature (adapted from Karsu and Morton (2015)):

- *The deviation from the mean* ($\sum_{i \in I}(x_i - \bar{x})$, where $\bar{x}$ is the mean value). This index measures the total deviation from the mean. In some applications, the mean of the outcome distribution is often unknown, especially at the time when the decisions leading to the outcome x are being made. Thus, the mean $\bar{x}$ is often approximated based on expert knowledge or derived endogenously in the model. Some studies use the total absolute deviation from the mean (i.e., $\sum_{i \in I} |x_i - \bar{x}|$). Note that the mean absolute deviation (MAD) disregards how these deviations are distributed. Thus, MAD does not provide an incentive to minimize the gap among high values of the outcome (i.e., above or equal to the average) and among low values of the outcome (i.e., below or equal to the average). Other studies use the mean squared deviation, the maximum component-wise deviation from mean, or only the positive or negative deviation from the mean as a measure of inequity. Mathematically, if the mean is (assumed) known, then using MAD in the objective function or a constraint may yield a linear optimization problem.
- *The range or difference between the minimum and maximum levels of outcomes* ($\max_i x_i - \min_i x_i$). Some studies also minimize this range normalized by the minimum outcome or enforce a constraint that ensures that $\frac{\min_i x_i}{\max_i x_i} \geq \beta$, where

β is an equity or fairness parameter (Chang et al., 2006). This index is used in many applications owing to its being simple and easy to understand. However, as pointed out by Karsu and Morton (2015), by considering the two extremes (e.g., most and least deprived), this index is rather crude as it fails to distinguish allocations with the same values of extremes but different levels of other values. Thus, the range is sensitive to extreme values and ignores the interior of the distribution. Note that the normalized range may lead to a nonlinear and complex formulation.

- *The variance or standard deviation.* A small variance means a low dispersion of the outcome. Both variance and standard deviation typically result in nonlinear optimization problems.
- *The Gini coefficient* ($\sum_{i\in I}\sum_{j\in I}|x_i - x_j|/2|I|\sum_{i\in I} x_i$). The Gini coefficient (and indices derived based on it) is one of the most widely used measures of income inequity in an economy that satisfies the Pigou-Dalton principle of transfers (PD), which states that any transfer from a poorer person to a richer person, other things remaining the same, should always lead to a less equitable allocation. Note that the Gini index is a dimensionless quantity, and thus, it cannot often be incorporated in a natural way with other terms in a multicriteria problem (Gutjahr & Fischer, 2018). Moreover, this measure has the disadvantage of being highly nonlinear, possibly making the resulting optimization problem extremely complex. The Gini index will always assume a value between 0 (indicating total equity) and 1 (indicating total inequity).
- *Sum of pairwise (absolute) differences* ($\sum_{i\in I}\sum_{j\in I}|x_i - x_j|$) and sum of squared deviations between all pairs. In contrast to MAD, the sum of pairwise absolute differences (SAD) does consider the spread of the outcome. Like MAD, the SAD term can be linearized.
- *The deviation from a predefined target*: If the predefined target is the best possible output, for example, then satisfying it indicates perfect equity. Thus, the larger the deviation, the larger the inequity. Minimizing this deviation is often referred to as minimizing regret. Related measures include minimizing maximum regret, minimizing absolute regret, etc.

Remark 1 Selecting one of the above (or other) indices implies a particular assumption on the decision-maker's or optimizer's attitude to equity.

2.3 Approaches Based on Inequity-Averse Aggregation Functions

Karsu and Morton (2015) propose approaches based on inequity-averse aggregation functions, which use the aggregation function of the distribution vector in the model that would encourage equitable distributions. Unlike an inequity index, which only focuses on the inequity in a distribution, an inequity-averse aggregation function reflects concerns for both equity and efficiency. There are several ways to capture equity in this approach. For example, some studies use aggregation functions that have convenient mathematical properties such as convexity.

Marín et al. (2010) use *ordered median functions*—weighted total cost functions in which the weights are rank-dependent—as objective functions of flexible discrete location problems. We refer to Karsu and Morton (2015) for more details on inequity-averse aggregation functions.

3 Equity Versus Uncertainty

Uncertainty is intrinsic to many HCF location problems since various key input parameters such as demand, costs, and travel times are often unpredictable. While inequity-averse optimization in a deterministic context is conceptually relatively simple, though often computationally nontrivial and demanding, location decisions under uncertainty represented by suitable stochastic models introduce additional challenges for the following primary reasons. First, because of uncertainty, the value estimates are often not perfectly accurate, whereby the ex post realized values of the impact (e.g., access to care, percentage of satisfied demand across geographical locations) and alternative decisions rarely coincide with their ex ante estimated values.

Second, in most real-world applications such as HCF location, it is unlikely that we can accurately infer the actual distributions of random parameters and thus quantify the impacts of decisions on equity, especially with limited data or no information during the planning stage. Even when historical data is available, the quality of such data may not be sufficient to estimate the distribution of uncertain factors accurately, and future uncertainty is often not distributed as the past. Various studies show that different distributions can typically explain raw data of uncertain parameters, indicating distributional ambiguity (i.e., uncertainty in distribution type (Esfahani & Kuhn, 2018; Vilkkumaa & Liesiö, 2021)).

Third, incorporating equity measures and addressing both uncertainty and distributional ambiguity may increase the overall complexity of HCF location problems. However, ignoring uncertainty and equity–uncertainty interaction may lead to devastating costs and health outcomes. Adverse outcomes associated with poor HCF location decisions include increased costs, disparities in service, and increased illness or death. For example, a hard-to-access healthcare facility is likely to be associated with increased morbidity (disease) and mortality (death).

In this section, we demonstrate through a simple example that uncertainty and equity interact in ways that should not be ignored—the decision one should make in the presence of both considerations is often different from the decision under either consideration in isolation. As detailed in Ahmadi-Javid et al. (2017) and Daskin and Dean (2005), and in Sect. 4, most of the existing HCF location models are extensions of basic discrete location problems (in which facilities can be established only at candidate locations), including p-median, p-center, covering-based, and fixed charge models. For brevity and illustrative purposes, herein, we next analyze extensions of

Table 1 General notation

Parameters and sets	
p	Number of facilities
I	Set of locations
$d_{i,j}$	Distance/travel time between any pair of nodes $i \in I$ and $j \in I$
w_i	Demand at node i
First-stage decision variables	
x_j	$\begin{cases} 1, & \text{if a facility is open at candidate location } j, \\ 0, & \text{otherwise.} \end{cases}$
$y_{i,j}$	$\begin{cases} 1, & \text{if demand point } i \text{ is assigned to a facility at candidate location } j, \\ 0, & \text{otherwise.} \end{cases}$

the p-median and p-center using some of the linear equity metrics discussed in the previous section. Specifically, we define different equity objectives based on the sum or maximum of pairwise (absolute) differences in distances or demand-weighted distances.

Table 1 summarizes the general notation we use in all formulations. The decision variables listed in the table are first-stage decision variables; we will introduce the second-stage variables shortly, when we discuss stochastic models. We assume that all demand nodes are also potential facility locations, but this assumption is straightforward to relax if necessary. Using the notation in Table 1, we define the following common feasible set among all formulations:

$$\mathcal{X} = \left\{ x, y : \begin{array}{ll} \text{(C1)} & \sum_{j \in I} x_j = p \\ \text{(C2)} & \sum_{j \in I} y_{i,j} = 1, \forall i \in I \\ \text{(C3)} & y_{i,j} \leq x_j, i \in I, j \in I \\ \text{(C4)} & y_{i,j} \in \{0, 1\}, \; x_j \in \{0, 1\}, \forall i \in I, j \in I \end{array} \right\}$$

Constraint (C1) specifies the total number of facilities to be established. Constraints (C2) ensure that each demand point is assigned to exactly one facility, and constraints (C3) limit assignments to open facilities. Constraints (C4) are integrality constraints. Table 2 presents several deterministic formulations for locating p facilities, including both classical facility location problems such as the p-median and p-center, as well as equity-based formulations. We do not claim that the equity-based approaches listed here are the best models for considering equity. Rather, we chose these measures because they are used frequently and because they are useful for illustrating the interaction between uncertainty and equity.

In the next section, we compare the optimal solutions obtained using the deterministic formulations in Table 2 and their stochastic programming (SP) counterparts. In the SP, travel time and demand are modeled as random variables that follow fully known probability distributions. The objective is to minimize the expectation of the objective function, where the expectation is taken with respect to an assumed known distribution. We approximate solutions to the SP models using

Table 2 Optimization models

Model name	Formulation
Median	$\min\left\{\sum_{i\in I}\sum_{j\in I} w_i d_{i,j} y_{i,j} : (x,y)\in\mathcal{X}\right\}$
Center	$\min\left\{\max_{i\in I}\sum_{j\in I} d_{i,j} y_{i,j} : (x,y)\in\mathcal{X}\right\}$
Total unweighted distance	$\min\left\{\sum_{i\in I}\sum_{j\in I} d_{i,j} y_{i,j} : (x,y)\in\mathcal{X}\right\}$
Total unweighted deviation	$\min\left\{\sum_{i\in I}\sum_{j\neq i\in I} \lvert z_i - z_j\rvert : (x,y)\in\mathcal{X}, z_i = \sum_{j\in I} d_{i,j} y_{i,j},\ \forall i\in I\right\}$
Max unweighted deviation	$\min\left\{\max_{i,j} \lvert z_i - z_j\rvert : (x,y)\in\mathcal{X},\ z_i = \sum_{j\in I} d_{i,j} y_{i,j},\ \forall i\in I\right\}$
Total weighted deviation	$\min\left\{\sum_{i\in I}\sum_{j\neq i\in I} \lvert z_i - z_j\rvert : (x,y)\in\mathcal{X},\ z_i = \sum_{j\in I} w_i d_{i,j} y_{i,j},\ i\in I\right\}$
Max unweighted deviation	$\min\left\{\max_{i,j} \lvert z_i - z_j\rvert : (x,y)\in\mathcal{X},\ z_i = \sum_{j\in I} w_i d_{i,j} y_{i,j},\ i\in I\right\}$
Max sum of unweighted deviations	$\min\left\{\max_{i\in I}\sum_{j\in I} \lvert z_i - z_j\rvert : (x,y)\in\mathcal{X},\ z_i = \sum_{j\in I} d_{i,j} y_{i,j},\ \forall i\in I\right\}$
Max sum of weighted deviations	$\min\left\{\max_{i\in I}\sum_{j\in I} \lvert z_i - z_j\rvert : (x,y)\in\mathcal{X},\ z_i = \sum_{j\in I} w_i d_{i,j} y_{i,j},\ \forall i\in I\right\}$
Sum of maximum unweighted deviations	$\min\left\{\sum_{i\in I}\max_{j\in I} \lvert z_i - z_j\rvert : (x,y)\in\mathcal{X},\ z_i = \sum_{j\in I} d_{i,j} y_{i,j},\ \forall i\in I\right\}$
Sum of maximum weighted deviations	$\min\left\{\sum_{i\in I}\max_{j\in I} \lvert z_i - z_j\rvert : (x,y)\in\mathcal{X},\ z_i = \sum_{j\in I} w_i d_{i,j} y_{i,j},\ \forall i\in I\right\}$

their sample average approximation (SAA). That is, we generate a sample of N scenarios (each scenario consists of a vector of realizations of demand and travel time which are drawn independently from the distributions corresponding to each node and pair of nodes, respectively) and then optimize the sample average of the objective. (The technical details of SAA are out of the scope of this chapter, and we refer the reader to Kim et al. (2015), Kleywegt et al. (2002), Mak et al. (1999), Shapiro et al. (2021) for a thorough discussion.) For example, the SAA of the p-center model is

$$\min\left\{\sum_{n=1}^{N}\frac{1}{N}z^n : (x,y)\in\mathcal{X},\quad z^n \geq \sum_{j\in I} d^n_{i,j} y_{i,j},\quad \forall i\in I,\ n\in N\right\} \tag{1a}$$

Here, the $d^n_{i,j}$ values are the travel times (distances) under the nth sample, and z^n is the optimal objective function value of the (deterministic) p-center problem under the nth sample.

3.1 Example: Where to Locate a New Hospital in Lehigh County?

In this subsection, we consider locating a single hospital (i.e., $p = 1$) in a service region based on Lehigh County, that is located in the Lehigh Valley region of the US state of Pennsylvania. We consider a dataset consisting of 20 selected nodes: the 20 largest communities in Lehigh County according to the 2010 census, plus the city of Easton[1] (see Fig. 1.) We calculated the distance and travel time between each pair of nodes using the Google API. We use these travel times as the $d_{i,j}$ values for the deterministic models. For the stochastic models we set both the average travel time ($\mu^d_{i,j}$) and the standard deviation of travel times ($\sigma^d_{i,j}$) between each pair of nodes (i, j) equal to the calculated travel time, for all (i, j). We use the population estimate in each county based on the 2010 US census (see Table 3) to construct the following demand structure. We use the population percentage (weight) at each node to generate the mean (average) demand at each node $i \in I$ as $\mu^w_i = \texttt{population\%} \times 1000$ (i.e., total demand of 1000). To a certain extent, this structure reflects what may be observed in real life, that is, locations with more population typically create greater demand. We set the standard deviation as $\sigma^w_i = 0.5\mu^w_i$, for all $i \in I$.

We generate the following two sets of N data samples for the parameters w and d.

- Set 1: $w_i \sim$ lognormal (LogN) with mean μ^w_i and standard deviation σ^w_i, and $d_{i,j} \sim U[\mu^d_{i,j} - \Delta, \mu^d_{i,j} + \Delta]$, where $\Delta = 10$ minutes.
- Set 2: $w_i \sim$ LogN with mean μ^w_i and standard deviation σ^w_i, and $d_{i,j} \sim$ LogN with mean $\mu^d_{i,j}$ and standard deviation $\sigma^d_{i,j}$.

We solve the SAA counterparts of the deterministic models in Table 2 with the generated data samples. In addition, we solve the deterministic formulations with one scenario ($N = 1$). Tables 4 and 5 present the optimal solutions under Sets 1 and 2. We make the following observation from these tables. First, the optimal location can change when we consider uncertainty *or* equity, which is expected and not new. Second, incorporating both uncertainty and equity result in a solution that is often different from incorporating either one. For example, the deterministic and SAA solutions of the p-median problem respectively locate the hospital at Catasauqua and Fountain Hill under Set 1. In contrast, the deterministic and SAA solutions of the total weighted deviation problem, which minimizes the sum of the absolute deviations in demand-weighted travel time, respectively locate the hospital at Allentown and Dorneyville. Third, different measures of inequity aversion under uncertainty can result in different optimal solutions. Fourth, the SAA solutions may be different under different distributions, motivating the need for distribution-free

[1] Easton is not in Lehigh County; we included it because it is the third largest city in the Lehigh Valley.

Fig. 1 Map of nodes in Lehigh Valley dataset

Table 3 Lehigh County nodes and their population based on the 2010 census of Lehigh County

City/town/etc.	Pop	Pop%	Avg. demand
Allentown	118,032	40.9%	409
Bethlehem	74,982	26.0%	260
Emmaus	11,211	3.9%	39
Ancient Oaks	6661	2.3%	23
Catasauqua	6436	2.2%	22
Wescosville	5872	2.0%	20
Fountain Hill	4597	1.6%	16
Dorneyville	4406	1.5%	15
Slatington	4232	1.5%	15
Breinigsville	4138	1.4%	14
Coplay	3192	1.1%	11
Macungie	3074	1.1%	11
Schnecksville	2935	1.0%	10
Coopersburg	2386	0.8%	8
Alburtis	2361	0.8%	8
Cetronia	2115	0.7%	7
Trexlertown	1988	0.7%	7
Laurys Station	1243	0.4%	4
New Tripoli	898	0.3%	3
Slatedale	751	0.3%	3
Easton	26,800	9.3%	93
Total	288,310		

Table 4 Optimal locations yielded by each model under Set 1. Notation: DET is the deterministic model solved with 1 scenario, and SAA is the SAA counterpart of each solved with $N = 50$

Model	DET	SAA
Median	Catasauqua	Fountain Hill
Center	Dorneyville	Catasauqua
Total unweighted distance	Dorneyville	Cetronia
Total unweighted deviation	Coplay	Cetronia
Max unweighted deviation	Allentown	Catasauqua
Total weighted deviation	Allentown	Dorneyville
Max unweighted deviation	Catasauqua	Dorneyville
Max sum of unweighted deviations	Allentown	Catasauqua
Max sum of weighted deviations	Catasauqua	Dorneyville
Sum of maximum unweighted deviations	Catasauqua	Catasauqua
Sum of maximum weighted deviations	Catasauqua	Fountain Hill

Table 5 Optimal locations yielded by each model under Set 2

Model	DET	SAA
Median	Cetronia	Allentown
Center	Catasauqua	Cetronia
Total unweighted distance	Wescosville	Dorneyville
Total unweighted deviation	Catasauqua	Cetronia
Max unweighted deviation	Catasauqua	Cetronia
Total weighted deviation	Cetronia	Allentown
Max unweighted deviation	Cetronia	Allentown
Max sum of unweighted deviations	Catasauqua	Cetronia
Max sum of weighted deviations	Cetronia	Allentown
Sum of maximum unweighted deviations	Catasauqua	Catasauqua
Sum of maximum weighted deviations	Cetronia	Allentown

models. For example, the SAA solution for max unweighted deviation locates the hospital at Catasauqua and Cetronia under Set 1 and Set 2, respectively.

4 Is the Stochastic HCFL Literature Inequity Averse?

In this section, we provide a high-level analysis of recent stochastic approaches for HCF location, focusing on studies proposing inequity-averse approaches published between 2004 and 2017. By inequity averse, we mean any approach that considers one or more of the considerations mentioned above or other equity-related objectives or constraints. Our goal is to bring attention to a fundamental and timely question: *Is the stochastic HCF location literature inequity averse?*. We next analyze the limited literature considering equity and uncertainty, highlighting existing equity-related objectives or constraints and the challenges of incorporating these.

For a comprehensive survey on the HCF location–allocation literature, we refer to (Rahman & Smith, 2000; Daskin & Dean, 2005; Li et al., 2011; Güneş et al., 2019; Cissé et al., 2017; Gutiérrez & Vidal, 2013; Grieco et al., 2020; Ahmadi-Javid et al., 2017). The recent survey by Ahmadi-Javid et al. (2017) provides a thorough classification of HCF location problems, models, and solution methods in the last decade, identifying gaps and possible future directions. They first provide a framework to classify different types of non-emergency and emergency HCFs. Then, they analyze the literature on HCF location problems along ten descriptive dimensions (e.g., uncertainty, single or multi-period settings, etc.). Next, we dive deeper into this literature, highlighting those considering equity and/or uncertainty. Note that mathematical differences between optimization models for each type (and sub-type) of HCFs are due to, for example, the nature of the service provided (and thus different objectives, constraints, and random factors), nature of the operation (mobile vs. static, e.g., mobile primary care clinic vs. a primary clinic in a hospital), decision-maker perspective, case study, etc.

4.1 Non-emergency HCF Location

Non-emergency health services include medical treatment, observation, prevention, testing, and other healthcare services provided to patients whose conditions are not urgent or considered an emergency. Next, we review some of the literature on major non-emergency HCFs.

4.1.1 Primary Care Facilities

Primary care facilities (PCF) is a class of HCF that provide primary care to the public, including early diagnosis and first-contact care. Most PCFs (e.g., hospitals, clinics) are open for 24 hours, and patients often tend to visit the nearest one. The optimal location of these facilities has received significant attention from the operations research community. Basic location models used include set covering, maximal covering, p-median, and fixed-charge. Within this literature, most papers proposed deterministic models. Mitropoulos et al. (2006) proposed a bi-objective mathematical programming model for locating hospitals and primary healthcare centers. The two objectives in this model are (1) minimization of the distance between patients and facilities and (2) equitable distribution of the facilities among citizens. Güneş et al. (2014) also considered minimizing the maximum travel distance in designing a primary care facility network. Other (deterministic) studies minimize deviations from a standard distance (Smith et al., 2013).

Oliveira and Bevan (2006) proposed two location–allocation models to redistribute hospital supply using different objective functions and assumptions about the utilization behavior of patients. The first model optimizes equity by minimizing variations between predicted and normative utilization (according to need) by small area. The second model optimizes equity by minimizing utilization flows between small areas and hospitals and a utilization flows target (defined flows of patients using closest and central hospitals).

Rahmaniani et al. (2014) proposed a multi-objective two-stage stochastic nonlinear integer programming model for the location–allocation of hospitals. Uncertainty considered includes the fixed cost of opening a facility, travel time (distance) between nodes, the capacity of facilities, and demand. They used the expected demand weighted travel time as a measure of random accessibility. Due to the challenges of solving the model exactly, Rahmaniani et al. (2014) proposed a heuristic solution algorithm based on variable neighborhood search (VNS).

Mestre et al. (2015) proposed two location–allocation models for handling demand uncertainty in the strategic design of a hospital network. In addition to operational objectives, both models aim to maximize access by minimizing the expected travel time to reach hospital services weighted by demand. McCoy et al. (2014) proposed equitable allocation strategies for motorcycle trips facilitating access to healthcare in rural areas. Ares et al. (2016) used a coverage score to model equity in terms of access to healthcare among the different populations. Beheshtifar and Alimoahmmadi (2015) proposed a new definition for equity by minimizing the variability of access distance to a healthcare clinic, where variability was measured in terms of the standard deviation of distances from the place of demand points to the related open site.

4.1.2 Blood Banks

A blood bank is an HCF that collects blood samples from donors and then stores and prepares them for transfusion to recipients. One of the main challenges to blood bank location is that human blood is scarce, perishable, and often in high demand. Furthermore, both demand and supply of blood are stochastic and subject to various disruptions.

Jabbarzadeh et al. (2014) presents a robust location–allocation model for dynamic supply chain network design for the supply of blood in disasters. Uncertainty considered includes the capacity of a temporary blood facility, capacity of a permanent blood facility, and maximum blood supply of each donor group. The objective considers both cost and resilience to disruptions. Fahimnia et al. (2017) also proposed a stochastic bi-objective supply chain design model for efficient (cost-minimizing) and effective (delivery time-minimizing) blood supply in disasters. Uncertainty considered includes demands and various costs. Although both studies considered uncertainty, they did not incorporate any equity objective or constraints.

4.1.3 Organ Transplant Centers

Organ transplant centers (OTC) are the main components of organ transplantation programs in most healthcare systems. The demand for organs is a major random factor that is often larger than the supply, which is also random. As a result, organ transplants suffer from long waiting lists. The time between the request for an organ and transplantation, transportation time of organs from donors' locations (e.g., hospital) to OTC, and transportation time of recipients to OTC are vital in the process of organ donation/transplantation and subject to a high degree of uncertainty. Zahiri et al. (2014a) present a robust probabilistic programming approach to multi-period location–allocation of OTCs. They demonstrate that solutions derived using their model are robust when taking into account uncertain conditions in the form of small yearly demand changes. Zahiri et al. (2014b) propose a multi-period location–allocation bi-objective mathematical programming model for designing an organ transplant transportation network under uncertainty. The model minimizes total cost and time, including waiting time in the queue for the transplant operation while considering organs' priorities. Uncertainty considered includes inter-arrival times of organs entering the transplant centers, arrivals of patients to the transplant centers, among others. These studies did not consider any equity objective or constraints.

4.1.4 Detection and Prevention Centers

Detection and prevention centers are HCFs that provide healthcare services defined based on local or national detection and prevention programs (Ahmadi-Javid et al., 2017). The common aim of these HCFs is to reduce the likelihood and severity of potentially life-threatening illnesses by protection and early detection. Therefore, the level of participation in preventive healthcare programs is crucial in their effectiveness and efficiency, so most studies considered participation maximization objectives. Other objectives include minimizing travel distance or time to increase accessibility and thus participation.

Zhang et al. (2009) use the total travel time, waiting time, and service time required for receiving the preventive service as a proxy for accessibility of a healthcare facility and assume that each patient would seek the facility's services with a minimum expected total of these metrics. To capture the congestion level, they modeled each facility as an M/M/1 queue. They show that the expected number of participants from each population zone decreases with the expected total time. They included a constraint that ensures equity among the people living in a population zone in terms of access to preventive health services. The model in Zhang et al. (2009) is highly nonlinear, so they propose a heuristic.

Zhang et al. (2010) proposed a bilevel model for preventive healthcare facility network design with congestion. They formulate the lower-level problem, which determines the allocation of clients to facilities, as a variational inequality, while the upper level is a facility location and capacity allocation problem. Major random factors considered include the demand rate at each zone, travel times, and service times. They incorporate congestion at the facilities in the model and assume that clients patronize the facility with the minimum expected total time. Thus, they considered the total time needed to receive preventive services at a facility as a proxy for its accessibility and did not include equity objectives or constraints.

Vidyarthi and Kuzgunkaya (2015) analyzed the impact of system-optimal (i.e., directed) choice on the design of the preventive healthcare facility network under congestion. The problem is set up as a network of spatially distributed M/G/1 queues and formulated as a nonlinear mixed-integer programming model. The model simultaneously determines the location and size of the facilities and the allocation of clients to these facilities to minimize the weighted sum of the total travel time and the congestion associated with waiting and service delay at the facilities. They linearize the model and present a cutting plane-based approach to solve the reformulation. The model in Vidyarthi and Kuzgunkaya (2015) does not capture the seasonality of the demand, nor does it include any equity objectives or constraints. Aboolian et al. (2016) focus on designing facility networks in the public sector to maximize the number of people benefiting from their services. They propose an analytical framework for the maximal accessibility network design problem that involves determining the optimal number, locations, and capacities of a network of public sector facilities. They assume that the time spent for receiving the service from a facility is a good proxy for its accessibility. Aboolian et al. (2016) did not consider equity objectives or constraints.

4.1.5 Medical Laboratories

A medical laboratory is an HCF where tests are carried out on clinical specimens from the patient to aid in diagnosis, treatment, and disease prevention. Although medical laboratories are critical for public health, there is a lack of research on the location of these HCFs. Notably, Saveh-Shemshaki et al. (2012) propose a p-median-based model for designing a network of tuberculosis testing laboratories to reduce transportation times and thereby decrease overall test turnaround time. They use the travel time from any region to any laboratory as a measure of equity. Accordingly, they included a constraint that allows decision-makers to specify an upper bound for origin–destination transportation time. Their results suggest that the optimal locations and capacities are not sensitive to this additional equity constraint.

4.1.6 Long-Term Care Centers

Long-term (nursing) care is an HCF that provides rehabilitative, restorative, and ongoing nursing care to patients or residents who need assistance with their health or daily living activities. The location of long-term care facilities is crucial to provide the best and most equitable possible services to aged people who represent the major demand group needing social and medical services. Given that this type of HCF provides medical care and social services to inpatients, the simultaneous determination of location, optimal capacity levels (e.g., number of beds), inventory levels, and locations are essential. Cardoso et al. (2015) proposed a fixed charge facility location-based model for planning a long-term care network, which considers demand uncertainty, multiple services, and various forms of equity (access, utilization, socioeconomic, and geographical equities) constraints. Cardoso et al. (2016) consider equity of access, geographical equity, and socioeconomic equity in long-term care (LTC) and network design decisions. They use minimization of total travel time for individuals accessing institutional LTC services to ensure equity of access, minimization of unmet need in the geographical area with the highest level of unmet need to ensure geographical equity, and minimization of unmet need for the lower-income groups to ensure socioeconomic equity in their model.

4.1.7 Other Non-emergency HCFs

Home healthcare (HHC) services provide healthcare services to people in their homes. HHC emerged in the early 50s to reduce the cost of care and health systems and improve patients' quality of life. The home service industry, especially home healthcare, has been rapidly growing worldwide due to emergent changes in family structures, work obligations, aging populations, and the outspread of chronic and infectious diseases. Hence, the operations research community investigated different challenges of the HHC services, such as routing, appointment scheduling, staffing, and various resource allocation issues. However, as pointed out by Ahmadi-Javid et al. (2017), no studies exist for HHC center locations. It follows that there are no studies considering equity in HHC center locations. Similarly, the location of rehabilitation HCF (i.e., HCF devoted to the physical rehabilitation of patients with, e.g., neurological, orthopedic, other medical conditions), doctors' offices, and drugstores are not studied as other types of HCFs. An analysis of equity concerns related to, for example, the distribution of these HCFs and access to them under uncertainty is an important future research area.

4.2 Emergency HCF Location

Life-threatening emergencies, such as a severe injury, stroke, or heart attack, require the services of emergency HCFs. In addition, emergency HCFs provide service for patients with an injury or illness that does not appear to be life-threatening, but the treatment of such patients cannot wait until the next day or for a primary care doctor to see them. Ahmadi-Javid et al. (2017) classify emergency HCFs according to whether they perform under permanent or temporary emergencies. *Permanent* emergency HCFs provide service regularly, including emergency off-site public access devices, trauma centers, and ambulance stations. In contrast, *temporary* emergency HCFs are constructed to respond to unexpected health (e.g., infectious disease outbreak) and other situations (e.g., disaster). Next, we analyze the literature on emergency and trauma centers, ambulance stations, and temporary medical centers.

4.2.1 Emergency and Trauma Centers

Emergency departments or emergency centers are permanent emergency facilities that provide medical and surgical care to both patients arriving in need of immediate care or walk-ins (unscheduled patients). They can be part of a hospital or free-standing. Locating these facilities has not received as much attention as other emergency facilities such as ambulance stations. Therefore, there is a need for inequity-averse stochastic models for the location of emergency centers, particularly those associated with hospitals and clinics or ambulance stations. Silva and Serra

(2008) is one of the early studies that presented a priority queuing-based covering location problem for locating emergency services considering different service priority levels. Silva and Serra (2008) did not focus on equity.

Trauma centers are hospitals that provide specialized medical and nursing care to patients suffering from major traumatic injuries (e.g., falls, motor vehicle collisions and accidents, gunshot wounds, etc.). Patients are typically transported to trauma centers via helicopters or ambulances. To account for air transportation, some studies consider a joint location problem of trauma centers and helicopters under some budget constraints. Most existing studies for locating trauma centers employ maximal coverage location or fixed charge location models. Ahmadi-Javid et al. (2017) propose a maximal backup coverage model (BACOP) for the joint location problem of trauma centers and helicopters with budget constraints. Although trauma centers are rife with uncertainty (especially in demand), none of the papers reported in Ahmadi-Javid et al. (2017) has incorporated uncertainty. In addition, equity and equity–uncertainty interaction has not been considered.

4.2.2 Ambulance Stations

Emergency medical services, more commonly known as EMS, is a system that provides out-of-hospital acute medical care and transfers patients to emergency centers/departments within or outside hospitals and trauma centers for definitive care. The OR community has paid significant attention to the location of ambulance stations, the deployment (location, relocation, fleet sizing) of ambulances in the stations, and the dispatch of ambulances to the demand points or emergency sites.

Most of the existing models for this type of HCF are extensions of the basic maximum or set covering location, maximum expected coverage location (Daskin, 1982, 1983), and p-center location models. Note that by focusing on maximizing the demand that can be covered, traditional covering models favor locating ambulances in more densely populated areas, resulting in longer response times for patients in more rural areas. That is, traditional covering models may lead to solutions in which the coverage pattern is quite good for those nodes counted as covered but extremely poor for those not covered, highlighting the need for equity-based models.

Most existing models for EMS are stochastic due to the stochastic nature of EMS operations. Random factors include, but are not limited to, the busy fraction of ambulances, demand or service requests, and travel time (Beraldi et al., 2004; Beraldi & Bruni, 2009; Gendreau et al., 2006; Ingolfsson et al., 2008; Rajagopalan et al., 2008; McLay, 2009; Rajagopalan & Saydam, 2009; Sorensen & Church, 2010; Noyan, 2010; Rajagopalan et al., 2011; Naoum-Sawaya & Elhedhli, 2013; Zhang & Li, 2015; Yoon et al., 2021). Noyan (2010) considers an EMS system design problem with stochastic demand. They proposed a capacitated fixed charge facility-like location model to locate the emergency response facilities and vehicles to ensure target levels of coverage, which are quantified using risk measures on random unmet demand. The model considers target service levels for each demand site and also for the entire service area. Noyan (2010) argues that considering the individual target service levels may be regarded as an alternative approach to model the coverage equity. To present risk preferences, they develop two types of stochastic optimization models involving alternate risk measures: integrated chance constraints (ICCs) and ICCs with a stochastic dominance constraint.

Chanta et al. (2011) proposed a minimum *p-envy* facility location model, aiming to find optimal locations for EMS facilities to balance customers' perceptions of equity in receiving service. Specifically, to deal with the issue of equity, they assigned an envy function (a function of the distance from a demand zone to its closest EMS station and the distance from a demand zone to its backup EMS stations weighted by priority of the serving stations and weighted by the proportion of demand) to each pair of demand nodes, for each level of priority. This value indicates the dissatisfaction level of a demand node with its serving station in comparison with other demand nodes that have the same level of priority.

To address the issue of fairness in semi-rural/semi-urban communities, Chanta et al. (2014) propose a bi-objective covering location model for locating EMS ambulances at preexisting rescue stations that balances efficiency (i.e., maximizing expected coverage) and equity. Specifically, they propose the following alternative objective functions for improving fairness in rural areas: minimize the maximum distance between uncovered demand zones and their closest opened station, minimize the number of uncovered rural demand zones, and minimize the number of uncovered demand zones. Chanta et al. (2014) use the ϵ-constraint method to solve their multi-objective model. Khodaparasti et al. (2016) studied balancing efficiency and equity in a location–allocation EMS model under uncertainty using data envelopment analysis. Enayati et al. (2019) proposed a multicriteria optimization approach to study the trade-offs in equity and efficiency for simultaneously optimizing location and multipriority dispatch of ambulances.

4.2.3 Temporary Medical Centers

Temporary medical centers (TMCs) provide healthcare services to victims of large-scale and catastrophic disasters. Examples of TMCs include Red Cross medical tents, casualty collection points, and any temporary HCF established before the disaster to play a short-term role in the immediate aftermath. TMC location has some stochastic characteristics similar to those we see in humanitarian logistics that should be considered. Despite the importance of TMC, as pointed out by Ahmadi-Javid et al. (2017), only a few papers addressed TMC location problems, and surprisingly under deterministic conditions. None of these studies incorporated equity objectives or constraints.

5 Future Directions

While great research efforts have been reported to improve stochastic HCF location theory, much work is still needed to incorporate equity and derive inequity-averse stochastic HCF location approaches and insights. We discuss a few critical research opportunities for the future in the following subsections.

5.1 *Analyzing Equity Measures*

The World Health Organization defines health equity as *the absence of unfair and avoidable or remediable differences in health among population groups defined socially, economically, demographically, or geographically* (WHO, 2021). Accordingly, when making HCF location–allocation decisions, one should ensure equal distributions of HCF, equal access to HCF, and equal utilization of healthcare services/HCFs across geographical areas, social groups, demographics, socioeconomic groups, and identified/non-identified groups under normal and abnormal (e.g., disaster, conflict, etc.) conditions. However, to ensure this, one should first find the right measure of equity. But is the obvious objective the right one? Are the obvious and classical constraints the right ones? Is the impact of decisions obtained from classical models with the obvious objectives equitable when compared across various people or groups of people?

To date, we do not have a formal analysis of equity objectives and constraints under uncertainty. Thus, the first step toward developing tractable and realistic stochastic inequity-averse HCF location approaches is to rigorously analyze the mathematical similarities and differences between different mathematical representation of a measure of equity under uncertainty, including their mathematical properties and their suitability for inclusion in optimization models.

5.2 *Capturing Uncertainty: Optimizer's Curse and Trade-Offs*

Acknowledging the inevitable uncertainty and the uncertainty–equity interaction in HCF location settings is crucial to devising inequity-averse models for emerging real-life HCF location problems. One possibility for hedging against uncertainty is to capture it in the parameters underlying optimization models to support the decision-making process. There are three main stochastic optimization (SO) frameworks: stochastic programming (SP), robust optimization (RO), and distributionally robust optimization (DRO). (In DRO, we assume that the probability distribution of the uncertain parameters is unknown but belongs to a certain set, and we wish to optimize a system by hedging against the worst-case distribution within that set.) The main distinguishing feature among these frameworks concerns the knowledge of the probability distribution of the underlying random vector. Hence, adopting one of these modeling frameworks depends mainly on the available information regarding uncertainty and its distribution.

If we know the uncertainty distribution or have sufficient and high-quality data to represent it, we can use SP to optimize the expected performance. However, the SP decision problem evaluates the objective and optimizes the decisions only for the given training sample (which may come from a biased distribution). As a result, decisions obtained from an SP model can be biased, that is, sensitive to the distribution or sample data employed in the SP, and hence perform poorly in the out-of-sample tests (under unseen data). This phenomenon is known as the *optimizers' curse* (Smith & Winkler, 2006). Disappointing consequences in healthcare include, but are not limited to, disparities in healthcare service and distribution of services, poor access to care, increased mortality and morbidity of the vulnerable population, and increased costs. Unfortunately, as shown in Sect. 4, most existing studies employ SP or other "sample-based" approaches.

RO models do not make strong distributional assumptions as in SP. Instead, one only needs to calibrate the so-called "*uncertainty set*" of possible outcomes of random parameters (Ben-Tal et al., 2015; Bertsimas & Sim, 2004; Soyster, 1973). Then, optimization is based on the worst-case scenario occurring within the uncertainty set, which may inevitably lead to *overconservatism* and suboptimal decisions for the other more likely scenarios (Chen et al., 2020; Rahimian & Mehrotra, 2019).

DRO is an alternative approach to model uncertainty that addresses the optimizer's curse and the concerns of overconservatism. In DRO, we optimize a system by hedging against the worst-case distribution within the predefined ambiguity set. DRO is known to offer several benefits, such as mitigating the optimizer's curse, relaxing strong assumptions on distributions, and offering tractable reformulations and approximations.

This suggests that it is worthwhile to consider DRO as an alternative approach for modeling uncertainty in HCF location. To date, there are no inequity-averse DRO approaches for facility location and other application domains.

These gaps and challenges call for more efforts toward rigorous analyses of equity under different scenarios of information availability and ambiguity. Formal analyses are also needed to answer questions about *when* to use each type of modeling approach (SP, DRO, or a trade-off between them) and about *what* is the value of adopting each for inequity-averse HCF location.

5.3 Dynamic Mapping and Databases

For many decades, locations or places have played an important role in understanding health patterns, disease patterns, and healthcare service distributions. Historically, maps have been the primary source for storing and communicating spatial information. In the past two decades, the advancement in computer power and the emergence of Geographic Information Systems (GIS; a system that captures, creates, stores, manages, analyzes, and maps all types of data) has allowed for better mapping and more widespread, complex, and comprehensive analyses than previously. Such advances have made it possible for governments, researchers, and others to seek answers to previously overly complex and computationally impractical questions.

Data accuracy, correctness, and completeness are crucial elements affecting our ability to use GIS to analyze health and equity issues effectively. Unfortunately, despite the significant improvement in technologies to obtain geographic data, we often need to geocode specific data such as patient data (volume, disease, etc.) and HCF utilization by different groups to undertake a particular geographical analysis. As pointed out by Lyseen et al. (2014), in the absence of standardized data collection methods or databases, this process imposes a significant challenge for health information systems in collecting data with adequate granularity, ensuring reliability and validity of the relevant health data, and maintaining appropriate privacy and security for the collected data. Moreover, even when the needed data has been collected, it is often unavailable to researchers due to privacy laws regarding patients' medical information (Lyseen et al., 2014).

Several governments and research groups publish interactive GIS maps of HCF locations with basic information. However, these are individual efforts, with no integrated or standardized databases even within the same country. Therefore, policies and strategies for developing standards-compliant and reliable databases on HCF and the ability to integrate these in GIS dynamically as information unfolds are important research directions and a prerequisite to developing data-driven inequity-averse HCF location models. Such databases may include, for example, real-time data on the location, type, characteristics, and utilization of existing HCFs, characteristics of people living in each geographical area, and demand and access for healthcare and health services across geographical areas and demographic categories.

5.4 *Multicriteria Approaches*

The specific approach or criteria to capture inequity is context-dependent and depends on the decision-maker's perception of equity. One can, of course, consider a multicriteria objective and a method to optimize multiple equity criteria (e.g., equity of access, socioeconomic equity, geographical equity) and efficiency criteria (e.g., transportation cost, operational costs). However, to the best of our knowledge, and according to a recent survey on HCF location (Ahmadi-Javid et al., 2017) and our analysis in Sect. 4, within the limited literature considering equity in stochastic HCF location, most studies only consider one equity-related policy objective, and only a few multi-objective models exist for the joint attainment of different equity objectives under uncertainty.

When we seek both equity (e.g., equitable allocation of HCF across urban and rural areas) and efficiency (e.g., minimum operational cost), there is the additional challenge of mathematically integrating them in a tractable model. As we mentioned earlier, considering equity may degrade efficiency. Thus, questions arise in integrating equity and efficiency; for example, how should one regulate the trade-off between the two under uncertainty? What is the price of equity (i.e., the efficiency difference between selecting an inequity-averse approach and not using an inequity-averse approach)?

In theory, multicriteria optimization problems such as HCF location with equity and efficiency concerns can be modeled and solved using multi-objective optimization (MOO) approaches (see, e,g., Ehrgott (2005)). However, multicriteria location models assume that decision-makers can articulate their objectives or have a well-defined metric for each objective. In most real-world applications such as HCF location, the decision-maker's priorities, goals, and equity concerns are not easy to articulate and could vary over time and among different decision-makers. Moreover, we often need to incorporate the (potentially conflicting) perspective of many stakeholders. For example, locating hospitals is a process that must take into consideration many different stakeholders (Burkey et al., 2012), including patients who need access to the hospital, clinical staff who want an attractive and easy-to-reach workplace, taxpayers who want value for their dollars, politicians who want to demonstrate their ability to deliver a better quality of life and healthcare services, and more.

The ability to quantify the trade-off between efficiency and equity would help decision-makers make informed and better location decisions (Karsu & Morton, 2015). In particular, to make decisions, a decision-maker needs to understand (a) what the efficiency loss might be and (b) what the equity loss might be for a specific solution. Although various studies analyze the price of fairness or price of equity (see, e.g., Bertsimas et al. (2012), Bertsimas et al. (2013)), this price has not been adequately analyzed in HCF location contexts under uncertainty.

Analyzing the robustness of solutions obtained from different inequity-averse HCF location approaches under uncertainty is another avenue for future research. Incorporating multiple equity measures and an analysis of the impact of uncertainty on each and the trade-off among them could shed some light on the question of how similar or different equity measures are. There are some attempts to analyze the commonality and differences of different equity measures in general facility location models (Batta et al., 2014; Karsu & Morton, 2015; Mulligan, 1991; López-De-Los-Mozos & Mesa, 2003). However, such analyses have not been conducted in the HCF location literature.

5.5 *Mobile HCF*

A mobile healthcare facility (MHCF) is a *facility-like vehicle* that can serve patients in a way similar to a static HCF when stationary but can also move from one place to another to provide health services to communities (Attipoe-Dorcoo et al., 2020; Malone et al., 2020; McGowan et al., 2020; Shehadeh, 2022; Santa González et al., 2020). They offer a wide variety of (prevention, testing, diagnostic) health services and are often staffed by a combination of physicians, nurses, community health workers, and other health professionals. As long-standing community-based service delivery models, mobile facilities have the potential to help underserved communities overcome common barriers to accessing healthcare, including availability, time, geography, and trust (Clark et al., 2011; Guruge et al., 2010; Sommers, 2015), and have demonstrated improvements in health outcomes and reductions in cost (Brown-Connolly et al., 2014; Oriol et al., 2009; Song et al., 2013).

MHCF also offer an alternative healthcare delivery option when a disaster, conflict, or other event causes stationary HCFs (e.g., hospitals) to close or stop operations (Blackwell & Bosse, 2007; Du Mortier & Coninx, 2007; Fox-Rushby & Foord, 1996; Gibson et al., 2011). Classical stochastic mobile facility deployment, routing, and scheduling models (see, e.g., Halper and Raghavan (2011), Shehadeh (2022), Lei et al. (2014), Lei et al. (2016)) do not incorporate equity objectives or constraints and thus may produce such inequitable decisions.

Developing inequity-averse MHCF location models is more challenging than the static HCF for the following reasons. First, there is limited literature on mobile facilities as compared to stationary facilities (Ahmadi-Javid et al., 2017). Second, in static HCF problems, we usually consider opening facilities at fixed locations. In contrast, MHCF location problems have the added challenge of determining a *routing plan* and a *time schedule* for each MHCF in the fleet (i.e., the node that each MHCF is located at in each time period). Mathematically, stochastic routing and scheduling are challenging stochastic combinatorial optimization problems. Thus, integrated MHCF deployment (e.g., determining the number of mobile vaccination clinics) and their capacity, routing, and scheduling problems with equity and efficiency criteria under uncertainty represent a new class of complex combinatorial, multicriteria, stochastic optimization problems. To date, optimization (or other) tools have not been employed to analyze and address these problems.

6 Conclusion

In this chapter, we focused on the issue of equity in stochastic HCF locations. We recognize that uncertainty is an intrinsic property of HCF location and explore how uncertainty and equity interact in ways that should not be ignored. The primary goal is to bring attention to equity issues and provide an understanding of the existing recent effort to incorporate equity measures in stochastic HCF locations. Our key findings are:

- There is limited literature considering equity in HCF location and equity–uncertainty interaction.

- Within the limited literature, most studies use the distance that a patient must travel or the travel time to HCF to measure equity and accessibility, with minimizing the maximum distance to the nearest facility as the most frequent measure. Some other studies account for factors such as spatial accessibility, multi-modal travel, temporal service availability, competition, multiple and hierarchical services, socioeconomic and personal factors, etc. in order to more realistically reproduce users' behaviors (Bruno et al., 2020; Higgs et al., 2019; Jin et al., 2019; Mathon et al., 2018; Mayaud et al., 2019; Lin et al., 2018; Shin & Lee, 2018; Yin et al., 2018).
- Most studies, especially in EMS, use covering-based models to ensure equity. However, covering-based models favor locating HCFs in more densely populated areas, potentially resulting in longer travel or response times for patients in more rural areas. In addition, traditional covering models may lead to solutions in which the coverage pattern is quite good for those nodes counted as covered but extremely poor for those not covered.
- Some studies investigate the trade-off between one or more equity objectives and multiple efficiency objectives. However, no study has analyzed the equity–equity trade-off (i.e., the trade-off between equity metrics) or the equity–efficiency trade-off with multiple equity metrics.
- Existing stochastic optimization (SO) models for HCF location assumes that the distribution of uncertainty is fully known or there is sufficient data to model it. To date, there are no data-driven and distribution-free inequity-averse HCF location models.
- To date, there are no inequity-averse mobile healthcare facility deployment, routing, and scheduling models.
- No studies rigorously analyze the mathematical similarities, differences, and complexity of equity objectives and constraints. In addition, there are no standardized and open-access databases for HCF locations that incorporate all the geographical, socioeconomic, and other relevant data that one needs to derive and test inequity-averse models.

We discuss and provide new directions for future research opportunities by recognizing these challenges and gaps in the literature. Our main recommendations for future research include:

- Conducting formal analyses of equity metrics and their mathematical complexity under uncertainty and distributional ambiguity
- Developing data-driven, distribution-free, and tractable multicriteria inequity-averse static and mobile HCF location approaches
- Developing standardized, granular, and dynamic HCF databases and integrate these with recent GIS technology
- Developing robust and resilient inequity-averse stochastic HCF location approaches

Declarations

Not applicable

Acknowledgments We thank all researchers who have contributed significantly to the facility location, equity, and optimization under uncertainty literature. Dr. Karmel S. Shehadeh dedicates her effort in this paper to every little dreamer in the whole world who has a dream so big and so exciting. Believe in your dreams and do whatever it takes to achieve them—the best is yet to come for you.

References

Aboolian, R., Berman, O., & Verter, V. (2016). Maximal accessibility network design in the public sector. *Transportation Science, 50*(1), 336–347.

Ahmadi-Javid, A., Seyedi, P., & Syam, S. S. (2017). A survey of healthcare facility location. *Computers & Operations Research, 79*, 223–263.

Ares, J. N., De Vries, H., & Huisman, D. (2016). A column generation approach for locating roadside clinics in africa based on effectiveness and equity. *European Journal of Operational Research, 254*(3), 1002–1016.

Attipoe-Dorcoo, S., Delgado, R., Gupta, A., Bennet, J., Oriol, N. E., & Jain, S. H. (2020). Mobile health clinic model in the covid-19 pandemic: lessons learned and opportunities for policy changes and innovation. *International Journal for Equity in Health, 19*(1), 1–5.

Batta, R., Lejeune, M., & Prasad, S. (2014). Public facility location using dispersion, population, and equity criteria. *European Journal of Operational Research, 234*(3), 819–829.

Beheshtifar, S., & Alimoahmmadi, A. (2015). A multiobjective optimization approach for location-allocation of clinics. *International Transactions in Operational Research, 22*(2), 313–328.

Ben-Tal, A., Den Hertog, D., & Vial, J.-P. (2015). Deriving robust counterparts of nonlinear uncertain inequalities. *Mathematical Programming, 149*(1-2), 265–299.

Beraldi, P., & Bruni, M. E. (2009). A probabilistic model applied to emergency service vehicle location. *European Journal of Operational Research, 196*(1), 323–331.

Beraldi, P., Bruni, M. E., & Conforti, D. (2004). Designing robust emergency medical service via stochastic programming. *European Journal of Operational Research, 158*(1), 183–193.

Bertsimas, D., & Sim, M. (2004). The price of robustness. *Operations Research, 52*(1), 35–53.

Bertsimas, D., Farias, V. F., & Trichakis, N. (2012). On the efficiency-fairness trade-off. *Management Science , 58*(12), 2234–2250.

Bertsimas, D., Farias, V. F., & Trichakis, N. (2013). Fairness, efficiency, and flexibility in organ allocation for kidney transplantation. *Operations Research, 61*(1), 73–87.

Blackwell, T., & Bosse, M. (2007). Use of an innovative design mobile hospital in the medical response to hurricane katrina. *Annals of Emergency Medicine, 49*(5), 580–588.

Brown-Connolly, N. E., Concha, J. B., & English, J. (2014). Mobile health is worth it! economic benefit and impact on health of a population-based mobile screening program in new mexico. *Telemedicine and e-Health, 20*(1), 18–23.

Bruno, G., Cavola, M., Diglio, A., & Piccolo, C. (2020). Improving spatial accessibility to regional health systems through facility capacity management. *Socio-Economic Planning Sciences, 71*, 100881.

Burkey, M. L., Bhadury, J., & Eiselt, H. A. (2012). A location-based comparison of health care services in four us states with efficiency and equity. *Socio-Economic Planning Sciences, 46*(2), 157–163.

Cardoso, T., Oliveira, M. D., Barbosa-Póvoa, A., & Nickel, S. (2015). An integrated approach for planning a long-term care network with uncertainty, strategic policy and equity considerations. *European Journal of Operational Research, 247*(1), 321–334.
Cardoso, T., Oliveira, M. D., Barbosa-Póvoa, A., & Nickel, S. (2016). Moving towards an equitable long-term care network: A multi-objective and multi-period planning approach. *Omega, 58*, 69–85.
Chakravarty, S. R. (1999). Measuring inequality: the axiomatic approach. In: *Handbook of income inequality measurement* (pp. 163–186). Springer.
Chang, K.-N., Lee, K.-D., & Kim, D. (2006). Optimal timeslot and channel allocation considering fairness for multicell cdma/tdd systems. *Computers & Operations Research, 33*(11), 3203–3218.
Chanta, S., Mayorga, M. E., Kurz, M. E., & McLay, L. A. (2011). The minimum p-envy location problem: a new model for equitable distribution of emergency resources. *IIE Transactions on Healthcare Systems Engineering, 1*(2), 101–115.
Chanta, S., Mayorga, M. E., & McLay, L. A. (2014). Improving emergency service in rural areas: a bi-objective covering location model for ems systems. *Annals of Operations Research, 221*(1), 133–159.
Chen, Z., Sim, M., & Xiong, P. (2020). Robust stochastic optimization made easy with RSOME. *Management Science, 66*(8), 3329–3339.
Cissé, M., Yalçındağ, S., Kergosien, Y., Şahin, E., Lenté, C., & Matta, A. (2017). Or problems related to home health care: a review of relevant routing and scheduling problems. *Operations Research for Health Care, 13*, 1–22.
Clark, C. R., Soukup, J., Govindarajulu, U., Riden, H. E., Tovar, D. A., & Johnson, P. A. (2011). Lack of access due to costs remains a problem for some in massachusetts despite the state's health reforms. *Health Affairs, 30*(2), 247–255.
Daskin, M. S. (1982). Application of an expected covering model to emergency medical service system design. *Decision Sciences, 13*, 416–439.
Daskin, M. S. (1983). A maximum expected covering location model: Formulation, properties and heuristic solution. *Transportation Science, 17*(1), 48–70.
Daskin, M. S., & Dean, L. K. (2005). Location of health care facilities. *Operations Research and Health Care*, 43–76.
Du Mortier, S., & Coninx, R. (2007). *Mobile health units in emergency operations: a methodological approach.* Humanitarian Practice Network, Overseas Development Inst.
Ehrgott, M. (2005). *Multicriteria optimization* (vol. 491). Springer.
Enayati, S., Mayorga, M. E., Toro-Díaz, H., & Albert, L. A. (2019). Identifying trade-offs in equity and efficiency for simultaneously optimizing location and multipriority dispatch of ambulances. *International Transactions in Operational Research, 26*(2), 415–438.
Esfahani, P. M., & Kuhn, D. (2018). Data-driven distributionally robust optimization using the wasserstein metric: Performance guarantees and tractable reformulations. *Mathematical Programming, 171*(1-2), 115–166.
Fahimnia, B., Jabbarzadeh, A., Ghavamifar, A., & Bell, M. (2017). Supply chain design for efficient and effective blood supply in disasters. *International Journal of Production Economics, 183*, 700–709.
Fox-Rushby, J. A., & Foord, F. (1996). Costs, effects and cost-effectiveness analysis of a mobile maternal health care service in west kiang, the gambia. *Health Policy, 35*(2), 123–143.
Güneş, E. D., Yaman, H., Çekyay, B., & Verter, V. (2014). Matching patient and physician preferences in designing a primary care facility network. *Journal of the Operational Research Society, 65*(4), 483–496.
Güneş, E. D., Melo, T., & Nickel, S. (2019). Location problems in healthcare. In: *Location science* (pp. 657–686). Springer.
Gendreau, M., Laporte, G., & Semet, F. (2006). The maximal expected coverage relocation problem for emergency vehicles. *Journal of the Operational Research Society, 57*(1), 22–28.

Gibson, J., Deng, X., Boe-Gibson, G., Rozelle, S., & Huang, J. (2011). Which households are most distant from health centers in rural china? evidence from a gis network analysis. *GeoJournal, 76*(3), 245–255.

Grieco, L., Utley, M., & Crowe, S. (2020). Operational research applied to decisions in home health care: a systematic literature review. *Journal of the Operational Research Society, 72*, 1–32.

Guruge, S., Hunter, J., Barker, K., McNally, M. J., & Magalhaes, L. (2010). Immigrant women's experiences of receiving care in a mobile health clinic. *Journal of Advanced Nursing, 66*(2), 350–359.

Gutiérrez, E. V., & Vidal, C. J. (2013). Home health care logistics management problems: A critical review of models and methods. *Revista Facultad de Ingeniería Universidad de Antioquia, 68*, 160–175.

Gutjahr, W. J., & Fischer, S. (2018). Equity and deprivation costs in humanitarian logistics. *European Journal of Operational Research, 270*(1), 185–197.

Halper, R., & Raghavan, S. (2011). The mobile facility routing problem. *Transportation Science, 45*(3), 413–434.

Hawthorne, T. L., & Kwan, M.-P. (2013). Exploring the unequal landscapes of healthcare accessibility in lower-income urban neighborhoods through qualitative inquiry. *Geoforum, 50*, 97–106.

Higgs, G., Langford, M., Jarvis, P., Page, N., Richards, J., & Fry, R. (2019). Using geographic information systems to investigate variations in accessibility to 'extended hours' primary healthcare provision. *Health & Social Care in the Community, 27*(4), 1074–1084.

Ingolfsson, A., Budge, S., & Erkut, E. (2008). Optimal ambulance location with random delays and travel times. *Health Care management science, 11*(3), 262–274.

Jabbarzadeh, A., Fahimnia, B., & Seuring, S. (2014). Dynamic supply chain network design for the supply of blood in disasters: a robust model with real world application. *Transportation Research Part E: Logistics and Transportation Review, 70*, 225–244.

Jin, C., Cheng, J., Lu, Y., Huang, Z., & Cao, F. (2015). Spatial inequity in access to healthcare facilities at a county level in a developing country: a case study of deqing county, zhejiang, china. *International Journal for Equity in Health, 14*(1), 1–21.

Jin, M., Liu, L., Tong, D., Gong, Y., & Liu, Y. (2019). Evaluating the spatial accessibility and distribution balance of multi-level medical service facilities. *International Journal of Environmental Research and Public Health, 16*(7), 1150.

Karsu, Ö., & Morton, A. (2015). Inequity averse optimization in operational research. *European Journal of Operational Research, 245*(2), 343–359.

Khodaparasti, S., Maleki, H. R., Bruni, M. E., Jahedi, S., Beraldi, P., & Conforti, D. (2016). Balancing efficiency and equity in location-allocation models with an application to strategic ems design. *Optimization Letters, 10*(5), 1053–1070.

Kim, S., Pasupathy, R., & Henderson, S. G. (2015). A guide to sample average approximation. In: *Handbook of simulation optimization* (pp. 207–243). Springer.

Kleywegt, A. J., Shapiro, A., & Homem-de-Mello, T. (2002). The sample average approximation method for stochastic discrete optimization. *SIAM Journal on Optimization, 12*(2), 479–502.

Kostreva, M. M., Ogryczak, W., & Wierzbicki, A. (2004). Equitable aggregations and multiple criteria analysis. *European Journal of Operational Research, 158*(2), 362–377.

López-De-Los-Mozos, M. C., & Mesa, J. A. (2003). The sum of absolute differences on a network: algorithm and comparison with other equality measures. *INFOR: Information Systems and Operational Research, 41*(2), 195–210.

Lei, C., Lin, W.-H., & Miao, L. (2014). A multicut l-shaped based algorithm to solve a stochastic programming model for the mobile facility routing and scheduling problem. *European Journal of Operational Research, 238*(3), 699–710.

Lei, C., Lin, W.-H., & Miao, L. (2016). A two-stage robust optimization approach for the mobile facility fleet sizing and routing problem under uncertainty. *Computers & Operations Research, 67*, 75–89.

Levinson, D. (2010). Equity effects of road pricing: a review. *Transport Reviews, 30*(1), 33–57.

Li, X., Zhao, Z., Zhu, X., & Wyatt, T. (2011). Covering models and optimization techniques for emergency response facility location and planning: a review. *Mathematical Methods of Operations Research, 74*(3), 281–310.

Lin, Y., Wan, N., Sheets, S., Gong, X., & Davies, A. (2018). A multi-modal relative spatial access assessment approach to measure spatial accessibility to primary care providers. *International Journal of Health Geographics, 17*(1), 1–22.

Lyseen, A.-K., Nøhr, C., Sørensen, E.-M., Gudes, O., Geraghty, E., Shaw, N. T., Bivona-Tellez, C., Group, I. H. G. W., et al. (2014). A review and framework for categorizing current research and development in health related geographical information systems (gis) studies. *Yearbook of Medical Informatics, 23*(01), 110–124.

Mak, W.-K., Morton, D. P., & Wood, R. K. (1999). Monte Carlo bounding techniques for determining solution quality in stochastic programs. *Operations Research Letters, 24*(1), 47–56.

Malone, N. C., Williams, M. M., Fawzi, M. C. S., Bennet, J., Hill, C., Katz, J. N., & Oriol, N. E. (2020). Mobile health clinics in the united states. *International Journal for Equity in Health, 19*(1), 1–9.

Marín, A., Nickel, S., & Velten, S. (2010). An extended covering model for flexible discrete and equity location problems. *Mathematical Methods of Operations Research, 71*(1), 125–163.

Mathon, D., Apparicio, P., & Lachapelle, U. (2018). Cross-border spatial accessibility of health care in the north-east department of Haiti. *International Journal of Health Geographics, 17*(1), 1–15.

Mayaud, J. R., Tran, M., Pereira, R. H., & Nuttall, R. (2019). Future access to essential services in a growing smart city: the case of surrey, British Columbia. *Computers, Environment and Urban Systems, 73*, 1–15.

McCoy, J. H., & Lee, H. L. (2014). Using fairness models to improve equity in health delivery fleet management. *Production and Operations Management, 23*(6), 965–977.

McGowan, C. R., Baxter, L., Deola, C., Gayford, M., Marston, C., Cummings, R., & Checchi, F. (2020). Mobile clinics in humanitarian emergencies: a systematic review. *Conflict and Health, 14*(1), 4.

McLay, L. A. (2009). A maximum expected covering location model with two types of servers. *IIE Transactions, 41*(8), 730–741.

Mestre, A. M., Oliveira, M. D., & Barbosa-Póvoa, A. P. (2015). Location–allocation approaches for hospital network planning under uncertainty. *European Journal of Operational Research, 240*(3), 791–806.

Mitropoulos, P., Mitropoulos, I., Giannikos, I., & Sissouras, A. (2006). A biobjective model for the locational planning of hospitals and health centers. *Health Care Management Science, 9*(2), 171–179.

Mostajabdaveh, M., Gutjahr, W. J., & Sibel Salman, F. (2019). Inequity-averse shelter location for disaster preparedness. *IISE Transactions, 51*(8), 809–829.

Mulligan, G. F. (1991). Equality measures and facility location. *Papers in Regional Science, 70*(4), 345–365.

Naoum-Sawaya, J., & Elhedhli, S. (2013). A stochastic optimization model for real-time ambulance redeployment. *Computers & Operations Research, 40*(8), 1972–1978.

Noyan, N. (2010). Alternate risk measures for emergency medical service system design. *Annals of Operations Research, 181*(1), 559–589.

Oliveira, M. D., & Bevan, G. (2006). Modelling the redistribution of hospital supply to achieve equity taking account of patient's behaviour. *Health Care Management Science, 9*(1), 19–30.

Oriol, N. E., Cote, P. J., Vavasis, A. P., Bennet, J., DeLorenzo, D., Blanc, P., & Kohane, I. (2009). Calculating the return on investment of mobile healthcare. *BMC Medicine, 7*(1), 1–6.

Panzera, D., & Postiglione, P. (2020). Measuring the spatial dimension of regional inequality: an approach based on the gini correlation measure. *Social Indicators Research, 148*(2), 379–394.

Rahimian, H., & Mehrotra, S. (2019). Distributionally robust optimization: a review. Preprint. arXiv:1908.05659.

Rahman, S.-u., & Smith, D. K. (2000). Use of location-allocation models in health service development planning in developing nations. *European Journal of Operational Research, 123*(3), 437–452.
Rahmaniani, R., Rahmaniani, G., & Jabbarzadeh, A. (2014). Variable neighborhood search based evolutionary algorithm and several approximations for balanced location–allocation design problem. *The International Journal of Advanced Manufacturing Technology, 72*(1-4), 145–159.
Rajagopalan, H. K., & Saydam, C. (2009). A minimum expected response model: Formulation, heuristic solution, and application. *Socio-Economic Planning Sciences, 43*(4), 253–262.
Rajagopalan, H. K., Saydam, C., & Xiao, J. (2008). A multiperiod set covering location model for dynamic redeployment of ambulances. *Computers & Operations Research, 35*(3), 814–826.
Rajagopalan, H. K., Saydam, C., Setzler, H., Sharer, E., et al. (2011). Ambulance deployment and shift scheduling: an integrated approach. *Journal of Service Science and Management, 4*(01), 66.
Rawls, J. (1999). *A theory of justice: Revised Edition*. Harvard University Press.
Santa González, R., Cherkesly, M., Crainic, T. G., & Rancourt, M.-È. (2020). Mobile clinics deployment for humanitarian relief: a multi-period location-routing problem, CIRRELT. https://www.cirrelt.ca/documentstravail/cirrelt-2020-39.pdf
Saveh-Shemshaki, F., Shechter, S., Tang, P., & Isaac-Renton, J. (2012). Setting sites for faster results: Optimizing locations and capacities of new tuberculosis testing laboratories. *IIE Transactions on Healthcare Systems Engineering, 2*(4), 248–258.
Shapiro, A., Dentcheva, D., & Ruszczynski, A. (2021). *Lectures on stochastic programming: modeling and theory*. SIAM.
Shehadeh, K. S. (2022). Distributionally robust optimization approaches for a stochastic mobile facility fleet sizing, routing, and scheduling problem. *Transportation Science, 57*(1), 197–229.
Shin, K., & Lee, T. (2018). Improving the measurement of the Korean emergency medical system's spatial accessibility. *Applied Geography, 100*, 30–38.
Silva, F., & Serra, D. (2008). Locating emergency services with different priorities: the priority queuing covering location problem. *Journal of the Operational Research Society, 59*(9), 1229–1238.
Smith, J. E., & Winkler, R. L. (2006). The optimizer's curse: Skepticism and postdecision surprise in decision analysis. *Management Science, 52*(3), 311–322.
Smith, H. K., Harper, P. R., & Potts, C. N. (2013). Bicriteria efficiency/equity hierarchical location models for public service application. *Journal of the Operational Research Society, 64*(4), 500–512.
Sommers, B. D. (2015). Health care reform's unfinished work—remaining barriers to coverage and access. *New England Journal of Medicine, 373*, 2395–2397.
Song, Z., Hill, C., Bennet, J., Vavasis, A., & Oriol, N. E. (2013). Mobile clinic in massachusetts associated with cost savings from lowering blood pressure and emergency department use. *Health Affairs, 32*(1), 36–44.
Sorensen, P., & Church, R. (2010). Integrating expected coverage and local reliability for emergency medical services location problems. *Socio-Economic Planning Sciences, 44*(1), 8–18.
Soyster, A. L. (1973). Convex programming with set-inclusive constraints and applications to inexact linear programming. *Operations Research, 21*(5), 1154–1157.
Vidyarthi, N., & Kuzgunkaya, O. (2015). The impact of directed choice on the design of preventive healthcare facility network under congestion. *Health Care Management Science, 18*(4), 459–474.
Vilkkumaa, E., & Liesiö, J. (2021). What causes post-decision disappointment? Estimating the contributions of systematic and selection biases. *European Journal of Operational Research, 296*(2), 587–600.
Wang, F. (2012). Measurement, optimization, and impact of health care accessibility: a methodological review. *Annals of the Association of American Geographers, 102*(5), 1104–1112.
WHO. (2021). Social Determinants of Health. Available online. https://www.who.int/health-topics/social-determinants-of-health#tab=tab_3. Accessed 30 Aug 2021

Yin, C., He, Q., Liu, Y., Chen, W., & Gao, Y. (2018). Inequality of public health and its role in spatial accessibility to medical facilities in china. *Applied Geography, 92*, 50–62.

Yoon, S., Albert, L. A., & White, V. M. (2021). A stochastic programming approach for locating and dispatching two types of ambulances. *Transportation Science, 55*(2), 275–296.

Zahiri, B., Tavakkoli-Moghaddam, R., & Pishvaee, M. S. (2014a). A robust possibilistic programming approach to multi-period location–allocation of organ transplant centers under uncertainty. *Computers & Industrial Engineering, 74*, 139–148.

Zahiri, B., Tavakkoli-Moghaddam, R., Mohammadi, M., & Jula, P. (2014b). Multi-objective design of an organ transplant network under uncertainty. *Transportation Research Part E: Logistics and Transportation Review, 72*, 101–124.

Zhang, Z.-H., & Li, K. (2015). A novel probabilistic formulation for locating and sizing emergency medical service stations. *Annals of Operations Research, 229*(1), 813–835.

Zhang, Y., Berman, O., & Verter, V. (2009). Incorporating congestion in preventive healthcare facility network design. *European Journal of Operational Research, 198*(3), 922–935.

Zhang, Y., Berman, O., Marcotte, P., & Verter, V. (2010). A bilevel model for preventive healthcare facility network design with congestion. *IIE Transactions, 42*(12), 865–880.

Part IV
Methods and Approaches Location Models with Uncertainty

Hub Location Models Under Uncertainty

Gita Taherkhani and Sibel A. Alumur

Abstract In this chapter, we review models and methods that incorporate uncertainty in hub location problems. In particular, we present stochastic and robust optimization models to formulate different sources of uncertainty including demand and costs and confer when each approach is best suited for use. We further describe and discuss hybrid modeling approaches and other extensions. We also review the common solution methods that are used to solve hub location models under uncertainty.

Keywords Hub location · Uncertainty · Stochastic programming · Robust optimization

1 Introduction

Hub facilities perform several functions that might involve sorting, switching, consolidation, or break-bulk to efficiently distribute the traffic and flow of freight, passengers, or data between many origins and destinations. Hub location problems address the decisions on how to optimally locate hub facilities, design the hub network, and determine the routes of flow on this network.

Several different versions of hub location problems have been studied in the literature for various applications including, but not limited to, the design of airline freight and passenger networks, less-than-truckload and truckload transportation, postal operations, express shipment, cargo delivery, liner shipping, public transit, and telecommunication networks. The interested reader can refer to a number of

G. Taherkhani
Quinlan School of Business, Loyola University, Chicago, IL, USA
e-mail: gtaherkhani@luc.edu

S. A. Alumur (✉)
Department of Management Sciences, University of Waterloo, Waterloo, ON, Canada
e-mail: sibel.alumur@uwaterloo.ca

H. A. Eiselt, V. Marianov (eds.), *Uncertainty in Facility Location Problems*,
International Series in Operations Research & Management Science 347,
https://doi.org/10.1007/978-3-031-32338-6_13

reviews and surveys written in this area, for example, Alumur and Kara (2008), Campbell and O'Kelly (2012), Farahani et al. (2013), Contreras and O'Kelly (2019), and Alumur et al. (2021).

Uncertainty is involved in every decision problem of the world. Since hub location can be considered a long-term strategic decision, incorporating uncertainty in hub location models is a natural extension to be able to design resilient hub networks. In particular, demand between origin-destination (O-D) pairs, transportation costs, and travel times, which are among the most common parameters used in hub location models, should be modeled under uncertainty as location decisions may have a long-lasting effect and their implementation can take a considerable amount of time. However, it is not always easy to incorporate and model uncertainty; it makes hub location models, which are already difficult in nature, more complicated, and even more difficult to solve.

Optimization under uncertainty generally consists of two streams of approaches: stochastic and robust optimization. In stochastic optimization, known probability distributions describe the behavior of uncertain parameters, and these distributions can be used to optimize the expected value of the objective function (e.g., Sim et al. (2009), Contreras et al. (2011), Mohammadi et al. (2014), Sadeghi et al. (2015), Correia et al. (2018), Correia and Saldanha-da Gama (2019), Peiró et al. (2019), and Taherkhani et al. (2020)). Another stochastic modeling approach is to use chance constraints, where a subset of constraints is defined to address the impact of uncertainty. The chance constraints are not required to always hold and the decision-maker is satisfied if they are held for a given probability (see, e.g., Snyder (2006) and Correia and Saldanha-da Gama (2019)).

In robust optimization, on the other hand, no probabilistic information is available for the uncertain parameters. In this case, uncertainty can be described by using a finite set of scenarios or can be modeled assuming that the values of the uncertain parameters can change within predefined intervals (e.g., Meraklı and Yaman (2016), Zetina et al. (2017), de Sá et al. (2018), Ghaffarinasab (2018)).

It is also possible to combine stochastic and robust optimization approaches to model uncertainty in hub location problems (see, e.g., Alumur et al. (2012) and Taherkhani et al. (2021)).

This chapter aims to review models and methods that incorporate uncertainty in hub location problems. As mentioned above, there are several applications of hub location problems, and each application setting, for example, whether it is airline passenger travel or less-than-truckload freight transportation, has its own criteria and requirements. For the sake of generality, we demonstrate different modeling approaches incorporating uncertainty on a classical hub location problem setting, with the aim that the formulations we present in this chapter can be used as a basis to model uncertainty in more complex problem settings.

The organization of this chapter is as follows. Before presenting the models with uncertainty, in the next section, we introduce the uncapacitated hub location problem with multiple allocation and present a deterministic model for this setting. The third section of the chapter studies hub location models under stochastic demand and stochastic costs and further discusses extensions with stochastic models. The fourth section describes robust optimization approaches under uncertain demand, uncertain transportation costs, uncertain setup costs, and some extensions with robust models. The fifth section presents hybrid models and other approaches including robust-stochastic models, distributionally robust models, and simulation-optimization. The sixth section reviews common solution methods that are used to solve hub location models under uncertainty. Finally, we present some concluding remarks in the last section.

2 Deterministic Model

In this section, we explain the assumptions, define the notation, and present the mathematical formulation of the deterministic uncapacitated hub location problem with multiple allocation.

In this classical hub location setting, it is assumed that the demand of all O-D pairs must be fully satisfied by using at least one hub facility, and hence, direct connections between two (non-hub) demand nodes are not allowed. Moreover, there are no connection costs for building the network links between the hub nodes as well as between non-hub nodes and hub nodes. The objective is to minimize the total cost that includes transportation costs as well as setup costs for establishing hubs. When the transportation costs are proportional to the distance and when distances satisfy the triangle inequality, each hub pair can be directly connected, that is, the hub-level network will be a complete subgraph induced by the hub nodes, as there are no fixed costs for building the network connections. It is assumed that there are economies of scale during transportation between hubs and that the unit transportation costs are lower between the hub facilities.

Consider a directed complete graph $G = (N, A)$, where N is the set of nodes and A is the set of arcs representing all the possible direct links between each pair of nodes. The sets of potential hub locations and commodities are denoted by $H \subseteq N$ and $K \subseteq N \times N$, respectively. Each $k \in K$ indicates a unique O-D pair whose origin and destination points belong to N. The parameter w_k represents the demand for commodity $k \in K$ to be routed from origin $o(k) \in N$ to destination $d(k) \in N$. Moreover, f_i represents the setup cost for establishing a hub located at node $i \in H$ and $c_{ij} = \zeta d_{ij}$ presents a transportation cost from node $i \in N$ to node $j \in N$, where d_{ij} represents the distance and ζ is the resource cost per unit distance.

Given the complete graph and assuming that distances satisfy the triangle inequality, every path between an origin $o(k)$ and a destination $d(k)$ will contain at least one and at most two hubs represented by $P_{ak} = (o(k), a_1, a_2, d(k))$, where $a = (a_1, a_2) \in A$ is a hub arc with the ordered pair of hubs a_1 and a_2, which are assigned to $o(k)$ and $d(k)$, respectively. Accordingly, the unit transportation cost of routing commodity k along path P_{ak} is expressed as $C_{ak} = \chi c_{o(k)a_1} + \alpha c_{a_1 a_2} + \delta c_{a_2 d(k)}$, where χ, α, δ are the collection, transfer, and distribution cost factors along the path. To reflect economies of scale between hubs, it is assumed that $\alpha < \chi$ and $\alpha < \delta$.

Using the above notation and following Hamacher et al. (2004), the deterministic uncapacitated hub location (DUHL) problem with multiple allocation can be formulated as:

$$\text{DUHL} \quad \min \sum_{k \in K} \sum_{a \in A} C_{ak} w_k x_{ak} + \sum_{i \in H} f_i y_i \tag{1}$$

$$\text{s.t.} \quad \sum_{a \in A} x_{ak} = 1 \qquad k \in K, \tag{2}$$

$$\sum_{a \in A: i \in a} x_{ak} \leq y_i \qquad i \in H, k \in K, \tag{3}$$

$$x_{ak} \geq 0 \qquad k \in K, a \in A, \tag{4}$$

$$y_i \in \{0, 1\} \qquad i \in H. \tag{5}$$

In this formulation, y_i is a binary variable that equals 1 if a hub is located at node $i \in H$, and 0 otherwise; x_{ak} is a continuous variable determining the fraction of commodity $k \in K$ that is satisfied through a path with hub arc $a \in A$.

The objective function (1) minimizes total cost. Constraints (2) ensure that the demand of all commodities must be satisfied. Constraints (3) guarantee that the demands of the commodities are satisfied only through open hubs. Constraints (4) and (5) indicate the domains of the decision variables.

The next sections of the chapter describe and detail how to incorporate uncertainty in various problem parameters for this classical hub location problem.

3 Stochastic Models

In this section, we present hub location models with stochastic demand and stochastic transportation costs and discuss further possible extensions with the stochastic models.

3.1 Stochastic Demand

In many applications of hub location problems, the demands of the commodities are not known in advance. Hence, the values of the parameter w_k, which represents the demand for commodity $k \in K$ to be routed from origin $o(k) \in N$ to destination $d(k) \in N$, may not be known with certainty. When the decision-maker has prior knowledge of the distribution of the demand, for example, through past observations and data, and can describe it by a known probability distribution, stochastic optimization can be used to incorporate the uncertainty associated with demand.

To model stochasticity, we can use stochastic programming with recourse, also known as two-stage stochastic programming, in which the stages determine the problem's informational context. The first stage, known as *ex ante*, refers to the decisions that need to be made without defining the stochastic parameters, while the recourse decisions, known as the second stage decision or *ex post*, can be determined after the stochastic parameters are known.

To model the uncapacitated hub location problem with stochastic demand, let $w_k(\psi)$ denote the random variables describing the future demand for commodity $k \in K$ under realization $\psi \in \Psi$, where Ψ is the support of ψ. In this stochastic setting with random demands, the strategic location decisions can be considered in the first stage, whereas the tactical decisions, including the allocations and the optimal routes of flows through the network, are determined in the second stage. The two-stage stochastic program for the uncapacitated hub location problem with stochastic demand (UHLSD) is modeled as follows:

$$\text{UHLSD} \quad \min \quad \sum_{i \in H} f_i y_i + \mathbb{E}_{\psi}\left[Q(\mathbf{x}, \psi)\right] \tag{6}$$

$$\text{s.t.} \quad (5), \tag{7}$$

where $\mathbb{E}_{\psi}$ denotes the expectation with respect to ψ and $Q(\mathbf{x}, \psi)$ represents the optimal value of the following problem for each realization of $\psi \in \Psi$:

$$\min \quad \sum_{k \in K} \sum_{a \in A} C_{ak} w_k(\psi) x_{ak}(\psi) \tag{8}$$

$$\text{s.t.} \quad \sum_{a \in A} x_{ak}(\psi) = 1 \qquad k \in K \tag{9}$$

$$\sum_{a \in A: i \in a} x_{ak}(\psi) \leq y_i \qquad i \in H, k \in K \tag{10}$$

$$x_{ak}(\psi) \geq 0 \qquad k \in K, a \in A. \tag{11}$$

In this formulation, $x_{ak}(\psi)$ is a continuous variable determining the fraction of commodity $k \in K$ that is satisfied through a path with hub arc $a \in A$ for realization $\psi \in \Psi$ and the objective of the problem is to minimize the total cost.

Contreras et al. (2011) showed that the uncapacitated hub location with stochastic demand is equivalent to the *expected value problem*, in which the deterministic model is solved by replacing the random variables with their expected values. However, this equivalence with the expected value problem is not true for the capacitated versions of the problem. The capacitated hub location problem with stochastic demand can be formulated by adding the following constraints to the above model:

$$\sum_{k \in K} \sum_{a \in A: i \in a} w_k(\psi) x_{ak}(\psi) \leq \Gamma_i y_i \qquad i \in H, \psi \in \Psi, \tag{12}$$

where Γ_i is the available capacity for a hub located at node $i \in H$. Observe that by adding this constraint, we cannot replace $w_k(\psi)$ variables by their expected value as the optimal solution of the second stage depends on the particular realization of the random vector ψ. Hence, the capacitated version no longer has the equivalency with the expected value problem and can be solved as a two-stage stochastic problem.

3.2 Stochastic Cost

In some real-life applications, the transportation costs may not be known with certainty. When there is prior knowledge on the behavior of the uncertain costs and it can be described by a known probability distribution, stochastic optimization can be used to incorporate the uncertainty associated with costs. In such a setting, the unit transportation cost from node $i \in N$ to node $j \in N$ can be denoted with a random variable $c_{ij}(\zeta)$ under realization $\zeta \in Z$, where Z is the support of ζ. The transportation cost associated with each path P_{ak} can then be defined as $C_{ak}(\zeta) = \chi c_{o(k)a_1}(\zeta) + \alpha c_{a_1 a_2}(\zeta) + \delta c_{a_2 d(k)}(\zeta)$.

To model the problem with stochastic costs, similar to UHLSD, the location decisions can be handled in the first stage, while the allocation decisions and the optimal routes of flows through the network are considered in the second stage. Accordingly, the hub location problem with stochastic transportation costs (UHLSC) can be modeled with a two-stage stochastic program.

In the previous section, under demand uncertainty, we elaborated the first and second stage decisions by writing two separate formulations. An alternative notation would be to model the first and second stage decisions within a single stochastic formulation as demonstrated in the following model with cost uncertainty:

$$\text{UHLSC} \quad \min \quad \mathbb{E}_\zeta \Big[\sum_{k \in K} \sum_{a \in A} C_{ak}(\zeta) w_k x_{ak}(\zeta) \Big] + \sum_{i \in H} f_i y_i \tag{13}$$

$$\text{s.t.} \quad \sum_{a \in A} x_{ak}(\zeta) = 1 \qquad k \in K \tag{14}$$

$$\sum_{a \in A: i \in a} x_{ak}(\zeta) \leq y_i \qquad i \in H, k \in K \tag{15}$$

$$x_{ak}(\zeta) \geq 0 \qquad k \in K, a \in A \tag{16}$$

$$y_i \in \{0, 1\} \qquad i \in H. \tag{17}$$

Unlike the equivalency with the expected value problem as shown for the UHLSD, note that $C_{ak}(\zeta)$ variables cannot be replaced by their expected values in the above formulation as the optimal values of the second-stage decisions are subject to the particular realization of the random vector ζ.

In addition to transportation costs, the setup costs can also be considered as stochastic parameters (i.e., $f_i(\zeta)$). However, as the optimal values of the second-stage decisions do not depend on any realization of ζ, the $f_i(\zeta)$ variables can be replaced by their expected value counterparts and thus the problem can be equivalent to the deterministic problem if the setup costs are the only stochastic parameters in the formulation.

3.3 Extensions

A more general variant of the problem would be the case where demand and cost are jointly stochastic. This problem can be also formulated using a two-stage stochastic program with recourse, where the strategic location decisions are *ex ante* and the routing decisions are *ex post* (recourse). The main challenge of this extension, compared to the cases where demand and costs are separately stochastic, would be the possible dependency between these two stochastic parameters.

Several extensions of stochastic hub location problems have been addressed in the literature. In particular, Contreras et al. (2011) study stochastic uncapacitated multiple allocation hub location problem in which demand or transportation costs are uncertain. Alumur et al. (2012) model single and multiple allocation hub location problems jointly under two sources of uncertainty: uncertain setup costs and stochastic demand. Correia et al. (2018) develop a two-stage stochastic modeling framework for multi-period capacitated multiple allocation hub location problem under uncertain demands. Peiró et al. (2019) present a stochastic uncapacitated r-allocation p-hub median problem with direct connections in which demand and transportation costs are associated both with uncertainty. In this problem setting, exactly p hubs need to be opened, nodes can be allocated to at most r hubs, and direct connections between non-hub nodes are allowed.

Taherkhani et al. (2020) model a two-stage stochastic program for the profit maximizing capacitated hub location problem under stochastic demand from different segments of commodities. Taherkhani et al. (2021) extended this setting where demand and revenue are jointly stochastic. They consider and model three separate cases depending on the interdependence between revenue and demand characterized by a linear revenue-demand function. Ghaffarinasab and Kara (2022) study risk-averse stochastic multiple allocation p-hub median problem in which demand is associated with uncertainty.

The effect of congestion at hubs can be incorporated into hub location models using stochastic parameters. Azizi et al. (2018) model hub network design under stochastic demand and congestion, where they model hubs as spatially distributed M/G/1 queues and congestion is captured using the expected queue lengths.

As mentioned in the introduction section, stochastic programming with chance constraints is another modeling approach that incorporates uncertainty into the problem setting, where uncertain parameters are introduced without explicitly defining different stages or recourse decisions. Instead, a subset of constraints is defined to address the impact of uncertainty and those constraints are to be held only for a given probability. Marianov and Serra (2003) model hubs as M/D/c queuing systems and use such chance constraints to limit the number of airplanes that can queue at a hub airport.

Travel times can be another stochastic parameter in modeling hub location problems. Sim et al. (2009) introduce the stochastic p-hub center problem and employ a chance-constrained formulation to model the minimum service-level requirement. They take travel time as stochastic parameters in their hub network design model where the objective is to minimize the maximum travel time through the network. Mohammadi et al. (2014) study a multi-objective stochastic hub location problem with uncertain demand. They use chance constraints along with a fuzzy multi-objective programming approach to formulate their problem. Sadeghi et al. (2015) propose a chance-constrained formulation for modeling a reliable p-hub covering location problem in which the links of the network are capacitated and their capacities are subject to stochastic degradations.

4 Robust Models

In some cases, no information is available on the distributions of the uncertain parameters. In that case, uncertainty can be introduced in the models by considering a finite set of scenarios or it can be modeled assuming that the uncertain parameters can take values independently within an interval. As such, this section discusses modeling of uncertainty in hub location problems using robust optimization techniques.

Depending on the attitude of the decision-maker, a min-max (i.e., optimizing for the objective under the worst-case scenario) or a min-max regret (i.e., mitigating the opportunity loss from not selecting the best scenario) criteria can be used for formulating the problems. As mentioned in Aissi et al. (2009), each criterion (i.e., min-max and min-max regret) fits best for a conservative decision-maker. With the min-max criterion, one can model uncertainty using either a continuous interval or a discrete set of scenarios. Using interval uncertainty for max-min, however, addresses a more general setting. For the min-max regret, on the other hand, only scenario-based uncertainty can be modeled as it requires obtaining the optimal solution for each realization. In this section, both of these approaches are demonstrated for hub location problems.

4.1 Uncertain Demand

Let us first model a robust hub location problem with uncertain demand by using the min-max criterion with a budget of uncertainty. As noted above, the min-max criterion optimizes the objective against the worst-case scenario to obtain a robust solution for a conservative decision-maker.

Let's assume that an interval uncertainty for demand can be used in which each parameter w_k for $k \in K$ takes values in $[\bar{w}_k, \bar{w}_k + \hat{w}_k]$, where $\bar{w}_k$ is the nominal value of demand (i.e., w_k) and $\hat{w}_k \geq 0$ represents the deviation from the nominal value. Furthermore, let us define a subset of commodities $U_w \subseteq K$ and a parameter $\gamma_w \in [0, |K|]$ with an integer value for the budget of uncertainty, which represents the maximum number of demand parameters w_k whose value is allowed to deviate from its nominal value. This parameter γ_w controls the level of conservatism in the objective. The robust hub location problem with uncertain demand (RHLUD) can then be modeled as:

$$\text{RHLUD} \quad \min_{(\mathbf{x},\mathbf{y})\in\mho} \sum_{k\in K}\sum_{a\in A} C_{ak}\bar{w}_k x_{ak} + \nu(\mathbf{x}) + \sum_{i\in H} f_i y_i, \tag{18}$$

where $\mho = \{(\mathbf{x}, \mathbf{y}) : \text{(2)–(5) are satisfied}\}$ and $\nu(\mathbf{x})$ is defined as follows:

$$\nu(\mathbf{x}) = \max_{U_w\subseteq K:|U_w|\leq\gamma_w} \sum_{k\in U_w}\sum_{a\in A} C_{ak}\hat{w}_k x_{ak}. \tag{19}$$

In the above formulation, $\nu(\mathbf{x})$ determines the worst-case deviation from the total demand over all possible demand scenarios for a given solution vector x. As noted in Bertsimas and Sim (2003), this approach finds an optimal solution that optimizes against all scenarios under which a number γ_w of the demand coefficients can vary in such a way so as to maximally influence the objective.

Note that in the extreme cases when $\gamma_w = 0$ or $\gamma_w = |K|$ (alternatively, when $U_w = \emptyset$ or $U_w = K$, respectively), the problem can be reduced to the deterministic model and it has trivial solutions such that for all commodities k, in the former case, $w_k = \bar{w}_k$, whereas in the latter case, $w_k = \bar{w}_k + \hat{w}_k$, where these cases represent the least and highest levels of conservatism, respectively.

The $\nu(\mathbf{x})$ can be reformulated by defining a binary variable z_k that determines whether or not commodity $k \in K$ is subject to uncertainty; that is, $z_k = 1$ if $k \in U_w$, and 0 otherwise:

$$\nu(\mathbf{x}) = \max \sum_{k\in K} \left(\hat{w}_k \sum_{a\in A} C_{ak} x_{ak} \right) z_k \tag{20}$$

$$\text{s.t.} \quad \sum_{k\in K} z_k \leq \gamma_w \tag{21}$$

$$z_k \in \{0, 1\} \qquad k \in K. \tag{22}$$

Since γ_w is integer, $\nu(\mathbf{x})$ simply sorts the commodities k in the nonincreasing order of their $\hat{w}_k \sum_{a\in A} C_{ak}x_{ak}$ values and selects the first γ_w of them. Hence, without losing integrality, constraint (22) can be replaced by its linear relaxation counterpart.

Observe that model (18) is nonlinear. Bertsimas and Sim (2003) show that by using the dual of problem $\nu(\mathbf{x})$, (18) has an equivalent linear formulation. Accordingly, we can define μ and λ_k as the dual variables associated with constraints (21) and the linear relaxation of (22), respectively. The dual of problem (20)–(22) can then be written as:

$$\nu(\mathbf{x}) = \min \ \gamma_w \mu + \sum_{k \in K} \lambda_k$$

$$\text{s.t.} \quad \mu + \lambda_k \geq \hat{w}_k \sum_{a \in A} C_{ak} x_{ak} \qquad k \in K \tag{23}$$

$$\lambda_k, \mu \geq 0 \qquad k \in K. \tag{24}$$

With this formulation of $\nu(\mathbf{x})$, the mathematical model RHLUD (18) can be reformulated as:

$$\min_{(\mathbf{x},\mathbf{y}) \in \mho} \quad \sum_{k \in K} \sum_{a \in A} C_{ak} \bar{w}_k x_{ak} + \gamma_w \mu + \sum_{k \in K} \lambda_k + \sum_{i \in H} f_i y_i$$

$$\text{s.t.} \quad (23), (24). \tag{25}$$

The above model is a linear mixed-integer programming formulation that can be solved using general purpose solvers.

4.2 *Uncertain Transportation Costs*

Let us now model the case where the transportation costs are uncertain and their values can change within an interval again using the min-max criterion. Under interval uncertainty, each parameter c_{ij} for $i \in N$ and $j \in N$ takes values in $[\bar{c}_{ij}, \bar{c}_{ij} + \hat{c}_{ij}]$, where $\bar{c}_{ij}$ is the nominal value and $\hat{c}_{ij} \geq 0$ its deviation. In this situation, each coefficient of the routing variable x_{ak} contains up to three uncertain parameters. $C_{ak} = \chi c_{o(k)a_1} + \alpha c_{a_1 a_2} + \delta c_{a_2 d(k)}$. Accordingly, the transportation cost associated with each path P_{ak} can be written as:

$$C_{ak} = \sum_{(i,j) \in P_{ak}} \tau_{ak}^{ij} c_{ij} = \sum_{(i,j) \in A} \tau_{ak}^{ij} c_{ij}$$

where $\tau_{ak}^{ij} = \chi$ if $(i, j) = (o(k), a_1) \in P_{ak}$, $\tau_{ak}^{ij} = \alpha$ if $(i, j) = (a_1, a_2) \in P_{ak}$, $\tau_{ak}^{ij} = \delta$ if $(i, j) = (a_2, d(k)) \in P_{ak}$, otherwise $\tau_{ak}^{ij} = 0$. Accordingly, robust hub location problem with uncertain transportation costs (RHLUC) and min-max criterion can be modeled as follows:

$$\text{RHLUC} \quad \min_{(\mathbf{x},\mathbf{y})\in\mho} \quad \sum_{k\in K}\sum_{a\in A} \bar{C}_{ak} w_k x_{ak} + \upsilon(\mathbf{x}) + \sum_{i\in H} f_i y_i, \tag{26}$$

where $\upsilon(\mathbf{x})$ is defined as:

$$\upsilon(\mathbf{x}) = \max_{U_c\subseteq A:|U_c|\leq\gamma_c} \sum_{k\in K}\sum_{a\in A}\sum_{(i,j)\in P_{ak}\cap U_c} \hat{c}_{ij}\tau_{ak}^{ij} w_k x_{ak}. \tag{27}$$

In this formulation, $U_c \subseteq A$ denotes a subset of arcs and $\gamma_c \in [0, |A|]$ represents the budget of uncertainty for the maximum number of transportation costs whose value are allowed to deviate from their nominal value. For a given solution x, $\upsilon(\mathbf{x})$ specifies the worst-case deviation from the total transportation cost over all possible cost scenarios. The $\upsilon(\mathbf{x})$ can be reformulated by defining a binary variable h_{ij}, which identifies whether or not arc $(i, j) \in A$ is under uncertainty; that is, $h_{ij} = 1$ if $(i, j) \in U_c$, and 0 otherwise.

$$\upsilon(\mathbf{x}) = \max \sum_{(i,j)\in P_{ak}\cap U_c} \left(\sum_{k\in K}\sum_{a\in A} \hat{c}_{ij}\tau_{ak}^{ij} w_k x_{ak}\right) h_{ij} \tag{28}$$

$$\text{s.t.} \quad \sum_{(i,j)\in A} h_{ij} \leq \gamma_c \tag{29}$$

$$h_{ij} \in \{0, 1\} \qquad (i, j) \in A. \tag{30}$$

Because of the same arguments we detailed for the RHLUD, constraint (30) can be replaced by its linear relaxation counterpart. Accordingly, we define μ and λ_{ij} as the dual variables associated with constraints (29) and the linear relaxation of (30), respectively. The dual of problem (28)–(30) can then be written as:

$$\upsilon(\mathbf{x}) = \min \ \gamma_c\mu + \sum_{(i,j)\in A} \lambda_{ij}$$

$$\text{s.t.} \quad \mu + \lambda_{ij} \geq \sum_{k\in K}\sum_{a\in A:(i,j)\in P_{ak}} \hat{c}_{ij}\tau_{ak}^{ij} w_k x_{ak} \qquad (i, j) \in A \tag{31}$$

$$\lambda_{ij}, \mu \geq 0 \qquad (i, j) \in A. \tag{32}$$

With this formulation of $\upsilon(\mathbf{x})$, mathematical model (26) can be reformulated as:

$$\min_{(\mathbf{x},\mathbf{y})\in\mho} \quad \sum_{k\in K}\sum_{a\in A} \bar{C}_{ak} w_k x_{ak} + \gamma_c\mu + \sum_{(i,j)\in A} \lambda_{ij} + \sum_{i\in H} f_i y_i$$

$$\text{s.t.} \quad (31), (32). \tag{33}$$

Similar to RHLUD, we end up with a mixed-integer programming model that can be solved using general purpose solvers.

4.3 Uncertain Setup Cost

We next model the case with uncertain setup costs. For this case, let us assume that a finite set of scenarios describe the uncertainty associated with the setup costs and this time let us use a min-max regret objective. As discussed above, the min-max regret criterion mitigates the opportunity loss by not selecting the best strategy (Aissi et al., 2009).

Let S_f describe the set of scenarios for uncertain setup costs and f_i^s the amount of setup cost for establishing a hub located at node $i \in H$ under scenario $s \in S_f$. The problem for each scenario $s \in S_f$ can then be formulated as:

$$Z_s^* = \min \quad \sum_{k \in K} \sum_{a \in A} C_{ak} w_k x_{ak} + \sum_{i \in H} f_i^s y_i \tag{34}$$

$$\text{s.t.} \quad (2)\text{–}(5),$$

where Z_s^* represents the lowest cost (i.e., optimal value) under scenario $s \in S_f$.

The regret of a solution (x, y) under setup cost scenario $s \in S_f$ is defined as the difference between the optimal cost that can be obtained under that scenario (i.e., Z_s^*) and the total cost associated with solution (x, y). With this definition, the robust hub location problem with uncertain setup cost (RHLUSC) is formulated as follows:

$$\text{RHLUSC} \quad \min \max_{s \in S_f} \Big\{ \sum_{k \in K} \sum_{a \in A} C_{ak} w_k x_{ak} + \sum_{i \in H} f_i^s y_i \Big\} - Z_s^* \tag{35}$$

$$\text{s.t.} \quad (2)\text{–}(5).$$

The above formulation can be linearized by replacing the inner maximization with a continuous variable V as follows:

$$\min \quad V \tag{36}$$

$$\text{s.t.} \quad (2)\text{–}(5)$$

$$V \geq \sum_{k \in K} \sum_{a \in A} C_{ak} w_k x_{ak} + \sum_{i \in H} f_i^s y_i - Z_s^* \qquad s \in S_f. \tag{37}$$

In this fashion, we can linearly model uncertainty in setup cost using the min-max regret criterion.

4.4 Extensions with Robust Models

An immediate extension with the robust models would be to consider a more general case of the problem by taking demand and costs simultaneously under uncertainty. There are a few studies in the literature modeling different robust hub location problems. Meraklı and Yaman (2016, 2017) study the uncapacitated and capacitated multiple allocation p-hub median problem under polyhedral demand uncertainty with two different uncertainty sets, hose and hybrid, and adopt a min-max criterion. In the hose model, they assume that the only available information is the upper limit on the total flow adjacent to each node, while in the hybrid model, they additionally impose lower and upper bounds on each pairwise demand. Zetina et al. (2017) address robust hub location problem with uncertain demand and transportation costs. They take interval uncertainty with a min-max objective and linearize the formulation by defining an additional continuous variable.

de Sá et al. (2018) consider an incomplete hub location problem in which a hub network can be partially interconnected by hub arcs and where both demand and transportation costs are under uncertainty. More recently, Taherkhani et al. (2021) take revenue under uncertainty in a profit maximizing hub location problem and formulate the problem with a max-min criterion and a min-max regret objective, considering both interval uncertainty and a discrete set of scenarios. They evaluate the level of robustness and conservatism of each metric in addressing the uncertainty associated with revenue, and for a particular case, they prove that the robust-stochastic version with max-min criterion can be viewed as a special case of the min-max regret stochastic model.

5 Hybrid Models and Other Approaches

In this section, we first describe a robust-stochastic model, in which two different types of uncertainty, including stochastic demand and uncertain transportation costs, are simultaneously incorporated into the problem. Then, we briefly discuss distributionally robust optimization and simulation-optimization techniques.

Let us start by modeling a robust-stochastic hub location problem. As explained in Sect. 3.1, the uncapacitated hub location problem with stochastic demand is equivalent to its expected value counterpart. Therefore, in this setting, we consider the capacitated version of the problem under stochastic demand. Assume that uncertain transportation costs can change within an interval and let us use the min-max criterion for formulating this problem.

Let $\Omega(\mathbf{y}, \psi)$ denote the set of all feasible decision variables $x_{ak}(\psi)$ under demand realization ψ that comply with the given location decisions $\mathbf{y}$. In particular,

$$\Omega(\mathbf{y}, \psi) = \{x_{ak}(\psi) : (9)\text{–}(12) \text{ are satisfied}\}. \tag{38}$$

The robust-stochastic capacitated hub location (RSCHL) problem can then be formulated as follows:

$$\text{RSCHL} \quad \min_{(\mathbf{x},\mathbf{y})\in\Omega} \quad \mathbb{E}_{\psi}\Big[\sum_{k\in K}\sum_{a\in A_k}\bar{C}_{ak}w_k(\psi)x_{ak}(\psi)+\rho_{\psi}(\mathbf{x})\Big]+\sum_{i\in H}f_iy_i, \tag{39}$$

where $\Omega = \{(\mathbf{x},\mathbf{y}) : \mathbf{x}(\psi) \in \Omega(\mathbf{y},\psi)\ \forall\psi \in \Psi, y_i \in \{0,1\}\ \forall i \in H\}$ is the set of solutions $(\mathbf{x},\mathbf{y})$ that are feasible under all realizations of ψ, and $\rho_{\psi}(\mathbf{x})$ is defined as follows:

$$\rho_{\psi}(\mathbf{x}) = \max \sum_{(i,j)\in P_{ak}\cap U_c}\left(\sum_{k\in K}\sum_{a\in A}\hat{c}_{ij}\tau_{ak}^{ij}w_k(\psi)x_{ak}(\psi)\right)h_{ij} \tag{40}$$

$$\text{s.t.} \quad (29)\text{–}(30).$$

By defining $\mu(\psi)$ and $\lambda_{ij}(\psi)$ as the dual variables associated with constraints (29) and the linear relaxation of (30), respectively, and taking the dual of the problem, the mixed-integer programming formulation of the robust-stochastic model can be written as:

$$\min_{(\mathbf{x},\mathbf{y})\in\Omega} \quad \mathbb{E}_{\psi}\Big[\sum_{k\in K}\sum_{a\in A}\bar{C}_{ak}w_k(\psi)x_{ak}(\psi)+\gamma_c\mu(\psi)+\sum_{(i,j)\in A}\lambda_{ij}(\psi)\Big]+\sum_{i\in H}f_iy_i \tag{41}$$

$$\text{s.t.} \quad \mu(\psi)+\lambda_{ij}(\psi) \geq \sum_{k\in K}\sum_{a\in A:(i,j)\in P_{ak}}\hat{c}_{ij}\tau_{ak}^{ij}w_k(\psi)x_{ak}(\psi) \quad (i,j)\in A \tag{42}$$

$$\lambda_{ij}(\psi), \mu(\psi) \geq 0 \quad (i,j)\in A. \tag{43}$$

Alternatively, if there is a discrete set of scenarios that can describe the behavior of uncertain transportation costs, one can also consider minimizing the regret incurred by the lack of perfect information. In this case, the resulting model under stochastic demand would be a min-max regret type robust-stochastic model.

The distributionally robust optimization technique is another approach in modeling uncertainty, where a parameter assumes only a partially known random distribution. The distributionally robust models are again formulated using min-max and min-max regret criteria as detailed above. However, compared to pure robust optimization, the advantage of this approach is that it is less conservative as it utilizes distribution information. Interested reader can refer to, for example, Wiesemann et al. (2014) and Chen et al. (2019) for more information on distributionally robust optimization. We would like to note that although both types of uncertainty are considered in the above hybrid robust-stochastic RSCHL formulation, it is different than the distributionally robust optimization approach by assuming that one set of parameters (i.e., demand) have a fully known distribution, while there is no known distribution for the other set of parameters (i.e., transportation costs).

Wang et al. (2020) adopt the distributionally robust optimization approach for addressing uncertainty involving both demand and costs in uncapacitated and capacitated hub location problems. In their setting, the joint distribution of demand and cost is allowed to be ambiguous and is only partially known defined by an ambiguity set. The objective is to minimize the worst-case expected cost over member distributions arising from this ambiguity set.

Uncertainty in hub location problems can also be addressed by integrating optimization with simulation. For example, Janschekowitz et al. (2023) study uncapacitated hub network design problems under various types and levels of uncertainty by implementing an iterative scenario-based hybrid simulation-optimization approach to obtain estimated global optimal solutions. They define a metric referred to as the value of simulation to compare the solutions obtained by using deterministic data or the expected value problem, with the solutions obtained under uncertainty. The advantage of using simulation coupled with optimization is that simulation can easily handle more complex settings, such as nonlinear transportation cost functions, reliability of hubs, congestion, and even user behavior.

6 Solution Methods

Hub location problems are usually very challenging to solve as they belong to a class of NP-hard problems involving joint location and network design decisions. Their main difficulty stems from the inherent interrelation between two levels of the decision process. The first level considers the selection of a set of nodes to locate hub facilities, whereas the second level deals with the design of the hub network, by selecting the links to connect origins, destinations and hubs, as well as the routing of flows through the network (Contreras & O'Kelly, 2019). Incorporating uncertainty makes these problems even more difficult to solve to optimality, particularly by general purpose solvers. Even with a finite set of scenarios, one often ends up with a large-scale mixed-integer linear problem with huge numbers of constraints and variables. In such cases, developing exact or heuristic solution approaches are required to be able to obtain a solution to the problems.

In this section, we overview the common approaches employed for solving stochastic and robust hub location problems in the literature. We would like to remark here that the complexity of the problems in this domain depends on the problem setting and its corresponding formulation. Hence, we cannot compare the problem complexity, in general, based on the employed modeling approaches (e.g., stochastic vs. robust).

One main difficulty in solving stochastic problems arises when the number of scenarios is infinite, and hence, the evaluation of the expected value of the objective function is challenging. To overcome this issue, a Monte Carlo simulation-based method known as sample average approximation (SAA) scheme can be used as suggested by Kleywegt et al. (2002). The main idea of this method is to reduce the size of the problem by generating a random sample and approximating the second-stage expectation value by the sample average function. This procedure is then replicated, and the overall average value is considered as the approximation of the optimal value of the stochastic problem. This approach has been successfully applied to several stochastic supply chain design as well as hub location problems with a large number of scenarios (see, e.g., Santoso et al. (2005), Schütz et al. (2009), Contreras et al. (2011), and Taherkhani et al. (2020), Taherkhani et al. (2021)).

Note that with a finite number of second-stage realizations, the full deterministic equivalent linear program becomes quite large. Accordingly, as customarily done in the literature, developing an exact (e.g., decomposition-based algorithm) or a heuristic algorithm (e.g., variable neighborhood search, tabu search, and GRASP) for solving the SAA counterpart of the stochastic problem is required. For example, Taherkhani et al. (2020, 2021) propose Benders decomposition algorithms coupled with SAA for profit maximizing capacitated hub location problem with stochastic demands. They further develop acceleration techniques to improve the convergence of the algorithms. Ghaffarinasab and Kara (2022) develop exact algorithms based on Benders decomposition for solving risk-averse stochastic p-hub median problem under demand data uncertainty represented by a finite set of scenarios. Peiró et al. (2019) propose a heuristic procedure for an uncapacitated r-allocation p-hub median problem with nonstop services in which demand and transportation costs are stochastic and stochasticity is captured by a finite set of scenarios.

The L-shaped algorithm is another approach for solving two-stage stochastic integer programs. The main idea of this algorithm is to approximate the nonlinear term in the objective including a solution of all second-stage linear programs and use the structure of the stochastic linear program for building a master problem and a subproblem (Van Slyke & Wets, 1969). Recently, Rostami et al. (2021) consider the single allocation hub location problem under demand uncertainty and develop a customized solution approach based on cutting planes that computationally outperforms the standard L-shaped method.

Similar to stochastic programming, the scientific community researches for efficient exact approaches to solve robust versions of the hub location problems. Meraklı and Yaman (2016, 2017) develop two Benders decomposition-based algorithms for robust uncapacitated and capacitated multiple allocation p-hub median problem. Zetina et al. (2017) propose a branch-and-cut algorithm to solve the robust uncapacitated hub location problem with uncertain demand and transportation cost. de Sá et al. (2018) devise a Benders decomposition algorithm for an incomplete multiple allocation hub location problem where the hubs are not necessarily fully interconnected. They assume that both demands and the fixed costs associated with the hubs are under uncertainty. Taherkhani et al. (2021) develop exact algorithms based on Benders decomposition integrated with SAA for solving robust-stochastic models for profit maximizing hub location problems in which revenue is uncertain. Exploiting the repetitive nature of SAA scheme, they propose generic acceleration methodologies to enhance the performance of the algorithms enabling them to solve large-scale intractable instances of the problem, in particular for the min-max regret version, to optimality.

7 Conclusion

In this chapter, we addressed essential aspects related to modeling hub location problems under uncertainty. We presented several stochastic and robust optimization models for a basic hub location problem setting and discussed further extensions with hybrid approaches.

Taking uncertainty into account in hub location models is a challenging field of research. Hence, despite the reported studies, the existing literature incorporating various uncertainty aspects into hub location problems, in terms of both modeling and solution methodology, is still rather limited. Nonetheless, to be able to provide resilient networks and routing solutions for the decisions made in the hub location domain, it is certainly very important to incorporate uncertainty into the problem settings.

References

Aissi, H., Bazgan, C., & Vanderpooten, D. (2009). Min–max and min–max regret versions of combinatorial optimization problems: a survey. *European Journal of Operational Research, 197*(2), 427–438.

Alumur, S., & Kara, B. Y. (2008). Network hub location problems: the state of the art. *European Journal of Operational Research, 190*(1), 1–21.

Alumur, S. A., Campbell, J. F., Contreras, I., Kara, B. Y., Marianov, V., & O'Kelly, M. E. (2021). Perspectives on modeling hub location problems. *European Journal of Operational Research*, 291:1–17.

Alumur, S. A., Nickel, S., & Saldanha-da Gama, F. (2012). Hub location under uncertainty. *Transportation Research Part B: Methodological, 46*(4), 529–543.

Azizi, N., Vidyarthi, N., & Chauhan, S. S. (2018). Modelling and analysis of hub-and-spoke networks under stochastic demand and congestion. *Annals of Operations Research, 264*(1), 1–40.

Bertsimas, D., & Sim, M. (2003). Robust discrete optimization and network flows. *Mathematical Programming, 98*(1–3), 49–71.

Campbell, J. F., & O'Kelly, M. E. (2012). Twenty-five years of hub location research. *Transportation Science, 46*(2), 153–169.

Chen, Z., Sim, M., & Xu, H. (2019). Distributionally robust optimization with infinitely constrained ambiguity sets. *Operations Research, 67*(5), 1328–1344.

Contreras, I., Cordeau, J.-F., & Laporte, G. (2011). Stochastic uncapacitated hub location. *European Journal of Operational Research, 212*(3), 518–528.

Contreras, I., & O'Kelly, M. E. (2019). Hub location problems. In *Location science* (pp. 327–363). Springer.

Correia, I., Nickel, S., & Saldanha-da Gama, F. (2018). A stochastic multi-period capacitated multiple allocation hub location problem: formulation and inequalities. *Omega, 74*, 122–134.

Correia, I., & Saldanha-da Gama, F. (2019). Facility location under uncertainty. In *Location science* (pp. 185–213). Springer.

de Sá, E. M., Morabito, R., & de Camargo, R. S. (2018). Benders decomposition applied to a robust multiple allocation incomplete hub location problem. *Computers & Operations Research, 89*, 31–50.

Farahani, R. Z., Hekmatfar, M., Arabani, A. B., & Nikbakhsh, E. (2013). Hub location problems: a review of models, classification, solution techniques, and applications. *Computers & Industrial Engineering, 64*(4), 1096–1109.

Ghaffarinasab, N. (2018). An efficient matheuristic for the robust multiple allocation p-hub median problem under polyhedral demand uncertainty. *Computers & Operations Research, 97*, 31–47.

Ghaffarinasab, N., & Kara, B. Y. (2022). A conditional β-mean approach to risk-averse stochastic multiple allocation hub location problems. *Transportation Research Part E: Logistics and Transportation Review, 158*, 102602.

Hamacher, H. W., Labbé, M., Nickel, S., & Sonneborn, T. (2004). Adapting polyhedral properties from facility to hub location problems. *Discrete Applied Mathematics, 145*(1), 104–116.

Janschekowitz, M., Taherkhani, G., Alumur, S. A., & Nickel, S. (2023). An alternative approach to address uncertainty in hub location. *OR Spectrum, 45*, 359–393.

Kleywegt, A. J., Shapiro, A., & Homem-de Mello, T. (2002). The sample average approximation method for stochastic discrete optimization. *SIAM Journal on Optimization, 12*(2), 479–502.

Marianov, V., & Serra, D. (2003). Location models for airline hubs behaving as M/D/c queues. *Computers & Operations Research, 30*(7), 983–1003.

Meraklı, M., & Yaman, H. (2016). Robust intermodal hub location under polyhedral demand uncertainty. *Transportation Research Part B: Methodological, 86*, 66–85.

Meraklı, M., & Yaman, H. (2017). A capacitated hub location problem under hose demand uncertainty. *Computers & Operations Research, 88*, 58–70.

Mohammadi, M., Torabi, S., & Tavakkoli-Moghaddam, R. (2014). Sustainable hub location under mixed uncertainty. *Transportation Research Part E: Logistics and Transportation Review, 62*, 89–115.

Peiró, J., Corberán, Á., Martí, R., & Saldanha-da Gama, F. (2019). Heuristic solutions for a class of stochastic uncapacitated p-hub median problems. *Transportation Science, 53*(4), 1126–1149.

Rostami, B., Kämmerling, N., Naoum-Sawaya, J., Buchheim, C., & Clausen, U. (2021). Stochastic single-allocation hub location. *European Journal of Operational Research, 289*(3), 1087–1106.

Sadeghi, M., Jolai, F., Tavakkoli-Moghaddam, R., & Rahimi, Y. (2015). A new stochastic approach for a reliable p-hub covering location problem. *Computers & Industrial Engineering, 90*, 371–380.

Santoso, T., Ahmed, S., Goetschalckx, M., & Shapiro, A. (2005). A stochastic programming approach for supply chain network design under uncertainty. *European Journal of Operational Research, 167*(1), 96–115.

Schütz, P., Tomasgard, A., & Ahmed, S. (2009). Supply chain design under uncertainty using sample average approximation and dual decomposition. *European Journal of Operational Research, 199*(2), 409–419.

Sim, T., Lowe, T. J., & Thomas, B. W. (2009). The stochastic p-hub center problem with service-level constraints. *Computers & Operations Research, 36*(12), 3166–3177.

Snyder, L. V. (2006). Facility location under uncertainty: a review. *IIE Transactions, 38*(7), 547–564.

Taherkhani, G., Alumur, S. A., & Hosseini, M. (2020). Benders decomposition for the profit maximizing capacitated hub location problem with multiple demand classes. *Transportation Science, 54*(6), 1446–1470.

Taherkhani, G., Alumur, S. A., & Hosseini, M. (2021). Robust stochastic models for profit-maximizing hub location problems. *Transportation Science, 55*(6), 1322–1350.

Van Slyke, R. M., & Wets, R. (1969). L-shaped linear programs with applications to optimal control and stochastic programming. *SIAM Journal on Applied Mathematics, 17*(4), 638–663.

Wang, S., Chen, Z., & Liu, T. (2020). Distributionally robust hub location. *Transportation Science, 54*(5), 1189–1210.

Wiesemann, W., Kuhn, D., & Sim, M. (2014). Distributionally robust convex optimization. *Operations Research, 62*(6), 1358–1376.

Zetina, C. A., Contreras, I., Cordeau, J.-F., & Nikbakhsh, E. (2017). Robust uncapacitated hub location. *Transportation Research Part B: Methodological, 106*, 393–410.

On Risk Management of Multistage Multiscale FLP Under Uncertainty

Laureano F. Escudero and Juan F. Monge

Abstract A tight mixed integer linear programming modeling framework is presented for the multistage multiscale facility location multiproduct allocation network expansion planning under uncertainty. Two types of decisions are considered, namely, the strategic and the operational ones. The strategic decisions are the selection of facility locations in a network as well as the related capacity dimensioning and expansion along a time horizon. A comprehensive literature overview on the problem is performed. Two types of uncertain parameters are considered, namely, strategic and operational ones, to be represented in multistage and two-stage scenario trees, resp. By using the special structure of the facility location problem, the coherent time-consistent risk-averse measure to consider is the expected conditional second-order stochastic dominance. Given the intrinsic problem's difficulty and the huge instances' dimensions, it is unrealistic to seek an optimal solution. A specialization of the matheuristic algorithm SFR3 is presented to obtain a (hopefully good) feasible solution in reasonable time as well as a lower bound to assess the solution quality. The performance of the overall approach is computationally validated by considering a dynamic supply network design problem with 100 raw material, 50 products, 30 candidate facilities (10 plants and 20 distribution centers), 31 strategic scenario nodes in the time horizon, and 4 operational ones per stage.

Keywords Facility location · Multistage multiscale stochastic mixed integer linear programming · Strategic and operational uncertainties · Coherent time-consistent risk-averse measures · Matheuristics

---

L. F. Escudero (✉)
Area of Statistics and Operations Research, Universidad Rey Juan Carlos, URJC, Móstoles, Madrid, Spain
e-mail: laureano.escudero@urjc.es

J. F. Monge
Center of Operations Research, Universidad Miguel Hernández, UMH, Elche, Alicante, Spain
e-mail: monge@umh.es

H. A. Eiselt, V. Marianov (eds.), *Uncertainty in Facility Location Problems*, International Series in Operations Research & Management Science 347,
https://doi.org/10.1007/978-3-031-32338-6_14

1 Introduction and Motivation

The optimization of real-life facility location planning (FLP) frequently requires strong MILP[1] modeling for problem-solving. Those dynamic problems are hard to solve with the additional difficulty of considering capacity expansion planning along a time horizon. Some examples of industrial sectors are energy and petrochemical networks expansion; transportation (aircraft fleet and rapid transit) network design; supplying, manufacturing, and distribution network management; forest harvesting planning; natural disaster relief preparedness resource allocation; and flow distribution through hub networks, just to name a few. Two types of time scaling are very frequently encountered in those problems, namely, a long one named stage and another shorter time unit.[2] Two types of uncertain parameters are considered, namely, strategic and operational ones. The strategic decisions (facility location, dimensioning, and expansion) and the realization of the uncertain strategic parameters take place at the nodes in the scenarios through the stages in a given time horizon.[3,4] The operational decisions and the realization of the uncertain operational parameters take place at the shorter time units.[5] The uncertainty in the strategic parameters is stagewise-dependent.[6] The uncertainty in the operational parameters is only stage-dependent.[7,8] Examples of the former are the facility building costs and residual values, and examples of the latter are the product demand, raw material costs, and facility capacity disruptions. Both types of uncertainties are considered in this work in an interlinked way. The capacitated facility location multiproduct allocation and extension planning (CFLEP) to deal with consists of selecting a given number of locations for a set of facilities at (the beginning of) the stages to manufacture products or producing services at the shorter time units. The goal is to minimize the expected cost of facility building,

---

[1] *MILP*, mixed integer linear programming.

[2] *Stage*, set of consecutive time units (say semesters, years), such that a facility location decision can be made at say (the beginning of) a semester and its operation is performed at shorter time units (say hours, days, weeks).

[3] *Strategic scenario*, node set where the realization of the strategic uncertain parameters is represented in a multistage tree from the first stage to the last one along the time horizon, so that only one node per stage is considered; see Fig. 1.

[4] *Strategic node* node in the multistage scenario tree where the realization of the strategic uncertain parameters take place for a given stage along he time horizon.

[5] *Operational scenario*, node where the realization of the operational uncertain parameters is represented in a two-stage tree in the time horizon; it belongs to the time units of a stage; see Fig. 1.

[6] *Stagewise-dependency*, property of the strategic uncertainty, so that it depends on the stage and also the uncertainty in the previous stages.

[7] *Stage-dependency*, property of the operational uncertainty, so that it does only depend on the stage.

[8] *The operational two-stage trees* are hanging from the multistage tree, being rooted at the strategic nodes, so that the second stage nodes represent the operational scenarios; see Fig. 1.

capacitated module installation, and expected cost of supplying, manufacturing, and distribution network management so that the solution is feasible under the strategic and operational scenarios along the time horizon. This policy is called risk neutral (RN)[9] in the absence of measures that control the impact in the objective function of the black swan scenarios. However, given the high stochasticity of the problem, a risk-averse measure (RAM) should be considered.[10] That stochasticity is due to the high uncertainty in the main strategic and operational parameters along the time horizon.

The coherent time-consistent RAM that is proposed in this work is the so-named expected conditional second-order stochastic dominance (ECSD); see Escudero and Monge (2018).[11]

To the best of our knowledge, a multistage multiscale stochastic approach for general CFLEP problems under uncertainty has not yet been considered in the literature, much less when the RAM ECSD is instrumented. Given the intrinsic problem's difficulty and the huge instances' dimensions (due to the network size of realistic instances as well as the cardinality of the strategic scenario tree and operational ones), it is unrealistic to seek an optimal solution. The matheuristic algorithm SFR3 is considered; see Escudero and Monge (2021). It obtains a (hopefully good) feasible solution in reasonable time as well as a lower bound of the optimal solution value to assess the solution quality.

The remainder of the work is organized as follows: Sect. 2 is devoted to a literature review on FLP variants under uncertainty to take benefit from by the main contributions of this work. Section 3 presents the main features of the stochastic version of CFLEP, so-named S-CFLEP, to deal with in this work. For completeness purposes and setting some notation to be used throughout the work, Sect. 4 reviews the main concepts of strategic multistage scenario trees and operational two-stage ones. Section 5 presents a strong MILP model for a comprehensive multistage multiscale S-CFLEP, where a dynamic supply, production, and distribution general facility location and expansion planning is considered. Section 6 specializes the decomposition matheuristic algorithm SFR3 to solving S-CFLEP, where the RAM ECSD is considered. Section 7 reports the main results of the proposal with a commercial state-of-the-art solver as a benchmark. Section 8 draws the main conclusions.

---

[9] *Risk neutral (RN)*, policy in problems under uncertainty where the impact of the so-named black swan scenarios in the objective function value is balanced with the impact of the good ones. Note: A black swan scenario has a low probability of occurrence, and it may have a high negative impact in the objective function depending upon the solutions.

[10] *Risk-averse measure (RAM)*, set of constraints and, perhaps, objective function elements that prevent solutions with high negative implications of the occurrence of black swan scenarios in the expected value of the function.

[11] *ECSD*, a risk-averse measure that prevent solutions according to the coherent properties presented in Artzner et al. (2007) and the time consistency ones presented in, for example, in Carpentier et al. (2012), Escudero and Monge (2018), Escudero et al. (2018b), Homem-de-Mello and Pagnoncelli (2016), and Werner et al. (2013), by considering a set of cost function thresholds that violations are upper bounded.

2 Literature Overview

See a recent comprehensive overview in Escudero and Monge (2021) for hub location problems, where the routing has a special importance for flow transporting between node pairs in a network. It includes works on uncapacitated and capacitated deterministic and Robust Optimization and static, two-stage, multistage, and multistage multiscale stochastic optimization. In any case, an overview on stochastic general FLP is as follows:

Static Stochastic Most of the works in the literature on static stochastic capacitated FLP (CFLP) are usually related to network infrastructure stochastic disruptions, due to natural disasters and terrorist attacks. See in Yu and Zhang (2018), Mohammadi et al. (2019) two comprehensive reviews on the subject. In particular, Yu and Zhang (2018) present a mixed integer nonlinear programming (MINLP) model for the uncapacitated FL under stochastic facility disruption, where the related risk is dealt with by considering the RAM Conditional Value-at-Risk (CVaR). Aghezaaf (2005) presents a Robust Optimization model for a strategic CFLP, where the facilities are plants and warehouses in the supply network design (SND) area under uncertainty on the demand. A Lagrangean relaxation approach is considered where the multipliers are constructed from the LP dual variables. Pagès-Bernaus et al. (2019) present a large-scale stochastic e-commerce CFLP for SND with uncertain demand. A mixture of a metaheuristic algorithm and simulation so-named simheuristic is introduced.

Two-Stage Stochastic The key parameters in FLP are frequently uncertain at the decision-making process in real-life problems. The realization of the uncertain parameters in mathematical optimization can usually be structured in a finite set of scenarios along a time horizon. Traditionally, special attention has been given to optimizing the Deterministic Equivalent Model (DEM),[12] in this case, by minimizing the expected facility network location and operation costs in the scenarios. The parameters' uncertainty in this field has been studied since the sixties; see Louveaux (1993) for a classical review and, more recently, Snyder (2006), Correia and Saldanha-da-Gama (2019), and Gago et al. (2022), among others. Most of the works in the literature deal with static two-stage models, and related algorithms for problem-solving.

A selected review of stochastic approaches for CFLP is presented in Correia and Melo (2021); see also Crainic et al. (2021), where additionally deterministic settings are reviewed. Most of the works deal with two-stage single-period MILP models, where the uncertainty (usually in the product demand along the time horizon) is represented by a set of finite scenarios. The FL decisions (designed here as "strategic") are considered as first-stage variables in the models, where no subordination is made to any single scenario but all of them are taken into account. On the other hand, the decisions on the allocation of products to facilities

[12] *Deterministic Equivalent Model (DEM)*, model equivalent to the stochastic one to represent in the deterministic setting the objective function and constraints in the scenarios; see Wets (1966).

and demand satisfaction (here called "operational") are considered as second-stage variables in the scenarios. Ntaimo and Sen (2005) present a very interesting approach for CFLP by considering a binary linear optimization (BLO) model for the two-stage stochastic server location problem, where the first-stage variables are the servers' location and the second stage ones are the allocation of clients to servers in the scenarios. The model is strengthened by generating valid inequalities. The divide-and-conquer exact algorithm so-named D^2, see Sen et al. (2002), is considered for solving an instance with over one million binary variables.

Chen et al. (2006) deal with a coherent RAM for single-period FLPs under uncertainty, although it can be easily generalized to other types of problems. It is presented in a two-stage stochastic BLO model, where the uncertainty is captured in a set of scenarios. It is assumed that a modeler-driven number of facilities is selected by considering all scenarios but without subordinating it to any of them (i.e., the selected set of facilities is a first-stage decision). On the other hand, each customer is allocated to a facility under each scenario, that is, the customer allocation is a second-stage decision. The aim is to minimize the expectation of the regret associated with the scenarios. Note: That regret for a scenario is measured as the difference between the demand weighted distance of the facility locations and their allocated customers in the proposed solution and the optimal one for the scenario alone. Albareda et al. (2011) present a BLO model for two-stage stochastic FLP with the following particularity: The location of the facilities as well as the assignment of customers to facilities (only one per customer) are first-stage decisions, where the number of assignments have a lower bound for each facility. The uncertainty relies on whether a customer has demand or not; it is assumed that it follows the Bernoulli distribution. If the number of customers with demand among of those assigned "a priori" to a facility is higher than a preset upper bound, then the service cost has an additional so-named outsourcing one. The aim is to select the facility location and the facility-customer assignment to minimize the facility location fixed cost plus the expected servicing one. Ivanov and Akmaeva (2021) consider an MILP model for two-stage stochastic CFLP with profit maximization, where the facilities can be open at the first stage and also under each scenario in the second stage, for a given value of the reliability level. The particularities mainly lie on: (a) The customer preferences for their facility allocation are taken into account. (b) Since the stochastic demand must be served, the quantile of losses is also considered in the solution's feasibility.

Dehghan et al. (2021) offer a review of the CFL-routing problem under uncertainty. Additionally, a two-stage MILP model is presented for the capacitated depot location-vehicle routing problem, where the simultaneous pickup and delivery is performed in a supply chain distribution network. The uncertainty on the depot disruption is modeled as a set of finite second-stage yes-no scenarios where the routing is performed. Several metaheuristics are tailored and computationally compared with the classical genetic algorithm.

Wang et al. (2021) provide a comprehensive review of emergency FLP, and most of the works intend to obtain in a static approach the optimal number of locations and the appropriate sites for the emergency facilities. Binary and general integer models are reviewed, mostly in the deterministic setting together with recent works that deal with the uncertainty related to services demand, site and resources availability, and cost and traffic congestion, among others. The uncertainty is dealt with by considering three methods, namely, chance constrained (by bounding the probability of not reaching certain levels of demand service covering), Distributionally Robust Optimization (DRO),[13] and classical scenario-based two-stage stochastic optimization. There is a broad variety of decomposition methodologies for emergency FLP solving, such as exact ones, matheuristics, metaheuristics, etc. Wang et al. (2021a) present a mixed integer second-order cone model for multiperiod CFLP to delivering relief supplies to affected areas in post-disaster humanitarian logistics. The ambiguity set is considered by the ∞-Wasserstein distance in chance constraints to assure a high probability of on time delivery when facing uncertain demand. Boonmee et al. (2017) present a survey on determining locations for emergency response facilities, such as distribution centers, warehouses, shelters, debris removal sites, and medical centers. The BLO models are static and multiperiod deterministic, two-stage stochastic and Robust Optimization in a set of application studies. A two-stage stochastic BLO model is presented in Gago et al. (2022) for ambulance location-allocation in an emergency medical service. The first stage decisions are the location of ambulance stations and fleet sizing, and the second stage decisions are the assignments of ambulances to emergencies in a given area under the scenarios. Recently, Zhu et al. (2022) present several cost-related two-stage Robust Optimization MILP models for delivering first aid products to the demand points. The first stage consists of locating warehouses and allocating available drones to them. The second stage assigns drones to demand points under the scenarios. A mixture of a column-and-constraint generation method and Benders Decomposition (BD) is considered.

Rahmaniani et al. (2018) present a two-stage stochastic single-period multi-product CFLP with stochastic demands in the origin-destination node pairs of the network. An MILP model is presented where first-stage strategic decisions are the subset of arcs to be located in the network, and the scenario-based second-stage operational decisions are given by the traffic flow through the located arcs in the network. The goal is the minimization of the arc investment cost plus the expected transportation cost in the scenarios. A set of valid inequalities is introduced as well as an accelerated BD methodology. Conde and Leal (2021) deal with the uncapacitated version of the two-stage stochastic FLP just presented above, by introducing a BLO robust model. The uncertainty is modelled by considering polyhedral sets so that the aim is to minimize a maximum regret total cost, by considering another tightened BD scheme. Mendoza-Ortega et al. (2021) present a two-stage stochastic MILP model for the single-period uncapacitated FLP. An MILP model is introduced where the first-stage strategic decisions are related to

[13] *Distributionally Robust Optimization (DRO)*, a Robust Optimization framework for problem-solving, where the uncertainty lies in an ambiguity set that is composed by some probabilistic distributions.

the facility locations. The scenario-based second-stage decisions are related to a multiple product distribution from producers to demand distributors. An agrofood case study is also presented, where the uncertainty on the crop yields per hectare is modeled.

An MILP model is presented in Alonso-Ayuso et al. (2003) for two-stage stochastic multiperiod CFLP. Its aim consists of production topology selection, facility sizing, product selection and allocation among plants, and vendor selection for raw material. The objective is the maximization of the expected benefit given by the product net profit over the time horizon minus the investment depreciation and operation costs. The main uncertain parameters are the product net price and demand, the raw material supply cost, and the production cost. The strategic decisions are made in the first stage. The tactical decisions are made in the second stage in the scenarios along the time horizon, being represented by continuous variables. Another MILP model is presented in Alonso-Ayuso et al. (2005) for two-stage stochastic multiperiod CFLP. It does product selection and plant location and dimensioning to maximize a mixture of the expected profit and the risk-averse reduction in the scenarios along a time horizon. Two alternative RAMs are considered. The first one maximizes the probability of reaching a profit in the scenarios satisfying a given threshold. The second one maximizes the VaR profit in the scenarios so that the expected number of scenarios that do not satisfy it is not higher than a given probability threshold (i.e., chance-constrained approach). An algorithm for problem-solving based on a splitting variable modeling object[14] via scenarios is considered. It uses the Branch-and-Fix Coordination (BFC) algorithm[15] introduced in Alonso-Ayuso et al. (2003a). An approach is presented in Ravi and Sinha (2006), where the facility locations could be in any of the two stages. An MILP model for a general two-stage stochastic multiperiod covering CFLP is presented in Marín et al. (2018). Its aim is to cover the demand of all nodes in a network up to a predefined distance threshold from the facility locations (with penalization for exceeding it). It is claimed that most of the works in the literature (even those with a deterministic subject) are particular cases of the proposed new formulation, where the facilities can be opened and closed at the periods. The goal consists of minimizing the expected cost in a finite set of scenarios for the demand and covering capabilities. Another two-stage stochastic multiperiod approach has been recently presented in Correia and Melo (2021), where two MILP models are introduced. In the first one, the strategic decisions on CFL are performed at the first periods before the uncertainty on the customers' demand is unveiled. In any way, the capacity can be modified (increasing and reducing it) in the periods along the

[14] *Splitting variable modeling object*, a constraint where a variable is equated to their copies such that, for example, its dualization allows the model's Lagrangean decomposition.

[15] *Branch-and-Fix Coordination (BFC)*, an algorithm where the coordination of the selection of the branching nodes and branching variables in the scenario subproblems is jointly done for those B&C nodes that share the same path on their common variables from the root node to each of them in their own branching process.

time horizon by considering the 'here and now' policy,[16] therefore, independently of the scenarios (i.e., first-stage decisions). The second model considers that those facility capacity modifications in the periods are performed under the scenarios (i.e., in the second stage). In both models, the operational decisions on the product flow from facilities to customers along the time horizon is carried out at the second stage (i.e., under the scenarios). Baptista et al. (2019) present a mixed integer bilinear optimization model for the two-stage stochastic multiperiod multiproduct close-loop chain design problem. The recovered end-of-life products from customers are evaluated in disassembly centers, and accordingly, either they are sent back to facilities for remanufacturing or leaving the network. In the latter case, either they are sold to third parties or sent to disposal. Typical uncertain parameters are cost investment on facility location as well as product demand, production cost, and returned product pricing, among others. Therefore, the stochastic optimization addresses different topology decisions on the location and capacity of some facilities (factories and distribution and sorting centers) at the first stage while some others (in particular some centers) at the second-stage periods in the scenarios. The goal is to maximize the net present value of the expected total profit along the time horizon. A mixture of chance-constrained and second-order stochastic dominance coherent time-inconsistent RAMs is considered for risk management at intermediate periods of the time horizon. A variant of the Fix-and-Relax (FR) methodology[17] is also presented. For assessing the computational validation of the approach, pilot cases are taken from a real-life glass supply chain that main features are retained.

Liu et al. (2019) present a two-stage DRO model for an emergency capacitated service station location problem with joint chance constrained to lower bound the probability of medical servicing demand. The problem is converted in a second-order cone mixed integer optimization model. Another two-stage DRO MILP model for CFLP is presented in Gourtani et al. (2020), where semi-infinite and semi-definite programming approaches are considered for problem-solving. A data-driven DRO model for CFLP is studied in Saif and Delage (2021), which distributional ambiguity set is represented as a Wasserstein ball around a small sample of the uncertain parameters, from where a set of scenarios is generated. Two MILP models are presented, one for the two-stage version and the other for the static one. Column generation iterative algorithms are proposed. A static CFLP with uncertainty in customer demands is presented in Ryu and Park (2021), where a cardinality-constrained uncertainty set is assumed for the robust problem. A mixture of the Dantzig-Wolfe decomposition and a branch-and-price algorithm is considered. A

[16] *'Here and now' policy*, a popular one in stochastic optimization so that the decisions are made at the stages (the first one in two-stage settings and at any stage in multistage settings), where the scenarios are taken into account but without subordinating to any of them.

[17] *Fix-and-Relax (FR)*, a matheuristic decomposition methodology for solving dynamic deterministic MILP problems, introduced in Dillenberger et al. (1994). It consists of solving a sequential series of subproblems, such that the variables in each one are partitioned in three subsets. The value of the variables in the first one is fixed, the integer variables in the second subset are kept integer, and the integrality of the variables in the third subset is relaxed.

two-stage DRO MILP model is presented in Basciftci et al. (2021), where the customer demand uncertainty is endogenous on the location of the facilities. A computational study is conducted to assess its added value by comparing it with the deterministic model as well as the classical stochastic and two-stage DRO ones with exogenous demand uncertainty. Another Robust Optimization approach is presented in Valtsa and Jayaswal (2021) for CFLP, where a two-stage stochastic multiperiod MILP problem is introduced by considering the uncertainty in the facility capacity. As a pilot case, medical doctors are to be assigned in a primary health center location-allocation problem. The approach minimizes the maximum regret between the optimal solution for each scenario and the one provided by the model, where the facility location is performed at the first stage. Several scenario dominance rules are introduced for reducing the model's dimensions, and a BD refinement is considered for problem-solving.

Given a network with a set of supplying (i.e., origin) nodes of different products and a set of receiver (i.e., destination) nodes of those products, but on smaller quantities, a cross-dock entity may serve as a consolidation point. The origin nodes can deliver the products to the cross-dock so that after being classified by type and destination, it can be transported to the destination nodes. As it is pointed out in Goodarzi et al. (2020), "cross-docking helps to accelerate the flow of parts and material, reduces the number of vehicles, and diminishes inventory costs. The main purpose of cross-docking in almost all companies is to collect various supply products in the form of pallets, consolidate them into a collection of mixed pallets of products with the same destination, and drop them off at the consume point (manufacturing/assembly plants/end user), according to the orders." A cross-dock has a number of receiving doors (strip doors) and a number of exiting (stack) doors, each of them with a pallet handling capacity during a given time period. The cross-docking assignment optimization is a young discipline. Most of the literature has been published during the last 25 years. See a comprehensive review in Goodarzi et al. (2020) focused on deterministic models and metaheuristic algorithms, but also considering the uncertainty on disruption, reliability, and reallocation of the facilities. However, the literature on cross-dock design under uncertainty is very scarce.

Multistage Stochastic for Capacitated Facility Location, Dimensioning and Expansion Planning, S-CFLEP As it can be observed in the approaches reviewed above (static stochastic, two-stage single-period stochastic, and two-stage multi-period stochastic), most of the works in the literature on stochastic CFLEP consider that the uncertainty is revealed only at a single moment. However, Current et al. (1998), one of the first works on stochastic FLP in a multistage setting, point out that "facility location decisions are frequently long-term in nature. Consequently, there may be considerable uncertainty regarding the way in which relevant parameters in the location decision will change over time." As Correia and Saldanha-da-Gama (2019) recently noted, there are many cases where the uncertainty is revealed along a time horizon and where the realization of the uncertainty is frequently stagewise-dependent. In those cases, the two-stage stochastic multiperiod scenario setting is not appropriate, since that approach is a relaxation of the multistage scenario tree where the nonanticipativity principle[18] is violated. However, there is not a broad literature on multistage stochastic FLP under uncertainty with respect to its deterministic counterpart. Among the very few works in the literature dealing with S-CFLEP, Hernández et al. (2012) present an MILP model and an algorithmic approach to location of prison facilities under uncertainty; it is applied to the Chilean prison system. The problem consists of finding locations and sizes of a preset number of new jails, and determining where and when to increase the capacity of both new and existing facilities over a time horizon, while minimizing the expected costs of the prison system location. The large-scale instances are solved via a heuristic mixture of BFC and B&C schemes to satisfy the constraints in the scenarios. Nickel et al. (2012) present an MILP model for S-CFLEP in the SND area, where the uncertainty in the customer demand and interest rates is represented.

Escudero et al. (2018a) present two matheuristric algorithms for providing lower and upper bounds for a BLO model that is introduced for large-scale S-CFLEP. Both algorithms consider, at each iteration, the solution where the variables' integrality related to the later stages is relaxed in the subproblems supported by the subtrees rooted at the nodes of a given stage in the multistage scenario tree. The first one, so-named CLD-LH (Cluster split-variable Lagrangean Decomposition and Lazy Heuristic),[19,20] is an iterative Lagrange multipliers updating scheme based on a scenario cluster split-variable Lagrangean decomposition; see Escudero et al. (2017). It is intended for obtaining strong (lower, in case of minimization) bounds of the solution value. The second scheme is the Fix-and-Relax Coordination (FRC) matheuristic,[21] presented in Albareda-Sambola et al. (2013) for S-CFLEP. It also works with the integrality relaxation of a subset of variables for different levels of the problem so that a lower-bound chain is generated from the LP relaxation up to the integer solution value. Additionally, a lazy heuristic scheme, based on the solutions of the relaxed problems, is considered in both procedures to obtain a (hopefully good) feasible solution of the original problem by fixing to 0 or 1 the fractional values of the variables based on different criteria. Another multistage stochastic MILP model is presented in Quezada et al. (2020) for the uncapacitated lot-sizing problem with uncertainty in demand and costs. An extension of the stochastic dual dynamic integer programming (SDDiP) algorithm, see Zou et al.

---

[18] *Nonanticipativity principle*: The scenarios in a two-stage tree or a multistage one that have a unique realization in the stages up to a given node should have the same solution; see Wets (1966).

[19] *Cluster split-variable Lagrangean Decomposition (CLD)*, scheme for performing Lagrangean Decomposition (LD) where the modeler-driven scenarios in a given cluster are not subjected to LD and, on the other hand, the split-variable constraints among the scenario clusters are dualized. Those constraints equate copies of the variables in the nodes of the scenario tree that belong to the first stages.

[20] *Lazy algorithm*, a scheme to fix variables in a model to the values retrieved from the solution of another algorithm.

[21] *Fix-and-Relax Coordination (FRC)*, an algorithm based on a specialization of the BFC methodology.

(2019), is proposed, by exploiting the polyhedral structure of the stochastic uncapacitated lot-sizing problem. Several Lagrangean relaxation approaches are considered in Taghavi and Huang (2020) for S-CFLEP, where temporary and permanent facility capacity types are allowed.

A three-stage stochastic mixed binary quadratic optimization (BQO) model is presented in Escudero et al. (2018, 2020) to determine a preparedness resource CFL and emergency good rescue allocation for managing natural disaster mitigation. Two types of uncertainty are considered: exogenous one due to the lack of full knowledge about the probability and intensity of the disaster for each point in a given network and endogenous uncertainty[22] that is based on the decision-maker's investment to obtain greater accuracy on the occurrence of the disaster to reinforce the network infrastructure. Additionally, the risk-averse measure ECSD is presented. Several scenario cluster-variants of Progressive Hedging Algorithm (PHA)[23] are benchmarked. See also some risk-averse two-stage stochastic approaches for the same natural disaster relief in Noyan (2012), Rawls and Turnquist (2012). A computational comparison between two-stage and multistage MILP models with CVaR variants is carried out in Yu et al. (2021). The classical time-inconsistent variant is considered for the two-stage models, and the coherent time-consistent Expected CVaR (ECVaR)[24] is done for the multistage version.

The cross-dock design under uncertainty along a time horizon is basically considered for SND, where the details of the cross-dock capacity design are not taken into account. Two works that are representative of the state-of-the art on the issue are Soanpet (2012) and Mousavi et al. (2014). In particular, Soanpet (2012) presents two stochastic models on cross-dock design for a multiproduct SND with origin and destination nodes, where the cross-docks are located to consolidate the products and saving transportation and handling costs. The first model is a chance-constrained BQO model to locate a given number of cross-dock centers, each one with a predefined capacity, as well as to assigning nodes and vehicles to the centers. The uncertainty lies in the capacity (assumed to follow the Normal Distribution) of each cross-dock center. The goal is to minimize the total cost, provided that the facility location cost and transportation and handling operations allow that the probability of satisfying the capacity of each facility is not smaller than a given threshold. The second model is a two-stage stochastic BQO one, where the first stage deals with the infrastructure of the cross-dock facilities. And the second stage deals with the supply network operations (i.e., assigning nodes and vehicles to the facilities) for each scenario, where the uncertainty lies in the cross-dock facility

---

[22] *Endogenous uncertainty*, also named decision-dependent one, is the uncertainty that results from the modification of the exogenous uncertainty by the decisions made in the model.

[23] *Progressive Hedging Algorithm (PHA)*, introduced in Rockafellar and Wets (1991), for providing the solution of stochastic problems with continuous variables. It was specialized in Gade et al. (2016) for obtaining (hopefully good) feasible solutions in stochastic MILP problems. An extension is presented in Boland et al. (2018) as a mixture of PHA and a Frank-Wolfe Simplicial decomposition for obtaining stronger lower bounds.

[24] *ECVaR*, Expected Conditional Value-at-Risk, a popular risk-averse measure.

capacity disruption. The goal is to minimize the expected total cost of FLP and product rooting and handling. For comparison purposes, the deterministic version of each of the two models is also presented. Mousavi et al. (2014) considers the location of multiple cross-docking facilities; the assignment of origin and destination nodes to the facilities; the product entering, stocking, and exiting; and the vehicle routing scheduling for supplying and delivering along a time horizon. Two MILP models are proposed as well as their integration. A hybrid of fuzzy possibilistic and stochastic optimization approaches is introduced to cope with the uncertainty in critical parameters, such as product supplying and demand, volume capacity of vehicles, required time for each vehicle to move between nodes, and transportation and operating costs.

3 Strategic Stagewise-Dependent and Operational Stage-Dependent Scenarios in S-CFLEP. The Subject of the Current Work

In real-life S-CFLEP problems, two types of decisions are to be considered, namely, the strategic and the operational ones. As Alumur et al. (2021) point out, "There is also a need to better integrate hub [here, facility] location models with service network design research to bridge the strategic and tactical [here, operational] decision models,...to bridge long- and short-term decisions, requiring managing the time scale differences between the different decisions." Thus, two types of time scaling are considered, namely, a long one (viz., semesters, years) and the other scale where the timing is shorter. The strategic decisions are the facility location in a network, and its capacity dimensioning and expansion along a time horizon. The operational decisions are the raw material supplying, its processing for manufacturing products, the customers satisfaction in the available CFL infrastructure, and the related transportation. Two types of uncertain parameters are also considered, namely, strategic and operational ones. Examples of strategic parameters are the costs of facility building and the cost of the installation of initial capacity modules and additional ones along the time horizon. The uncertainty of that type of parameters is usually stagewise-dependent, that is, it varies depending on the uncertainty in the previous stages. In fact, the parameters in a given stage may have different realizations for each set in the previous stage. On the other hand, the uncertainty of the operational parameters is only stage-dependent, being represented in a two-stage scenario tree, where the nodes in the second stage give the realizations (so-named operational scenarios) of those uncertain parameters. Examples of this type of parameters are the cost of the raw material supplying, transportation to and processing in the facilities, product demand, facility partial or full disruption, etc.

See in Escudero and Monge (2018) a scheme related to the partition of uncertain parameters into strategic and operational ones and, in case, tactical parameters. Basically, it consists of considering that the strategic decisions should not depend on *individual* operational uncertainties in the previous stages. By contrary, strategic decisions should depend on the realizations of the strategic uncertain parameters in the stages as well as on the set of realizations *as a whole* of the operational parameters in the next stages. Therefore, while dealing with uncertainty in multistage scenario trees, that observation is translated into considering that the strategic nodes in the tree should *not* be successors of the individual operational scenarios.

The multistage multiscale approach is considered in Werner et al. (2013), Kaut et al. (2014), and Escudero and Monge (2018), although none of these works present any algorithmic approach for empirically validating the proposals for large-scale instances. Moreover, the approach is also considered in rapid transit network design (see Cadarso et al. (2018)), hub network design (see Escudero & Monge (2021)), and SND (see (Castro et al., 2023)), where the algorithmic developments are empirically validated. In a different context, a multistage forest stand harvesting selection planning is presented in Alonso-Ayuso et al. (2020), where a multiperiod "tactical" activity replaces the two-stage "operational" one. Note that it is related to storable production as opposed to the service one as in the current work, where no stocking for later stages is considered. Then, that activity *does* influence, as an *expected* one, on the decisions to be made in the successor strategic nodes. See also strong multistage multiscale-based formulation in Glanzer and Pflug (2020), and on the other hand, Maggioni et al. (2020) present a scheme for obtaining lower and upper bounds on this type of stochastic problems.

The S-CFLEP problem addressed in this work involves decisions to be taken for FLP, initial capacity dimensioning and facility, and capacity expansion in a multistage multiscale stochastic setting. The representation of the uncertain data affects the type of decision variables to consider and, then, the type of model and decomposition methodologies for problem-solving. Therefore, the quality of the solution to offer to the decision-making process is also affected by the type of scenario trees to generate. While dealing with large time horizon problems as in the current work, there are two types of optimization submodels, namely, the strategic and the operational ones. They are very different in all aspects and intrinsically interrelated while embedded in a usually large-scale model for real-life problem-solving.

Contributions of the Work

To the best of our knowledge, no work in the literature considers multiple-allocation S-CFLEP in a time horizon setting with a multistage multiscale approach for modeling and problem-solving. The main contributions are as follows:

- The framework for representing the strategic stagewise-dependent uncertainty in S-CFLEP is based on the strategic multistage stochastic tree. Its nodes are rooting the two-stage stochastic trees where the operational stage-dependent scenarios are represented in the second-stage nodes.

- The investment on the facilities and capacity dimensioning and expansion is assumed to be made at the strategic nodes of the stages along the time horizon. The operation of the available facilities in the network is carried out under the operational scenarios in the stages. It is considered that the facilities can have a total or partial capacity disruption depending on the type of events to occur (say sabotage, misuse, etc.) that could diminish their capacity in a stochastic way. The objective function to minimize is the expected facility investment cost plus the expected operational one in the scenarios, minus the expected residual value of the facilities at the end of the time horizon. Therefore, an MILP model is introduced where binary, general integer, and continuous variables are considered. A key element in the model is the *step variable* modeling object[25] for representing the facility building in strategic nodes.
- Very few works deal with RAMs in multistage CFLP for reducing the drawbacks of the risk neutral (RN) function. As examples, see Yu et al. (2021) for multistage non-multiscale CFLP, and Alonso-Ayuso et al. (2020) for multistage multiscale CFLEP in the forestry stand harvesting sector; both approaches deal with the risk-averse measure ECVaR. By contrast, the current work deals with the impact of ECSD on S-CFLEP modeling and problem-solving.
- The large-scale feature of the instances in S-CFLEP is due to the facility network cardinality and number of capacity modules in realistic cases as well as the cardinality of the strategic scenario tree and operational ones. It renders unrealistic to problem-solving up to optimality by straightforward use of MILP solvers and, probably, any other current means. So a variant of the constructive decomposition matheuristic SFR3 (see (Escudero & Monge, 2021)) is also provided for obtaining feasible solutions with optimality gap. The *step variable* modeling object is a key one for the good performance of the matheuristic. It allows that state variables only link pairs of consecutive stages; in fact, the linking is performed between a strategic node and its immediate ancestor one. SFR3 is specialized to consider ECSD. The validity of the proposed approach is computationally analyzed.

[25] *Step variable* modeling object, a scheme for modeling (usually integer) variables of which values may last for several consecutive periods; it is tighter than its impulse variable counterpart. As an example, let us assume that a facility can only be made available once, if any, in a time horizon, being $t = 1, 2, 3$ the time periods. Let also the so-named impulse binary variable modeling object be such that the value 1 for variable x_t means that the facility is built *at* that stage t and otherwise, 0. So the constraint is $x_1+x_2+x_3 \leq 1$. Observe that the solution, say, $x_1 = 0.6$, $x_2 = 0.0$, $x_3 = 0.4$ is a valid one for that constraint in the LP relaxation in the B&C scheme. However, it is not accepted by the *step variable* modeling object $x_{t-1} \leq x_t$, where x_t has the value 1 if the facility is made available *by* period t and otherwise, 0.

4 Strategic Multistage Operational Two-Stage Stochastic Scenario Trees

4.1 Strategic Multistage Stochastic Tree

Let the notation taken from our work Escudero and Monge (2021). The information about the strategic nodes and scenarios can be visualized in the tree depicted in Fig. 1, where each root-to-leaf path represents a specific scenario and, then, it corresponds to a realization of the whole set of the uncertain parameters. Let us point out that it is beyond the scope of this work to present a methodology for multistage scenario tree generation and reduction; see, for example, Heitsch and Römisch (2009), Pflug and Pichler (2014, 2015), Leövey and Römisch (2015), Li and Floudas (2016), and Henrion and Römisch (2022), among others.

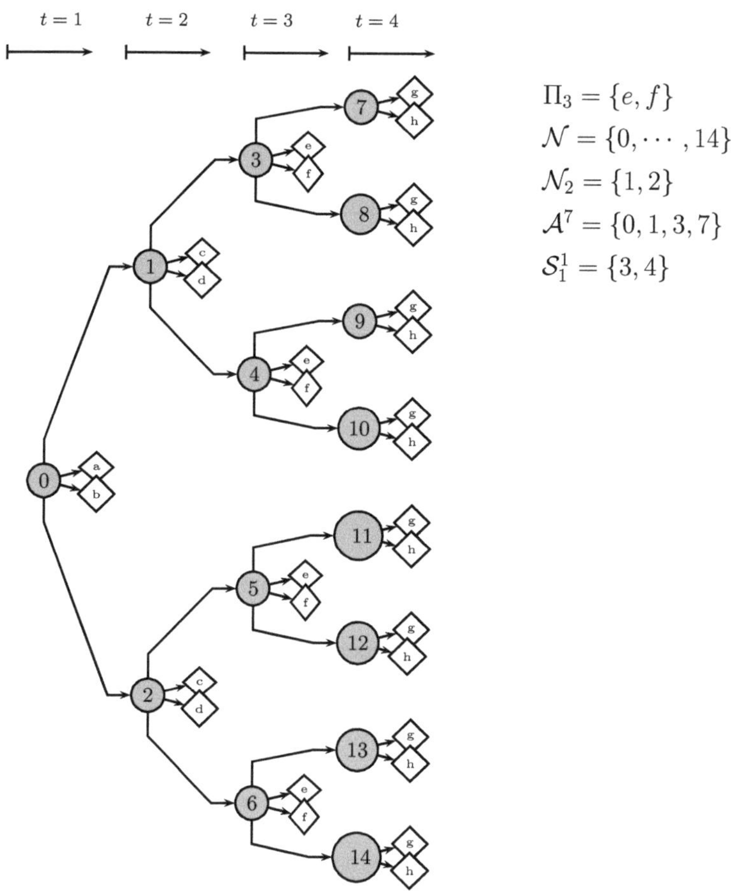

Fig. 1 Strategic multistage scenario tree with operational two-stage scenario trees

4.1.1 Lexicographically Ordered Sets in the Strategic Tree

$\mathcal{T}$, stages, where $T = |\mathcal{T}|$.
$\mathcal{N}$, strategic nodes in the scenario tree, such that $\mathcal{N} = \{0, \cdots, N-1\}$, where $N = |\mathcal{N}|$.
$\mathcal{N}_t$, nodes that belong to stage t, where $\mathcal{N}_t \subset \mathcal{N}$, for $t \in \mathcal{T}$. Note: By construction, $|\mathcal{N}_1|$=1.
Ω, strategic scenarios. Each one is included by the nodes in the Hamiltonian path from root node 0 to a node, namely, ω in the last stage, through the stages in set $\mathcal{T}$. Note: For convenience, a scenario has traditionally been denoted by its last node in the path; therefore, $\omega = n \in \mathcal{N}_T$.
$\mathcal{A}^n$, node n and its ancestors, for $n \in \mathcal{N}$. Note that $\mathcal{A}^1$ is only included by node 0, where $0 \in \mathcal{N}_1$.
$\Omega^n \subset \Omega$, scenarios having one-to-one correspondence with node n
$\mathcal{S}^n$, successors of node n, for $n \in \mathcal{N}$. Note: $\mathcal{S}^n = \emptyset$, for $n \in \mathcal{N}_T$.
$\mathcal{S}_1^n \subset \mathcal{S}^n$, immediate successors of node n, for $n \in \mathcal{N}$.

4.1.2 Other Elements in the Strategic Scenario Tree

w^n, weight factor representing the likelihood that is associated with node n, for $n \in \mathcal{N}$. Note: $w^n = \sum_{\omega \in \Omega^n} w^\omega$, where w^ω gives the modeler-driven likelihood associated with scenario ω, such that $\sum_{\omega \in \Omega} w^\omega = 1$.
σ^n, immediate ancestor of node n, for $n \in \mathcal{N}$. Note: It is assumed that $\sigma^0 = -1$, where -1 is the numbering of a null node that represents the existence of the facility network before the beginning of the time horizon under consideration.
t^n, stage to which node n belongs to, therefore, $n \in \mathcal{N}_{t^n}$.[26]

4.2 *Operational Two-Stage Trees Rooted at Strategic Nodes*

The operational uncertainty is represented in a finite set of stage-dependent operational scenarios in each stage t, for $t \in \mathcal{T}$. Therefore, it is assumed that the operational uncertainty is stage-independent of the strategic one. It is structured in two-stage trees rooted at the strategic nodes, so the operational realizations (i.e., scenarios) are visualized in the nodes of the second stage. Let the following additional notation.

Π_t, set of operational scenarios in stage t, for $t \in \mathcal{T}$; see Fig. 1.

[26] By construction, the scenarios in set Ω^n have a unique solution up to stage t^n, according with the non-anticipativity principle. Observe that the scenarios in Ω^n have the same realizations of the strategic uncertain parameters up to stage t^n, since they share the nodes in set $\mathcal{A}^n$.

w^{π}, weight factor representing the likelihood that is associated with operational scenario π, for $\pi \in \Pi_t$, such that $\sum_{\pi \in \Pi_t} w^{\pi} = 1$, for $t \in \mathcal{T}$.

Remark For the unlikely case where the strategic nodes are also stagewise-dependent on the operational ones, then, instead of the tree depicted in Fig. 1, the full combination of strategic and operational scenarios results in a gigantic multistage scenario tree.[27]

5 Multistage Multiscale Stochastic Assembly Plants and Distribution Centers Location Design and Expansion Planning

The aim is to select the locations of a plant subset in set, namely, $\mathcal{P}$ to manufacture a product set, namely, $\mathcal{J}$ and to select the locations of a distribution center (DC) subset in set, namely, $\mathcal{C}$, by blending raw material from supply set, namely, $\mathcal{I}$, along a time horizon. It is assumed that once a plant or a DC is available at any stage, then it continues being so until the end of the time horizon, although the plant capacity can be temporarily diminished due to different types of disruptions. The main uncertain strategic parameters are the cost of the plants and DCs building, the costs of initial and additional capacity modules in the plants along the time horizon, and the residual value of the plants and DCs. The main operational uncertain parameters are the cost of the raw material supplying, transportation to and manufacturing in the plants, the processing coefficients of the raw material, the cost of plant maintenance, the cost of transporting the product from the plans to the DCs, and the product demand in the DCs. The strategic variables are related to the location (and, then, the availability) of plants and DCs as well as the plant capacity module expansion. The operational variables are related to the raw material supplying, transportation to and manufacturing in the available plants, the product volume and its transportation to the DCs, plus the product demand shortfall at those DCs. For illustrative purposes, given the strategic multistage operational two-stage tree depicted in Fig. 1, let a representation of the strategic decisions as depicted in Fig. 2 under strategic scenario $\omega = 5$, say set $\mathcal{A}^{\omega}$, composed by the strategic node set $\{0, 2, 5\}$. The figure shows the plants initial availability and expansion, module capacity expansion, and DC availability. It also depicts a sketch of the raw material and product's traffic as operational decisions in a given operational scenario for each node in the strategic set $\{0, 2, 5\}$.

[27] Escudero and Monge (2021) report that for a case with $T = 5$, $|\mathcal{N}| = 31$, $|\mathcal{S}_1^n| = 2\,\forall n \in \mathcal{N} : t^n < T$, the scenario tree is composed of 23, 405 nodes and $|\Omega| = 16,384$ scenarios for $|\Pi_t| = 4$, and 629, 145 nodes and 524, 288 scenarios for $|\Pi_t| = 8\,\forall t \in \mathcal{T}$. Therefore, it is useful to consider the proposed approximation type in order to have an affordable structure for the case where the strategic uncertainty in a stage also depends on the operational one at the previous stage.

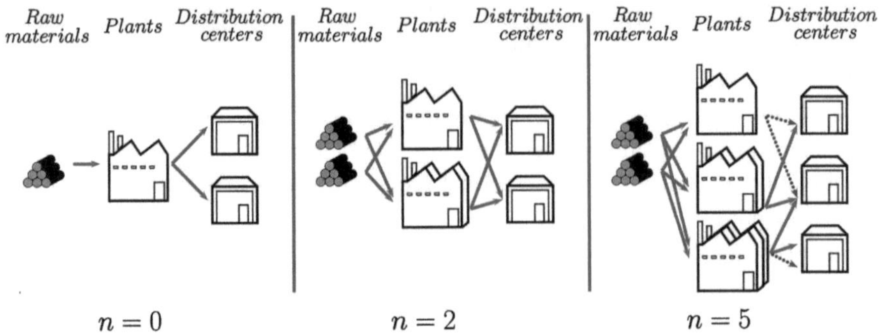

Fig. 2 Supplying, manufacturing plants, and distribution center network design and extension

5.1 Strategic Multistage Operational Two-Stage-Based Model

Let the notation of the elements, where capital letters and the symbol $\overline{(.)}$ usually denote data, and lowercase and Greek letters usually denote variables.

5.2 Additional Sets

$\mathcal{I}_j \subseteq \mathcal{I}$, raw material to blend for manufacturing product j, for $j \in \mathcal{J}$.
$\mathcal{J}_p \subseteq \mathcal{J}$, candidate products to be manufactured in plant p, for $p \in \mathcal{P}$.
A, candidate arcs (pc) for transporting product from plant p to DC c, for $p \in \mathcal{P}$, $c \in \mathcal{C}$.

5.3 Deterministic Data

$\hat{\gamma}_p^{\mathcal{P},-1}$ and $\hat{\gamma}_c^{\mathcal{C},-1}$, binary data on the current existing capacity of plant p, for $p \in \mathcal{P}$, and current existing DC c, for $c \in \mathcal{C}$, resp., at the beginning of the time horizon. Note: $\hat{\gamma}_p^{\mathcal{P},-1} = 0$ (resp., $\hat{\gamma}_c^{\mathcal{C};-1} = 0$) means that plant p (resp., DC c) is anew.
$\hat{\delta}_p^{-1}$, number of capacity modules already installed in plant p, for $p \in \mathcal{P}$. Note: For simplification purposes, it is assumed that the modules are identical in any plant.
$\overline{\gamma}^{\mathcal{P}}$, $\overline{\gamma}^{\mathcal{C}}$, maximum number of plants and DCs centers that can be available, resp., at any stage.
$\overline{\overline{\delta}}$, maximum number of capacity modules that are allowed to be installed (i.e., built) in any plant at any stage.

$\overline{\delta}$, maximum number of capacity modules that are allowed in any plant at any stage, independently of their stage installation.

K_p, reference module capacity in plant p, for $p \in \mathcal{P}$.

B_t, budget available for plant (either initial capacity or expansion) investment at stage t, for $t \in \mathcal{T}$.

$\overline{x}_i$, maximum volume of raw material i that can be supplied at any stage (i.e., under any operational scenario), for $i \in \mathcal{I}$.

$\overline{x}^t_{ip}$, upper bound on the supplying and transportation of raw material i to plant p at stage t, for $i \in \mathcal{I},\ p \in \mathcal{P},\ t \in \mathcal{T}$.

$\overline{y}^t_{jpc}$, upper bound on the flow of product j from plant p to DC c at stage t, for $j \in \mathcal{J},\ p \in \mathcal{P},\ c \in \mathcal{C} : (pc) \in A,\ t \in \mathcal{T}$.

Q_{jc}, product j demand shortfall unit penalization for DC c at any stage, for $j \in \mathcal{J},\ c \in \mathcal{C}$.

5.4 Uncertain Strategic Data in Node n, for $n \in \mathcal{N}$

$C_p^{\mathcal{P},n}$ and $C_p^{\delta,n}$, cost of plant p building and unit cost of capacity module installation, resp., for $p \in \mathcal{P}$. Note: $\mathcal{P}$ and δ stand for plant and capacity module, resp., to make a difference with the other C^n-costs.

$C_c^{\mathcal{C},n}$, cost of DC c building, for $c \in \mathcal{C}$. Note 1: $\mathcal{C}$ stand for DC. Note 2: For simplification purposes, it is assumed that a DC has a fixed capacity, if any, that is enough for product handling to satisfy product demand.

5.5 Uncertain Strategic Data in Node n, for $n \in \mathcal{N}_T$

$V_p^{\mathcal{P},n}$ and $V_p^{\delta,n}$, residual value of the investment on plant p and one capacity module, resp., at the end of the time horizon, for $p \in \mathcal{P}$.

$V_c^{\mathcal{C},n}$, residual value of the investment on DC c at the end of the time horizon, for $c \in \mathcal{C}$.

5.6 Uncertain Operational Data Under Scenario π, for $\pi \in \Pi_t,\ t \in \mathcal{T}$

C^{π}_{ij}, unit cost of raw material i supplying, transportation to and manufacturing in plant j, for $i \in \mathcal{I},\ j \in \mathcal{P}$, and unit cost of product transportation from plant i to DC j, for $i \in \mathcal{P},\ j \in \mathcal{C} : (ij) \in A$.

$M_p^{\mathcal{P},\pi}$ and $M_c^{\mathcal{C},\pi}$, maintenance cost of plant p and DC c, resp., for $p \in \mathcal{P},\ c \in \mathcal{C}$.
$M_p^{\delta,\pi}$, maintenance cost of a capacity module in plant p, for $p \in \mathcal{P}$.
U_{ijp}^{π}, volume requirement of raw material i for manufacturing a unit of product j in plant p, for $i \in \mathcal{I}_j,\ j \in \mathcal{J}_p,\ p \in \mathcal{P}$.
R_{jp}^{π}, capacity requirement of plant p for manufacturing a unit of product j, for $j \in \mathcal{J}_p,\ p \in \mathcal{P}$.
D_{jc}^{π}, product j demand in DC c, for $j \in \mathcal{J},\ c \in \mathcal{C}$.
ρ_p^{π}, fraction disruption of the capacity in plant p, in case that it is available, for $p \in \mathcal{P}$.

5.7 Strategic Variables in Node n, for n ∈ $\mathcal{N}$

$\gamma_p^{\mathcal{P},n}$, step binary variable of which value 1 means that plant p has been made available (jointly with some initial capacity modules) for product manufacturing *by* strategic node n and otherwise, 0.[28]
δ_p^n, step general integer variable that gives the total number of capacity modules installed in plant p *by* strategic node n, and otherwise, 0.[29]
$\gamma_c^{\mathcal{C},n}$, step binary variable which value 1 means that DC c has been made available for product distribution *by* strategic node n and otherwise, 0.
$\gamma_p^{\mathcal{P},-1}$, δ_p^{-1} and $\gamma_c^{\mathcal{C},-1}$, variables that represent the plant-DC network status already available at the beginning of the time horizon.

As an example, Fig. 2 can be interpreted as $|\mathcal{J}| = 1$ with the following facility locations along the stages:

- In $n = 0$: $\gamma_1^{\mathcal{P},n} := 1$ then set $|P| := 1$, $\delta_1^n := 1$, $\gamma_1^{\mathcal{C},n} := 1$ and $\gamma_2^{\mathcal{C},n} := 1$ then set $|\mathcal{C}| := 2$.
- In $n = 2$: $\gamma_2^{\mathcal{P},n} := 1$ then update $|\mathcal{P}| := 2$, $\delta_2^n := 2$.
- In $n = 5$: $\gamma_3^{\mathcal{P},n} := 1$ then update $|\mathcal{P}| := 3$, $\delta_3^n := 3$, $\gamma_3^{\mathcal{C},n} := 1$ then update $|\mathcal{C}| := 3$.

[28] $\gamma_p^{\mathcal{P},n} - \gamma_p^{\mathcal{P},\sigma^n} = 1$ means that plant p has been made available (i.e., built) *at* strategic node n. On the other hand, $\gamma_p^{\mathcal{P},n} - \gamma_p^{\mathcal{P},\sigma^n} = 0$ means that plant p has been made available by strategic node σ^n for $\gamma_p^{\mathcal{P},\sigma^n} = 1$, and it has not yet been made available for $\gamma_p^{\mathcal{P},n} = 0$. Note: $\gamma_p^{\mathcal{P},\sigma^n} \le \gamma_p^{\mathcal{P},n}$.
[29] $\delta_p^n - \delta_p^{\sigma^n} > 0$ gives additional number of capacity modules that are installed *at* node n. Therefore, observe that δ_p^n is the result of the cumulated number of capacity modules that has been installed *at* the previous strategic nodes back to stage $t = 1$, including node n (i.e., set $\mathcal{A}^n$).

5.8 Operational Variables Under Scenario π in Strategic Node n, for $\pi \in \Pi_{t^n}$, $n \in \mathcal{N}$

$x_{ip}^{n,\pi}$, volume of raw material i to be supplied and transported to plant p, for $i \in \mathcal{I}$, $p \in \mathcal{P}$.
$y_{jpc}^{n,\pi}$, product j volume to be transported from plant p to DC c, for $j \in \mathcal{J}_p$, $p \in \mathcal{P}$, $c \in \mathcal{C} : (pc) \in A$.
$s_{jc}^{n,\pi}$, slack variable that gives the product j demand shortfall in DC c, for $j \in \mathcal{J}$, $c \in \mathcal{C}$.

5.9 Elements of the Coherent Time-Consistent Risk-Averse Measure ECSD

- Sets.

  $\underline{\mathcal{T}} \subseteq \mathcal{T}$, stages of which strategic scenario nodes in the multistage tree have one-to-one correspondence with the scenario groups where ECSD is to be considered for.
  $\mathcal{B}^n$, profiles on strategic and operational costs, for $n \in \mathcal{N}_t$, $t \in \underline{\mathcal{T}}$.

- Parameters in profile b for strategic scenario group Ω^n, for $b \in \mathcal{B}^n$, $n \in \mathcal{N}_t$, $t \in \underline{\mathcal{T}}$.

  ϕ^b, cost threshold in profile b as a target for any scenario ω in the group, being composed by the strategic cost plus the expected operational one minus the expected residual value.
  $\overline{r}^b$, upper bound on the surplus over cost threshold ϕ^b in any scenario ω in the group.
  $\overline{\overline{r}}^b$, upper bound on the *expected* surplus over cost threshold ϕ^b in the group.
  w'^{ω}, weight factor representing the likelihood that is associated with scenario ω in the group. It can be expressed as $w'^{\omega} = w^{\omega} / \sum_{\omega' \in \Omega^n} w^{\omega'}$.

- $r^{\omega,b}$, continuous variable that gives the surplus over cost threshold ϕ^b under scenario ω in strategic group Ω^n, for $n \in \mathcal{N}_t$, $t \in \underline{\mathcal{T}}$.

5.10 Model for the Strategic Multistage Operational Two-Stage S-CFLEP

Let F^n denote the strategic and related operational cost for node $n \in \mathcal{N}$. It can be expressed

$$F^n = \sum_{p \in \mathcal{P}} \left[C_p^{\mathcal{P},n} \left(\gamma_p^{\mathcal{P},n} - \gamma_p^{\mathcal{P},\sigma^n} \right) + C_p^{\delta,n} \left(\delta_p^n - \delta_p^{\sigma^n} \right) \right] \tag{1a}$$

$$+\sum_{c\in\mathcal{C}} C_c^{\mathcal{C},n}\left(\gamma_c^{\mathcal{C},n}-\gamma_c^{\mathcal{C},\sigma^n}\right) \tag{1b}$$

$$+\sum_{\pi\in\Pi_{t^n}} w^\pi\left[\sum_{p\in\mathcal{P}}\left(M_p^{\mathcal{P},\pi}\gamma_p^{\mathcal{P},n}+M_p^{\delta,\pi}\delta_p^n\right)+\sum_{c\in\mathcal{C}} M_c^{\mathcal{C},\pi}\gamma_c^{\mathcal{C},n}\right. \tag{1c}$$

$$\left.+\sum_{p\in\mathcal{P}}\left(\sum_{i\in\mathcal{I}} C_{ip}^\pi x_{ip}^{n,\pi}+\sum_{c\in\mathcal{C}:(pc)\in A} C_{pc}^\pi\sum_{j\in\mathcal{J}_p} y_{jpc}^{n,\pi}\right)\right] \tag{1d}$$

Expression (1a) gives the plant building costs in strategic scenario n, including the related capacity modules installation costs. Expression (1b) gives the DC building costs in strategic scenario n. Expression (1c) gives the maintenance expected cost in the operational scenarios for strategic node n. And expression (1d) gives the raw material supplying, transportation and blending expected costs, plus product transportation expected costs from the plants to DCs.

Let V^ω denote the residual value of the facility network assets investment at the end of the time horizon under strategic scenario ω, for $\omega\in\Omega$. It can be expressed

$$V^\omega=\sum_{p\in\mathcal{P}}\left(V_p^{\mathcal{P},\omega}\gamma_p^{\mathcal{P},\omega}+V_p^{\delta,\omega}\delta_p^\omega\right)+\sum_{c\in\mathcal{C}} V_c^{\mathcal{C},\omega}\gamma_c^{\mathcal{C},\omega}. \tag{2}$$

Recall that $\omega=n\in\mathcal{N}_T$.

Let P^n denote the penalization of the product demand shortfall, for $n\in\mathcal{N}$, composed by the penalization of the overall demand shortfall $\sum_{j\in\mathcal{J}}\sum_{\pi\in\Pi_{t^n}} Q_{jc}D_{jc}^\pi$ in the operational scenarios for the unavailable DCs, and the penalization of the demand shortfall $\sum_{j\in\mathcal{J}}\sum_{\pi\in\Pi_{t^n}} Q_{jc}s_{jc}^\pi$ in the operational scenarios for the available DCs. It can be expressed

$$P^n=\sum_{j\in\mathcal{J}}\sum_{c\in\mathcal{C}}\sum_{\pi\in\Pi_{t^n}} Q_{jc}\left(D_{jc}^\pi(1-\gamma_c^{\mathcal{C},n})+s_{jc}^{n,\pi}\right) \tag{3}$$

Given the RHS of constraint (4o), it is easy to show that the elements $D_{jc}^\pi(1-\gamma_c^{\mathcal{C},n})$ and $s_{jc}^{n,\pi}$ are exclusive of each other.

The DEM can be expressed

$$\min\sum_{n\in\mathcal{N}} w^n(F^n+P^n)-\sum_{\omega\in\Omega} w^\omega V^\omega \tag{4a}$$

$$\text{s.t. }\gamma_p^{\mathcal{P},n}\in\{0,1\},\ \gamma_p^{\mathcal{P},\sigma^n}\le\gamma_p^{\mathcal{P},n}\qquad\forall p\in\mathcal{P},\ n\in\mathcal{N} \tag{4b}$$

$$\gamma_c^{\mathcal{C},n}\in\{0,1\},\ \gamma_c^{\mathcal{C},\sigma^n}\le\gamma_c^{\mathcal{C},n}\qquad\forall c\in\mathcal{C},\ n\in\mathcal{N} \tag{4c}$$

$$\delta_p^n\in\mathbb{Z}^+,\ \delta_p^{\sigma^n}\le\delta_p^n\qquad\forall p\in\mathcal{P},\ n\in\mathcal{N} \tag{4d}$$

$$\gamma_p^{\mathcal{P},n} \leq \delta_p^n \leq \overline{\delta}\gamma_p^{\mathcal{P},n} \qquad \forall p \in \mathcal{P},\ n \in \mathcal{N} \tag{4e}$$

$$\sum_{p\in\mathcal{P}} \gamma_p^{\mathcal{P},n} \leq \overline{\gamma}^{\mathcal{P}} \qquad \forall n \in \mathcal{N} \tag{4f}$$

$$\sum_{c\in\mathcal{C}} \gamma_c^{\mathcal{C},n} \leq \overline{\gamma}^{\mathcal{C}} \qquad \forall n \in \mathcal{N} \tag{4g}$$

$$\delta_p^n - \delta_p^{\sigma^n} \leq \overline{\overline{\delta}} \qquad \forall p \in \mathcal{P},\ n \in \mathcal{N} \tag{4h}$$

$$\sum_{p\in\mathcal{P}} \left[C_p^{\mathcal{P},n}(\gamma_p^{\mathcal{P},n} - \gamma_p^{\mathcal{P},\sigma^n}) + C_p^{\delta,n}(\delta_p^n - \delta_p^{\sigma^n})\right] \leq B_{t^n} \ \forall n \in \mathcal{N} \tag{4i}$$

$$\gamma_p^{\mathcal{P},-1} = \hat{\gamma}_p^{\mathcal{P},-1},\ \delta_p^{-1} = \hat{\delta}_p^{-1} \qquad \forall p \in \mathcal{P} \tag{4j}$$

$$\gamma_c^{\mathcal{C},-1} = \hat{\gamma}_c^{\mathcal{C},-1} \qquad \forall c \in \mathcal{C} \tag{4k}$$

$$\sum_{j\in\mathcal{J}_p} R_{jp}^{\pi} \sum_{c\in\mathcal{C}:(pc)\in A} y_{jpc}^{n,\pi} \leq (1-\rho_p^{\pi})K_p\delta_p^n \qquad \forall p \in \mathcal{P},\ \pi \in \Pi_{t^n},\ n \in \mathcal{N} \tag{4l}$$

$$x_{ip}^{n,\pi} = \sum_{j\in\mathcal{J}_p:i\in\mathcal{I}_j} U_{ijp}^{\pi} \sum_{c\in\mathcal{C}:(pc)\in A} y_{jpc}^{n,\pi} \qquad \forall i \in \mathcal{I},\ p \in \mathcal{P},\ \pi \in \Pi_{t^n},\ n \in \mathcal{N} \tag{4m}$$

$$\sum_{p\in\mathcal{P}} x_{ip}^{n,\pi} \leq \overline{x}_i \qquad \forall i \in \mathcal{I},\ \pi \in \Pi_{t^n},\ n \in \mathcal{N} \tag{4n}$$

$$\sum_{p\in\mathcal{P}:(pc)\in A} y_{jpc}^{n,\pi} + s_{jc}^{n,\pi} = D_{jc}^{\pi}\gamma_c^{\mathcal{C},n} \qquad \forall j \in \mathcal{J},\ c \in \mathcal{C},\ \pi \in \Pi_{t^n},\ n \in \mathcal{N} \tag{4o}$$

$$0 \leq x_{ip}^{n,\pi} \leq \overline{x}_{ip}^{t^n} \ \forall i \in \mathcal{I},\ p \in \mathcal{P},\ \pi \in \Pi_{t^n},\ n \in \mathcal{N} \tag{4p}$$

$$0 \leq y_{jpc}^{n,\pi} \leq \overline{y}_{jpc}^{t^n} \ \forall j \in \mathcal{J}_p,\ p \in \mathcal{P},\ c \in \mathcal{C} : (pc) \in A,\ \pi \in \Pi_{t^n},\ n \in \mathcal{N} \tag{4q}$$

$$0 \leq s_{jc}^{n,\pi} \qquad \forall j \in \mathcal{J},\ c \in \mathcal{C},\ \pi \in \Pi_{t^n},\ n \in \mathcal{N} \tag{4r}$$

$$\sum_{n'\in\mathcal{A}^{\omega}} F^{n'} - V^{\omega} - r^{\omega,b} \leq \phi^b \ \forall \omega \in \Omega^n,\ b \in \mathcal{B}^n,\ n \in \mathcal{N}_t,\ t \in \underline{\mathcal{T}} \tag{4s}$$

$$0 \leq r^{\omega,b} \leq \overline{r}^b \qquad \forall \omega \in \Omega^n,\ b \in \mathcal{B}^n,\ n \in \mathcal{N}_t,\ t \in \underline{\mathcal{T}} \tag{4t}$$

$$\sum_{\omega\in\Omega^n} w'^{\omega} r^{\omega,b} \leq \overline{\overline{r}}^b \qquad \forall b \in \mathcal{B}^n,\ n \in \mathcal{N}_t,\ t \in \underline{\mathcal{T}}. \tag{4u}$$

The objective function (4a) to minimize consists of the expected cost (1) minus the expected residual value (2) of the facilities in the strategic scenarios, plus the DC product demand's expected shortfall penalization (3).

The strategic constraint system (4b)–(4d) introduce the step variable modeling object for plant and DC building as well as for the initial and expansion capacity module installation in the strategic nodes. (It is assumed that the investment in the plants is performed at the beginning of the stages.) The strategic constraints (4e) force that one capacity module is installed, at least, at the same strategic node where the related plant is built. Additionally, an upper bound is imposed on the modules that are allowed in any plant. The strategic constraints (4f)–(4g) upper bound the cardinality of the subsets of plants and DCs locations that are available at the stages. The strategic constraints (4h) impose an upper bound on the number of capacity modules to install in each plant at the stages. The strategic constraints (4i) impose budget limitations on the investment for plant building, initial capacity and expansion at the stages. A zero-value for the constraints (4j)–(4k) means that the facility network is anew.

The operational constraint system (4l)–(4r) refers to the system under the operational scenarios for each strategic node. Constraints (4l) bound the product manufacturing volume to the capacity of the available plants. Constraints (4m) define the raw material requirements by the product manufacturing in each plant. Constraints (4n) bound the raw material volume to cover the manufacturing needs in the plants. Observe that it is assumed the zero-stock policy at the end of the stages. The operational constraints (4o) balance the product j manufactured in the plant set for each available DC c and its demand and, therefore, defining the product demand shortfall in case $\gamma_c^{\mathcal{C},n} = 1$ (i.e., DC c is available) in node n.

The minimization of function (4a) is constrained by the ECSD constraint system (4s)–(4u) so that the negative impact of the cost related to the black swan scenarios on the expected overall cost is kept under preset limits. Constraints (4s) define the cost surplus over the thresholds in the policy profiles under the scenarios in the strategic groups that have been selected based on the modeler chosen stages. Constraints (4t) and (4u) bound the scenario cost shortfall and the overall expected cost shortfall in the scenario groups, resp.

6 Specialization of SFR3, a Decomposition Matheuristic Algorithm

Given the dimensions of the large-scale instances of the S-CFLEP model (4), straightforward use of a state-of-the-art MILP optimizer requires very high computational effort; see Sect. 7. There is a broad literature on exact and inexact decomposition algorithms for stochastic MILP problem-solving; see Escudero et al. (2017) for a literature overview. However, the model solving up to optimality even for a single scenario along a time horizon could be unaffordable. Therefore, a Lagrangean-based approach, in principle, would have no practical interest for solving interconnected scenario-based submodels. As we know, a type of algorithms that could be considered is the Stochastic Nested Decomposition methodology (see Zou et al. (2019); Escudero et al. (2020a); Ahmed (2022)), among others, where single strategic nodes-based submodels are very attractive in the location framework for the risk neutral variant of the problem. However, the computing time could be expensive even in that case.

This section briefly presents the specialization of the constructive matheuristic algorithm SFR3 (see (Escudero & Monge, 2021)), to solve S-CFLEP model (4). SFR3 stands for Scenario variables Fixing and constraints and variables' integrality iteratively Randomizing Relaxation Reduction. It provides feasible solutions with good optimality bounds for medium- and large-scale instances. It is based on the Fix-and-Relax (FR) methodology, where the partition of the variables results from an ordering that is established *a priori*, and the variables declared integer in each subproblem define the so-named FR *level*. In particular, the variables that are fixed at level ℓ are the variables fixed at level $\ell - 2$ plus the variables of which values are retrieved from the solution in level $\ell - 1$. The approach has given good results while solving large-scale dynamic MILP real-life problems; see, for example, Escudero and Salmerón (2005), Baena et al. (2015), and Escudero and Pizarro (2017) for deterministic settings, and Alonso et al. (2000), Albareda-Sambola et al. (2013), and Escudero and Monge (2021) for stochastic ones. However, as a matter of fact, the relaxation of the integrality in a sizable subset of variables in the original model (4) prevents to take benefit of that integrality feature when solving the submodels in classical FR. Note that the knowledge of the variables' integrality by any solver strongly helps to model's tightening by performing probing, fixing variables, redundant constraints elimination, and new cuts appending. Therefore, the computing effort could be unaffordable for problem-solving by straightforward use of the solver as well as by using classical FR in the presence of a high number of integer variables in the instances.

Some of those FR drawbacks can be reduced in SFR3. It starts by partitioning the set of stages into modeler-driven so-named *stage blocks*; each one is a disjoint subset of consecutive stages, thus creating a collection of subtrees rooted at the strategic nodes $\{n\}$ that belong to the first stage in the block. It is worth to point out that SFR3 independently solves relaxations of the submodels of model (4) supported by the scenario subtrees rooted at those strategic nodes $\{n\}$. Those submodels are composed of the constraints and variables in the root nodes $\{n\}$ and successors in its own stage block and, in a partial *relaxation*, the constraints and variables in the other successors up to the leaf ones in the scenario tree. That relaxation is *randomly reduced* in the successive iterations of the algorithm. So two of the main modeler-driven SFR3 parameters are $\alpha_{t'}$ and $\beta_{t'}$, the assigned *probabilities* of operational scenario π in set $\Pi_{tn'}$ and strategic node n' in $\mathcal{S}^n \cap \mathcal{N}_{t'}$, resp., *not* to be relaxed in the n-submodel, where t' is a stage that does not belong to the given stage block. At a given iteration, the solution of the n-submodels is retrieved for node n and its successors that belong to its own block. That partial solution is fixed in the n'-

submodels at the next iteration, for $n' \in \mathcal{S}_1^{n''}$, which node set $\{n''\}$ is composed of the nodes that belong to the last stage in the block. Figure 3 depicts the supporting scenario tree rooted at node n, for $n = 0$ (then, first SFR3 iteration), which related stage block is $\{1, 2\}$. As an illustration, the node set is composed of the *non-relaxed* nodes (i.e., the non-shadowed ones). Observe that, by construction, the nodes that belong to the stage block that node n belongs to are not relaxed. The rationale behind that scheme consists of keeping the submodels' dimensions within affordable limits until obtaining a (hopefully good) feasible solution.

Note that, by construction, none of the strategic nodes in a *non*-relaxed scenario can be relaxed. On the other hand, the number of non-relaxed scenarios in each group $\Omega^{\underline{n}}$, for $\underline{n} \in \bigcup_{\underline{t} \in \underline{\mathcal{T}}} \mathcal{N}_{\underline{t}}$, should be kept above a given threshold in order to keep its representativeness in the stochastic dominance constraints (4s)–(4u) for ECSD. As an illustration, let us assume $\underline{t} = 2$, $\underline{n} = 2$ for the case depicted in Fig. 3, where $T = 4$, $n = 0$. Note that the strategic nodes in the scenarios given by the paths $\{0, 1, 3, 7\}$ and $\{0, 2, 5, 12\}$ in Fig. 3 are non-relaxed ones.

7 Computational Results

7.1 *Introduction. Computational Environment*

This section reports the main computational results that have been obtained while experimenting with a test bank composed of a variety of instances from medium one up to large-scale sizing. The computational experiment was conducted on a PC with a 2.9 gigahertz dual-core Intel Core i5 processor, 8 gigabyte of RAM, and operating system OS X 10.12.1. The modeling approach as well as SFR3 have been implemented in a C++ experimental code. The default options of CPLEX v20.1.0 are used for the full model (4) solving as well as for the n-submodels to be optimized in the matheuristic. However, given the difficulty of the problem, the optimality tolerance has been set to 1%.

Table 1 shows the S-CFLEP dimensions in the experiment. These instances contain both strategic and operational uncertainties. The scenario tree dimensions are $T = 5$ stages, $|\mathcal{N}| = 31$ strategic nodes, $|S_1^n| = 2$ immediate successors of node n, for $n \in \mathcal{N} : t^n < T$, and $|\Pi_t| = 4$, for $t \in \mathcal{T}$; see Fig. 1. For illustrative purposes on the strength of the risk-averse measure ECSD, let us consider $|\underline{\mathcal{T}}| = 1$, $\underline{\mathcal{T}} = \{1\}$, $0 \in \mathcal{N}_1$, $|\mathcal{B}^0| = 1$ and $1 \in \mathcal{B}^0$, such that ECSD is to be performed on the whole set of scenarios.

The data for the experiment have been randomly generated according to the following distributions: $C_p^{\mathcal{P},n} = 1000 + 100.(\sigma^n) + 100000.n$, $C_p^{\delta,n} = 100 + 100.(\sigma^n) + 10000.n$ and $C_c^{\mathcal{C},n} = 10000 + 1000.(\sigma^n) + 10000.n$, being the maintenance costs 25%, 15% and 25%, and the residual value 45%, 35% and 30%

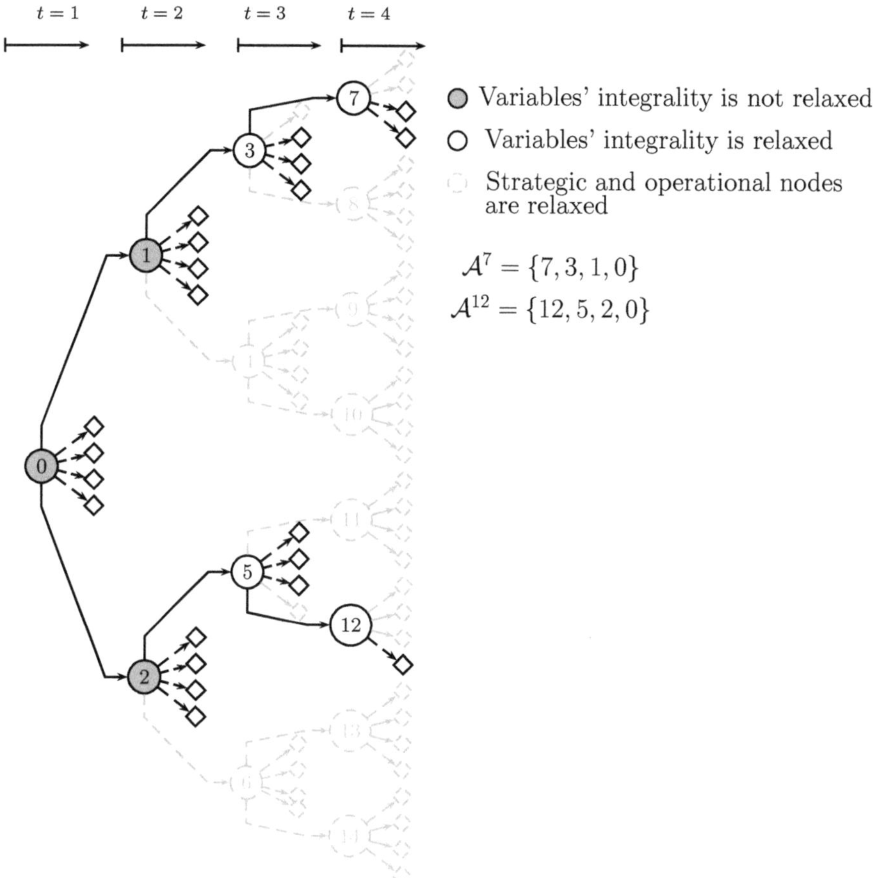

Fig. 3 Strategic multistage scenario tree with operational two-stage scenario trees. SFR3 nodes relaxation

Table 1 S-CFLEP dimensions

inst	$\lvert\mathcal{I}\rvert$	$\lvert\mathcal{J}\rvert$	$\lvert\mathcal{P}\rvert$	$\lvert\mathcal{C}\rvert$	$\overline{\gamma}^{\mathcal{P}}$	$\overline{\gamma}^{\mathcal{C}}$	$\overline{\delta}$
i1	20	5	5	10	3	8	3
i2	20	10	10	20	5	15	5
i3	50	20	10	20	5	15	5
i4	100	50	10	20	5	15	5

of the investments, resp., for $p \in \mathcal{P}$, $c \in \mathcal{C}$, $n \in \mathcal{N}$; $C^{\pi}_{ip} := U(4, 10)$ and $C^{\pi}_{pc} := U(6, 12)$; $U^{\pi}_{ijp} = U(2, 4)$; and $D^{\pi}_{jc} = U(40.t^n.(\pi + 1)/|\Pi_t| + 80.(t^n + 1)/T,\ 50 + 40.t^n.(\pi + 1)/|\Pi_t| + 80.(t^n + 1)/T))$, for $i \in \mathcal{I}$, $j \in \mathcal{J}$, $p \in \mathcal{P}$, $c \in \mathcal{C}$, $\pi \in \Pi_t$, $t \in \mathcal{T}$. (The data and other results are available from the authors under request).

Tables 2 and 3 show the main results obtained by CPLEX straightforward use (Sect. 7.2) and SFR3 (Sect. 7.3), resp., to solve model (4). The common headings to both approaches use the following result:

- $z^{\omega}_{(.)}$, cost under scenario ω related to function $(F - V)$ in the solution value of variant (.) of model (4), for $\omega \in \Omega$, where $(.) = RN$ is the risk neutral (4a)–(4r) and $(.) = SD$ is the risk- averse measure ECSD (4a)–(4u). It is computed as $\sum_{n \in \mathcal{A}^{\omega}} F^n - V^{\omega}$, where F^n and V^{ω} are given in (1) and (2), resp.

The headings are as follows:

- $\nabla\phi$, modeler-driven fraction of $z^{\overline{\omega}}_{RN}$ to consider as the cost threshold in ECSD, where $\overline{\omega} = argmax_{\omega \in \Omega}\{z^{\omega}_{RN}\}$ (i.e., the scenario with the highest cost in the RN variant of model (4)).
- $a(F - V)$, expected value of function $(F - V)$ to be computed as $\sum_{\omega \in \Omega} w^{\omega} z^{\omega}_{(.)}$, for $(.) \in \{RN, SD\}$.
- $d(F - V)$, expected value of the absolute deviation of the z-cost in the scenarios with respect to $a(F - V)$, to be computed as $\sum_{\omega \in \Omega} w^{\omega} |z^{\omega}_{(.)} - a(F - V)|$, for $(.) \in \{RN, SD\}$.

7.2 CPLEX Straightforward Use. Results

Table 2 shows the dimensions of model (4) and the CPLEX main results. The new headings are as follows: m, $n01$, ngi and nc, number of constraints, binary, general integer, and continuous variables, resp.; $dens\%$, density of the constraint matrix nonzero elements; $\underline{z}_{CPX}$, lower bound of the solution value (i.e., value of the best node in the B&B tree up to the optimization's interruption); z_{CPX} and t_{CPX}, incumbent MILP solution value (i.e., the smallest expected cost (4a)), and its computing time (in seconds, as for all experiments), resp.; and GAP_{CPX}, optimality gap of the incumbent solution, being computed as $100.\frac{z_{CPX} - \underline{z}_{CPX}}{z_{CPX}}$. Observe that for the large instances, there are $ngi = 300$ general integer variables of which the range is {0, 5}; it highly increases the difficulty of the instances i3 and i4 solving. Note: The dimensions' difference of the variants of a model only lies on the number of the new continuous variables that are required by ECSD.

Table 2 has two blocks of results for each instance ix, for $x = 1, 2, 3, 4$. The first block, say ix.0, has only one line; it shows the results for the RN variant. The second block has one line for each value of the risk averse parameter $\nabla\phi$, say 0.90 for ix.1, 080 for ix.2 and 0.70 for ix.3, so that the cost threshold in (4s) of ECSD functional is given by $\phi = \nabla\phi.z^{\overline{\omega}}_{RN}$. A 4h computing time limit is imposed for all instances in the test bank, but for i4 which limit is 8h.

It can be observed in the table that CPLEX straightforward use provides good quality results for instances i1 and i2. The cost threshold ϕ constraining implies an increase on the expected objective function value z_{CPX} and a decrease on the expected cost $a(F - V)$. Note that ECSD constraint system is also impacting on the

Table 2 S-CFLEP model (4). Dimensions and CPLEX results

$inst$	m	$n01$	ngi	nc	$dens\%$	$\underline{z}_{CPX}$	z_{CPX}	t_{CPX}	GAP_{CPX}	$\nabla\phi$	$a(F-V)$	$d(F-V)$
i1.0	29,078	465	155	50,236	0.06	192,137	192,644	60	0.26	–	1773	289
i1.1	29,111	465	155	50,252	0.06	207,802	208,582	69	0.37	0.90	1250	341
i1.2	29,111	465	155	50,252	0.06	214,300	214,932	88	0.29	0.80	1113	289
i1.3	29,111	465	155	50,252	0.06	221,583	222,177	95	0.27	0.70	974	268
i2.0	80,368	930	310	298,856	0.03	869,886	875,092	652	0.59	–	11,228	534
i2.1	80,401	930	310	298,872	0.03	924,666	930,451	1321	0.62	0.90	9530	656
i2.2	80,401	930	310	298,872	0.03	971,140	979,605	3662	0.86	0.80	8471	591
i2.3	80,401	930	310	298,872	0.03	1,025,824	1,031,838	1567	0.58	0.70	7412	514
i3.0	170,888	930	310	608,856	0.03	1,775,830	1,776,492	5995	0.04	–	69,717	490
i3.1	170,921	930	310	608,872	0.03	1,853,388	1,877,419	14,400	1.28	0.90	62,959	437
i4.0	387,888	930	310	1,489,256	0.02	4,573,136	4,581,987	28,478	0.19	–	359,957	4673
i4.1	387,921	930	310	1,489,272	0.02	5,024,331	–	28,800	–	0.90	–	–

Table 3 S-CFLEP model (4). SFR3 matheuristic results

$inst$	$\underline{z}_{FR3}$	$\underline{t}_{FR3}$	z_{FR3}	$\tilde{z}_{FR3}$	t_{FR3}	GAP_{FR3}	GR_{FR3}	$\nabla\phi$	$a(F\text{-}V)$	$d(F\text{-}V)$
i1.0	187,852	13	193,795	193,852	68	3.06	1.006	–	1693	292
i1.1	203,033	66	209,005	209,074	77	2.86	1.002	0.90	1251	339
i1.2	209,550	58	215,475	215,543	73	2.75	1.003	0.80	1113	294
i1.3	216,880	44	222,812	222,902	77	2.66	1.003	0.70	974	267
i2.0	842,754	208	872,083	873,780	491	3.36	0.997	–	11,420	553
i2.1	898,581	317	938,215	946,705	591	4.22	1.008	0.90	9530	654
i2.2	945,513	802	977,833	979,434	618	3.31	0.998	0.80	8471	588
i2.3	1,001,405	553	1,033,139	1,037,309	626	3.07	1.001	0.70	7412	514
i3.0	1,722,785	905	1,780,296	1,780,948	2285	3.23	1.002	–	69,674	589
i3.1	1,810,564	5534	1,868,183	1,874,516	2873	3.08	0.995	0.90	62,422	606
i4.0	4,441,440	4409	4,576,514	4,583,393	15,450	2.95	0.999	–	359,475	5258
i4.1	4,916,226	7418	5,043,965	5,052,222	15,347	2.53	–	0.90	290,816	7116

computing time from, say, 652s for i2.0 to 1567s for i2.3 ($\nabla\phi = 0.70$). The large instance i3 has also small optimality gaps. Anyway, it reaches the time limit for i3.1; see also the increase on z_{CPX} with respect to i3.0. The performance on instance i4.0 is very good (note that $GAP_{CPX} = 0.19\%$), but the computing time is very high. However, no solution is provided for i4.1 in the time limit.

7.3 SFR3 Matheuristic. Results

The section reports the main results for solving the original model (4) by using SFR3. For lack of space only, the results of the following strategy are shown: $\overline{e} = 5$ executions, $\alpha_t = 0.50,\ \beta_t = 0.50\,\forall t \in \mathcal{T}$. The computing time limit has been set to 2h for each n-submodel solving.

Table 3 reports the main results for the best of the executions of SFR3 in the variants RN and ECSD of model (4). As in Table 2, it is also organized in two blocks for each instance. The new headings are as follows: $\underline{z}_{FR3}$ and $\underline{t}_{FR3}$, lower bound of the optimal solution value (see below) and computing time, resp.; z_{FR3}, incumbent solution value in the scenarios; GAP_{FR3}, optimality gap of the incumbent solution, being computed as $100.\frac{z_{FR3}-\underline{z}_{FR3}}{z_{FR3}}$; and GR_{FR3}, goodness ratio of the SFR3 incumbent solution z_{FR3} over the CPLEX one, being computed as $\frac{z_{FR3}}{z_{CPX}}$, where the smaller $GR_{FR3} < 1$, the better performance of SFR3 versus CPLEX straightforward use. Additionally, t_{FR3} gives the computing time that is required by the whole set of the $\overline{e}$ executions, and $\tilde{z}_{FR3}$ gives the median of the expected cost in the set of those executions.

The lower bound $\underline{z}_{FR3}$ is independently obtained by considering the strategy $\overline{e} = 1,\ \alpha_t = 1.00,\ \beta_t = 1.00\,\forall t \in \mathcal{T}$ for node $n = 0$, where the variables' integrality is relaxed at all stages but $t = 1$ (i.e., it is the first *level* of the classical FR approach; see Sect. 6).

It can be observed in Table 3 that SFR3 provides a solution for all of the instances in the two variants, requiring a reasonable computing time, anyway, much smaller than the CPLEX one. Additionally, the goodness ratio GR_{FR3} is very close to 1 for the cases where CPLEX obtains a solution, and on the other hand, GAP_{FR3} is reasonable (not higher than 4.22%). It is worth to point out the small 2.53% optimality gap for the most difficult instance, i4.1, where CPLEX does not find any solution. The overall SFR3 computing time in the $\overline{e} = 5$ executions in any variant of any instance is much smaller than the one required by CPLEX. Even adding the computing times $\underline{t}_{FR3}$ and t_{FR3}, the total time is smaller.

8 Conclusions

In this work, a comprehensive literature review is performed on FLP under uncertainty. Additionally, an MILP model is presented for the difficult problem where the capacity and location expansion planning under uncertainty is considered along a multiscale time horizon. It has been named S-CFLEP. Two variants are considered, namely, the risk neutral (RN) and its counterpart risk management. In this work, the coherent time-consistent expected conditional second-order stochastic dominance risk-averse functional ECSD is considered. Given the time scaling nature of the problem, difficult types of decisions have to be made, namely, strategic and operational ones. It has been proved with empirical validation; first, the importance of having two-stage scenario trees to represent the operational uncertainty, being rooted at the nodes of the multistage strategic scenario tree; second, the efficiency of the step variable modeling object for representing the state variables; and, third, the usefulness of ECSD in order to have a balance between the negative impact of the occurrence of black swan scenarios in the objective function value and its RN minimization. On the other hand, large-scale S-CFLEP instances are difficult to be straightforward solved even by state-of-the-art optimizers, as CPLEX. The matheuristic decomposition algorithm SFR3 has been proved to be very efficient on providing solutions with small optimality gap, and being very close to the CPLEX ones for the instances where the latter gives a solution.

Perspectives of the Current Work The structure of model S-CFLEP (4) can be considered on other types of system or network expansion planning, where risk management should be considered in a multistage multiscale framework. Examples are those policies where the goal is to maximize the expected overall profit in the presence of uncertainty on the main parameters as market uncertainty (product price and demand), strategic and operational costs, facility disruptions due to accidental and non-accidental events, etc. Note that the risk-averse functional considered in this work allows to prevent, up to some extent, solutions that do not consider certain thresholds on other functions as well.

Acknowledgments This research has been partially supported by the projects RTI2018-094269-B-I00 (L.F. Escudero) and PID2019-105952 GB-I00 funded by Ministerio de Ciencia e Innovación/, Agencia Estatal de Investigación/https://doi.org/10.13039/501100011033 (Juan F. Monge).

References

Aghezaaf, E. (2005). Capacity planning and warehouse location in supply chains with uncertain demands. *Journal of Operational Research Society, 56*, 453–462.

Ahmed, S., Goulart Cabral, F., & Freitas Paulo da Costa, B. (2022). Stochastic Lipschitz dynamic programming. *Mathematical Programming, 191*, 755–793.

Albareda-Sambola, M., Alonso-Ayuso, A., Escudero, L. F., Fernández, E., & Pizarro, C. (2013). Fix-and-relax-coordination for a multi-horizon location-allocation problem under uncertainty *Computers and Operations Research, 40*, 2878–2892.
Albareda-Sambola, M., Fernández, E., & Saldanha-da-Gama, F. (2011). The facility location problem with Bernoulli demands. *Omega, 39*, 335–345.
Alonso-Ayuso, A., Escudero, L. F., Garín, A., Ortuño, M. T., & Pérez, G. (2003). A Stochastic 0–1 program based approach for strategic supply chain planning under uncertainty. *Journal of Global Optimization, 26*, 97–124.
Alonso-Ayuso, A., Escudero, L. F., Garín, A., Ortuño, M. T. & Pérez, G. (2005). On the product selection and plant dimensioning problem under uncertainty. *Omega, The International Journal of Management Science, 33*, 307–318.
Alonso-Ayuso, A., Escudero, L. F., Guignard, M., & Weintraub, A. (2020). On dealing with strategic and tactical decision levels in forestry planning under uncertainty. *Computers and Operations Research, 115*, 104836.
Alonso, A., Escudero, L. F. & Ortuño, M. T. (2000). Stochastic 0–1 program based approach for air traffic management. *European Journal of Operational Research, 120*, 47–62.
Alonso-Ayuso, A., Escudero, L. F. & Ortuño, M. T. (2003a). BFC, a Branch-and-Fix Coordination algorithmic framework for solving some types of stochastic pure and mixed 0–1 programs. *European Journal of Operational Research, 151*, 503–519.
Alumur, S. A., Campbell, J. F., Contreras, I., Kara, B. Y., Marianov, V., & O'Kelly, M. E. (2021). Perspectives on modelling hub location problems. *European Journal of Operational Research, 291*, 1–17.
Artzner, P., Delbaen, F., Eber, L., Health, D., & Ku, H. (2007). Coherent multi-period risk adjusted values and Bellman's principle. *Annals of Operations Research, 152*, 5–22.
Baena, D., Castro, J., & González, J. A. (2015). Fix-and-Relax approaches for controlled tabular adjustment. *Computers and Operations Research, 58*, 41–52.
Baptista, S., Barbosa-Povoa, A. P., Escudero, L. F., Gomes, M. I., & Pizarro, C. (2019). On risk management for a two-stage stochastic mixed 0–1 model for designing and operation planning of a closed-loop supply chain. *European Journal of Operational Research, 274*, 91–107.
Basciftci, B., Ahmed, S., & Shen, S. (2021). Distributionally robust facility location problem under decision-dependent stochastic demand. *European Journal of Operational Research, 292*, 548–561.
Boland, N., Christiansen, J., Dandurand, B., Eberhard, A., Linderoth, J., Luedtke, J., & Oliveira, F. (2018). Combining Progressive Hedging with a Frank-Wolfe method to compute Lagrangian dual bounds in stochastic mixed-integer programming. *SIAM Journal on Optimization, 28*, 1312–1336.
Boonmee, Ch., Arimura, M., & Takumi Asada, T. (2017). Facility location optimization model for emergency humanitarian logistics. *International Journal of Disaster Risk Reduction, 24*, 485–498.
Cadarso, L., Escudero, L. F., & Marín, A. (2018). On strategic multistage operational two-stage stochastic 0–1 optimization for the Rapid Transit Network Design problem. *European Journal of Operational Research, 271*, 577–593.
Carpentier, P., Chancelier, J. P., Cohen, G., de Lara, M., & Girardeau, P. (2012). Dynamic consistency for stochastic optimal control problems. *Annals of Operations Research, 200*, 247–263.
Castro, J., Escudero. L. F., & Monge, J. F. (2023). On solving large-scale multistage stochastic optimization problems with a new specialized interior-point approach. *European Journal of Operational Research, 310*, 268–285.
Chen, G., Daskin, M. S., Max-Shen, Z. J., & Uryasev, S. (2006). The a-reliable mean-excess regret model for stochastic facility location modeling. *Naval Research Logistics, 53*, 617–626.
Conde, E., & Leal, M. (2021). A robust optimization model for distribution network design under a mixed integer set of scenarios. *Computers and Operations Research*. https://doi.org/10.1016/j.cor.2021.105493.

Correia, I., & Melo, T. (2021). Integrated facility location and capacity planning under uncertainty. *Computational and Applied Mathematics, 40*, 175.

Correia, I., & Saldanha-da-Gama, F. (2019). Facility location under uncertainty. In G. Laporte, S. Nickel, & F. Saldanha-da-Gama (Eds.), *Location science* (pp. 185–213, 2nd ed.). Springer.

Crainic, T. G., Gendreau M., & Gendron B. (Eds.) (2021). *Network Design with Applications to Transportation and Logistics*. Springer.

Current, J., Ratick, S., & ReVelle, C. (1998). Dynamic facility location when the total number of facilities is uncertain: A decision analysis approach. *European Journal of Operational Research, 110*, 597–609.

Dehghan, M., Hejazil, S. R., Karimi-Mamaghan, M., Mohammadi, M., & Pirayesh, A. (2021). Capacitated location routing problem with simultaneous pick and delivery under the risk of disruption. *RAIRO Operations Research, 55*, 1371–1399.

Dillenberger, Ch., Escudero, L. F., Wollensak, A., & Zhang, W. (1994). On practical resource allocation for production planning and scheduling with period overlapping setups. *European Journal of Operational Research, 75*, 275–286.

Escudero, L. F., Garín, M. A., Monge, J. F. & Unzueta, A. (2018). On preparedness resource allocation planning for natural disaster relief under endogenous uncertainty with time-consistent risk-averse management. *Computers & Operations Research, 88*, 84–102.

Escudero, L. F., Garín, M. A., Monge, J. F. & Unzueta, A. (2020). On multistage stochastic mixed 0–1 bilinear optimization based on endogenous uncertainty and time consistent stochastic dominance risk management. *European Journal of Operational Research, 285*, 988–1001.

Escudero, L. F., Garín, M. A., Pizarro, C., & Unzueta, A. (2018a). On efficient matheuristic algorithms for multi-period stochastic facility location-assignment problems. *Computational Optimization and Applications, 70*, 865–888.

Escudero, L. F., Garín, A., & Unzueta, A. (2017). Cluster Lagrangean decomposition for risk averse in multistage stochastic optimization. *Computers & Operations Research, 85*, 154–171.

Escudero, L. F., & Monge, J. F. (2018). On capacity expansion planning under strategic and operational uncertainties based on stochastic dominance risk averse management. *Computational Management Science, 15*, 479–500.

Escudero, L. F., & Monge, J. F. (2021). On multistage multiscale stochastic capacitated multiple allocation hub network expansion planning. *Mathematics, 9*, 3177.

Escudero, L. F., Monge, J. F., & Rodríguez-Chía, A. M. (2020a). On pricing-based equilibrium for network expansion planning. A multi-period bilevel approach under uncertainty. *European Journal of Operational Research, 287*, 262–279.

Escudero, L. F., Monge, J. F., & Romero-Morales, D. (2018b). On time-consistent stochastic dominance risk averse measure for tactical supply chain planning under uncertainty. *Computers & Operations Research, 100*, 270–286.

Escudero, L. F., & Pizarro, C. (2017). On solving a large-scale problem on facility location and customer assignment with interaction costs along a time horizon. *TOP, 25*, 601–622.

Escudero, L. F., & Salmerón, J. (2005). On a Fix-and-Relax framework for large-scale resource-constrained project scheduling. *Annals of Operations Research, 140*, 163–188.

Gade, D., Hackebeil, G., Ryan, S. M., Watson, J.-P., Wets, R.J.-B., & Woodruff, D. L. (2016). Obtaining lower bounds from the Progressive Hedging Algorithm for stochastic mixed-integer programs. *Mathematical Programming, 157*, 47–67.

Gago, I., Aldasoro, U., Ceberio J., & Merino, M. (2022). A stochastic optimization model for ambulance (re)location-allocation under equitable coverage and multi-layer response time. SSRN. https://ssrn.com/abstract=4283397.

Glanzer, M., & Pflug, G. C. (2020). Multiscale stochastic optimization: modeling aspects and scenario generation. *Computational Optimization and Applications, 75*, 1–34.

Goodarzi, A. H., Zegordi, S. H., Alpan, G., Kamalabadi, I. N., & Kashan, A. H. (2020). Reliable cross-docking location problem under the risk of disruptions. *Operational Research*. https://doi.org/10.1007/s12351-020-00583-5.

Gourtani, A., Nguyen, T.-D., & Xu, H. (2020). A distributionally robust optimization approach for two-stage facility location problems. *EURO Journal on Computational Optimization, 8*, 141–172.

Heitsch, H., & Römisch, W. (2009). Scenario tree reduction for multistage stochastic programs. *Computational Management Science, 6*, 117–133.

Henrion, H., & Römisch, W. (2022). Problem-based optimal scenario generation and reduction in stochastic programming. *Mathematical Programming, 191*, 347–380.

Hernández, P., Alonso-Ayuso, A., Bravo, F., Escudero, L. F., Guignard, M., Marianov, V., & Weintraub, A. (2012). Prison facility site selection under uncertainty. *Computers & Operations Research, 29*, 2232–2241.

Homem-de-Mello, T., & Pagnoncelli, B. K. (2016). Risk aversion in multistage stochastic programming: A modeling and algorithmic perspective. *European Journal of Operational Research, 249*, 188–199.

Ivanov, S. V., & Akmaeva V. N. (2021). Two-stage stochastic facility location model with quantile criterion and choosing reliability level. *Vestnik YuUrGU Seriya Matematicheskoe Modelirovanie i Programmirovanie, 21*, 5–17.

Kaut, M., Midthun, K. T., Werner, A. S., Tomasgard, A., Hellemo, L., & Fodstad, M. (2014). Multi-horizon stochastic programming. *Computational Management Science, 11*, 179–193.

Leövey, H., & Römisch, W. (2015). Quasi-Monte Carlo methods for linear two-stage stochastic programming. *Mathematical Programming, 151*, 314–345.

Liu, K., Li, Q., & Zhang, Z. H. (2019). Distributionally robust optimization of an emergency medical service station location and sizing problem with joint chance constraints. *Transportation Research Part B: Methodological, 119*, 79–101.

Li, Z., & Floudas, Ch. (2016). Optimal scenario reduction framework based on distance of uncertainty distribution and output performance: II. Sequential reduction. *Computers and Chemical Engineering, 84*, 599–610.

Louveaux, F. V. (1993). Stochastic location analysis. *Location Science, 1*, 127–154.

Maggioni, F., Allevi, E., & Tomasgard, A. (2020). Bounds in multi-horizon stochastic programs. *Annals of Operations Research, 292*, 605–625.

Marín, A., Martínez, L. I., Rodríguez-Chía, A. M., & Saldanha-da-Gama, F. (2018). Multiperiod stochastic covering location problems: Modelling framework and solution approaches. *European Journal of Operational Research, 268*, 432–449.

Mendoza-Ortega, G. P., Soto M., Ruiz-Meza J., Salgado R., & Torregroza A. (2021). Scenario-based model for the location of multiple uncapacitated facilities: case study in an agro-food supply chain. In J. C. Figueroa-García, Y. Díaz-Gutierrez, E. E. Gaona-García, & A. D. Orjuela-Cañón (Eds.). *Applied Computer Sciences in Engineering. Communications in Computer and Information Science* (Vol. 1431). Springer.

Mohammadi, M., Jula, P., & Tavakkoli-Moghaddam, R. (2019). Reliable single-allocation hub location with disruptions. *Transportation Research Part E: Logistics & Transportation Review, 62*, 89–115.

Mousavi, S. M., Behnam Vahdani, B., Tavakkoli-Moghaddam, R., & Hashemi, H. (2014). Location of cross-docking centers and vehicle routing scheduling under uncertainty: A fuzzy possibilistic–stochastic programming model. *Applied Mathematical Modeling, 38*, 2249–2264.

Nickel, S., Saldanha-da-Gama, F., &. Ziegler, H. P. (2012). A multi-stage stochastic supply network design problem with financial decisions and risk management. *Omega, 40*, 511–524.

Noyan, N. (2012). Risk-averse two-stage stochastic programming with an application to disaster management. *Computers & Operations Research, 39*, 541–559.

Ntaimo, L., & Sen, S. (2005). The million-variable 'march' for stochastic combinatorial optimization. *Journal of Global Optimization, 32*, 385–400.

Pagès-Bernaus, A., Ramalhinho, H., Juan, A. A., & Calvet, L. (2019). Designing e-commerce supply chains: A stochastic facility-location approach. *International Transactions in Operational Research, 26*, 507–528.

Pflug, G. Ch., & Pichler, A. (2014). *Multistage stochastic optimization*. Springer.

Pflug, G. Ch., & Pichler, A. (2015). Dynamic generation of scenario trees. *Computational Optimization and Applications, 62*, 641–668.

Quezada, F., Gicquel, C., & Kedad-Sidhoum, S. (2020). Combining polyhedral approaches and stochastic dual dynamic integer programming for solving the uncapacitated lot-sizing problem under uncertainty. 2020. hal-02868707.

Rahmaniani, R., Crainic, T. G., Gendreau, M., & Rey, W. (2018). Accelerating the Benders Decomposition method: Application to stochastic network design problem *SIAM Journal on Optimization, 28*, 875–903.

Ravi, R., & Sinha, A. (2006). Hedging uncertainty: Approximation algorithms for stochastic optimization problem. *Mathematical Programming, 108*, 97–114.

Rawls, C. G., & Turnquist, M. A. (2012). Pre-positioning and dynamic delivery planning for short-term response following a natural disaster. *Socio-Economic Planning Sciences, 46*, 46–54.

Rockafellar, R. T., & Wets, R.J-B. (1991). Scenario and policy aggregation in optimisation under uncertainty. *Mathematics of Operations Research, 16*, 119–147.

Ryu, J., & Park, S. (2021). A branch-and-price algorithm for the robust single-source capacitated facility location problem under demand uncertainty. arXiv, 2103-13010v1.

Saif, A., & Delage, E. (2021). Data-driven distributionally robust capacitated facility location problem *European Journal of Operational Research, 291*, 995–1007.

Sen, S., Higle, J. L. & Ntaimo, L. (2002). A summary and illustration of disjunctive decomposition with set convexification. In D. L. Woodruff (ed.) *Stochastic integer programming and network interdiction models* (pp. 105–125). Kluwer Academic Press.

Snyder, L. V. (2006). Facility location under uncertainty: A review. *IIE Transactions, 38*, 537–554.

Soanpet, A. (2012). Optimization models for locating cross-docks under capacity uncertainty. Graduate Theses, Dissertations, and Problem Reports. 582, West Virginia University, VI, USA. https://researchrepository.wvu.edu/etd/582.

Taghavi, N., & Huang, K. A. (2020). A Lagrangian relaxation approach for stochastic network capacity expansion with budget constraints. *Annals of Operations Research, 284*, 605–621.

Valtsa, A. K., & Jayaswal, S. (2021). Capacitated multi-period maximal covering location problem with server uncertainty. *European Journal of Operational Research, 289*, 1107–1126.

Wang, W., Wu, S., Wang, S., Zhen, L., & Qu, X. (2021). Emergency facility location problems: Status and perspectives. *Transportation Research-E: Logistics & Transportation Review, 154*, 102465.

Wang, Z., You, K., Wang, Z., & Liu, K. (2021a). Multi-period facility location and capacity planning under ∞-Wasserstein joint chance constraints in humanitarian logistics. arXiv2111.15057.

Werner, A. S., Pichler, A., Midthun, K. T., Hellemo, L., & Tomasgard, A. (2013). Risk measures in multihorizon scenarios tree. In R. Kovacevic, G. Ch. Pflug, & Vespucci, M. T. (Eds.). *Handbook of risk management in energy production and trading* (pp. 177–201). Springer.

Wets, R.J.-B. (1966). Programming under uncertainty: The equivalent convex program. *SIAM Journal on Applied Mathematics, 14*, 89–105.

Yu, G., & Zhang, J. (2018). Multi-dual decomposition for risk-averse facility location problem. *Transportation Research-E: Logistics & Transportation Review, 116*, 70–89.

Yu, X., Siqian Shen, S., & Ahmed, S. (2021). On the value of multistage stochastic facility location with risk aversion. arXiv2105.11005.

Zhu,T., Boyles, S. D., & Unnikrishnan, A. (2022). Two-stage robust facility location problem with drones. *Transportation Research Part C, 137*, 103563.

Zou, J., Ahmed, S., & Sun, X. A. (2019). Stochastic dual dynamic integer programming. *Mathematical Programming, 175*, 461–502.

Some Heuristic Methods for Discrete Facility Location with Uncertain Demands

Maria Albareda-Sambola, Elena Fernández, and Francisco Saldanha-da-Gama

Abstract In this chapter, we consider discrete facility location problems with uncertainty and focus on the case where the service demand of each customer follows a Bernoulli distribution. This problem can be modeled as a two-stage stochastic programming problem where the first stage determines a set of facilities to open together with a tentative allocation of customers to open facilities, and the second stage builds the actual assignment of customers to open plants for each possible realization of the customers' demands. The objective is to minimize the sum of the cost of the first-stage decision plus the expected cost of the recourse action. Given that, in practice, the exact evaluation of the recourse function becomes computationally unaffordable, we illustrate the application of possible heuristics. We discuss GRASP and Path Relinking as the building blocks of a heuristic solution method for the considered problem. We also present mathematical programing formulations for the case where uncertainty is expressed by means of a given set of scenarios, which can be embedded in a Sample Average Approximation algorithm. Numerical results from computational experiments are discussed and analyzed.

Keywords Discrete facility location · Uncertainty · Bernoulli demand · Heuristics

M. Albareda-Sambola
Universitat Politècnica de Catalunya, Department d'Estadística i Investigació Operativa, Terrassa, Spain
e-mail: maria.albareda@upc.edu

E. Fernández (✉)
Universidad de Cádiz, Departamento de Estadística e Investigación Operativa, Puerto Real, Spain
e-mail: elena.fernandez@uca.es

F. Saldanha-da-Gama
Universidade de Lisboa, Faculdade de Ciências, Department Estatística e Investigação Operacional e Centro de Investigação Operacional, Lisboa, Portugal
e-mail: faconceicao@fc.ul.pt

H. A. Eiselt, V. Marianov (eds.), *Uncertainty in Facility Location Problems*, International Series in Operations Research & Management Science 347, https://doi.org/10.1007/978-3-031-32338-6_15

1 Introduction

Discrete facility location problems (FLPs) determine one fundamental class of problems within location analysis (see, e.g., Eiselt & Marianov, 2011; Daskin, 2013; Laporte et al., 2019). Broadly speaking, input data consists of (i) a discrete set of potential locations for the facilities, with associated setup costs and capacities, (ii) a set of customers with associated demands, and (iii) transportation costs between the potential facilities and the customers. The goal is to establish what facilities to open and how to allocate the customers demand to open facilities so as to minimize the sum of setup plus transportation costs.

One common characteristic of many FLPs is the strategic nature of the location decisions, which should be *long-lasting*, in the sense that the selected facilities should operate for long periods of time. Since the external circumstances are likely to change during the planning horizon, but their actual evolution is typically unknown at the moment when decisions have to be made, very often uncertainty is present in FLPs. Uncertainty may involve setup costs, travel times and costs, availability of supply, demands, etc. Uncertainty may even affect the underlying setting of the problem, namely, the set of potential facilities or available connections between facilities and customers.

Under uncertainty, the use of a classical deterministic model for FLP may produce a solution that is no longer optimal or is even infeasible, when it has to be implemented. Thus, it may be necessary to build a new solution from scratch. It is thus better to model FLPs with uncertainty as two-stage problems (see, e.g., Birge & Louveaux, 2011; Klein Haneveld et al., 2020) in which the first-stage decision is to select a set of facilities (plants) to open together with a *tentative* (*a priori*) allocation of customers within the set of selected facilities. The second-stage solution is guided by the *recourse function*, which modifies the solution built in the first stage in order to render it feasible (or cheaper) once the uncertainty is revealed.

When no probabilistic information on the random parameters is available, the possible realizations of uncertain parameters are expressed by a set of scenarios. In principle, when probability information on the random parameters is available, uncertainty could be described using the corresponding probability distributions. Unfortunately, even when probabilistic information is available, it is most often the case that no algebraic expressions are available representing the involved probability distributions; thus, the probabilistic information cannot be embedded within a mathematical optimization model that can be handled with off-the-shelf solvers. In such a case, uncertainty must also be handled by means of a suitable set of scenarios.

As pointed out in Snyder (2006), an advantage of working with scenarios is that it allows more flexibility for modeling uncertain parameters. On the other hand, working with scenarios involves two main types of difficulties. One is to identify an appropriate set of scenarios and to assign suitable probabilities to them. The second is to find a trade-off between the number of scenarios (which for computational purposes should be relatively small) and the size of the set of decision states that are actually evaluated, which should represent sufficiently the whole search space. Moreover, even if the number of scenarios is relatively small, the computational burden for evaluating solutions with sufficient accuracy may be too high so as to resort to exact solution methods, and it can be the case that heuristic solution methods are the only realistic alternative.

In FLPs, very often the source of uncertainty is driven by the motivating application. Thereby, for problems focusing on potential applications related to natural disasters (Dönmez et al., 2021), it can be very hard to estimate in advance (i) what potential locations will be available for emergency facilities, (ii) what areas will require humanitarian relief, or (iii) what connections will be available to reach the damaged areas from the installed facilities. Similar sources of uncertainty emerge in FLPs arising in supply chain management aiming at reducing vulnerability due to disruptions (see, e.g., Snyder & Daskin, 2005). In the above contexts, it seems suitable to adopt a conservative robust optimization approach, focusing on the worst-case performance of the system by looking for first-stage solutions that are robust against all possible realizations (scenarios) in the second stage. Instead, the vulnerability of other FLPs arising in supply chain management may stem from seasonal demand or fluctuations in commodity prices, which do not have a direct effect on the transportation network to be used, but may impact strongly on location and allocation decisions. In such contexts, it seems suitable to consider an optimization model looking for first-stage solutions that minimize the expected value over all possible realizations in the second stage.

In this chapter, we illustrate the application of possible heuristic approaches for this latter type of FLPs with uncertainty using the facility location problem with Bernoulli Demand (FLBD). The FLBD is an FLP in which the only source of uncertainty is demand. The capacity of the facilities is expressed in terms of the maximum number of customers that they can actually serve, and it is assumed that the service demand of each customer follows a Bernoulli distribution. The FLBD can be modeled as a two-stage stochastic programming problem: the first-stage decision is to select a set of facilities to open together with a tentative *a priori* allocation of customers to open facilities, and the second stage builds the actual assignment of customers to open plants for each possible realization of the customers' demands. The objective is to minimize the sum of the cost of the first-stage decision plus the expected cost of the recourse action.

The FLBD models situations where a facility provides a service and demand refers to whether a customer requires to be served. Companies providing repair or maintenance services fit within this modeling framework. In this case, customers can be grouped (e.g., according to their location) and assigned to a facility that should handle the existing demand. The term "facility" should be considered in a generic way. For instance, it may refer to a worker or a team. Moreover, facilities may be mobile in the sense that service can be provided at the customer's locations. One particular example is elevator maintenance: each repair or maintenance team must assist (*serve*) a prespecified set of customers in case they call for service. If the actual demand turns out to be higher than the service capacity, then the service still has to be provided. This may call for temporary relocation of workers from other teams or simply for outsourcing the service to a third party. Mobile health clinics provide another potential application of the FLBD. In this type of application, "facilities" assist some specific area or region previously assigned to them. In case the occurring demand is higher than the service capacity, extra personnel is necessary, which may incur additional costs. Other examples of settings fitting the FLBD include target-oriented advertisement activities, door-to-door product demonstration, etc. In all these cases, potential customers are previously assigned to the facility and may or may not have actual demand.

We address a general version of the FLBD, where demand probabilities are not necessarily the same for all customers. Such assumption often holds in practice and reflects the fact that different customers usually have different demands. We further assume that customer's demands are uncorrelated. This is also a natural assumption, which holds when customers do not obey to some *common interest* and demand is not seasonal. Unfortunately, the generality gained when assuming that different customers may have different probabilities of demand comes at the expenses of some additional difficulties. To the best of our knowledge, it is not possible to express the joint probability distribution of the customer's demand by means of algebraic expressions that can be handled with off-the-shelf solvers. Then the exact evaluation of the recourse function (the expected costs of service plus outsourcing), essentially amounts to enumerate, for each open facility, all possible subsets of the set of customers assigned to it in the *a priori* solution that actually have demand. Such an enumeration becomes computationally unaffordable. Then either heuristic methods are used in which the value of solutions is *estimated*, for instance, with Monte Carlo sampling, or the use of optimization models is restricted to those in which uncertainty is expressed by means of a given set of scenarios. In this chapter, we discuss both possibilities although the latter is focused within the context of a Sample Average Approximation (SAA) algorithm. We also summarize the GRASP and Path Relinking heuristic proposed in Albareda-Sambola et al. (2017).

The chapter is organized as follows. Section 2 gives a short overview of some of the existing literature on problems related to the FLBD, and in Sect. 3, we introduce some notation that we will use and formally define the FLBD. In Sect. 4, we give a generic introduction to GRASP and Path Relinking, which are the building blocks of the FLBD heuristic of Albareda-Sambola et al. (2017) that is presented in Sect. 5. Section 6 focuses on the two outsourcing policies introduced in Albareda-Sambola et al. (2011) and presents the mathematical programming formulations for the case where uncertainty is expressed by means of a given set of scenarios. These formulations will be embedded in the SAA algorithm that can be used for problems in which uncertainty is expressed in a probabilistic way, which is presented in Sect. 7. Section 8 is dedicated to the computational experiments. The sets of test instances that we have used and their characteristics are described in Sect. 8.1 and some relevant implementation details in 8.2. The results of the GRASP + PR heuristic in Sect. 8.3, and those of the SAA algorithm, are presented in Sect. 8.4. The chapter ends in Sect. 9 with some conclusions and final comments.

2 Some Related Literature

There are several alternatives for addressing location problems with uncertainty in service demands, depending on their nature. Queuing location models have been used to model situations where the goal is to optimize the system performance (e.g., Marianov & ReVelle, 1996; Carrizosa et al., 1998; Fernández et al., 2005). Several references can be found where a certain level of service is guaranteed by means of probabilistic constraints (see, e.g., Toregas et al., 1971; Beraldi & Bruni, 2009).

Examples of uncertain location problems where robustness is measured by the cost associated with the most adverse scenario are Averbakh and Berman (1997, 2000, 2003), Carrizosa and Nickel (2003), Conde (2007), and Wagner et al. (2009) or, more recently, Alvarez-Miranda et al. (2015). Two-stage models for uncapacitated FLPs considering the minimization of the expected cost of the recourse function have been studied in Louveaux and Peeters (1992) and Laporte et al. (1994a).

Heuristic Methods for FLPs with Uncertainty Despite of the difficulty of FLPs with uncertainty, the literature on heuristic methods for problems of this type is scarce. Two examples are mentioned next.

Albareda-Sambola et al. (2013) propose a Fix-and-Relax-Coordination matheuristic for a multi-period location/allocation problem under uncertainty. Uncertainty is assumed on the costs and on some of the requirements along the planning horizon. A compact 0–1 formulation is proposed for the deterministic equivalent counterpart of the problem under two alternative strategies for the location decisions. The results of an extensive computational experience allow to compare the alternative modeling strategies and assess the effectiveness of the proposed approach versus the plain use of an off-the-shelf solver. Pagès-Bernaus et al. (2019) present two facility–location models to represent two alternative distribution policies in e-commerce (one based on outsourcing and another one based on in-house distribution). The authors propose two-stage mathematical-programming models and show that, because of the computational effort they involve, they are not useful for solving large-scale instances. Furthermore, a *simheuristic* is also introduced, to deal with large-scale instances in short computing times. Extensive computational experiments on benchmark instances illustrate the efficiency of the simheuristic.

More recently, Turkẽs et al. (2021) develop a matheuristic to solve a stochastic facility location problem under uncertainty. The latter is triggered by demands inventory spoilage and transportation network availability. The problem aims at determining the location and size of storage facilities, the quantities of various types of supplies stored in each facility, and the assignment of demand locations to the open facilities, which minimize unmet demand and response time in lexicographic order. The proposed matheuristic resorts to iterated local search to look for good location and inventory configurations, whereas optimal assignments for the selected configurations are obtained by means of a mathematical programming formulation.

Stochastic Combinatorial Problems with Bernoulli Demand Several stochastic combinatorial optimization problems with Bernoulli demands have been addressed in the literature in the context of the minimization of a recourse function. This is the case of the probabilistic traveling salesman problem. Jaillet (1988) introduces this problem and presents a closed expression for computing efficiently the expected value of the length of any given tour. Laporte et al. (1994b) propose a linear stochastic program, which is solved with a branch-and-cut approach. More recently, Bianchi and Campbell (2007) propose a heuristic approach for the same problem. The reader is referred to this latter work for additional references on this problem. Berman and Simchi-Levi (1988) study a single-vehicle location-routing problem with Bernoulli demands. The authors formulate the problem and develop a lower bound on the value of the optimal *a priori* tour. In Albareda-Sambola et al. (2007) study a location-routing problem with Bernoulli demands. The authors propose heuristics and lower bounds to minimize the expected value of the defined recourse function. The Stochastic Generalized Assignment Problem, where it is assumed that the customer's demands follow a Bernoulli distribution, is studied in Albareda-Sambola et al. (2006), where the authors propose an exact algorithm for minimizing the expected cost of a recourse function.

Unit-Demand FLPs Deterministic FLPs with unit-demand customers have been widely studied in the literature, motivated by different types of applications mostly in telecommunications (e.g., Fortz, 2015) and healthcare (e.g., Ahmadi-Javid et al., 2017). To the best of our knowledge, the stochastic counterparts of these problems have yet received little attention (Albareda-Sambola et al., 2011, 2017; Bieniek, 2015). Still, examples of non-deterministic unit demands abound in logistics-related discrete location problems where demand levels might change over time (postal services, distribution systems of goods with seasonal demand, airports, etc.)

The FLBD and Extensions The FLBD was motivated and introduced in Albareda-Sambola et al. (2011) for the particular case when the distributions of the customers' demands are independent, all with the same demand probability. Such an assumption reflects situation where there are no common underlying reasons for the change of demand, for example, seasonal, economic up- or downturns. In that work, two different outsourcing policies were considered, and closed forms for the corresponding recourse functions were presented. The obtained numerical results showed that the proposed methodology was computationally highly demanding as the sizes of the instances increased. In Albareda-Sambola et al. (2017), the same authors addressed the heterogeneous version of the problem where the demand probabilities are not necessarily the same, for the same two outsourcing policies, again under the assumption of independent demands. Several heuristics, based on GRASP and Path Relinking, were proposed for that case. In Albareda-Sambola et al. (2022), the previous work on the FLBD was extended in several ways. First, it was no longer assumed that the probability distributions of the customers' demands are independent. Second, the set of outsourcing policies considered in Albareda-Sambola et al. (2011, 2017) was extended with two additional strategies. Finally, an empirical comparison among the considered outsourcing policies was carried out in terms of both their computational performance as well as their capability of producing good quality solutions for the other policies.

The FLBD and some of its variants have attracted the attention of other researchers. A bi-objective version of the FLBD has been considered in Shiripour and Mahdavi-Amiri (2019). A recourse function is considered, which includes a penalty for unmet demand (instead of outsourcing the deficit of capacity). The first objective is to minimize the sum of the setup costs plus the expected cost of the recourse function. The second objective is to balance the number of customers allocated to activated facilities. Small problems are solved with the augmented ε-constraint method, whereas two metaheuristic solution algorithms are proposed for solving large problems.

Bieniek (2015) studies an FLP with uncertainty in demand, which extends the FLBD for the case of independent demands, as they are assumed to be independent and identically distributed with arbitrary distribution. The author studies the recourse functions for two outsourcing policies. In each case, a closed expression is given for the recourse function as well as a deterministic equivalent formulation. Numerical results from some computational experiments are also presented.

3 Definition of the Problem

Let I and J, with $n = |J|$, denote the set of indices for the potential locations of facilities and for customers, respectively. We assume that the demands for service of customers follow independent Bernoulli probability distributions, with probabilities p_j, $j \in J$. We denote by Ω the set of all possible scenarios, by π^ω the probability of scenario ω ($\sum_{\omega\in\Omega} \pi^\omega = 1$), and by $d_j^\omega \in \{0, 1\}$ the demand of customer $j \in J$ in scenario $\omega \in \Omega$. Since d_j^ω takes binary values, $D^\omega = \sum_{j\in J} d_j^\omega$ indicates the number of customers with (non-zero) demand in scenario ω. Following the terminology introduced in Albareda-Sambola et al. (2011), such customers will be referred to as *customers with demand* or just as *demand customers*.

We have the following additional data. For each potential location $i \in I$, f_i is the fixed setup cost for opening facility i; ℓ_i is a lower bound on the number of customers that have to be *assigned* to facility i if it is opened; and K_i is the maximum number of customers that can be *served* from facility i if it is opened. For each pair $i \in I$, $j \in J$, c_{ij} is the cost for serving customer j from facility i.

For a given scenario $\omega \in \Omega$, not all the customers need to have demand. Hence, we distinguish between the *assignment* of customers to open plants, which is done *a priori* and is independent of the potential realizations, and the *service* of customers from open plants, which is decided *a posteriori*, once the realization is known. An *a priori* solution is given by a set of *operating* (open) facilities together with an assignment of all the customers to these facilities, such that for any open plant i, the number of customers that are assigned to it is at least ℓ_i. Since K_i is an upper bound on the number of customers that can be served from an open plant, it does not affect the feasibility of *a priori* solutions. Let $i(j) \in I$ denote the facility to which customer $j \in J$ is assigned in the *a priori* solution and $J_i = \{j \in J : i(j) = i\}$, the set of customers assigned to facility i in the *a priori* solution.

Given an *a priori* solution, the *a posteriori* solution indicates the decisions to make once demand customers are known, that is, it describes the actual services to demand customers. Let $J_i^\omega = J_i \cap \{j \in J : d_j^\omega = 1\}$ denote the set of customers assigned to facility $i \in I$ with demand in scenario ω, and $\eta_i^\omega = |J_i^\omega|$ the number of such customers. If $\eta_i^\omega \leq K_i$, then in the *a posteriori* solution all customers indexed in J_i^ω receive service from plant i, each of them incurring a service cost c_{ij}, $j \in J$. Instead, when $\eta_i^\omega > K_i$, the *a posteriori* solution consists of serving K_i (out of η_i^ω) demand customers from facility i and outsourcing to some third party the remaining $\eta_i^\omega - K_i$. A penalty cost g_i is incurred for every outsourced demand customer. The way in which, for a realization, it is decided whether a demand customer assigned to a plant with $\eta_i^\omega > K_i$ is actually served from i or outsourced depends on the outsourcing policy that is applied (see Sect. 6). The recourse function is the expected cost of the *a posteriori* solution, over all possible realizations of the demand vector.

The FLBD consists of finding a set of facilities to open and an allocation of the customers to the opened facilities, such that the lower bounds ℓ_i are satisfied, and the sum of the fixed cost associated with the open facilities and the recourse function is minimized.

To formulate the FLBD, we define two sets of decision variables: For $i \in I$, y_i is a binary variable equal to one if and only if facility i is established; x_{ij} is a binary variable indicating whether costumer $j \in J$ is assigned (*a priori*) to facility $i \in I$.

The generic formulation for the FLBD proposed in Albareda-Sambola et al. (2011) is:

$$(P) \quad \min \quad \sum_{i \in I} f_i y_i + \mathcal{Q}(x), \tag{1}$$

$$s.t. \quad \sum_{i \in I} x_{ij} = 1, \qquad j \in J, \tag{2}$$

$$x_{ij} \leq y_i, \qquad i \in I,\ j \in J, \tag{3}$$

$$\ell_i y_i \leq \sum_{j \in J} x_{ij}, \qquad i \in I, \tag{4}$$

$$y_i \in \{0, 1\}, \qquad i \in I, \tag{5}$$

$$x_{ij} \in \{0, 1\}, \qquad i \in I,\ j \in J, \tag{6}$$

where the recourse function is $\mathcal{Q}(x) = \mathbb{E}\left[\text{Service cost} + \text{Penalty cost}\right]$. Constraints (2) assure that all customers will be assigned to (exactly) one facility while constraints (3) impose that these assignments are only done to operating facilities.

Constraints (4) state the minimum number of customers that must be assigned to each operating facility. Finally, (5)–(6) define the domain of the variables. The objective function (1) includes the fixed costs for opening the facilities and the recourse function. Needless to say, the precise definition of this function will depend on the outsourcing policy adopted. The choice of the outsourcing policy is a very relevant issue in these problems as it determines the criterion according to which the quality of *a priori* solutions will be evaluated. In particular, the choice of the outsourcing policy may have a notable impact on the specific location/allocation selections, given that different outsourcing policies may lead to different solutions. This issue will be further discussed in Sect. 6.

4 Two Well-Known Heuristics: GRASP and Path Relinking

In this section, we describe the main elements of GRASP and Path Relinking, two well-known heuristics, that we use as the building blocks of the heuristic solution algorithm for the FLBD that will be presented in Sect. 5.

GRASP is the acronym for *Greedy Randomized Adaptive Search Procedure* and was introduced in Feo and Resende (1995). Despite its simplicity, it has proven to be highly effective for different classes of difficult optimization problems. GRASP is an iterative procedure that combines at each iteration the two essential features of any heuristic method: a constructive phase followed by an improvement phase. The constructive phase builds a solution from scratch by incorporating step-by-step additional elements to the solution under construction. A randomized greedy criterion is used at each step to select the element to be incorporated in the current partial solution. This combines the rationale of a greedy search with the effect of a (partial) randomization, thus overcoming one of the major drawbacks of pure greedy methods by diversifying the outcome of the construction phase and (possibly) producing different solutions when the process is repeated. Typically, the improvement phase is a local search where one (or more) neighborhood is explored leading to a local optimum.

Algorithm 1 presents a template for a generic GRASP. The greedy criterion is *measured* through a given function φ and a Restricted Candidate List (RCL) is used at each step, which contains the elements that potentially could enter the solution under construction. In particular, $RCL = \{j \notin S : \varphi_j \leq \varphi_{min} + \alpha\,(\varphi_{max} - \varphi_{min})\}$, where (i) S denotes the current partial solution, (ii) φ_{min} and φ_{max} are, respectively, the best and worse values, relative to the greedy function φ, of the elements that do not belong to the partial solution S, and (iii) the parameter $0 \leq \alpha \leq 1$ determines the level of randomization of the search as it regulates the size of the RCL. Note that $\alpha = 0$ leads to a purely greedy method and $\alpha = 1$ to a totally randomized one. The current solution S is *extended* at each step by adding to it a new element randomly chosen from RCL.

The local search aims at iteratively improving the current solution by exploring a (prespecified) neighborhood until no further improvement can be obtained.

Algorithm 1 `GRASP Template`(α)

`stop` $\leftarrow$ `false`;
while (not `stop`) **do**
// Constructive Phase
    $S \leftarrow \emptyset$;
    while (S not solution) **do**
        Identify φ_{min} and φ_{max} as the best and worse values of the elements not in S;
        RCL $\leftarrow \{j \notin S : \varphi_j \leq \varphi_{min} + \alpha\,(\varphi_{max} - \varphi_{min})\}$;
        Randomly select $j^* \in RCL$;
        $S \leftarrow S \cup \{j^*\}$;
// Improvement Phase: Local Search on a neighborhood N
    while (not `stop`) **do**
        Select $S' \in N(S)$;
        if ($f(S') < f(S)$) **then**
            $S \leftarrow S'$;
        else
            `stop`$\leftarrow$`true`;
    Update best solution;

Path Relinking (PR) (see, e.g., Glover, 1997; Glover & Laguna, 1997) is a generalization of Scatter Search (Glover et al., 2000). Similarly to other evolutionary methods, it operates with a *reference set (RS)* obtained from a *pool* of solutions rather than with a single solution at a time. The initial RS can be obtained in multiple ways, and any procedure able to produce high-quality solutions together with *diverse* solutions can be used to generate it. Later in this chapter, we will use GRASP for this purpose. PR creates new solutions from paths connecting pairs of solutions of RS. To generate a path from a source solution S_c to a target one, S_t, it is only necessary to perform moves that progressively introduce in S_c attributes of S_t. It may be needed to modify the intermediate solutions of the paths so as to make them feasible. The feasible solutions are then improved with some intensification method and the RS updated according to the new solutions obtained. The procedure is repeated until the RS does not change. A template for a generic PR procedure is presented in Algorithm 2.

Indeed, Algorithms 1 and 2 are very general and can be applied nearly to any optimization problem, independently of whether or not it incorporates uncertain elements. The main difficulty that may arise in the case of problems with uncertainty is to have procedures for evaluating the objective function value of the different solutions that are tested at the different steps of the heuristic. For two-stage stochastic problems with recourse function, the exact evaluation of an *a priori* solution would involve to identify the *a posteriori* solution for each possible scenario, evaluate its corresponding value, and then compute the expectation over all these values. Since the computational burden of such an enumeration

Algorithm 2 `PR Template`

Generate a starting reference set RS;
`stop` ← `false`
while (not `stop`) **do**
    Select S_c, $S_t \in RS$;
        // Create new solutions from a path connecting S_c and S_t;
    while $S_c \neq S_t$ **do**
        $S \leftarrow$ `next`(S_c);   //Obtain S introducing in S_c some attribute of S_t
        if (S not feasible) **then**
            $S \leftarrow$ `feas`(S_c);   //Apply a restoration mechanism to make S feasible.
        $S \leftarrow$ `intens`(S_c);   //Apply intensification method to S.
        $S_c \leftarrow S$;
        Check if RS can be updated;
    if (RS has not been updated) **then**
        `stop`← `true`;

may be unaffordable, the exact evaluation of solutions must be substituted by some approximate evaluation based, for instance, on Monte Carlo sampling. In its turn, this may distort the performance of the heuristics, due to the lack of precision of the considered approximations. Finding a trade-off between these two difficulties increases substantially the difficulty (and reduces the reliability) of heuristic methods when applied to optimization problems with uncertainty as the FLBD.

5 GRASP with Path Relinking for the FLBD

In this section, we present the main elements of the two-phase heuristic for obtaining an *a priori* solution for the FLBD proposed in Albareda-Sambola et al. (2017), where the interested reader may find further details. The first phase consists of a GRASP Feo and Resende (1995), which produces two pools of solutions: the *elite pool* containing a certain number of the best solutions found and the *diverse pool* containing the most diverse solutions among the ones in both pools. Both pools have a limited size, denoted by *nElite* and *nDiverse*, respectively. When one pool is full and we want to insert a solution there, then the worst solution in the pool is removed.

When both pools have been built, the second phase starts. It is an intensification phase consisting of a PR. It repeatedly chooses a target solution from the *elite pool* and then, starting from a solution selected at random from the *diverse pool*, explores a path linking them in an attempt to find better feasible solutions.

The overall procedure is summarized in Algorithm 3. In the first phase (lines 1 to 17), the GRASP is executed *max_iter* times. In each execution, a solution S is constructed (line 5—function `GreedyRandomizedConstruction`(α, p^+)) and repaired (`RepairSolution`(S)) if it is infeasible (lines 6 and 7). A local search (`LocalSearch_1`(S)) is applied at the end. All moves are evaluated by estimating their impact in the solution cost.

Algorithm 3 Heuristic Framework for the FLBD

// initialization
1: $elite\_pool \leftarrow \emptyset$, $worstValueElite \leftarrow \infty$;
2: $diverse\_pool \leftarrow \emptyset$; $worstDiverseValue \leftarrow 0.0$;
3: `initializeConstruction()`
// phase 1: GRASP used for building two pools of solutions;
4: **for** $k = 1, \ldots, max\_iter$ **do**
5: $S \leftarrow$ `GreedyRandomizedConstruction`(α, p^+);
6: **if** S not feasible **then**
7: $S \leftarrow$ `RepairSolution`(S);
8: $S \leftarrow$ `LocalSearch_1`(S); // cost variation estimated in all moves;
"// solution cost estimated at the end of the local search;"
9: $f(S) \leftarrow$ `estimateCost`$(S, highPrecision)$;
10: **if** $S \notin elite\_pool$ and $f(S) < worstValueElite$ **then**
11: $elite\_pool \leftarrow$ `insertInElitePool`$(S, worstValueElite)$;
12: **else**
13: **if** $S \notin diverse\_pool$ **then**
14: dist(S) $\leftarrow$ `pool_distance`(S);
15: **if** dist(S) $> worstDiverseValue$ **then**
16: $diverse\_pool \leftarrow$ `insertInDiversePool`$(S, worstDiverseValue)$;
17: $S^* \leftarrow \arg\min_{S \in elite\_pool}\{f(S)\}$; // best solution so far;
// end of phase 1;
// phase 2: a PR procedure is used for improving the solution;
18: **for** $\ell = 1, \ldots, nRep$ **do**
19: **for all** $S \in elite\_pool$ **do**
20: $S_t \leftarrow S$;
21: $S_c \leftarrow$ `getRandom`$(diverse\_pool)$;
22: $S^* \leftarrow$ `pathRelinking`(S_c, S_t, S^*);
// end of phase 2;
23: **return** S^*.

Once a solution is built, we decide whether to insert it in one of the pools—lines 9 to 16. The value of the solution produced by the local search is estimated (see line 9, `estimateCost`$(S, highPrecision)$) and we check if its quality indicates that it should be inserted into the *elite pool*. In this case, we insert it in the pool (see in line 11 the call for `insertInElitePool`$(S, worstValueElite)$), updating if necessary the worst value among all solutions of the pool. Otherwise, we check whether the solution should be inserted into the *diverse pool*. If so, the solution enters that pool (function `insertInDiversePool(`S`,` $worstDiverseValue$`)`) and the worst diverse value among solutions in the pool is updated (if necessary). The implementation details of the function estimating the cost of a solution (`estimateCost(`S`,` $highPrecision$`)`) are provided in Sect. 8.2.1. At termination of GRASP (line 17), the best solution found in all executions of the GRASP is set as the incumbent solution (S^*).

In the PR procedure (lines 18–23), paths are explored for pairs of solutions, one from the *elite pool* and the other one from the *diverse pool*. The former is selected sequentially and is set as the target solution (S_t); the latter is randomly selected and is set as the current solution (S_c). After defining a target and a current solution, a path linking them is explored. The incumbent solution is updated every time a solution is found with a better cost estimate. The PR procedure is repeated $|elite\_pool| \times nRep$ times.

Below we detail the different functions invoked by this heuristic.

5.1 *The Greedy Randomized Procedure*

As usual, the construction phase of the GRASP is the randomization of a greedy algorithm. At each iteration, an element is randomly selected from a RCL, which contains the *best* elements that could be incorporated into the partial solution according to some prespecified greedy criterion. The selected element is then incorporated to the solution under construction and the procedure repeated.

In our case, a partial solution is given by a set of open facilities together with an assignment of customers to the current set of open facilities. The construction phase ignores constraints (4), imposing a lower limit on number of customers assigned to every open facility. Hence, since it may produce an infeasible solution, a feasibility restoration mechanism is applied at the end, if needed. The procedure terminates with a local search in an attempt to improve the feasible solution obtained. All these elements are detailed next.

5.1.1 Construction Phase

This phase starts by opening one single facility, randomly chosen from the RCL, and assigning all the customers to it. In a general step k, a new facility is opened and some customers are reassigned to it. The full procedure is detailed in Algorithm 4. We denote by I^k the set of open facilities at the end of step k, by $(\tilde{\mathbf{y}}^{\mathbf{k}}, \tilde{\mathbf{x}}^{\mathbf{k}})$ the current partial solution, and by $i(j)$ the plant to which customer $j \in J$ is currently assigned. Again we denote by J_i the set of customers assigned to i in the *a priori* solution.

In the first iteration (see line 2), each candidate facility is evaluated according to

$$\delta_i^0 \leftarrow \frac{1}{u_i}\left(f_i + \sum_{j \in J} c_{ij}\right), \tag{7}$$

where u_i is an *auxiliary assignment capacity* of facility i, which remains constant for all executions of the GRASP. In particular,

$$u_i = \max\left\{\ell_i, K_i, \left\lfloor \frac{n}{\overline{p}} \frac{K_i}{\sum_{t \in I} K_t} \right\rfloor\right\},$$

where $\overline{p}$ denotes the average of the demand probabilities p_j, $j \in J$.

In subsequent iterations, the incremental cost of facility i, δ_i^k (line 18), is estimated as

$$\delta_i^k \leftarrow f_i + \sum_{t=1}^{r_i} \sigma_{ij_{[t]}}$$

where r_i is the number of customers that would be assigned to facility i if it were open (determined in line 17).

For each $j \in J, \sigma_{ij}$ is the estimated variation in its service costs when reassigning customer j to facility i, computed as (line 15)

$$\sigma_{ij} = c_{ij} - c_{i(j)j} - p_j \times g(i(j))\frac{(r_{i(j)} - K_{i(j)})^+}{r_{i(j)}}. \tag{8}$$

The RCL is built with the non-open facilities that seem most promising if opened (line 21). It contains all closed facilities with an incremental cost within the interval $[\delta^{\min}, \delta^{k\min} + \alpha^k(\delta^{k\max} - \delta^{k\min})]$, where $\delta^{k\min}$ and $\delta^{k\max}$ respectively denote the smallest and largest nonpositive incremental costs δ_i^k. Based on preliminary testing, we used $\alpha^k = 2\alpha$ in the first iteration ($k = 0$), whereas in subsequent iterations ($k > 0$) $\alpha^k = \alpha$, where α is a given parameter. After the next facility to open is randomly chosen from the RCL (line 23), the set of customers assigned to each open facility is updated in the loop 25–27.

The termination criterion of the construction phase is not fixed in advance. Usually, the constructive phase would continue until all facilities are open or until all the non-open facilities have a nonnegative incremental cost. Preliminary computational testing, however, indicated that these criteria tend to produce solutions with too many open facilities. For this reason, an additional parameter p^+ is introduced, which denotes the probability with which the process terminates even if none of the above termination criteria are met (lines 10, 11).

Algorithm 4 `GreedyRandomizedConstruction`(α, p^+, u)

// choose the first facility to open
$k \leftarrow 0$;
Compute $\{\delta_i^k\}_{i \in I}$ using (7), $\quad \alpha^k \leftarrow 2\alpha$;
RCL $\leftarrow \left\{i \in I : \delta^{k\,\min} \leqslant \delta_i^k \leqslant \delta^{k\,\min} + \alpha^k(\delta^{k\,\max} - \delta^{k\,\min})\right\}$;
$i^k \leftarrow$ `RandomSelect(RCL);`
$I^k \leftarrow \{i^k\}$;
$i(j) \leftarrow i^k, j \in J$;
$J_{i^k} \leftarrow J, J_i \leftarrow \emptyset, i \in I \setminus \{i^k\}$;
// main loop
repeat
  $k \leftarrow k + 1$;
  $\beta \leftarrow$ `RandomSelect([0, 1]);`
  if $(\beta > p^+)$ **then** Stop;
  else
    for $(i \in I)$ **do**
      for $(j \in J)$ **do**
        Compute σ_{ij} according to (8);
      $\{j_{[1]}, \ldots, j_{[n]}\} \leftarrow$ `sort_`σ`_increasing(`J`);`
      $r_i \leftarrow \min\left\{u_i, \max\left\{\ell_i, \max\{q : \sigma_{ij_{[q]}} < 0\}\right\}\right\}$;
      $\delta_i^k \leftarrow f_i + \sum_{t=1}^{r_i} \sigma_{ij_{[t]}}, \quad \alpha^k \leftarrow \alpha$;
   $\delta^{k\,\min} \leftarrow \min\left\{0, \min\{\delta_i^k : i \in I\}\right\}, \quad \delta^{k\,\max} \leftarrow \min\left\{0, \max\{\delta_i^k : i \in I\}\right\}$;
    if $\delta^{k\,\min} < 0$ **then**
      RCL $\leftarrow \left\{i \in I : \delta^{k\,\min} \leqslant \delta_i^k \leqslant \delta^{k\,\min} + \alpha^k(\delta^{k\,\max} - \delta^{k\,\min})\right\}$;
      if RCL $\neq \emptyset$ **then**
        $i^k \leftarrow$ `RandomSelect(RCL);`
        $I^k \leftarrow I^{k-1} \cup \{i^k\}$;
        for $(j = [1], \ldots, [r_{i^k}])$ **do**
          $J_{i(j)} \leftarrow J_{i(j)} \setminus \{j\}; i(j) \leftarrow i^k$;
          $J_{i^k} \leftarrow J_{i^k} \cup \{j\}$;
until $\delta^{k\,\min} \geq 0$;
return I^k and $i(j), j \in J$.

5.1.2 Feasibility Restoration

When the outcome of the construction phase violates the lower limit ℓ_i on the number of assigned customers of some open facility ($|J_i| < \ell_i$), the following two-step feasibility restoration procedure is applied (Algorithm 3—function `RepairSolution(S)`):

Step 1: We check whether the current solution satisfies $\sum_{i \in I^k} \ell_i > n$. If this is the case, facilities are closed one at a time until $\sum_{i \in I^k} \ell_i \leq n$. In each iteration, the facility that is closed is

$$i^* \in \arg \max_{i \in I^k : \ell_i > |J_i|} \left\{ f(i) + \sum_{j \in J_i} c_{ij} \quad + g(i) \left(\overline{p}_i \times |J_i| - K_i \right)^+ \right\},$$

where $\overline{p}_i$ denotes the average demand probability of all customers assigned to facility i.

After closing facility i^*, all the customers of J_i are reassigned among the remaining open facilities. This is done in a greedy way. For each involved customer j, we estimate the increase in the solution cost for reassigning it to open facility i as

$$c_{ij} + \max\{|J_i| + 1 - K_i, 0\} p_j \frac{g_i \times K_i}{|J_i| \times (|J_i| + 1)}.$$

Then j is reassigned to the open facility i that contributes the least to the estimated increase in the solution cost.

Step 2: After ensuring that $\sum_{i \in I^k} \ell_i \leq n$, we check whether the constraint (4) associated with some open facility i is still violated, that is, $|J_i| < \ell_i$. In this case, some customers currently assigned to facilities i' such that $|J_{i'}| > \ell_{i'}$ are assigned to facility i. Again, this is done in a greedy way. First, the reassignment cost of j to i' is estimated as

$$p_j(c_{ij} - c_{i'j}) + \max\{|J_i| + 1 - K_i, 0\} p_j \frac{g_i \times K_i}{|J_i| \times (|J_i| + 1)}$$
$$- \max\{|J_{i'}| - K_{i'}, 0\} p_j \frac{g_{i'} \times K_{i'}}{|J_{i'}|} \times (|J_{i'}| - 1).$$

Then the reassignment that is chosen is the one that produces the least estimated increase in the solution cost.

5.1.3 Local Search

In the local search, the set of open facilities remains fixed so the considered neighborhoods only affect the assignment of customers within that set. Two different neighborhoods are considered: (i) those induced by *reassignments* of customers and (ii) those induced by *interchanges* of assignments between pairs of customers. Both types of neighborhoods are explored with a first improving policy relative to the estimated variation in the solution cost, computed as follows:

Reassignments
The estimated cost variation when reassigning customer $j \in J_{i_1}$ to i_2 is:

$$\sigma_{j,i_2} = p_j(c_{i_2 j} - c_{i_1 j}) + \max\{|J_{i_2}| + 1 - K_{i_2}, 0\} g_{i_2} p_j \times \sum_{s=K_{i_2}}^{|J_{i_2}|} \mathrm{Bin}(|J_{i_2}|, \bar{p}_{i_2}, s)$$

$$- \max\{|J_{i_1}| - K_{i_1}, 0\} g_{i_1} p_j \times \sum_{s=K_{i_1}+1}^{|J_{i_1}|} \mathrm{Bin}(|J_{i_1}|, \bar{p}_{i_1}, s),$$

where $\overline{p}_{i_1}$ and $\overline{p}_{i_2}$ denote the arithmetic average of the demand probabilities for the customers of J_{i_1} and J_{i_2}, respectively, and $\mathrm{Bin}(\zeta, p, s)$ the probability that a binomial random variable with parameters ζ and p takes value s; *that is* $\mathrm{Bin}(\zeta, p, s) = \binom{\zeta}{s} p^s (1-p)^{\zeta - s}$.

Feasibility is kept by considering only customers j with $|J_{i(j)}| > \ell_{i(j)}$.

Note that since cost variations are only estimated, cycles of reassignments could happen. This is avoided by estimating simultaneously the cost of a reassignment and the cost of its reverse move and only performing moves such that the direct move has a better cost estimate than the reverse one.

Interchanges
For the assignment interchange of two customers j_1, j_2, with $j_1 \in J_{i_1}, j_2 \in J_{i_2}$, we assume w.o.l.g. that $p_{j_1} > p_{j_2}$. Then the estimate of the cost variation is:

$$\sigma_{j_1 j_2} = c_{i_1 j_2} + c_{i_2 j_1} - c_{i_1 j_1} - c_{i_2 j_2}$$
$$+ g_{i_2} \left(p_{j_1} - p_{j_2}\right) \left[\frac{\left(|J_{i_1}| - K_{i_1}\right)^+}{|J_{i_1}|} - \frac{\left(|J_{i_2}| - K_{i_2}\right)^+}{|J_{i_2}|} \right].$$

For both, reassignments and interchanges, we set an upper limit for the number of times that any customer can be reassigned to the same facility. We alternate the exploration of the two neighborhoods until there are no further promising moves.

5.1.4 Diversity Measure

In the GRASP, when a solution S is not inserted in the *elite pool*, we check whether it should be inserted in the *diverse pool*. The diversity measure that is used (`pool_distance(S)`) evaluates the average number of noncoincident customer assignments over all the solutions in one of the pools, namely,

$$\frac{1}{n_e + n_d}\left(\sum_{S_e \in elite\ pool} |\{j \in J : i_{S_e}(j) \neq i_S(j)\}| \right.$$
$$\left. + \sum_{S_d \in diverse\ pool} |\{j \in J : i_{S_d}(j) \neq i_S(j)\}| \right),$$

where n_e (n_d) is the current number of *elite* (*diverse*) solutions and $i_{\mathrm{s}}(j)$ denotes the facility to which customer j is assigned in solution s.

5.2 *Path Relinking*

The second phase of the heuristic consists of a PR that explores paths linking pairs of current and target solutions, (S_{c}, S_{t}), as detailed in Algorithm 3. Recall also that S^* stands for the incumbent solution.

Algorithm 5 `pathRelinking`(S_{c}, S_{t}, S^*)

// initialization;
1: nOpenCurrent ← `countFacilitiesOpen`(S_{c});
2: nOpenTarget ← `countFacilitiesOpen`(S_{t});
// end of initialization;
3: β ← `RandomSelect`([0, 1]);
4: **while** ($\beta \leq p^{++}$) and (nOpenCurrent > nOpenTarget) **do**
5: S_{c} ← `closeFacility`(S_{c});
6: nOpenCurrent ← nOpenCurrent-1;
7: S^* ← `updateIncumbent`(S^*, S_{c});
8: β ← `RandomSelect`([0, 1]);
9: (S_{c}, S^*) ← `exchangeFacilities`(S_{c});
10: (S_{c}, S^*) ← `closingFacilities`(S_{c});
11: (S_{c}, S^*) ← `openingFacilities`(S_{c});
12: S^* ← `localSearch_2`(S_{c}, S^*);
13: **return** S^*.

In Albareda-Sambola et al. (2017), two versions of the PR were implemented. The first one, referred to as PR1, is sketched in Algorithm 5. The second one, referred to as PR2, is as PR1 but without the cycle in lines 4–8. Both versions apply three types of local moves for transforming the current solution, S_c, into the target, S_t: (*i*) closing a facility that is closed in S_t, (*ii*) opening a facility that is open in S_t, or (*iii*) interchanging a facility that is open in S_c but closed in S_t by a facility that is closed in S_c and open in the S_t. Each of the above moves includes the necessary reassignments of customers and updates the incumbent solution when the estimated value of S_c is better than that of S^*. The loop in lines 4–8 used in PR1 explores the *close-facility* neighborhood only and is regulated by a probability p^{++} of changing to other neighborhoods before the number of open facilities in S_c and S_t coincides. In both versions, when S_t has been reached, a local search is applied to the current incumbent, consisting of reassignments of customers only. Again, a check is applied at the end to see if S^* should be updated. Some details of Algorithm 5 are given below.

Incumbent Update As will be explained in Sect. 8.2.1.2, the cost estimation of a solution can be computed with different precisions. Since a higher precision requires a higher computing time, in both versions of the PR, the higher precision is only used when trying to update the incumbent. Algorithm 6 gives the details of function `updateIncumbent(`S^*`,` S_c`)`, where a tolerance factor is used as a threshold for the acceptance of a solution that is candidate to become the new incumbent. This function is called at the end of each of the functions called in lines 9–12 of Algorithm 5.

Algorithm 6 `updateIncumbent(`S^*`,` S_c`)`

$f(S_c) \leftarrow$ estimateCost(S_c, lowPrecision);
if $f(S_c) < (tolerance \times f(S^*))$ **then**
    $f(S_c) \leftarrow$ estimateCost(S_c, highPrecision);
    if $f(S_c) < f(S^*)$ **then**
        $S^* \leftarrow S_c$;
        $f(S^*) \leftarrow f(S_c)$;
return S^*.

Facility Moves We next detail the three facility moves that are considered in PR. In order to differentiate among customers assignments of different solutions, we will use $J_i^{\mathtt{S}}$ to denote the set of customers assigned to facility i in solution s.

Closing facilities: When some of the facilities that are open in S_c are closed in S_t, we choose one of them for closing. This is done in a greedy way. For each facility i candidate for closing, we estimate the cost increment in the current solution if i is closed as

$$-f_i + \sum_{j \in J_i^{S_c}} p_j (c_j^{\min} - c_{ij}),$$

where, for each customer $j \in J_i^{S_c}$, $c_j^{\min}$ is its minimum service cost among the set of open facilities in S_c (except for the facility i, which is being checked for closing).

When a facility is eventually closed, each customer of $J_i^{S_c}$ is taken in turn. If the facility to which it is assigned in S_t is open, it is directly assigned to that facility. Otherwise, it is assigned greedily using the function defined in the GRASP constructive procedure.

Opening facilities: Facilities that are open in S_t but are closed in S_c are the candidates for being opened. Again, the facility chosen to be opened is the one producing the least estimated increase in the solution cost. The estimate for a given facility i is

$$f_i + \sum_{j \in J_i^{S_t}} p_j(c_{ij} - c_{i(j),j}),$$

where $i(j)$ is the facility to which customer j is assigned in S_t.

When a facility is opened, all customers assigned to it in S_t are automatically reassigned to it. When doing so, infeasibilities with respect to the lower bound on the number of customers assigned to open plants may arise. In such a case, we use the repair mechanism introduced in Sect. 5.1.2.

Exchange facilities The goal of these moves is to swap pairs of facilities that do not have the same status in S_c and in S_t. Let i_1 be a facility that is open in S_t but closed in S_c. Let also i_2 be a facility that is closed in S_t but open in S_c. Then swapping i_1 and i_2 means that in S_c, facility i_1 is opened and i_2 closed. Furthermore, all the customers initially assigned to i_2 are reassigned to i_1. If $\ell_1 > \ell_2$ and there are less than ℓ_1 customers initially assigned to i_2, then the new solution will violate the lower bound constraint associated with i_1. In such a case, we apply the restoration mechanism explained in Sect. 5.1.2.

As for the selection of the pair to swap, for each candidate pair (i_1, i_2), we estimate the variation in the solution cost after the swap, taking into account that all customers currently assigned to i_2 in S_c should be reassigned if i_2 is closed. Thus, we consider the partition of $J_{i_2}^{S_c}$ given by $J_{i_2}^{S_c} = J_{i_2}^1 \cup J_{i_2}^2$, with $J_{i_2}^1 = J_{i_2}^{S_c} \cap J_{i_1}^{S_t}$ and $J_{i_2}^2 = J_{i_2}^{S_c} \setminus J_{i_2}^1$. The estimation of the variation in the solution cost is

$$f_{i_1} - f_{i_2} + \sum_{j \in J_{i_2}^1} p_j(c_{i_1,j} - c_{i_2,j}) + \sum_{j \in J_{i_2}^2} p_j(c_j^{\min} - c_{i_2,j}),$$

where $c_j^{\min}$ denotes the minimum service cost for customer j computed among the facilities that are open in $(S_c \setminus \{i_2\}) \cup \{i_1\}$ (except for i_2, which is the facility being checked for closing).

Customer Moves Algorithm 5—`LocalSearch_2(S)` performs a local search involving the customers. It consists of a reassignment procedure, different from the one used in the GRASP, which is based on the following estimation for the variation of the solution cost when reassigning customer j from plant $i_1 = i(j)$ to i_2:

$$c_{i_2 j} - c_{i_1 j} + g_{i_2} \frac{p_j}{|J_{i_2}| + 1} \max\{|J_{i_2}| + 1 - K_{i_2}, 0\} - g_{i_1} \frac{p_j}{|J_{i_1}|} \max\{|J_{i_1}| - k_{i_1}, 0\}.$$

6 Outsourcing Policies for the FLBD

As we already mentioned when introducing the FLBD, the choice of the outsourcing policy is on itself a strategic decision that must be made in advance, which may have a notable impact on the specific location/allocation selections, given that different outsourcing policies may lead to different solutions. Below, we describe the two outsourcing policies that will be used in the remainder of this chapter. They are called *facility outsourcing* (FO) and *customer outsourcing* (CO). For each outsourcing policy, we present a mixed-integer linear programming formulation, which allows to optimally solve the problem when uncertainty is expressed via a set of scenarios, and will be used in the SAA solution algorithm discussed in Sect. 7. FO and CO have been used in Albareda-Sambola et al. (2011, 2017); alternative outsourcing policies have been considered and compared with FO and CO in Albareda-Sambola et al. (2022).

6.1 *Facility Outsourcing*

With the facility outsourcing (FO) policy, under scenario ω, facility i takes delivery of the whole set J_i^ω. When $\eta_i^\omega > K_i$, then $\eta_i^\omega - K_i$ units of product are outsourced to a third party, at a unit cost g_i. Then facility i serves the overall demand of its assigned customers, η_i^ω, at the same cost c_{ij} that would be incurred if it were not outsourced.

To formulate the FO-FLBD, in addition to the y and x decision variables introduced above, we consider θ_i^ω denoting the number of demand customers outsourced at facility $i \in I$ under scenario $\omega \in \Omega$. Additionally, we use z^ω to represent the total penalty incurred under scenario $\omega \in \Omega$. The formulation is:

$$\text{FO} \quad \min \sum_{i \in I} f_i y_i + \sum_{i \in I} \sum_{j \in J} p_j c_{ij} x_{ij} + \sum_{\omega \in \Omega} \pi^\omega z^\omega, \tag{9}$$

$$s.t. \quad (2) - (4),$$

$$\theta_i^\omega \geqslant \sum_{j \in J} d_j^\omega x_{ij} - K_i y_i, \qquad i \in I, \omega \in \Omega, \tag{10}$$

$$z^\omega \geqslant \sum_{i \in I} g_i \theta_i^\omega, \qquad \omega \in \Omega, \tag{11}$$

$$z^{\omega} \geqslant 0, \theta_i^{\omega} \geqslant 0, \qquad i \in I, \omega \in \Omega, \tag{12}$$

$$y_i \in \{0, 1\}, \qquad i \in I, \tag{13}$$

$$x_{ij} \in \{0, 1\}, \qquad i \in I, j \in J. \tag{14}$$

The objective function (9) includes the costs for opening facilities plus the expected value of the service plus outsourcing costs. As explained, Constraints (2)–(4) guarantee the feasibility of the *a priori solution*. Constraints (10) force θ variables to take consistent values, and Constraints (11) compute the penalty cost of each scenario and will hold as equality in any optimal solution. Thus, they are not strictly needed, as their right-hand side could be substituted in the last term of the objective function instead of using the variables z^{ω}. The domain of the variables is defined by (12)–(14). Note that given the structure of the formulation, integrality constraints on the z and θ variables can be relaxed to nonnegativity constraints.

Formulation (9)–(14) uses $|I|(1 + |J|) + |\Omega|(|I| + 1)$ variables and has $|J|(1 + |I|) + |I| + |\Omega|(|I| + 1)$ constraints. Depending on the size of Ω, these numbers can be quite high, even for moderate numbers in terms of customers and facilities. Hence, enhancing the formulation can be very useful to decrease the computing time required to solve such model to proven optimality using an off-the-shelf solver. Inequalities (15) and (16) below have proven to give a good balance between the increase in the size of the formulation and the improvement obtained when solving the model.

$$\sum_{\omega \in \Omega} \pi^{\omega} \theta_i^{\omega} \geqslant \sum_{j \in J} p_j x_{ij} - K_i y_i, \qquad i \in I, \tag{15}$$

$$\sum_{i \in I} K_i y_i + \sum_{i \in I} \theta_i^{\tilde{\omega}} \geqslant D^{\tilde{\omega}}. \tag{16}$$

Inequality (15) states that the expected number of demand customers outsourced at facility i ($i \in I$) is at least the expected number of demand customers assigned to that facility minus the capacity of the facility. Note that these constraints are activated only if $y_i = 1$. In (16), $\tilde{\omega} \in \Omega$ is the scenario with the largest number of demand customers ($D^{\tilde{\omega}}$). This constraint ensures that the maximum number of customers that can be served by the open facilities plus the outsourced demand customers is never below the total demand. This constraint holds for every scenario, but adding such a constraint for all scenarios would increase considerably the size of the formulation, which explains why we consider solely (16).

6.2 Customer Outsourcing

With the customer outsourcing (CO) strategy, in the scenarios where the number of demand customers assigned to facility i exceeds its capacity K_i, that is, $\eta_i^{\omega} > K_i$, exactly K_i customers are served directly from facility i, whereas the remaining $\eta_i^{\omega} - K_i$ customers of J_i^{ω} are outsourced and receive service from an external third party. Service costs c_{ij} are incurred for the customers served from facility i, whereas a penalty g_i is incurred for each outsourced customer, which depends on the facility the customer is assigned to. Hence, to formulate the FLBD with a CO policy, additional decision variables are needed, to identify the outsourced customers. In particular, we define a binary variable, s_{ij}^{ω} equal to one if and only if customer $j \in J$ is served from facility $i \in I$ under scenario $\omega \in \Omega$.

In the CO policy that we use here, in the scenarios where $\eta_i^{\omega} > K_i$, the demand customers that are served from facility i are selected according to a FIFO policy, relative to the order in which requests of service have arrived. With this CO policy, a scenario $\omega \in \Omega$ is not fully characterized by its probability and demand customers, since the order in which calls for service from demand customers arrive must also be known. We use the notation $j' \prec^{\omega} j$ to indicate that customers $j, j' \in J$ have demand under scenario ω and j' requested service before j. A formulation for the CO is then:

$$\text{CO} \quad \min \sum_{i \in I} f_i y_i + \sum_{\omega \in \Omega} \sum_{i \in I} \sum_{j \in J} c_{ij} s_{ij}^{\omega} + \sum_{\omega \in \Omega} \pi^{\omega} z^{\omega}, \tag{17}$$

$$s.t. \quad (2)-(4),$$

$$\sum_{\omega \in \Omega} s_{ij}^{\omega} \leqslant \left(\sum_{\omega \in \Omega} d_j^{\omega} \right) x_{ij}, \qquad i \in I, j \in J, \tag{18}$$

$$\sum_{j \in J} d_j^{\omega} s_{ij}^{\omega} \leqslant K_i, \qquad i \in I, \omega \in \Omega, \tag{19}$$

$$\sum_{j \in J} d_j^{\omega} s_{ij}^{\omega} + \theta_i^{\omega} \geqslant \sum_{j \in J} d_j^{\omega} x_{ij}, \qquad i \in I, \omega \in \Omega, \tag{20}$$

$$z^{\omega} \geqslant \sum_{i \in I} g_i \theta_i^{\omega}, \qquad \omega \in \Omega, \tag{21}$$

$$K_i d_j^{\omega} (x_{ij} - s_{ij}^{\omega}) \leqslant \sum_{j' \prec^{\omega} j} d_{j'}^{\omega} s_{ij'}^{\omega}, \qquad i \in I, \omega \in \Omega, j \in J. \tag{22}$$

$$z^{\omega} \geqslant 0, \theta_i^{\omega} \geqslant 0, \qquad i \in I, \omega \in \Omega, \tag{23}$$

$$y_i \in \{0, 1\}, \qquad i \in I, \tag{24}$$

$$x_{ij} \in \{0, 1\}, \qquad i \in I, j \in J, \tag{25}$$

$$s_{ij}^{\omega} \in \{0, 1\}, \qquad i \in I, j \in J, \omega \in \Omega. \tag{26}$$

Again, Constraints (2)–(4) guarantee the feasibility of the *a priori* solution. The second-stage variables s_{ij}^{ω} are now used to compute the expected service cost. Constraints (18) ensure that service from open facilities is only provided according to the *a priori* assignments dictated by the x variables. Constraints (19)–(20) state the service capacities of the facilities and set the right value to the number of outsourced units at each facility, respectively. Finally, constraints (22) ensure that the FIFO policy is followed for selecting the customers that will be served from a given facility when $\eta_i^\omega > K_i$. Note that if this last set of constraints is relaxed, then the decision on the customers to serve is made solely based on a cost criterion (see Albareda-Sambola et al., 2022 for further details). Again, the structure of the problem allows to relax integrality constraints on θ variables and restrict them to be just nonnegative.

The number of variables of formulation CO has increased in $|I| \times |J| \times |\Omega|$ with respect to the number of variables of the FO formulation. Its number of constraints is also larger, as it has raised to $|J|(1 + 2|I|) + |I| + |\Omega|(2|I| + |I||J| + 1)$.

7 Sample Average Approximation

A common approach to deal with two-stage problems where the unknown parameters are described through their probability distributions is Sample Average Approximation (SAA) (Kleywegt et al., 2001). The basic idea of SAA is to use Monte Carlo simulation to estimate the value of the stochastic program. To this end, an iterative procedure is used. At each iteration, the recourse function is approximated by the average cost over a sample of possible scenarios generated according to the assumed probability distribution. The so-called sample average optimization problem is solved and both its optimal value and its solution are stored. The output of the algorithm is the average of the optimal values obtained in the different runs (different samples), which is taken as an estimate of the optimal value of the stochastic problem, and the best solution found along the process, denoted by x^*. Algorithm 7 illustrates a generic SAA implementation in pseudocode. In this pseudocode, $v(\Omega^t)$ stands for the optimal value of the sample average problem defined by the scenarios in Ω^t, and $f(x)$ stands for the actual value of x as a solution of the two-stage program.

Depending on the structure of the recourse function, the actual value of this solution can be exactly computed or estimated using again Monte Carlo simulation, typically with a larger sample than those used in the sample average problems. On the other hand, the average of the optimal values of the sequence of subproblems solved, $\overline{z} \leftarrow \frac{1}{t}\sum_{s=1}^{t} z^s$, produces an estimate of the optimal value of the stochastic problem, which provides a basis for assessing the quality of the best solution found. Moreover, its convergence gives a termination criterion.

Note that the only requirement needed for implementing SAA is to have an oracle for solving the sample average problem; that is, the stochastic problem restricted to a subset of scenarios, all with the same weight. Thus, mathematical programming formulations are needed within an SAA algorithm. Indeed, such formulations depend on the chosen recourse action. For the FLBD, we will use the FO and CO formulations presented in Sect. 6.

Algorithm 7 Generic SAA pseudocode

```
procedure SAA (max_it)
Set z* ← ∞,  z̄ ← ∞,  t ← 0;
while (t < max_it and z̄ has not converged) do
    t ← t + 1;
    Select a sample of possible scenarios Ω^t;
    z^t ← v(Ω^t),  x^t ∈ arg min v(Ω^t);
    if (f(x^t) < z*) then
        z* ← f(x^t);
        x* ← x^t;
    end;
    z̄ ← (1/t) Σ_{s=1}^{t} z^s;
end;
return z̄, z*, x*;
```

8 Computational Experiments

We next describe the computational experiments carried out to test the approximate methods described in this chapter and analyze the results that they produce.

8.1 Test Instances

The results that we present correspond to instances of two different classes: with homogeneous demand and with heterogeneous demand. For the homogeneous case, we selected some of the instances from Albareda-Sambola et al. (2011), and for the class with general demands, we took instances from those used by Albareda-Sambola et al. (2022). The main characteristics of the generation process are detailed next.

The basis for the homogeneous instances is the traveling salesman problem (TSP) instances taken from http://comopt.ifi.uni-heidelberg.de/software/TSPLIB95/: `berlin52`, `eil51`, `eil76`, `kroA100`, `kroB100`, `kroC100`, `kroD100`, `kroE100`, `pr76`, `rqt99`, and `st70`. From those TSP instances, FLBD instances of two sizes were generated: small instances, with $|I| = 15$ and $|J| = 30$, and large instances with $|I| = 20$ and $|J| = 60$. To do so, nodes were selected randomly from the original TSP instance. Naturally, the smaller TSP instances with less than 80 nodes were only used for generating small instances. For each combination of plants and customers, the remaining data were generated considering three different values for the probability of demand (0.1, 0.5, and 0.9) and two different capacity levels (low and high). In total, we considered a set of 306 homogeneous instances.

Each homogeneous instance is identified by a label of the form `name_id_pr_`γ`_h`, where `name` is the name of the original TSP instance (with the prefix `L` for large instances), `id` $\in \{1, 2, 3\}$ identifies the instance among the three generated from the same original TSP instance, `pr` $\in \{1, 5, 9\}$ indicates the homogeneous demand probability ("1" corresponds to 0.1, "5" corresponds to 0.5, and "9" corresponds to 0.9), $\gamma \in \{1, 4\}$ is the value of the parameter γ used for generating the capacities of the facilities, and finally, `h` stands for "homogeneous."

For the class of heterogeneous FLBD instances, we consider the 306 instances introduced by Albareda-Sambola et al. (2022). They were generated from the 306 homogeneous instances just described, with customers classified as low-, medium-, and high-probability demand customers. The demand probability of each customer was randomly taken from $U(0.10, 0.25)$, $U(0.40, 0.60)$, or $U(0.75, 0.90)$, depending on the customer class. Then three patterns for the demand are considered: In pattern 1 (PT1), there are 60% of low-probability demand customers, 20% medium, and 20% high. In pattern 2 (PT2), these percentages are 20%, 60%, and 20%, respectively, and finally, in pattern 3 (PT3), they are 20%, 20%, and 60%. The reader is referred to Albareda-Sambola et al. (2022) for further details.

Summarizing, for each combination of one of the two capacity levels and one of the three demand probabilities (resp. probability patterns), we have a set of 33 small instances and a set of 18 large instances of the homogeneous (resp. non-homogeneous) FLBD.

8.2 *Implementation Details*

All algorithms analyzed in this section were coded in C, using CPLEX 12.6 callable libraries for solving MIP formulations. All the tests were carried out on a Pentium(R) 4, 3.2 GHz, 1.0 GB of RAM. Next, we give some implementation details specific for the different methods.

8.2.1 Grasp + Path Relinking

The evaluation of each particular solution of the FLBD, especially in the non-homogeneous case, is a relevant issue in the implementation of the GRASP and Path Relinking methods. Therefore, exact evaluation is used only in some cases, whereas in other cases, the solution values are approximated. In this section, we provide the details of these evaluations/estimations. The final choices for parameter values used in the algorithm are summarized in Table 1.

Table 1 Values set for the parameters

Parameter	Usage	Description	Value
`nElite`	GRASP (Algorithm 3)	Maximum number of solutions in the *elite_pool*	15
`nDiverse`	GRASP (Algorithm 3)	Maximum number of solutions in the *diverse_pool*	15
`max_iter`	GRASP (Algorithm 3)	Number of GRASP iterations	500
α	GRASP (Algorithm 4)	Range for the RCL	0.15
p^+	GRASP (Algorithm 4)	Probability ruling the process of opening further facilities while the stopping criteria are not met	0.9
`nMoves`	GRASP (`LocalSearch_1(`S`)`)	Maximum number of times the same reassignment can be done in the interchange and reassignment moves	2
`rep`	Algorithm 3	Number of times each elite solution will be used	2
p^{++}	PR1	Probability of keeping closing facilities in the first phase of converting the current solution into the target	0.9
`lower precision`	PR1 and PR2 (Algorithm 6)	Lower precision for simulating the cost of an *a priori* solution	5E-4
`higher precision`	GRASP, PR1 and PR2 (Algorithms 3 and 6)	Higher precision for simulating the cost of an *a priori* solution	1E-5
`tolerance`	GRASP, PR1 and PR2 (Algorithms 3 and 6)	Tolerance used to exclude a given solution as a candidate to become an incumbent	1.05
`iterMin`	Antithetic estimator	Minimum number of random sequences to generate independently of the stopping criteria	500
`ntop`	Solution evaluation (customer outsourcing)	Maximum number of demand customers assigned *a priori* to one facility for which we compute the service cost exactly	20

8.2.1.1 Evaluating a Feasible Solution

Enumerating independently the sets of scenarios restricted to each J_i reduces notably the full enumeration of potential subsets of demand customers. Still, it can be computationally very demanding, particularly for instances with low-demand probabilities where sets J_i tend to be large. This can become unaffordable, even for one single evaluation. This becomes especially true when customer outsourcing is considered. Indeed, in this case, scenarios are defined not only by the actual set of demand customers but also by their calling sequence. Therefore, in this case, the exact evaluation would require to enumerate all possible call sequences for each possible subset of J_i. Thus, we simulate the expected service cost of an open facility when $|J_i|$ is beyond some threshold, $nMax$. Since the above type of evaluation cannot be used repeatedly due to unaffordable computing times, we decided to apply it only once, for the final solution returned by the heuristic.

8.2.1.2 Estimating the Cost of a Feasible Solution (General FLBD)

We apply Monte Carlo simulation and use as an estimate of this cost the average cost associated with a sample of scenarios. However, in the context of the PR heuristic, we do not fix the sample size; instead, we run the simulation until we achieve the convergence of the average:

$$100 \times \left| \frac{\text{previous average} - \text{current average}}{\text{previous average}} \right| < \text{precision}, \tag{27}$$

where "precision" is a small value defined exogenously.

Different strategies exist to accelerate the computation of this average. The use of antithetic estimators proved useful to accelerate this convergence. The basis of antithetic estimates (the reader is referred to Hammersley and Mauldon (1956) for the seminal paper on the topic) is to use couples of negatively correlated estimators instead of a single estimator at each iteration. In particular, at each iteration, a sequence $u_1, \ldots u_{|J_i|}$ of numbers is drawn from a $U(0; 1)$ distribution, one sample is built by setting $d_j^1 = 1$ if $u_j < p_j$ and $d_j^1 = 0$ otherwise, and a second one by setting $d_j^2 = 1$ if $1 - u_i < p_j$ and $d_j^2 = 0$ otherwise. The costs associated with these two scenarios are two estimates of the actual cost of the solution with negative correlation. Their average is used as the iteration estimate, which is averaged with the estimates of all the other iterations.

Regarding the convergence of the average stated in (27), two different precision levels were used in our implementation (see Algorithms 3 and 6). For evaluating most solutions, we used a low precision. Indeed, we only evaluated solutions systematically with high precision during the first phase of the heuristic (Algorithm 3, lines 1–17). In the second phase (Algorithm 3, lines 18–23), we use low precision most of the time and resort to high precision only when we find a solution that is potentially better than the incumbent. Therefore, we always have an accurate estimate of the value of incumbent solutions.

Preliminary experiments were carried out using different functions for the approximation of the solution costs such as (*i*) the exact evaluation of the cost function (see above), (*ii*) the exact evaluation function associated with the homogeneous case using the average demand probability, (*iii*) the exact cost evaluation assuming that all customers assigned to the same facility have an equal demand probability given by their average probability, (*iv*) the cost function simulated as above but always using the lower precision, and (*v*) the cost function simulated as above but always using the higher precision. The obtained results showed that the above-described simulation function is the most effective one, among those we tested.

Since for the homogeneous case an exact evaluation of the cost is possible by making use of the probabilities corresponding to a binomial distribution, when Algorithm 3 is applied to an instance of that problem, function `estimateCost`$(S, precision)$ is replaced by a function that returns the exact value.

8.2.2 Sample Average Approximation

A major issue in the design and implementation of an SAA method is the choice of the sample size in the definition of the sample average problems. In general terms, large sample sizes involve large computational times at each iteration, and on the other side, smaller samples require more iterations before convergence. After some preliminary tests on randomly generated instances, we set a sample size $|\Omega^k| = 180$ for both outsourcing policies. The algorithm terminates after five consecutive iterations with a modification of the z^* estimate below 1‰ or a maximum of 200 iterations. It is well known that under some technical conditions, the probability that $x^t \in \arg\min f(\Omega^t)$ is optimal to the original problem converges to 1 as $|\Omega^t|$ increases, and similarly, the set of ϵ-optimal solutions to Ω^t converges to the set of ϵ-optimal solutions to the original problem as the sample size increases. So in our executions, we did not force optimization of the SAA subproblems to optimality, but we accepted ϵ-optimal solutions with $\epsilon = 0.1$ aiming at converging to ϵ-optimal solutions of the FLBD.

For evaluating the solution obtained at each iteration, we have combined an exact and an approximate evaluation as described next. Note that for a given *a priori* assignment of customers to facilities, if service requests of different customers are independent, the cost associated with a particular facility i is independent from the possible requests for service of customers not assigned to it. Therefore, as in the case of GRASP and PR, to compute this cost, it is enough to enumerate all possible outcomes of the demand requests of customers in J_i, ignoring any other customer. Even if the size of this set of outcomes can be large, it will be in general much smaller than the full set of possible scenarios considering all customers. Thus, to evaluate the cost of a particular solution, we evaluate separately the costs associated with each of the facilities opened in that solution. According to our preliminary experiments, the exact evaluation of such costs is affordable for facilities with, at most, 15 assigned customers. The costs associated with facilities with more customers is approximated by Monte Carlo simulation. Since the convergence of the algorithm using these estimates was rather slow, again we used antithetic estimators as explained in 8.2.1 for the GRASP + PR (see Freimer et al., 2012) to reduce the variance of each estimate.

8.3 Results for GRASP + Path Relinking

For evaluating the heuristics based on GRASP and Path Relinking, we used both classes of instances: homogeneous and non-homogeneous ones. We performed five runs of the GRASP with PR for every instance. In the case of homogeneous instances, as a basis for comparison, we took the optimal solution values obtained in Albareda-Sambola et al. (2011) for the set of homogeneous instances considered in this chapter.

We start by presenting in Table 2 for the homogeneous instances the average percentage of optimal solutions found over all the runs on each instance. In this table, results are aggregated according to the original TSP they were generated from. Hence, each line shows the averages over the 18 instances of five-run averages. Separate results are provided for GRASP and the two variants of Path Relinking presented. This is done in columns 2–4 for facility outsourcing and in columns 5–7 for customer outsourcing. In this table, the major conclusion is that complementing the GRASP with Path Relinking leads to a significant increase in terms of the percentage of optimal solutions found. Regarding both versions of the Path Relinking, we also see that none of them dominates the other. A further difference between FO and CO that we can appreciate from the table is that the percentage of runs where the optimal solution was found is slightly higher in the former.

8.3.1 Homogeneous Instances

We continue our analysis on the homogeneous instances by analyzing the percentage gap of the feasible solutions obtained by the approximate algorithms (GRASP, PR1, and PR2) as well as the computing times required.

For FO, this is depicted in Figs. 1 (gap) and 2 (computing times in seconds). In each instances group, the first bar corresponds to the average value associated with executing the GRASP procedure alone. The two following bars correspond to the two Path Relinking variants. In the case of computing times, these bars refer only to the Path Relinking phase.

From Fig. 1, we conclude that in the case of homogeneous instances, GRASP alone already finds quite good solutions that are often under 1% away from the

Table 2 Homogeneous case | percentage of optimal solutions found

	Facility outsourcing			Customer outsourcing		
	GRASP	GRASP +PR1	GRASP +PR2	GRASP	GRASP +PR1	GRASP +PR2
berlin52	27.8	44.4	41.1	7.8	35.6	30.0
eil51	0.0	30.0	28.9	5.6	35.6	36.7
eil76	16.7	46.7	36.7	15.6	40.0	44.4
kroA100	20.0	46.7	45.6	15.6	47.8	45.6
kroB100	16.7	47.8	44.4	11.1	34.4	36.7
kroC100	16.7	40.0	37.8	16.7	41.1	40.0
kroD100	20.0	45.6	42.2	11.1	38.9	42.2
kroE100	10.0	47.8	48.9	12.2	40.0	40.0
pr76	16.7	35.6	30.0	16.7	37.8	26.7
rat99	2.2	46.7	45.6	11.1	40.0	35.6
st70	5.6	38.9	31.1	0.0	13.3	7.8
LkroA100	11.1	26.7	28.9	6.7	16.7	16.7
LkroB100	0.0	27.8	25.6	0.0	16.7	16.7
LkroC100	21.1	25.6	25.6	11.1	14.4	15.6
LkroD100	0.0	30.0	30.0	0.0	16.7	13.3
LkroE100	4.4	14.4	15.6	0.0	13.3	13.3
Lrat99	5.6	16.7	15.6	11.1	16.7	15.6

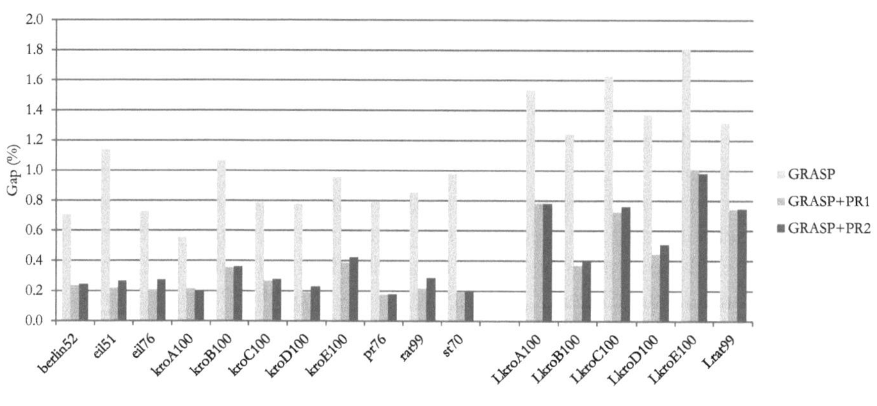

Fig. 1 Homogeneous case—facility outsourcing—gap (%)

optimal solution, being 1.81% the largest gap. Indeed, in many instances, it could already find the optimal solution. This is especially true for the set of smaller instances. These gaps are further reduced after applying any of the proposed PR

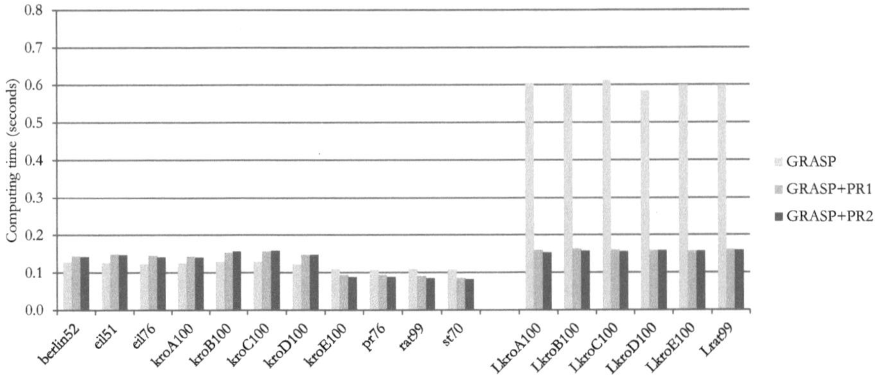

Fig. 2 Homogeneous case—facility outsourcing—computing time (sec.)

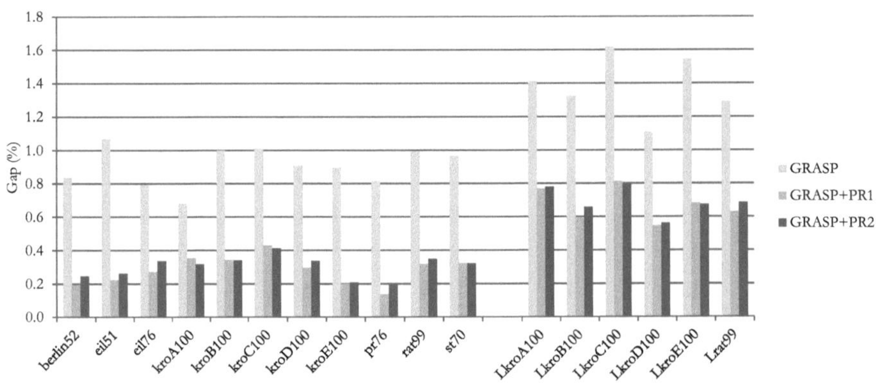

Fig. 3 Homogeneous case—customer outsourcing—gap (%)

procedures. In both cases, the reduction is quite significant, and there is not a clearly dominating variant.

As far as computing times are concerned, we can see that they are really small, even for the sets of larger instances. Although in those cases the GRASP computing time dominates the overall time of the heuristic, the total time is still very small. Adding the time of the GRASP and the time of any of the PR procedures, we do not reach one second of average in any instances group. As it happened with the solution quality, no PR variant dominates the other in terms of efficiency.

Following the structure of the previous results, Figs. 3 and 4 summarize the experiments made with GRASP and PR for the homogeneous instances when the CO outsourcing policy is considered.

In terms of solution quality, we observe that the behavior of GRASP is similar to what we could see for the FO policy, maybe with less differences between the quality of the solutions in small and larger instances. Percent gaps are again around 1%, being the largest average 1.62%. Again, both PR alternatives help improving

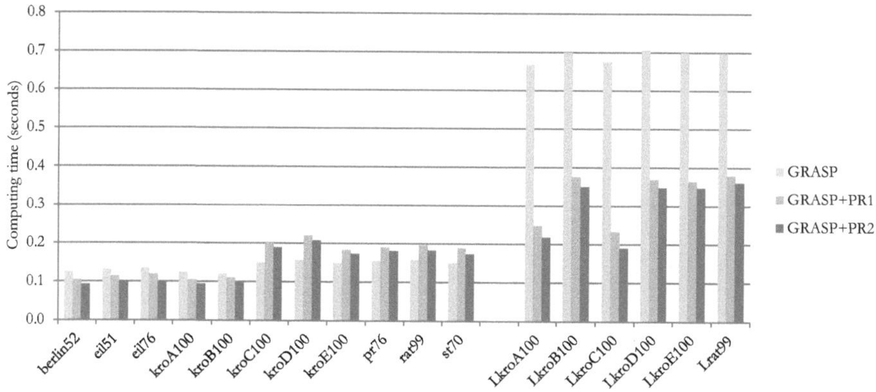

Fig. 4 Homogeneous case—customer outsourcing—computing time (sec.)

these figures yielding percent deviations from the optimum below 1% on average in all the groups of instances. The differences in solution quality between both PR variants are again very small.

The computational effort required for dealing for the CO outsourcing policy is similar to what we observed under FO. Again, times are extremely small. We can observe that as it happened with FO, the computing times for the GRASP procedure increase significantly with the instance size while those associated with PR do not. This is because the number of iterations and the size of the pools of solutions within the PR do not depend on the instance characteristics.

Summing up the information depicted in this subsection, we can say that the heuristic framework presented here is extremely effective for the homogeneous FLBD, under any of the considered outsourcing policies.

8.3.2 Results for the General Problem

As opposite to what happens for the homogeneous case, there is no exact algorithm available for solving the problem under non-homogeneous probability demands. Therefore, to assess the efficiency and efficacy of GRASP + PR in this case, we compare it with SAA. To this end, for each considered instance, we performed five runs of GRASP + PR with each PR variant, and one run of SAA, and recorded the best solution found over all runs. Then percent deviations from that best known solution are computed for each of the methods.

Figures 5 and 6 depict the average percent deviations under the FO and CO policies, respectively. As it was already explained in Sect. 8.4, in the case of the CO policy, SAA was only affordable for the small instances so the larger ones are not included in Fig. 6.

Looking into these figures, we observe that using SAA, the average deviation is above zero which is an indication that not always the best feasible solution was

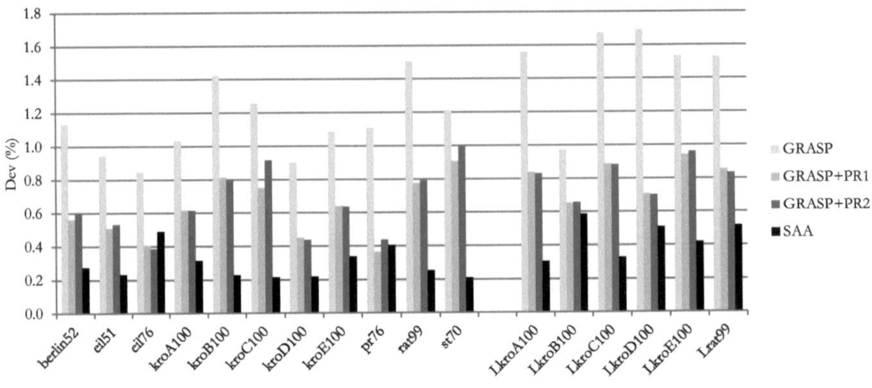

Fig. 5 Non-homogeneous case—facility outsourcing—dev (%)

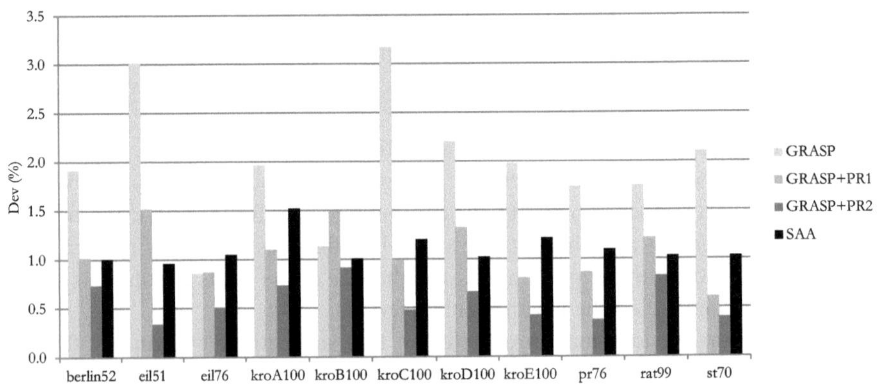

Fig. 6 Non-homogeneous case—customer outsourcing—dev (%)

obtained with SAA but with GRASP and PR. Indeed, when CO outsourcing policy is considered, the average deviations of the SAA solutions tend to be worse than those of GRASP + PR2 solutions. In fact, the deviations of the solutions provided by the GRASP method are already quite small, especially under FO.

Again, like in the homogeneous case, the solution improvements achieved by the PR procedures are quite significant. Now, additionally, we can observe a superior performance of PR2 as compared to PR1, at least under the CO policy (Fig. 6). We can see that the maximum percent deviation associated with GRASP + PR2 is approximately 1% for the FO policy and 0.9% under CO. The results of the computing times are depicted in Figs. 7 and 8. SAA times had to be represented relative to the second axis, to make it possible to appreciate the differences among the other times. In fact, the requirements of GRASP + PR are negligible as

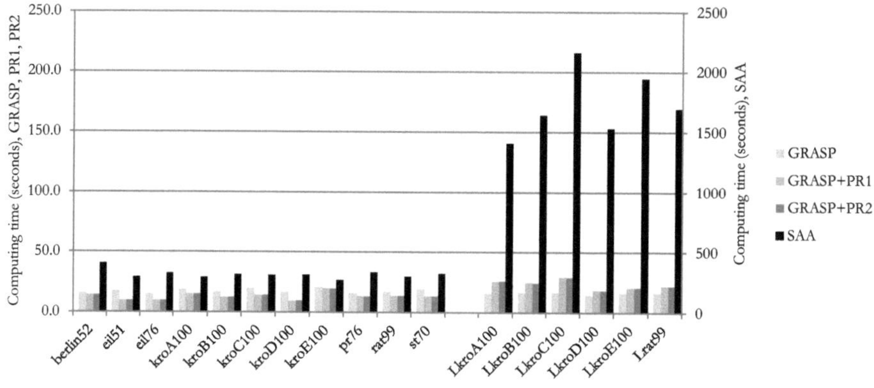

Fig. 7 Non-homogeneous case—facility outsourcing—computing time (sec.)

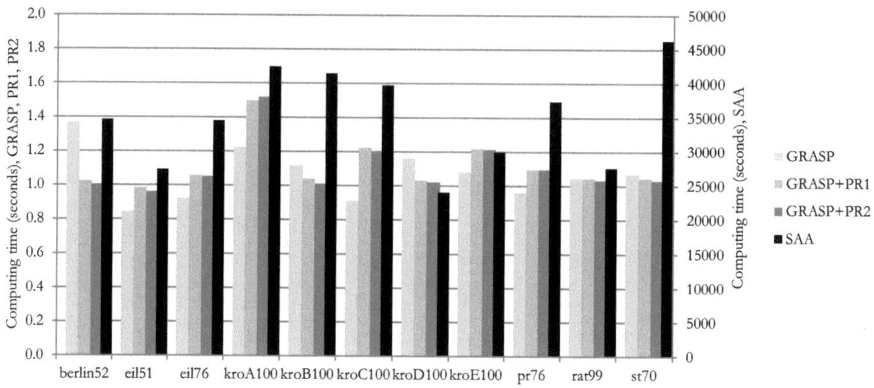

Fig. 8 Non-homogeneous case—customer outsourcing—computing time (sec.)

compared to those of SAA. In Fig. 7, we can see how, as observed before, the SAA time increases substantially with the instance size. In contrast, the time increase for GRASP and PR is moderate. In this regard, we observe that now PR times are more sensitive to instance size than GRASP times, as opposite to what happened for the homogeneous instances. We believe that this difference is due to the additional effort required for evaluating the candidate solutions. In general, times are quite homogeneous across instances of the same size. As for the comparison between PR1 and PR2, again, computation times are quite similar.

All the computations performed indicate that the heuristic framework proposed is robust as evidenced by the outcome of the five runs executed for each instance, which yielded similar results. We can also observe a slight superiority of *PR2* on average, even if *PR1* is able to find the optimal solution in more instances.

8.4 Results of the SAA

As mentioned in the introduction, the formulations presented in Albareda-Sambola et al. (2011) for the homogeneous case allow to solve the problem quite efficiently. On the other hand, the computational burden associated with the SAA is naturally considerable. Therefore, the natural application of the SAA presented in this work is on the heterogeneous case, for which there is no exact algorithm in the literature. For this reason, we have tested this algorithm only with the non-homogeneous instances presented above.

We first tested the SAA for the FO policy. The obtained results are depicted in Tables 3 and 4. In both tables, instances are grouped by capacity level and demand probability pattern. In Table 3, each of these groups contains 33 instances, whereas in Table 4, each group contains 18 large instances. For each group, we report the computing time, in seconds (minimum, average, and maximum), as well as the percent deviation of the value of the best solution found, z^*, with respect to the optimal value estimate, $\bar{z}$ (provided by the average of the optimal values of the

Table 3 SAA results for small instances with facility outsourcing

		Computing time			% dev. from estim.		
		Min	Avg.	Max	Min	Avg.	Max
PT1	$\gamma = 1$	99.2	242.5	667.6	−1.1	0.0	1.7
	$\gamma = 4$	66.8	136.1	211.2	−1.1	−0.1	1.2
PT2	$\gamma = 1$	348.1	688.6	3424.1	−0.7	0.1	0.9
	$\gamma = 4$	116.6	175.5	258.0	−4.1	−0.2	0.2
PT3	$\gamma = 1$	10.3	383.0	686.1	−2.5	−0.1	0.1
	$\gamma = 4$	5.2	165.3	432.2	−1.6	0.0	0.2
Total		5.2	298.5	3424.1	−4.1	−0.1	1.7

Table 4 SAA results for large instances with facility outsourcing

		Computing time			% dev. from estim.		
		Min	*Avg.*	Max	Min	Avg.	Max
PT1	$\gamma = 1$	1300.5	2484.7	4780.8	−1.6	0.8	5.3
	$\gamma = 4$	511.5	804.5	2036.8	−1.6	−0.2	0.9
PT2	$\gamma = 1$	1401.5	2477.8	6563.1	−0.3	0.1	1.4
	$\gamma = 4$	613.2	766.7	931.3	−2.0	−0.2	0.1
PT3	$\gamma = 1$	34.5	1277.8	2588.1	−2.7	−0.1	0.0
	$\gamma = 4$	18.3	574.4	967.8	−0.9	0.0	0.1
Total		18.3	1397.7	6563.1	−2.7	0.2	5.3

sample average subproblems): $100\frac{z^{*}-\bar{z}}{\bar{z}}$. Again, minimum, average, and maximum values, over the set of instances, are provided.

Comparing the two tables, we can appreciate that the instance size has a great impact on the computing times required to solve the instances. Indeed, while the solution times range between 5.2 seconds and 57 minutes for the small instances, the times required by the larger ones range between 18.3 seconds and 1.8 hours. Regarding the computing times, we also observed that they increase significantly with the proportion of medium demand customers and decreases with the capacity level. Indeed, the hardest instances to solve in both cases (small and large instances) were those with the largest proportion of medium-demand customers (PT2) and the lowest capacity level ($\gamma = 1$).

This behavior might seem counterintuitive. However, we think that since customers with high-demand probability will have demand in most scenarios and those with low-demand probability will seldom appear, most of the variability in the solutions is caused by the customers with medium-demand probability and the algorithm will take longer to converge for instances with more customers in this group. On the other side, instances with large facility capacities will require less facilities to be opened, which reduces the set of solutions with low costs.

Regarding the solution quality, we can appreciate that deviations of the obtained solution values from the SAA estimate of the optimal value are in general small. This makes us very confident on the quality of the obtained solutions. Note that the absolute value of the average deviations within the different groups of instances is at most 0.8% and it is below 0.2% in most cases. Here, there is not a clear influence of the parameters describing the instances on the values of these deviations.

Finally, Table 5 summarizes the results obtained by SAA for the CO policy on the small instances. As expected, instances are much harder to solve under this policy. For this reason, now, only one instance for each combination of original TSP instance, probability demand pattern, and capacity level was considered. Table 5 has the same structure as the two previous ones, excepting that, now, each cell of the table refers to 11 instances, and not 33 as in Table 3. We observe that the computing

Table 5 SAA results for small instances with order driven—customer outsourcing

		Computing time			% dev. from estim.		
		Min	Avg.	Max	Min	Avg.	Max
PT1	$\gamma = 1$	14,273.7	24,855.5	38,476.7	−5.1	1.0	4.3
	$\gamma = 4$	3986.1	5581.7	7013.6	−9.3	−3.2	1.7
PT2	$\gamma = 1$	50,331.3	94,692.7	175,558.6	−5.1	−0.4	3.4
	$\gamma = 4$	11,323.3	16,671.2	26,335.8	−9.7	−4.1	5.3
PT3	$\gamma = 1$	7377.0	53,809.1	103,566.0	−4.9	−1.8	0.6
	$\gamma = 4$	6083.9	14,312.6	30,016.5	−13.0	−5.3	2.6
Total		3986.1	34,987.1	175,558.6	−13.0	−2.3	5.3

times for these small instances range between 1.1 and 48.8 hours. For this reason, we did not test the SAA on the larger instances.

The effect of the proportion of medium demand probability is now more evident than in the FO case. Indeed, on the average, solving each instance in group PT2 took almost three times as much time as the corresponding instance in group P3 and more than 3.5 times the time it took to solve the corresponding instance in group PT1.

As far as the quality of the solutions is concerned, deviations of the value of the best solution found throughout the process from the optimum value estimate provided by the average of the problems solved at the different iterations are larger than in the FO policy. These deviations range now from -13% to 5.3%, and on the average, their absolute value is 3.5%. In 38% of the instances, it is below 2%, and in 24% of them, it is below 1%. Given the complexity of the problem, we believe that these results are reasonably good. Indeed, from Table 5, it might give the impression that in the case of the CO policy, the algorithm was often terminated before convergence because of the iterations limit. However, we can observe that the larger deviations do not correspond to the most demanding instance sets for any policy, which contradicts this argument. We believe that these larger deviations are due to a deterioration of the quality of the optimal value estimate in this last policy.

9 Conclusions

In this chapter, we studied the use of approximate methods for discrete facility location under uncertainty. We discussed several challenges and ways for overcoming them. The main challenge identified is the evaluation of the objective function, for which several strategies based on Monte Carlo simulation have been explored. The analysis was illustrated with a specific facility location problem, namely, that in which demands follow a Bernoulli distribution.

Two different approximate algorithms were revisited both having advantages and disadvantages when compared with each other. Sample Average Approximation is a method that relies on a mathematical model restricted to a subset of scenarios. This explains the computational effort it involves and its difficulty in tackling large instances of problems with more involved recourse functions. On the other hand, the GRASP with Path Relinking is an algorithm prepared to handle large-scale instances. Therefore, we cannot say promptly that one algorithm overcomes the other for the illustrative problem used in this chapter. Sample Average Approximation is certainly worth trying for small instances of the stochastic problem we considered, whereas when large instances are considered with more involved recourse functions, the GRASP with Path Relinking is certainly advisable.

The chapter highlights the need to develop further work in terms of the development of heuristics for stochastic facility location problems. This includes the application of existing heuristics to facility location problems other than the ones addressed in this chapter and also the development of especially tailored approximate algorithms (e.g., constructive procedures).

The focus of this chapter was put on two-stage stochastic facility location problems. In some settings, the problems may easily become multistage, as it happens when time-dependent and interrelated decisions are to be made. For such cases, existing approximate algorithms that have been successfully applied to multistage stochastic mixed-integer programming, such as the progressive hedging procedure, could be a promising avenue of research.

Acknowledgments This research has been partially supported by the *Spanish Ministry of Science and Innovation* and ERDF funds, through grants RED2018-102363-T and MTM2019-105824GB-I00, and by national funding from FCT — Fundação para a Ciência e Tecnologia, Portugal—UIDB/04561/2020. This support is gratefully acknowledged.

References

Ahmadi-Javid, A., Seyedi, P., & Syam, S. S. (2017). A survey of healthcare facility location. *Computers & Operations Research, 79*, 223–263.

Albareda-Sambola, M., Alonso-Ayuso, A., Escudero, L., Fernández, E., & Pizarro, C. (2013). Fix-and-relax-coordination for a multi-period location-allocation problem under uncertainty. *Computers & Operations Research, 40*(12), 2878–2892.

Albareda-Sambola, M., Fernández, E., & Laporte, G. (2007). Heuristic and lower bounds for a stochastic location routing problem. *European Journal of Operational Research, 179*, 940–955.

Albareda-Sambola, M., Fernández, E., & Saldanha-da-Gama, F. (2011). The facility location problem with Bernoulli demands. *Omega, 39*, 335–345.

Albareda-Sambola, M., Fernández, E., & Saldanha-da-Gama, F. (2017). Heuristic solutions to the facility location problem with general Bernoulli demands. *INFORMS Journal on Computing, 29*, 737–753.

Albareda-Sambola, M., Fernández, E., & Saldanha-da-Gama, F. (2022). Outsourcing policies for the facility location problem with Bernoulli demands. *In preparation.*

Albareda-Sambola, M., van der Vlerk, M., & Fernández, E. (2006). Exact solutions to a class of stochastic generalized assignment problems. *European Journal of Operational Research, 173*, 465–487.

Álvarez-Miranda, E., Fernández, E., & Ljubić, I. (2015). The recoverable robust facility location problem. *Transportation Research. Part B, Methodological, 79*, 93–120.

Averbakh, I., & Berman, O. (1997). Minimax regret p-center location on a network with demand uncertainty. *Location Science, 5*, 247–254.

Averbakh, I., & Berman, O. (2000). Minmax regret median location on a network under uncertainty. *INFORMS Journal on Computing, 12*, 104–110.

Averbakh, I., & Berman, O. (2003). An improved algorithm for the minmax regret median problem on a tree. *Networks, 41*, 97–103.

Beraldi, P., & Bruni, M. (2009). A probabilistic model applied to emergency service vehicle location. *European Journal of Operational Research, 196*, 323–331.

Berman, O., & Simchi-Levi, D. (1988). Finding the optimal a priori tour and location of a traveling salesman with non homogeneous customers. *Transportation Science, 22*, 148–154.

Bianchi, L., & Campbell, A. M. (2007). Extension of the 2-p-opt and 1-shift algorithms to the heterogeneous probabilistic traveling salesman problem. *European Journal of Operational Research, 176*, 131–144.

Bieniek, M. (2015). A note on the facility location problem with stochastic demands. *Omega, 55*, 53–60.

Birge, J. R., & Louveaux, F. V. (2011). *Introduction to stochastic programming* (2nd ed.). New York: Springer.

Carrizosa, E., Conde, E., & Muñoz, M. (1998). Admission policies in loss queueing models with heterogeneous arrivals. *Management Science, 44*, 311–320.

Carrizosa, E., & Nickel, S. (2003). Robust facility location. *Mathematical Methods in Operations Research, 58*, 331–349.

Conde, E. (2007). Minmax regret location-allocation problem on a network under uncertainty. *European Journal of Operational Research, 179*, 1025–1039.

Daskin, M. (2013). *Network and discrete location: Models, algorithms, and applications* (2nd ed.). Wiley.

Dönmez, Z., Kara, B. Y., Karsu, Ö., & Saldanha-da-Gama, F. (2021). Humanitarian facility location under uncertainty: Critical review and future prospects. *Omega, 102*, 102393.

Eiselt, H., & Marianov, V. (Eds.) (2011). *Foundations of location analysis*. Number 115 in International Series in Operations Research & Management Science (2nd ed.). Berlin: Springer.

Feo, T. A., & Resende, M. G. C. (1995). Greedy randomized adaptive search procedures. *Journal of Global Optimization, 6*, 109–133.

Fernández, E., Hinojosa, Y., & Puerto, J. (2005). Filtering policies in loss queuing-location problems. *Annals of Operations Research, 136*, 259–283.

Fortz, B. (2015). Location problems in telecommunications. In G. Laporte, S. Nickel, & Saldanha-da-Gama, F. (Eds.). *Location science* (pp. 537–554). Berlin: Springer.

Freimer, M. B., Linderoth, J. T., & Thomas, D. J. (2012). The impact of sampling methods on bias and variance in stochastic linear programs. *Computational Optimization and Applications, 51*, 51–75.

Glover, F. (1997). Tabu search and adaptive memory programing – advances, applications and challenges. In R. S. Barr, R. V. Helgasson, & J. L. Kennington (Eds.). *Interfaces in Computer Science and Operations Research* (pp. 1–75). Berlin: Springer.

Glover, F., & Laguna, M. (1997). *Tabu search*. Norwell, Massachusetts: Kluwer.

Glover, F., Laguna, M., & Martí, R. (2000). Fundamentals of scatter search and path relinking. *Control and Cybernetics, 29*(3), 653–684.

Hammersley, J. M., & Mauldon, J. G. (1956). General principles of antithetic variates. *Mathematical Proceedings of the Cambridge Philosophical Society, 52*, 476–481.

Jaillet, P. (1988). A priori solution of a traveling salesman problem in which a random subset of the customers are visited. *Operations Research, 36*, 929–936.

Klein Haneveld, W. K., van der Vlerk, M. H., & Romeijnders, W. (2020). *Stochastic programming: Modeling decision problems under uncertainty*. Switzerland: Springer Nature.

Kleywegt, A. J., Shapiro, A., & Homem-de-Mello, T. (2001). The sample average approximation method for stochastic discrete optimization. *SIAM journal on Optimization, 12*, 479–502.

Laporte, G., Louveaux, F., & Hamme, L. V. (1994a). Exact solution to a location problem with stochastic demands. *Transportation Science, 28*, 95–103.

Laporte, G., Louveaux, F., & Mercure, H. (1994b). An exact solution for the a priori optimization of the probabilistic traveling salesman problem. *Operations Research, 42*, 543–549.

Laporte, G., Nickel, S., & Saldanha-da-Gama, F. (Eds.) (2019). *Location science* (2nd edn.). Berlin: Springer.

Louveaux, F., & Peeters, D. (1992). A dual-based procedure for stochastic facility location. *Operations Research, 40*, 564–573.

Marianov, V., & ReVelle, C. (1996). The queueing maximal availability location problem: a model for the siting of emergency vehicles. *European Journal of Operational Research, 93*, 110–120.

Pagès-Bernaus, A., Ramalhinho, H., Juan, A., & Calvet, L. (2019). Designing e-commerce supply chains: A stochastic facility location approach. *International Transactions in Operational Research, 26*(2), 507–528.

Shiripour, S., & Mahdavi-Amiri, N. (2019). Bi-objective location problem with balanced allocation of customers and bernoulli demands: Two solution approaches. *Soft Computing, 23*(13), 4999–5018.

Snyder, L. V. (2006). Facility location under uncertainty: A review. *IIE Transactions, 38*, 547–564.

Snyder, L. V., & Daskin, M. (2005). Reliability models for facility location: The expected failure cost case. *Transportation Science, 39*(3), 400–416.

Toregas, C., Swain, R., & ReVelle, C. (1971). The location of emergency service facilities. *Operations Research, 19*, 1363–1373.

Turkěs, R., Sörensen, K., & Cuervo, D. (2021). A matheuristic for the stochastic facility location problem. *Journal of Heuristics, 27*, 649–694.

Wagner, M., Bhadury, J., & Peng, S. (2009). Risk management in uncapacitated facility location models with random demands. *Computers & Operations Research, 36*, 1002–1011.

MIX
Papier aus verantwortungsvollen Quellen
Paper from responsible sources
FSC® C105338

If you have any concerns about our products,
you can contact us on
ProductSafety@springernature.com

In case Publisher is established outside the EU,
the EU authorized representative is:
Springer Nature Customer Service Center GmbH
Europaplatz 3, 69115 Heidelberg, Germany

Printed by Libri Plureos GmbH
in Hamburg, Germany